Climate Change and Its Impact on Ecosystem Services and Biodiversity in Arid and Semi-Arid Zones

Ahmed Karmaoui
Southern Center for Culture & Sciences (SCCS), Morocco

A volume in the Advances in Environmental Engineering and Green Technologies (AEEGT) Book Series

IGI Global
DISSEMINATOR OF KNOWLEDGE

Published in the United States of America by
IGI Global
Engineering Science Reference (an imprint of IGI Global)
701 E. Chocolate Avenue
Hershey PA, USA 17033
Tel: 717-533-8845
Fax: 717-533-8661
E-mail: cust@igi-global.com
Web site: http://www.igi-global.com

Library of Congress Cataloging-in-Publication Data

Names: Karmaoui, Ahmed, 1982- editor.
Title: Climate change and its impact on ecosystem services and biodiversity in arid and semi-arid zones / Ahmed Karmaoui, editor.
Description: Hershey PA : Engineering Science Reference, [2019] | Includes bibliographical references.
Identifiers: LCCN 2018025306| ISBN 9781522573876 (hardcover) | ISBN 9781522573883 (ebook)
Subjects: LCSH: Climatic changes--Morocco. | Climatic changes--Government policy--Morocco. | Biodiversity--Climatic factors--Morocco. | Ecosystem services--Morocco.
Classification: LCC QC903.2.M8 C55 2019 | DDC 577.540964--dc23 LC record available at https://lccn.loc.gov/2018025306

This book is published in the IGI Global book series Advances in Environmental Engineering and Green Technologies (AEEGT) (ISSN: 2326-9162; eISSN: 2326-9170)

British Cataloguing in Publication Data
A Cataloguing in Publication record for this book is available from the British Library.

For electronic access to this publication, please contact: eresources@igi-global.com.

Advances in Environmental Engineering and Green Technologies (AEEGT) Book Series

Sang-Bing Tsai
University of Electronic Science and Technology of China Zhongshan Institute, China
Ming-Lang Tseng
Lunghwa University of Science and Technology, Taiwan
Yuchi Wang
University of Electronic Science and Technology of China Zhongshan Institute, China

ISSN:2326-9162
EISSN:2326-9170

Mission

Growing awareness and an increased focus on environmental issues such as climate change, energy use, and loss of non-renewable resources have brought about a greater need for research that provides potential solutions to these problems. Research in environmental science and engineering continues to play a vital role in uncovering new opportunities for a "green" future.

The **Advances in Environmental Engineering and Green Technologies (AEEGT)** book series is a mouthpiece for research in all aspects of environmental science, earth science, and green initiatives. This series supports the ongoing research in this field through publishing books that discuss topics within environmental engineering or that deal with the interdisciplinary field of green technologies.

Coverage

- Sustainable Communities
- Green Transportation
- Pollution Management
- Contaminated Site Remediation
- Biofilters and Biofiltration
- Renewable Energy
- Electric Vehicles
- Air Quality
- Radioactive Waste Treatment
- Industrial Waste Management and Minimization

IGI Global is currently accepting manuscripts for publication within this series. To submit a proposal for a volume in this series, please contact our Acquisition Editors at Acquisitions@igi-global.com or visit: http://www.igi-global.com/publish/.

The Advances in Environmental Engineering and Green Technologies (AEEGT) Book Series (ISSN 2326-9162) is published by IGI Global, 701 E. Chocolate Avenue, Hershey, PA 17033-1240, USA, www.igi-global.com. This series is composed of titles available for purchase individually; each title is edited to be contextually exclusive from any other title within the series. For pricing and ordering information please visit http://www.igi-global.com/book-series/advances-environmental-engineering-green-technologies/73679. Postmaster: Send all address changes to above address.

Titles in this Series

For a list of additional titles in this series, please visit: www.igi-global.com/book-series

Green Finance for Sustainable Global Growth
Sang-Bing Tsai (University of Electronic Science and Technology of China Zhongshan Institute, China & Civil Aviation University of China, China & WUYI University, China) Chung-Hua Shen (Shih-Chien University, Taiwan) Hua Song (Renmin University of China, China) and Baozhuang Niu (South China University of Technology, China)
Business Science Reference • copyright 2019 • 397pp • H/C (ISBN: 9781522578086) • US $215.00 (our price)

Building Sustainability Through Environmental Education
Lynn A. Wilson (SeaTrust Institute, USA) and Carolyn N. Stevenson (Purdue University Global, USA)
Engineering Science Reference • copyright 2019 • 335pp • H/C (ISBN: 9781522577270) • US $195.00 (our price)

Unique Sequence Signatures in Plant Lipolytic Enzymes Emerging Research and Opportunities
Nihed Ben Halima (University of Sfax, Tunisia)
Engineering Science Reference • copyright 2019 • 197pp • H/C (ISBN: 9781522574828) • US $145.00 (our price)

Handbook of Research on the Adverse Effects of Pesticide Pollution in Aquatic Ecosystems
Khursheed Ahmad Wani (Government Degree College Bijbehara, India) and Mamta (Jiwaji University, India)
Engineering Science Reference • copyright 2019 • 538pp • H/C (ISBN: 9781522561118) • US $295.00 (our price)

Harnessing Marine Macroalgae for Industrial Purposes in an Australian Context Emerging Research and Opportunities
Göran Roos (Economic Development Board of South Australia, Australia) Anthony Cheshire (Balance Carbon Pty Ltd, Australia) Sasi Nayar (South Australian Aquatic Sciences Centre, Australia) Steven M. Clarke (South Australian Research and Development Institute, Australia) and Wei Zhang (Flinders University, Australia)
Engineering Science Reference • copyright 2019 • 291pp • H/C (ISBN: 9781522555773) • US $195.00 (our price)

Advanced Methodologies and Technologies in Engineering and Environmental Science
Mehdi Khosrow-Pour, D.B.A. (Information Resources Management Association, USA)
Engineering Science Reference • copyright 2019 • 506pp • H/C (ISBN: 9781522573593) • US $295.00 (our price)

Transitioning Island Nations Into Sustainable Energy Hubs Emerging Research and Opportunities
Catalina Spataru (UCL Energy Institute, UK)
Engineering Science Reference • copyright 2019 • 172pp • H/C (ISBN: 9781522560029) • US $135.00 (our price)

Recent Challenges and Advances in Geotechnical Earthquake Engineering
T.G. Sitharam (Indian Institute of Science, India) and J. S. Vinod (University of Wollongong, Australia)
Engineering Science Reference • copyright 2019 • 315pp • H/C (ISBN: 9781522569480) • US $195.00 (our price)

IGI Global
DISSEMINATOR OF KNOWLEDGE

701 East Chocolate Avenue, Hershey, PA 17033, USA
Tel: 717-533-8845 x100 • Fax: 717-533-8661
E-Mail: cust@igi-global.com • www.igi-global.com

Table of Contents

Preface xvii

Acknowledgment xxiii

Section 1
Climate Change, Ecosystem, and Ecosystem Services

Chapter 1
The Need to Establish a Long-Term Ecological Research Network in Morocco as a Tool to Monitor Ecosystems Under Climate Change 1
Ahmed Karmaoui, Southern Center for Culture and Sciences, Morocco
Siham Zerouali, Southern Center for Culture and Sciences, Morocco

Chapter 2
Associations Between Climate, Ecosystems, and Ecosystem Services in the Pre-Sahara: Case Study of Tafilalet, Morocco 23
Abdelkrim Ben Salem, Cadi Ayyad University, Morocco
Souad Ben Salem, Cadi Ayyad University, Morocco
Mohammed Yacoubi Khebiza, Cadi Ayyad University, Morocco
Awatif Zine Elabidine, Kennesaw State University, USA

Chapter 3
Ecosystem Services Demand Management Under Climate Change Scenarios: Use of WEAP Software in Case of Water Demand in Ziz Basin, Morocco 45
Abdelkrim Ben Salem, Cadi Ayyad University, Morocco
Souad Ben Salem, Cadi Ayyad University, Morocco
Mohammed Khebiza Yacoubi, Cadi Ayyad University, Morocco
Mohammed Messouli, Cadi Ayyad University, Morocco

Section 2
Biodiversity and Ecosystem Services

Chapter 4
Monetary Value Change of Some Provisioning Ecosystem Services of Middle Draa Valley, South of Morocco 67
Siham Zerouali, Southern Center for Culture and Sciences, Morocco
Mohammed Yacoubi-Khebiza, Cadi Ayyad University, Morocco
Fadoua El Qorchi, Cadi Ayyad University, Morocco

Chapter 5
Potential Role of Endophytes for Sustainable Environment 78
Zubair A. Dar, Sher-e-Kashmir University of Agricultural Sciences and Technology of Kashmir, India
Bhat Rifat, University of Kashmir, India
Javeed I. A. Bhat, Sher-e-Kashmir University of Agricultural Sciences and Technology of Kashmir, India
Asma Absar Bhatti, Sher-e-Kashmir University of Agricultural Sciences and Technology of Kashmir, India
Shamsul Haq, Sher-e-Kashmir University of Agricultural Sciences and Technology of Kashmir, India
Azra Amin, Sher-e-Kashmir University of Agricultural Sciences and Technology of Kashmir, India
Shakeel Ahmad Dar, Sher-e-Kashmir University of Agricultural Sciences and Technology of Kashmir, India

Chapter 6
Health Effects of Air Pollution in Urban Environment 96
Banwari Dandotiya, Jiwaji University Gwalior, India

Section 3
Climate Change Impacts on Biodiversity and Ecosystem Services

Chapter 7
Climate Change Impacts on Biodiversity in Arid and Semi-Arid Areas: Biodiversity Under Climate Change 117
Hanane Boutaj, Cadi Ayyad University, Morocco
Aicha Moumni, Cadi Ayyad University, Morocco
Oumayma Nassiri, Cadi Ayyad University, Morocco
Abdelhak Ouled Aitouna, Cadi Ayyad University, Morocco

Chapter 8
Impacts of Climate Change on Biodiversity and Ecosystem Services: Current Trends 142
Vartika Singh, Amity University Noida, India

Chapter 9
Impact of Climatic Changes on Earth's Survival: An Analysis of Changing Environment and Its Consequences at a Global Level ... 160
Akash, Gurukula Kangri University, India
Navneet, Gurukula Kangri University, India

Chapter 10
Potential Impacts of Climate Change on Land Degradation and Desertification: Land Degradation and Climate Change ... 183
Dharumarajan S., National Bureau of Soil Survey and Land Use Planning (ICAR), India
Veeramani S., National Bureau of Soil Survey and Land Use Planning (ICAR), India
Kalaiselvi Beeman, National Bureau of Soil Survey and Land Use Planning (ICAR), India
Lalitha M., National Bureau of Soil Survey and Land Use Planning (ICAR), India
Janani N., National Bureau of Soil Survey and Land Use Planning (ICAR), India
Srinivasan R., National Bureau of Soil Survey and Land Use Planning (ICAR), India
Rajendra Hegde, National Bureau of Soil Survey and Land Use Planning (ICAR), India

Chapter 11
Potential Impacts of Climate Change on the Inland Fisheries of Arid and Semi-Arid Regions of Africa: Impacts of Climate Change on Inland Fisheries ... 196
Imefon Udo Udo, University of Uyo, Nigeria
Imekan Isaac Akpan, Akwa Ibom State University, Nigeria

Chapter 12
Climate Change and Its Impacts on Oases Ecosystem in Morocco ... 217
Mohamed Ait El Mokhtar, Cadi Ayyad University, Morocco
Raja Ben Laouane, Cadi Ayyad University, Morocco
Mohamed Anli, Cadi Ayyad University, Morocco
Abderrahim Boutasknit, Cadi Ayyad University, Morocco
Abdessamad Fakhech, Cadi Ayyad University, Morocco
Said Wahbi, Cadi Ayyad University, Morocco
Abdelilah Meddich, Cadi Ayyad University, Morocco

Section 4
Climate Change, Ecosystem Services, Food Sercurity, and Health

Chapter 13
Ecosystem Services, Climate Change, and Food Security ... 247
Samreen Siddiqui, Texas A&M University – Corpus Christi, USA

Chapter 14
Ecosystem Services-Climate-Health Associations: Water-Climate-Leishmaniasis Nexus in an Endemic Focus of Zoonotic Cutaneous Leishmaniasis 280
Ahmed Karmaoui, Southern Center for Culture and Sciences, Morocco
Siham Zerouali, Southern Center for Culture and Sciences, Morocco
Ashfaq Ahmad Shah, National University of Sciences and Technology (NUST), Pakistan
Mohammed Yacoubi-Khebiza, Cadi Ayyad University, Morocco
Fadoua El Qorchi, Cadi Ayyad University, Morocco

Chapter 15
Relationships Between Climate Parameters and the Density of *Phlebotomus papatasi*, the Main Vector of Zoonotic Cutaneous Leishmaniasis 291
Ahmed Karmaoui, Southern Center for Culture and Sciences, Morocco
Siham Zerouali, Southern Center for Culture and Sciences, Morocco

Section 5
Biodiversity and Ecosystem Services Adaptation to Climate Change

Chapter 16
Effectiveness of Local Community Policy Responses to Climate Change Impact on Ecosystem Services for Biodiversity Conservation in the Semi-Arid Zones 304
Wilson Truman Okaka, Kyambogo University, Uganda

Chapter 17
Adaptive Strategies of Small Family Farms in the Face of Climate Change: The Case of the Tleta Watershed in Northern Morocco 316
Mohamed El Amrani, National School of Agriculture of Meknes, Morocco

Compilation of References 331

About the Contributors 402

Index 406

Detailed Table of Contents

Preface..xvii

Acknowledgment..xxiii

Section 1
Climate Change, Ecosystem, and Ecosystem Services

Chapter 1
The Need to Establish a Long-Term Ecological Research Network in Morocco as a Tool to Monitor Ecosystems Under Climate Change.. 1
Ahmed Karmaoui, Southern Center for Culture and Sciences, Morocco
Siham Zerouali, Southern Center for Culture and Sciences, Morocco

Several countries are associated with the long-term ecological research (LTER) program, while others are in the process of joining the network. In Morocco, there is an urgent need to be a member of this network because the diversity of challenges associated with the various ecosystems requires multidisciplinary long-term studies. The chapter discusses this need by assessing the environmental vulnerability of Morocco basing on data from the environmental vulnerability index profile. Ecosystem change and vulnerability were investigated at the sites within the LTER program by exploring the criteria of the selected sites. The driving forces, pressures, states, impacts, responses framework would be a useful approach to study and explain the ecological changes of each selected site. The chapter highlights the purpose, significance, the mission, objectives, and the international cooperation of the proposed LTER network called Mo-LTER.

Chapter 2
Associations Between Climate, Ecosystems, and Ecosystem Services in the Pre-Sahara: Case Study of Tafilalet, Morocco 23
Abdelkrim Ben Salem, Cadi Ayyad University, Morocco
Souad Ben Salem, Cadi Ayyad University, Morocco
Mohammed Yacoubi Khebiza, Cadi Ayyad University, Morocco
Awatif Zine Elabidine, Kennesaw State University, USA

This chapter focuses on the richness and wide variety of Sub-Saharan ecosystems particularly in the region of Tafilalet, Morocco. The Tafilalet ecosystems provide an extensive list of goods (see below) to the region, as well as "ecosystem services" for the inhabitants. However, in the last 30 years, these ecosystems are threatened by conjugated impacts of climate change and human overuse of the nature resources. The primary objective of this chapter is to evaluate both the damage inflicted on Tafilalet ecosystems and the associated ecosystem services due to climate change. Adaptation to climate change is of paramount importance to keep the survival of the diversity of Tafilalet ecosystems including agro-ecosystems, forests, and inland waters. Adaptation is a process to reduce the vulnerability of people to prevent displacement and conflict for scarce ecosystem services. The findings of this chapter show that a large number of ecosystem services of the region are limited by climatic conditions that affect the resilience of the whole oasis of Tafilalet.

Chapter 3
Ecosystem Services Demand Management Under Climate Change Scenarios: Use of WEAP Software in Case of Water Demand in Ziz Basin, Morocco 45
Abdelkrim Ben Salem, Cadi Ayyad University, Morocco
Souad Ben Salem, Cadi Ayyad University, Morocco
Mohammed Khebiza Yacoubi, Cadi Ayyad University, Morocco
Mohammed Messouli, Cadi Ayyad University, Morocco

Water ecosystem service is the most important element that supports Tafilalet agro-ecosystems. In this region, drought frequency is increasing, which complicate the management groundwater reserves. The ephemeral flows of the rivers force people to use groundwater to meet the population demand. Consequently, water resource management is of significant importance the sustainability of this area. Water evaluation and planning (WEAP) is useful management software used to evaluate and trace the trend of water demand. This model was applied in case of Ziz basin in order to simulate and analyze the situation of water under different scenarios. The results show an increasing of demand for water irrigation and with introducing modern irrigation scenario. However, a decreasing trend in reservoir storage volume and groundwater storage was projected in Tafilalet.

Section 2
Biodiversity and Ecosystem Services

Chapter 4
Monetary Value Change of Some Provisioning Ecosystem Services of Middle Draa Valley, South of Morocco 67
Siham Zerouali, Southern Center for Culture and Sciences, Morocco
Mohammed Yacoubi-Khebiza, Cadi Ayyad University, Morocco
Fadoua El Qorchi, Cadi Ayyad University, Morocco

The desert oasis is a wetland agro-ecosystem more and more under pressure. This trend threatens the oasis ecosystem services (ES) and then the wellbeing of the local population. In order to quantify this trend in Middle Draa Valley, South of Morocco, the monetary values for a selection of ES have been estimated. In this chapter, the economic values of 28 provisioning ES in 2014 was examined and next compared with the economic values of 2009. These values were standardized to US$ per hectare per year. The total value of the studied provisioning ES in 2014 was found to be 8658.3 US$/ha/year for the arable area (26000ha) and 150 US$/ha/year for the whole Middle Draa Valley (1500000ha). Comparing these findings with the data of 2009, a slight increase of $1245.7 (in 26000ha) and $21.5 US$/ha/year (the whole area) was detected in the period 2009-2014.

Chapter 5
Potential Role of Endophytes for Sustainable Environment 78
Zubair A. Dar, Sher-e-Kashmir University of Agricultural Sciences and Technology of Kashmir, India
Bhat Rifat, University of Kashmir, India
Javeed I. A. Bhat, Sher-e-Kashmir University of Agricultural Sciences and Technology of Kashmir, India
Asma Absar Bhatti, Sher-e-Kashmir University of Agricultural Sciences and Technology of Kashmir, India
Shamsul Haq, Sher-e-Kashmir University of Agricultural Sciences and Technology of Kashmir, India
Azra Amin, Sher-e-Kashmir University of Agricultural Sciences and Technology of Kashmir, India
Shakeel Ahmad Dar, Sher-e-Kashmir University of Agricultural Sciences and Technology of Kashmir, India

Endophytes are symptomless fungal and bacterial microorganisms found in almost all living plants. They are vital components of plant microbiomes. Endophytes affect plant growth and plant responses to pathogens, herbivores, and environmental change by producing a range of natural products having antifungal, antibacterial, and insecticidal properties. Endophytes have shown particular promise in agriculture particularly as beneficial crop inoculants and are known to enhance abiotic and biotic plant stress tolerance by increasing tolerance to drought and water stress, as well as tolerance to high temperature and high salinity. A better understanding of their plant growth-promoting mechanisms could simplify higher production of energy crops in a more sustainable manner even on marginal land and feed stocks for industrial processes, thus contribute to avoiding conflicts between food and energy production Many endophytes can be exploited to improve the efficiency of phytoremediation as they are found to be resistant to heavy metals and capable of detoxifying organic contaminants.

Chapter 6
Health Effects of Air Pollution in Urban Environment........ 96
Banwari Dandotiya, Jiwaji University Gwalior, India

In the last few decades, urban development and growing industrialization has caused air pollution to become a major issue in urban areas of developing and developed countries. Urban area is more susceptible compare to other because of higher exposure time of urban residents, due to unbound interference of air pollutants in indoor environment. Exposure to air pollutants has been associated with increased mortality and hospital admissions due to respiratory and cardiovascular diseases. This chapter focused on sources and health effects of air pollution in urban areas in India. Most of the urban areas of India are suffering from higher concentrations of air pollutants including gaseous and particulates.

Section 3
Climate Change Impacts on Biodiversity and Ecosystem Services

Chapter 7
Climate Change Impacts on Biodiversity in Arid and Semi-Arid Areas: Biodiversity Under Climate Change............ 117
Hanane Boutaj, Cadi Ayyad University, Morocco
Aicha Moumni, Cadi Ayyad University, Morocco
Oumayma Nassiri, Cadi Ayyad University, Morocco
Abdelhak Ouled Aitouna, Cadi Ayyad University, Morocco

Considerable attention has been paid to climate change and its impacts on biodiversity. The climate change has caused several problems such as continuous ecosystem degradation and a resultant biodiversity decline. In addition, climate warming has a range of indirect effects through changes in vegetation type level and sea that affect physical and biological systems. This has also led to changes in the distribution of species, as well as reductions in the size of populations, or even local extinctions of these populations. Moreover, many species are disappearing with time due to climate change combined with the emergence of disease that develops and increases with time. These problems affect different biodiversity components that are close to collapse. This chapter explored the richness of biodiversity in arid and semi-arid zones. It is also illuminates the effects of climate change on distribution of biodiversity. The authors highlight the responses of biodiversity under climate change, in terms of species extinction, biodiversity loss, and the impacts of climate change to ecological tourism. Finally, the authors show how biodiversity can overcome the effect of climate change, by developing some systems that allow to them to survive and conservation of species and ecosystems.

Chapter 8
Impacts of Climate Change on Biodiversity and Ecosystem Services: Current Trends............ 142
Vartika Singh, Amity University Noida, India

The chapter illuminates the effect of climate change on biodiversity and ecosystem services. It also explains the present scenario of biological resources with biological diversity. The author describes the known biological resources and their uses with the help of flow chart and also shows the linkage among them. The author clearly explicates the inter-relationship between human activity, climate change, and biological diversity with charts. In the last section, the author talk about the effect on the paleo climate and impact of climate change on different parts of ecosystem services.

Chapter 9
Impact of Climatic Changes on Earth's Survival: An Analysis of Changing Environment and Its Consequences at a Global Level............ 160
Akash, Gurukula Kangri University, India
Navneet, Gurukula Kangri University, India

Every species' survival on earth is dependent on each other for their demand and dependent on the environment and various other sources. These resources include fresh food, clean drinking water, timber for construction, natural gas and coal for industries, fibers for clothing. All the human activity affects the environment severely in different ways. The biggest threats to the environment are climatic changes. Climate is an important factor that affects all survival on earth. The different pollutants, transport, dispersion, chemical transformation, as well as the deposition can be affected by meteorological variable such as humidity, wind, temperature. Climatic changes are expected to worsen the quality of air and water by changing the atmospheric processes and chemistry. Not only human beings but every aspect of the ecosystem is affected due to the changing climate. This chapter will explore the impacts of climatic changes on biodiversity by various activities of humans. Additionally, it will sketch how the impacts can be reduced by plants.

Chapter 10
Potential Impacts of Climate Change on Land Degradation and Desertification: Land Degradation and Climate Change 183
Dharumarajan S., National Bureau of Soil Survey and Land Use Planning (ICAR), India
Veeramani S., National Bureau of Soil Survey and Land Use Planning (ICAR), India
Kalaiselvi Beeman, National Bureau of Soil Survey and Land Use Planning (ICAR), India
Lalitha M., National Bureau of Soil Survey and Land Use Planning (ICAR), India
Janani N., National Bureau of Soil Survey and Land Use Planning (ICAR), India
Srinivasan R., National Bureau of Soil Survey and Land Use Planning (ICAR), India
Rajendra Hegde, National Bureau of Soil Survey and Land Use Planning (ICAR), India

Land degradation and desertification have been graded as a major environmental and social dispute in most of the emerging countries. Changes in temperature, wind speed, and precipitation patterns will influence plant biomass production, land use, land cover, soil moisture, infiltration rate, runoff and crop management, and ultimately, land degradation. Close relations between climate change and land degradation processes have been perceived in the past decades. Climate change models and land use models should be combined with hydrologic/erosion models to accurately compute or predict climate change impacts on land degradation. This chapter introduces the advancements in modeling of impact of climate changes in land degradation and need for the critical investigation to better understand and forecast the responses of land degradation processes to a changing climate in the future.

Chapter 11
Potential Impacts of Climate Change on the Inland Fisheries of Arid and Semi-Arid Regions of Africa: Impacts of Climate Change on Inland Fisheries 196
Imefon Udo Udo, University of Uyo, Nigeria
Imekan Isaac Akpan, Akwa Ibom State University, Nigeria

Inland fisheries of arid and semi-arid regions of Africa are seriously threatened by negative impacts of climate change. Literature and several models show increase in temperature of 1.1oC in some areas. Sea level rise is projected to increase to 0.8 m by the year 2100. Fish yields have increased almost linearly by around half a million metric tons per decade over the past 60 years, while clear cyclical variations in the residuals of about 20 years' periodicity above and below the trend line have been observed. Although fisher folks, their communities, and local institutions are already constantly adapting to various forms of change, flimsiness in the wider governance and macro-economic environment has weakened the overall adaptive capacity of these regions and fishers are vulnerable to projected climate change. For significant benefits of inland fisheries to be accomplished, planned adaptation at scales from the local to the regional is very necessary.

Chapter 12
Climate Change and Its Impacts on Oases Ecosystem in Morocco 217
Mohamed Ait El Mokhtar, Cadi Ayyad University, Morocco
Raja Ben Laouane, Cadi Ayyad University, Morocco
Mohamed Anli, Cadi Ayyad University, Morocco
Abderrahim Boutasknit, Cadi Ayyad University, Morocco
Abdessamad Fakhech, Cadi Ayyad University, Morocco
Said Wahbi, Cadi Ayyad University, Morocco
Abdelilah Meddich, Cadi Ayyad University, Morocco

Today, the international community agrees that climate change is one of the most important threats to the equilibrium of ecosystems. For its part, Morocco is not immune to this phenomenon because of climatic disturbances that the country has known for several years. Moroccan oases are among the most affected ecosystems by climate change, which has resulted in reduced yields of agricultural production, poverty, and deterioration of natural resources. The reports showed that in Morocco the area of palm groves has been significantly reduced during the last decades resulting from the upsurge of drought phenomena, soil salinity, and the spread of "Bayoud." Therefore, the deterioration of the socio-economic and environmental importance of the date palm is well established. Aware of these issues, many actions have been initiated by the Moroccan government to preserve this ecosystem. The aim of this chapter is to provide an overview of how climate change affects different levels of oases ecosystem in Morocco and to propose sustainable strategies to limit degradation of this vulnerable area.

Section 4
Climate Change, Ecosystem Services, Food Sercurity, and Health

Chapter 13
Ecosystem Services, Climate Change, and Food Security 247
Samreen Siddiqui, Texas A&M University – Corpus Christi, USA

Food security is a rising concern around the world, especially in developing countries within arid and semi-arid regions. Ecosystem provides different services to support living and human survival, which includes some major food sources around the world like agriculture, fisheries, and livestock. With advancing times, humans improved these services and produced enough food to support the rising population. However, with increasing greenhouse gases, a new problem came into existence, commonly known as climate change (CC), which accelerated issues like food security and safety. Under such issues, people don't have access to basic facilities and food supply to survive, and with future population growth estimates, it is becoming even more difficult. Some major food sources together with alternate sources are discussed in this chapter. Upcoming CC impacts are discussed in detail in relation to major food sources and supported by world maps to provide a better picture. Major actions, government initiatives, and some suggestions are also provided to overcome this global crisis.

Chapter 14
Ecosystem Services-Climate-Health Associations: Water-Climate-Leishmaniasis Nexus in an Endemic Focus of Zoonotic Cutaneous Leishmaniasis 280
Ahmed Karmaoui, Southern Center for Culture and Sciences, Morocco
Siham Zerouali, Southern Center for Culture and Sciences, Morocco
Ashfaq Ahmad Shah, National University of Sciences and Technology (NUST), Pakistan
Mohammed Yacoubi-Khebiza, Cadi Ayyad University, Morocco
Fadoua El Qorchi, Cadi Ayyad University, Morocco

Water is the main ecosystem service that supports the oasis system. Middle Draa Valley is an oasis zone located in the south of Morocco. The water availability in this area is the key element of vegetation cover change. This change added to other factors can cause some parasitic diseases. The zoonotic cutaneous leishmaniasis is one of these diseases. In this chapter, an analysis of the interaction between some key risk factors and the disease transmission was carried out. The outputs of this work rivaled that there is a very strong correlation between this disease and ecosystem services such as water stored and the dam outflow (directed to the oasis for the irrigation), and the groundwater availability. Regarding the correlation between this vector-borne disease and the cropping area, a strong correlation was recorded. However, for the relationship between leishmaniasis and the precipitation and the dam inflow, average correlations were found. Basically, in MDV, the water availability is the first element that affects an ensemble of processes that cause the disease infection.

Chapter 15
Relationships Between Climate Parameters and the Density of *Phlebotomus papatasi*, the Main Vector of Zoonotic Cutaneous Leishmaniasis 291
Ahmed Karmaoui, Southern Center for Culture and Sciences, Morocco
Siham Zerouali, Southern Center for Culture and Sciences, Morocco

This chapter presents an analysis of the seasonal activity of the Phlebotomus papatasi (leishmaniasis vector) through a correlation between its monthly activity and the monthly meteorological parameters. Three sites from three leishmaniasis endemic countries were used Morocco, Iran, and Greece. The high density of P. papatasi was observed in May in Greece, in July in Morocco and September in Iran. Before May and after September, an important decrease of P. papatasi activity was observed with a decline of maximum and minimum temperatures and an increase of precipitations. Secondly, strong associations between P. papatasi density and minimum and maximum temperatures were explored, third, no significant association have been detected between P. papatasi abundance and precipitations in the three sites. There is an average correlation between relative humidity and P. papatasi density for Isfahan and Creater Athens. Basically, P. papatasi is sensitive to low temperatures and high relative humidity, which makes these two variables better indicators for the possible emergence of cutaneous leishmaniasis.

Section 5
Biodiversity and Ecosystem Services Adaptation to Climate Change

Chapter 16
Effectiveness of Local Community Policy Responses to Climate Change Impact on Ecosystem Services for Biodiversity Conservation in the Semi-Arid Zones 304
Wilson Truman Okaka, Kyambogo University, Uganda

The Ugandan government has decentralized climate change policy strategies at the local government district development planning levels to build local community adaptive capacity for accelerated action for disaster resilience. This chapter focuses on the local community response by Isingiro local government district authority. The purpose is to provide a context of the significance of eco-services and natural biodiversity resources in the semi-arid district, located in the South-Western Uganda, elaborate on the district climate change action plan for community-based adaptation strategies, and the biodiversity conservation district development plan. For decades, most households and local communities have perpetually suffered from the severe climatic stress of galloping heatwaves, shocking floods, ferocious frequent whirlwinds and wind storms, catastrophic droughts, perennial food insecurity, malnutrition, migrations, and famines. The district has adopted strategies to enhance biodiversity conservation for eco-services for food security and sustainable community livelihoods.

Chapter 17
Adaptive Strategies of Small Family Farms in the Face of Climate Change: The Case of the Tleta Watershed in Northern Morocco 316
Mohamed El Amrani, National School of Agriculture of Meknes, Morocco

Climate change is now an accepted reality. It is felt globally through increased temperature and precipitation regime disruption and increased frequency of extreme events such as floods and droughts. In the absence of effective mitigation and adaptation actions, these changes could have significant negative impact on the sustainability of agriculture and the resilience of populations especially in areas with fragile ecology. However, these changes remain an issue that is difficult to grasp and still not well integrated into management strategies at the farm, sector, and territory levels. The objectives of this research are to describe the production systems, and to study the resilience strategies, perception, and adaptive practices of farms in the Tleta watershed in Northwest Morocco in the face of climate change and landscape dynamics. It describes farming systems and activities, attempts to analyze how farmers perceive global changes in their landscape, and adopts innovative strategies and practices to adapt to them. It also shows that the actions of institutional actors in the area that can contribute to the resilience of the populations are numerous but remain fragmentary and lack integration.

Compilation of References 331

About the Contributors 402

Index 406

Preface

The climate change concept refers mainly to the long-term change in precipitation and temperature. These two parameters are the main factors controlling the repartition of ecosystems, species, and ecosystem services. Since the publication of the first report of the Intergovernmental Panel on Climate Change (IPCC) in 1990, studies of the impact of climate change (CC) on ecosystems and communities have multiplied. The causes of this phenomenon are due to the natural processes of the Earth and the anthropogenic activities.

It is currently recognized that the industrial revolution of more than 150 years has contributed to the combustion of billions of tons of fossil fuels in order to produce energy; this in turn raises the carbon dioxide rate in air.

The threats due to climate change are large including air pollution, over-exploitation of resources, erosion of biodiversity, etc. In this book, we explore different aspects of CC in relation to ecosystem services and biodiversity. Additionally, we are briefly tracing the contours of climate change impacts in the arid and semi-arid zones. The climate change affects negatively more than six billion of people lives through different vital sectors; namely agriculture, economy, water, and health. There is a large number of variables that affects CC: soil, water, and air pollution, industrialization, urbanization, forest cutting and burning, and overuse of resources…

The most dangerous impacts of climate change include an increasing demand of ecosystem services and a decreasing trend of water supply, the accelerating of melting large glaciers, the raise of sea levels, climate disasters acceleration (flood and drought), soil fertility lose, agricultural yield decreasing, soil and biodiversity erosion, desertification, water born disease emergence, and vector borne disease occurrence.

This book tries to answer two main questions: First, what is the climate change impact on biodiversity and the associated ecosystem services? Second, what is the impact on economy, water and soil resources, well-being, and health in arid and semi-arid zones?

The main objectives of this book are to:

- Provide an essential resource for practitioners in the field, specifically, in ecosystem services research projects.
- Reflect on current trends in this field.
- Help to understand how climate change impacts the biodiversity and the associated ecosystem services and how it can be overcome at various strategic policy formulations.
- Provide relevant theoretical frameworks and the latest empirical research findings for professionals in this field.

- Highlight natural resources on which majority of people depend, such as water for drinking and irrigation, and soil for crop production

This book provides an opportunity to share experiences and methods for assessing ecosystem services and biodiversity. The focus of the edited Book is to link science and practice to use ecosystem services concept in conservation and restoration. According to the Millennium Ecosystem Assessment, ecosystem services are classified into four types: Provisioning, Regulating, Cultural, and Supporting services. As it is well established, it is important to understand the effects of climate change and the human pressure on these types of ecosystem services. It provides also a set of new tools and research trends, models, relevant literature, and analysis on this topic. These can be used for education, sensitization and scientific research, and then suggest strategies for different spatial temporal scales. It also explores several components related to ecosystem services and draw different directions to assess their vulnerability. Finally, it provides analyses of socio-ecological systems in relation to climate change, as well a multidisciplinary vision of ecosystem services.

This book is a collaborative effort of several experts across the world, with a focus on spatial-temporal levels. This collective book covers a wide range of domains and areas, including: water, land, air, agriculture, fisheries, health, and socioeconomic systems. The topics explored in this book are summarized hereafter.

Chapter 1 investigates the need to establish long-term ecological research network in morocco as a tool to monitor ecosystems under climate change. Several countries are associated with the Long-Term Ecological Research (LTER) program, while others are in the process of joining the network. In Morocco, there is an urgent need to be a member of this network because the diversity of challenges associated with the various ecosystems requires multidisciplinary long-term studies. The chapter discusses firstly, this need by assessing the environmental vulnerability of Morocco basing on data from the Environmental Vulnerability Index profile. Ecosystem change and vulnerability were investigated at the sites within the LTER program by exploring the criteria of the selected sites. The Driving Forces, Pressures, States, Impacts, Responses framework would be a useful approach to study and explain the ecological changes of each selected site. The chapter highlights secondly, the purpose, significance, the mission, objectives and the international cooperation of the proposed LTER Network called (Mo-LTER).

Chapter 2 explores the associations between climate, ecosystems, and ecosystem services in the Pre-Sahara. This chapter focuses on the richness and wide variety of Sub-Saharan ecosystems particularly in the region of Tafilalet, Morocco. The Tafilalet ecosystems provide an extensive list of goods to the region, as well as "ecosystem services" for the inhabitants. However, in the last 30 years, these ecosystems are threatened by conjugated impacts of climate change and human overuse of the natural resources. The primary objective of this chapter is to evaluate both the damage inflicted on Tafilalet ecosystems and the associated ecosystem services due to climate change. Adaptation to climate change is of paramount importance to keep the survival of the diversity of Tafilalet ecosystems; including agro-ecosystems, forests, and inland waters. Adaptation is a process to reduce the vulnerability of people to prevent displacement and conflict for scarce ecosystem services. The findings of this chapter show that a large number of ecosystem services of the region are limited by climatic conditions that affect the resilience of the whole oasis of Tafilalet.

Chapter 3 focuses on ecosystem services demand management under climate change scenarios, using WEAP Software in case of water demand in the arid zone of Morocco. In fact, water ecosystem service is the most important element that supports Tafilalet agro-ecosystems. In this region, drought frequency is increasing, which complicate the management groundwater reserves. The ephemeral flows of the rivers force people to use groundwater to meet the population demand. Consequently, water resource management is of significant importance the sustainability of this area. Water Evaluation and Planning (WEAP) is useful management software used to evaluate and trace the trend of water demand. This model was applied in case of Ziz basin in order to simulate and analyze the situation of water under different scenarios. The results show an increasing of demand for water irrigation and with introducing modern irrigation scenario. However, a decreasing trend in reservoir storage volume and groundwater storage was projected in this area.

Chapter 4 explores the monetary value change of some provisioning ecosystem services of an oasis zone of Morocco. The desert oasis is a wetland agro-ecosystem more and more under pressure. This trend threatens the oasis ecosystem services (ES) and then the well-being of the local population. In order to quantify this trend in Middle Draa Valley, south of Morocco, the monetary values for a selection of ES have been estimated. In this chapter, the economic values of 28 provisioning ES in 2014 was examined and next compared with the economic values of 2009. These values were standardized to US$ per hectare per year. The total value of the studied provisioning ES in 2014 was found to be 8658.3 US$/ha/year for the arable area (26000ha) and 150 US$/ha/year for the whole Middle Draa Valley (1500000ha). Comparing these findings with the data of 2009, a slight increase of $1245.7 (in 26000ha) and $21.5 US$/ha/year (the whole area) was detected in the period 2009-2014.

Chapter 5 highlights the potential role of endophytes for sustainable environment. Endophytes are symptomless fungal and bacterial microorganisms found in almost all living plants. They are vital components of plant microbiomes. Endophytes affect plant growth and plant responses to pathogens, herbivores, and environmental change by producing a range of natural products having antifungal, antibacterial and insecticidal properties. Endophytes have shown particular promise in agriculture particularly as beneficial crop inoculants, and are known to enhance abiotic and biotic plant stress tolerance by increasing tolerance to drought and water stress, as well as tolerance to high temperature and high salinity. A better understanding of their plant growth-promoting mechanisms could simplify higher production of energy crops in a more sustainable manner even on marginal land and feed stocks for industrial processes, thus contribute to avoiding conflicts between food and energy production Many endophytes can be exploited to improve the efficiency of phytoremediation as they are found to be resistant to heavy metals and capable of detoxifying organic contaminants.

Chapter 6 studies some health effects of air pollution in urban environment. In the last few decades, urban development and growing industrialization have caused air pollution to become a major issue in urban areas of developing and developed countries. The urban area is more susceptible compared to other because of higher exposure time of urban residents, due to unbound interference of air pollutants in the indoor environment. Exposure to air pollutants has been associated with increased mortality and hospital admissions due to respiratory and cardiovascular diseases. This chapter focused on sources and health effects of air pollution in urban areas in India. Most of the urban areas of India are suffering from higher concentrations of air pollutants including gaseous and particulates.

Chapter 7 analyses the Climate Change Impacts on Biodiversity in Arid and Semi-Arid Areas. Currently, considerable attention has been paid to climate change and its impacts on biodiversity. Climate change has caused several problems such as continuous ecosystem degradation and a resultant biodiversity decline. In addition, climate warming has a range of indirect effects through changes in vegetation type level and sea that affect physical and biological systems. This has also led to changes in the distribution of species, as well as reductions in the size of populations, or even local extinctions of these populations. Moreover, many species are disappearing with time due to climate change combined with the emergence of disease that develops and increases with time, affecting different biodiversity components that are close to collapse. This chapter is about exploring, how biodiversity has changed under climate change and looking into the possibility of how biodiversity may overcome climate change.

Chapter 8 reviews the impacts of climate change on biodiversity and ecosystem services, highlighting the current trends. It explains also the present scenario of biological resources with biological diversity. The author enlightens the known biological resources and their uses with the help of flowchart and also shows the linkage among them. The author clearly explicates the inter-relationship between human activity, climate change and biological diversity with the chart. In the last, the author talks about the paleoclimate.

Chapter 9 explores the impact of climatic changes on earth's survivals at global level. Every survival on earth is dependent on each other for their demand and depends on the environment various other sources. These resources include fresh food, clean drinking water, timber for construction, natural gas, coal for industries, and fibers for clothing. All the human activity affects the environment severely in different way. The biggest threats to the environment are climatic changes. Climate is an important factor which affects every survival on earth. The different pollutants, transport, dispersion, chemical transformation as well as the deposition can be affected by meteorological variable such as humidity, wind, temperature. On the other way climatic changes are expected to worsen the quality of air, water by changing the atmospheric processes and chemistry. Not only human beings, but every aspect of ecosystem gets affected due to the changing of climate. This chapter will explore about the impacts of climatic changes on biodiversity by various activity of human. Additionally, it will on an out sketch the impacts can be reduced by plants.

Chapter 10 focuses on the potential impacts of climate change on land degradation and desertification. These later have been graded as a major environmental and social dispute in most of the emerging countries. Changes in temperature, wind speed, and precipitation patterns will influence plant biomass production, land use land cover, soil moisture, infiltration rate, runoff and crop management, and ultimately land degradation. Close relations between climate change and land degradation processes have been perceived in the past decades. Climate change models and land use models should be combined with hydrologic/erosion models to accurately compute or predict climate change impacts on land degradation. This chapter introduces the advancements in the modeling of the impact of climate changes in land degradation and needs for the critical investigation to better understand and forecast the responses of land degradation processes to a changing climate in the future.

Chapter 11 investigates the potential impacts of climate change on the inland fisheries of arid and semi-arid regions of Africa. Inland fisheries of arid and semi-arid regions of Africa are seriously threatened by negative impacts of climate change. Literature and several models show an increase in temperature of 1.1°C in some areas. Sea level rise is projected to increase to 0.8 m by the year 2100. Fish yields have increased a most linearly by around half a million metric tonnes per decade over the past 60 years, while clear cyclical variations in the residuals of about 20 years' periodicity above and below the trend line have been observed. Although Fisher folks, their communities, and local institutions are already constantly adapting to various forms of change, flimsiness in the wider governance and macroeconomic environment has weakened the overall adaptive capacity of these regions and fishers are vulnerable to projected climate change. For significant benefits of inland fisheries to be accomplished, planned adaptation at scales from the local to the regional is very necessary.

Chapter 12 analyses the climate change impacts on oases ecosystem in Morocco. The international community agrees that climate change is one of the most important threats to the equilibrium of ecosystems. For its part, Morocco is not immune to this phenomenon because of climatic disturbances that the country has known for several years. Moroccan oases are among the most affected ecosystems by climate change, which has resulted in reduced yields of agricultural production, poverty and deterioration of natural resource. The reports showed that in Morocco the area of palm groves has been significantly reduced during the last decades resulting from the upsurge of drought phenomena, soil salinity and the spread of "Bayoud". Therefore, the deterioration of the socio-economic and environmental importance of the date palm is well established. Aware of these issues, many actions have been initiated by the Moroccan government to preserve this ecosystem. The aim of this chapter is to provide an overview of how climate change affects different levels of oases ecosystem in Morocco and to propose sustainable strategies to limit degradation of this vulnerable area.

Chapter 13 examines the association between climate change, ecosystem services and food security. Food security is a rising concern around the world, especially in developing countries within Arid and Semi-Arid Regions. Ecosystem provides different services to support living and human survival, which includes some major food sources around the world like agriculture, fisheries, and livestock. With advancing times human improved these services and produced enough food to support the rising population. However, with increasing greenhouse gases, a new problem came into existence, commonly known as climate change (CC), which accelerated issues like food security and safety. Under such issues, people don't have access to basic facilities and food supply to survive and with future population growth estimates; it is becoming even more difficult. Some major food sources together with alternate sources are discussed in this chapter. Upcoming CC impacts are discussed in detail in relation to major food sources and supported by world maps to provide a better picture. Major actions, government initiatives, and some suggestions are also provided to overcome this global crisis.

Chapter 14 analyses the Ecosystem Services-Climate-Health Associations, highlighting the Water-Climate-Leishmaniasis Nexus in an Endemic Focus of Zoonotic Cutaneous Leishmaniasis. Water is the main ecosystem services that support the oasis system. Middle Draa Valley is an oasis zone located in the south of Morocco. The water availability in this area is the key element of vegetation cover change. This change is added to other factors can cause some parasitic diseases. The Zoonotic cutaneous leishmaniasis is one of these diseases. In this chapter, an analysis of the interaction between some key risk factors and the disease transmission was carried out. The outputs of this work rivaled that there is a very strong correlation between this disease and ecosystem services such as water stored and the dam outflow (directed to the oasis for the irrigation), and the groundwater availability. Regarding the correlation between this vector-borne-disease and the cropping area, a strong correlation was recorded. However, for the relationship between leishmaniasis and the precipitation and the dam inflow, average correlations were found. Basically, in MDV, the water availability is the first element that affects an ensemble of processes that cause the disease infection.

Chapter 15 explores the relationships between climate parameters and the density of Phlebotomus papatasi, the main vector of Zoonotic Cutaneous Leishmaniasis. This chapter presents an analysis of the seasonal activity of the *Phlebotomus papatasi* (leishmaniasis vector) through a correlation between its monthly activity and the monthly meteorological parameters. Three sites from three leishmaniasis endemic countries were used Morocco, Iran, and Greece. The high density of *P. papatasi* (was observed in May in Greece, in July in Morocco and September in Iran. Before May and after September, an important decrease in *P. papatasi* activity was observed with a decline of maximum and minimum temperatures and an increase of precipitations. Secondly, strong associations between *P. papatasi* density and minimum and maximum temperatures were explored, third, no significant association have been detected between *P. papatasi* abundance and precipitations in the three sites. Basically, *P. papatasi* is sensitive to low temperatures and high relative humidity, which makes these two variables better indicators for the possible emergence of cutaneous leishmaniasis.

Chapter 16 focuses on the effectiveness of Decentralized Climate Change Policy to Climate Change Impact on Ecosystem Services in Ugandan Semi-Arid Zones. Ugandan government has decentralized climate change policy strategies at the local government district development planning levels to build local community adaptive capacity for accelerated action for disaster resilience. This chapter focuses on the local community response by Isingiro local government district authority. The purpose is to provide a context of significance of ecoservices and natural biodiversity resources in the semi-arid district, located in the South –Western Uganda, elaborate on the district climate change action plan for community based adaptation strategies, and the biodiversity conservation district development plan. For decades, most households and local communities have perpetually suffered from the severe climatic stress of galloping heat waves, shocking floods, ferocious frequent whirlwinds and wind storms, catastrophic droughts, perennial food insecurity, malnutrition, migrations, and famines. The district has adopted strategies to enhance biodiversity conservation for ecoservices for food security and sustainable community livelihoods.

Chapter 17 presents some adaptive strategies of small family farms in the face of climate Change in Northern Morocco. Climate change is now an accepted reality. It is felt globally through increased temperature and precipitation. In the absence of effective mitigation and adaptation actions, these changes could have a significant negative impact on the sustainability of agriculture and the resilience of populations especially in areas with fragile ecology. However, these changes remain an issue that is difficult to grasp and still not well integrated into management strategies at the farm, sector and, territory levels. The objectives of this research are to describe the production systems and to study the resilience strategies, perception and adaptive practices of farms in the Tleta watershed in the North West Morocco in the face of climate change and landscape dynamics. It describes farming systems and activities, attempts to analyze how farmers perceive global changes in their landscape, and adopt innovative strategies and practices to adapt to them.

Ahmed Karmaoui
Southern Center for Culture and Sciences (SCCS), Morocco

Acknowledgment

Climate change impacts on ecosystem services and biodiversity have become common in Ecology and in its associated domains; the socio-economic and biophysics. In order to assess these impacts, multidimensional research is needed. In this context, several researchers and experts have proposed empirical and theoretical approaches.

This book would not be possible without the contribution of the IGI Global team. I thank the international publisher IGI Global, for giving me this excellent way to realize and edit this book. Special thanks go to, book development editor, IGI Global, for the help, the patience, and support in the realization of this collective work. Specifically, I would like to thank the authors who contributed to the content of this book and the reviewers for their comments and suggestions.

I appreciate the support of members of IGI Global for the development of this edited book. If I could edit this book it is thanks to my little family, my wife SIHAM, my two children, Yassin and Amine. Many thanks are addressed to my dad MIMOUN for his continuous encouragement. There is also the time to say "thank you" to my friends and colleagues.

Section 1

Climate Change, Ecosystem, and Ecosystem Services

Chapter 1
The Need to Establish a Long-Term Ecological Research Network in Morocco as a Tool to Monitor Ecosystems Under Climate Change

Ahmed Karmaoui
https://orcid.org/0000-0003-3881-4029
Southern Center for Culture and Sciences, Morocco

Siham Zerouali
Southern Center for Culture and Sciences, Morocco

ABSTRACT

Several countries are associated with the long-term ecological research (LTER) program, while others are in the process of joining the network. In Morocco, there is an urgent need to be a member of this network because the diversity of challenges associated with the various ecosystems requires multidisciplinary long-term studies. The chapter discusses this need by assessing the environmental vulnerability of Morocco basing on data from the environmental vulnerability index profile. Ecosystem change and vulnerability were investigated at the sites within the LTER program by exploring the criteria of the selected sites. The driving forces, pressures, states, impacts, responses framework would be a useful approach to study and explain the ecological changes of each selected site. The chapter highlights the purpose, significance, the mission, objectives, and the international cooperation of the proposed LTER network called Mo-LTER.

DOI: 10.4018/978-1-5225-7387-6.ch001

INTRODUCTION

The international long-term ecological research Network (ILTER) is an international network of selected sites of different ecosystems throughout the world that study environmental changes (Nakano *et al.,* 2012). Kim, (2006) reported that the long-term ecological research is essential in providing understanding of spatial-temporal changes. Additionally, it provides knowledge required for sustainable development in global collaborations (Mirtl et al., 2018). The LTER program was initiated in the USA by the National Science Foundation (NSF) in 1980 to support research on ecological processes (Callahan, 1984). The purpose of this program is to study these ecological processes at different spatiotemporal scales (Karasti & Baker, 2008). The gathered information at long-term scale can give valuable finding on ecosystem structure and function (Holmes & Sherry, 2001; Holmes, 2011). Thus, ILTER's focus is on long-term. In fact, the understanding of the changes of ecosystems and resources were done on short-term rather than on long-term studies (Kohli *et al.,* 2005). Historically, the funding programs played an important role in the development of short-term than long-term researches (ILTER, 1998). The long-term studies became very interesting to understand the ecological process change. For example, the trees need several decadal years to develop, hurricanes returns in one site each 50 years, drought for decades; then, a long-term vision is required to understand the changes (Hobbies *et al.,* 2003).

The LTER network gathers several sites at different geographic scales. The importance of this network resides especially in the information management (Karasti & Baker, 2008). Since the beginning of LTER network, data was one of the important issues in its agenda (Baker *et al.,* 2000). The Need to Establish LTER Network was realized in the first time as mentioned above by the NSF in 1980, and its mission (Waide *et al.,* 1998) had been (between others):

- Understand ecological phenomenon at long-term scale;
- Constitution of a historical observation on long-term experiment;
- Beginning complex research efforts;
- Providing information on ecological questions to the society.

In 1980, the United States LTER was established, including five sites which have evolved to 26 sites. In 1993, the International LTER Network was constituted and today, it contains 44 active networks (Mirtl et al., 2018). The first is the Jordan that is working to establish its LTER Network. The second is the Norwegian long-term monitoring and research sites, is not yet a member of the International LTER. Four countries in southern Africa have joined the network, and none of the North African countries are scheduled to join the LTER network. In this chapter, we discuss the need to establish a Long-Term Ecological Research Network in Moroccan that we'll call Moroccan Long-Term Ecological Research (Mo-LTER).

The choice of Morocco is due to its ecological richness and the high environmental vulnerability. This creates the need to a long-term research and to a detailed vulnerability assessment. Morocco, like other countries in the Mediterranean region is experiencing, and may know important changes in land use after Millennium Ecosystem Assessment (2005) and whose changes are related to climate change, a socio-economic context and to high population dynamics (Messouli *et al.,* 2013). According to a report from the World Bank (World Bank, 2011), the climate change will influence the Moroccan unmet demand of water, and the shortage will increase from 10 to 20 km^3 in 2020-2030 period and 40 km^3 in 2040-2050 period. Elsewhere, agriculture plays an important role in Morocco, for the national economy. In fact, it contributes 19% to GDP (Kingdom of Morocco, 2013). According to Arrus and Rousset (2007), rainfed crops will suffer a regression of 10% in normal years to approximately 50% in a dry year in the horizon of 2020. In term of annual adaptation costs, the Moroccan government spends from $71 to $92 million for specific research needs especially in rural sector (roads and irrigation) (Rosegrant, 2009).

Through this chapter, we can make contribution by proposing and discussing the possibility to joining Morocco to the ILTER. Firstly, we examine and discuss environmental vulnerability of Morocco and secondly, the need of having long-term monitoring network in the country. The topic is contemporary and the importance of the monitoring network is understandable for the country to take appropriate measures for sustainable management of country's resources and environment.

The methodology used combines data extracted from the Environmental Vulnerability Index profile developed by the Stockholm environment institute (SEI) at national scale, and also DPSIR model at regional scale (pilot sites), for the main three selected sites and then highlights and expresses the significance and the need for establishing LTER in Morocco that we have called Mo-LTER. This paper highlights also the criteria of pilot sites choice, mission, objectives and the possible international cooperation of this Network.

MATERIALS AND METHODS

Study Area

The Kingdom of Morocco is located on the north-western edge of North Africa, bordering the North Atlantic Ocean and the Mediterranean Sea (Figure 1). It has common borders with Algeria and Mauritania.

Morocco's geography is diverse. The northern and western regions are the most fertile lowlands and also the most developed in the country (Speth et al., 2010). However, the mountainous system and the south of the Atlas Mountain are less developed and marginalized (Speth et al., 2010).

In this country, the biodiversity is very rich, effectively including 3913 of flora taxa, with 1298 subspecies (981 families) (Rankou *et al.*, 2015). Fennane & Ibn Tattou (2012) reported 7000 species of flora, including 1350 endemic. However, the fauna includes about 24, 000 species (92 species of terrestrial mammals, 334 species of birds and about 15 300 species of terrestrial invertebrates); the marine fauna includes more than 7100 species, and continental fauna, includes 1575 species (SECE, 1997). This richness is threatened by climate change and anthropogenic impacts. In fact, people's lives are closely linked to climate and its fluctuations. The economy is very dependent on water resources, agriculture, tourism, and coastal. In the last thirty years, the frequency of extreme flood and drought has increased in Morocco (Agoumi, 2003). These extreme climatic events are a real threat to the socio-economic development and will accelerate the environmental vulnerability. Thus, because of climate change and anthropogenic impacts various forms of environmental degradation were recorded in the country as reported in DOEC (1999):

- Forest loss of about 31 000 ha per year;
- Direct land clearing for cultivation (about 4500 ha / year);
- Fires ravaged about 4,500 ha / year;
- Water and wind erosion of 22 000 ha / year;
- 90% of Moroccan soils are threatened by desertification problems;
- Degradation following mining and quarrying.

Figure 1. Morocco location
Source: (Authors)

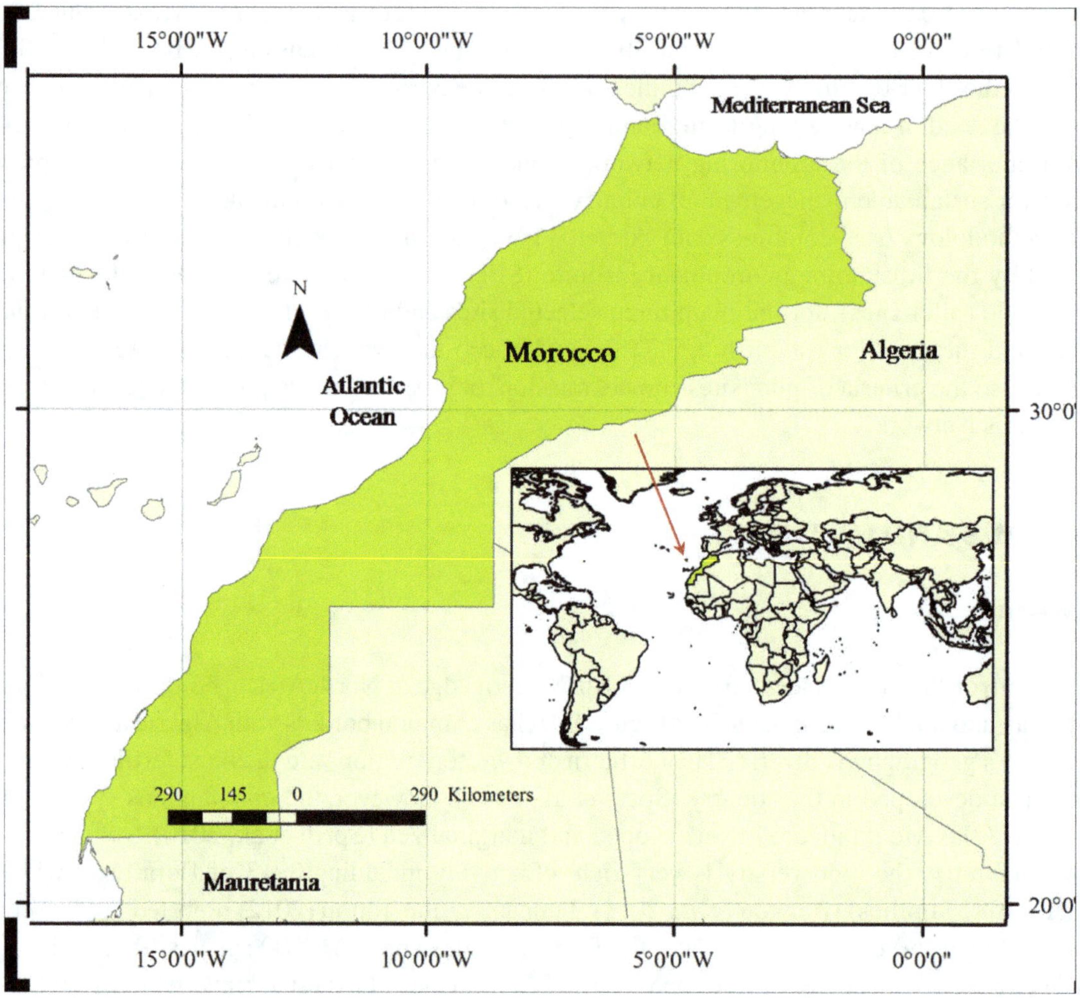

Over-exploitation of resources (over-fishing, over-exploitation of algae, over-grazing, loss of habitat through deforestation, urbanization, loss of wetlands) threatens many species with extinction. Indeed, 1670 species of flora, 610 species of fauna are threatened (SECE, 1997).

The need to join the international LTER network is justified by the ecological importance (the diversity of climate, the relief, and the rich biodiversity) and the environmental vulnerability.

Methodology

The current chapter was split into two stages. The first deals with the current environmental conditions (drivers and responses) in Morocco and the second presents a logical and government-supported LTER design in view of the vulnerability assessment.

Protecting the environment requires a global vision; it requires the mobilization of financial, human and technical capital. The environment of Morocco is threatened at numerous levels: air and water pollution, waste recycling difficulties, and massive deforestation. The environmental vulnerability index (EVI) is a tool that leads to a global vision of the environmental vulnerability state (citation needed). The index could serve as a basis for improving strategies for sustainable development. With five-year evaluations repeated, it could become a monitoring tool for monitoring policy responses.

In order to assess the environmental vulnerability, we used information extracted from the profile of the EVI developed by the Stockholm environment institute. The notion of vulnerability has a global recognition for the sustainable development of countries (Kaly et al., 2003). Analysis of environmental vulnerability is performed using the results collected from this index profile. In this chapter, we extracted, compiled and analyzed the indicators scores of the EVI profile of Morocco. The EVI uses 50 indicators for estimating the vulnerability of the environment of a country. The integrated data of these indicators is located in a scale that ranges from 1 to 7, where 1 indicates low vulnerability and 7 is extreme vulnerability. The position of the Moroccan environmental vulnerability have been compared also with the other four countries of North Africa, namely, Algeria, Tunisia, Libya and Egypt. This comparison was carried out based on total scores of these countries extracted and compiled from their environmental vulnerability profiles. In order to describe the environmental vulnerability of Morocco, a selection of indicators with high vulnerability scores was done.

The LTER established should identify some sites representing the relative common region of the country (with the same geographic, socioeconomic, and climatic characteristics). The proposed sites must fulfill the basic requirements considered essential to ensure the continuation of long-term research activities. These sites will have a scientific body to conduct the research projects (Sprott, 1998). To determine these sites, several criteria were adopted and the DPSIR frameworks were used, basing on climatic, geographic, life-style, water resources and environmental degradation stress factors. The DPSIR (Driving Force - Pressure - State - Impact – Response) is applied as an integrated evaluation approach to collect the change factors. This approach is useful approach to study and explain ecosystem change and presents a method to evaluate the ecosystems degradation. All the DPSIR factors help to simplify the complexities and the interaction of all components of the system. The result of this assessment can be used as a baseline for data collection and decision makers. Basing on the outputs (indicators, drivers and responses), we have presented a government-supported LTER design in view of the first stage.

In this stage, we discuss the possibility to joining the International LTER in order to express and defend this need and facilitate the establishment of the related sites. In order to joining Morocco to the International LTER Network, the procedure established by this international network was followed.

RESULTS

Assess the Moroccan Environmental Vulnerability Through Environmental Vulnerability Index (EVI)

Comparing the vulnerability of five countries of the North Africa, Morocco scores 315, which classes it as the most vulnerable (Table 1) than Algeria, Tunisia, Libya and Egypt to climate change and anthropogenic impacts.

The data of the Moroccan environmental vulnerability was used and compiled in Figure 2 (the heading graph). We selected indicators scored four or above that means vulnerable to extremely vulnerable to climate change and anthropogenic pressure.

Table 1. Environmental vulnerability of Northern Africa, with the scores and classification of each county

Score	315	275	306	256	298
Classification	Vulnerable	Vulnerable	Vulnerable	At risk	Vulnerable
Countries	Morocco	Algeria	Tunisia	Libya	Egypt

Source: (Sopac, n.d.)

Figure 2. Environmental vulnerability indicators of Morocco.
Source: (Authors' extraction and compilation from Environmental vulnerability Index profile).

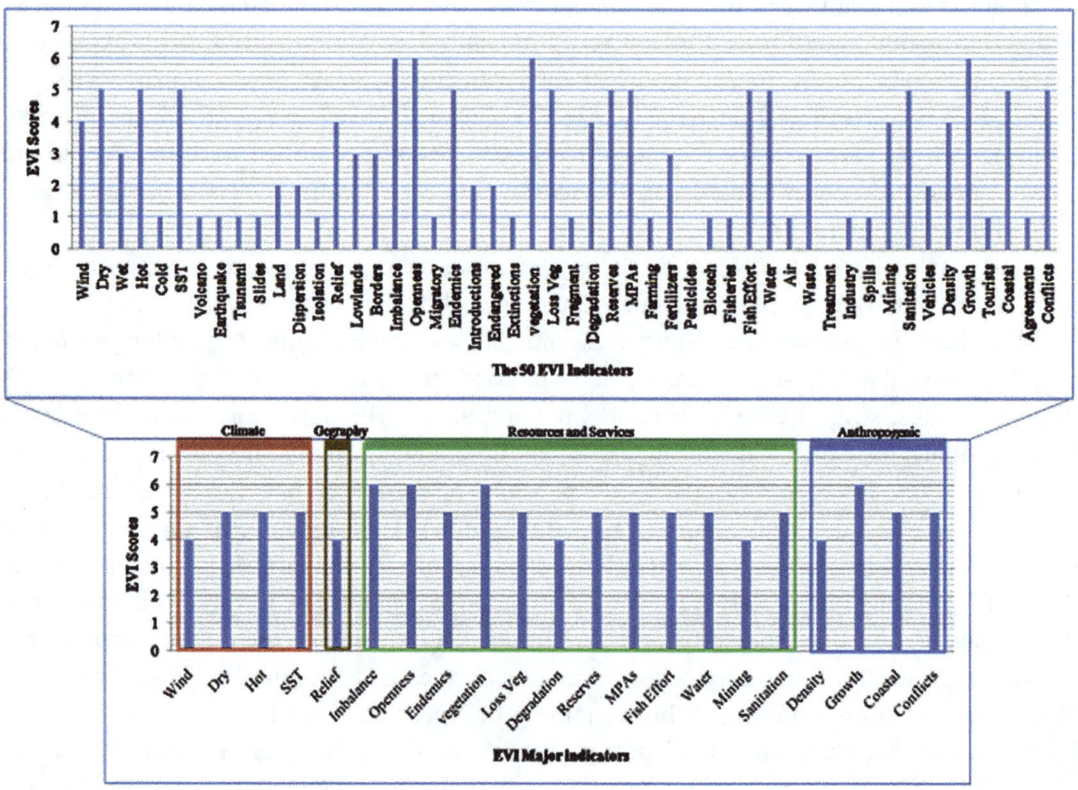

From Figure 2, Morocco is vulnerable on indicators of (*Imbalance, Openness, Endemics, Vegetation, Loss Veg., Degradation, Reserves, MPAs, Fish Effort, Water, Mining and Sanitation, Hot, SST, Dry, Relief, Coastal, Growth and Conflicts*). The Moroccan profile shows also no vulnerability in geologic indicators.

This means that Morocco is vulnerable to extremely vulnerable in terms of 21 indicators from 50. Population and industrial overload causes a multitude of environmental degradation forms. The indicators listed in (Figure 2), show increasing vulnerability for water and soil services. These services enter in the process of the majority of ecosystem services. In fact, our well-being depends critically upon the ecosystem services.

Dynamic Change by Pilot Sites Using DPSIR Model

Criteria for Selection of LTER Sites

What will be the criteria for site selection?

1. Based on Climatic characteristics, Knippertz *et al.,* (2003) classify Morocco in three homogeneous rainfall regions, the northern and western parts of Morocco (Atlantic region: ATL) including the Atlantic coastline region, northeastern Morocco and northwestern Algeria close to the Mediterranean coast (Mediterranean region: MED), and the Moroccan and Algerian stations south of the Atlas Mountains (Atlas region: SOA), including the oasis regions.
2. The presence of ecological, biological and historical high interest sites;
3. Even for administrative regionalization, there is a cultural similarity and common historical characteristics;
4. Regional division of the National direction of meteorology (NDM project: scheduled);
5. Natural resources (water, types and soil quality, biodiversity, and forest types);
6. The different topography of a site to another;
7. Great diversity of ecosystems: Ocean, Sea and Sahara, Mountains, Plateaus, wide plain, Lakes, large rivers…

The Proposed Pilot Sites of the Moroccan LTER

Figure 3 summarizes the approximate location of the three selected sites based on the mentioned criteria. The criteria help to differentiate three main sites, the Ocean (Atlantic) site in the west, the Sea (Mediterranean) in the north and the Sahara in south and south eastern. The Mediterranean site is bounded on the north by the Mediterranean Sea, west by the Atlantic Ocean, and south by the Gharb region. The Atlantic site is located in the Argan Biosphere Reserve. The Saharan site is at the border of the hyper-arid deserts, oasis those participate in maintenance of transitional spaces that are buffers.

The proposed network can contain ecosystems like Dry Valleys in the Sahara zone (Arid) and some very complex ecosystems such as the Atlantic forest in the Atlantic zone (Semi Arid) and the Mediterranean zone (Humid). Table 2 gives a comparative description of these sites.

Using the same indicators as mentioned above and following the DPSIR classification (Driving Force - Pressure - State - Impact – Response), we can aggregate the degradation factors as shown in table 4.

After determining the three sites, we have explored the dynamic of each site following the DPSIR model. Based on the DPSIR classification using available information, we gathered the main indicators of change that impact each site (Table 3).

Table 3 depicts each one of the model components. Therefore, socioeconomic drivers, environmental pressures, the environment state of the selected sites are explored; and thereafter impacts on local population. Several changes in the selected sites have been recorded in the 50 last years. Human impact on biodiversity has been documented on all national and local projects about the alteration of the ecosystems. An example of these impacts is the advanced desertification. Agriculture, livestock breeding and urbanization, are the principal socioeconomic drivers that exert various pressures on well-being of the local population. Measures can be taken to protect and restore the different component of the sites. State response should focus upon this issue.

Figure 3. Geographic situation of the three pilot sites of the Mo-LTER; 1, the Mediterranean, 2, the Atlantic, and 3, the Saharan site. For the reserves: Intercontinental Biosphere Reserve of the Mediterranean (IBRM), Argan Biosphere Reserve (ABR), and Biosphere Reserve of the south (oases) or BRS
Source: (Authors)

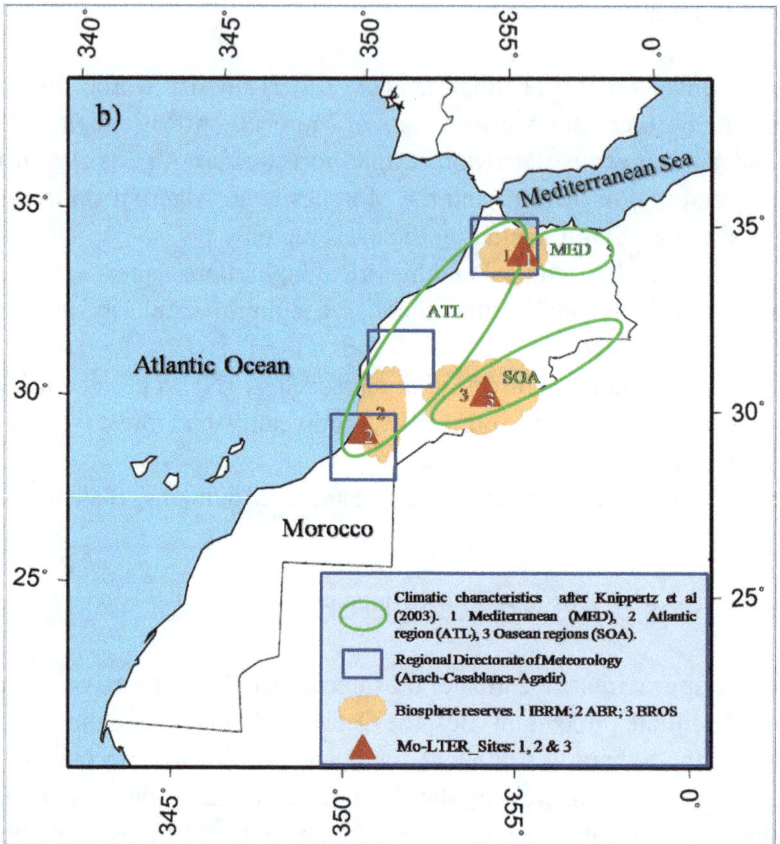

DISCUSSION

Moroccan Environmental Vulnerability

Morocco has a set of ecosystems with remarkable biodiversity (El Fasskaoui, 2009), like in other vulnerable countries, climate change is impacting this biodiversity (IPCC, 2007). Additionally, at global scale, the conjugated climate change and human pressure lead to the six mass extinction of biodiversity (Pereira *et al.,* 2010). The rate of species extinction is now about 1000 times the natural rate (IUCN, 2005). In addition, 4161species (aquatic and terrestrial) are threatened by climate change, 33% at risk because of habitat shifts and climate change, 29% threatened by extreme temperatures, and 28% by drought (Staudinger *et al.*, 2012). In North Africa, the data show an increase of 1°C in the 20th century, with an important trend in the last 30 years century (CHANGE, 2003). These data also indicate an increase in the frequency of climatic extreme events. In fact, we passed from a drought each 10 years from the beginning of the century to five-six in the last decade (Agoumi, 2003). At Moroccan scale, the simulation of climate (Babqiqi, 2008) gives the following information:

Table 2. Description of the three selected pilot sites

	Site 1: Mediterranean Site	Site 2: Atlantic Site	Site 3: The Saharan Site
Climatic characteristics (Knippertz *et al.*, 2003)	Northeastern Morocco and northwestern Algeria close to the Mediterranean coast	Northern and western parts of Morocco (Atlantic region) including the Atlantic coastline region	Moroccan and Algerian stations south of the Atlas Mountains (Atlas region), including the oasean regions.
Biosphere Reserve (BR)	Intercontinental BR of the Mediterranean (IBRM) (UNESCO/2006), located in the heart of Tingitane peninsula and Andalusia (South Spain) *.	Argan BR, original Moroccan forest, located in semi arid region, declared as a biosphere reserve of southern Morocco (UNESCO/1998).	Moroccan BR of the south (oases) was classified (UNESCO/2000). Indeed it plays an important socio-economic and ecological role.
Socio-economy	PDES, 2004: • The urbanization rate (Tanger-Tétouan) amounted to 55.9% in 1994 • 383.000 ha of agricultural area. • 56% cereals • 44 cooperatives. • 5 fishing ports. • 241 hotel establishments (16% National accommodation capacity).	• Over one hundred cooperatives framing over 4,000 women between 1999 and 2007 (Lybbert *et al.*, 2011). • Most of the accommodation (guesthouses) is concentrated on the coastal part (Aboutayeb, 2014).	• The economy of this oasean region closely linked with the agricultural sector. • This sector is mostly for subsistence and local consumption (karmaoui et *al.*, 2015a)

Source: (Authors)

Table 3. The Driving Force - Pressure - State - Impact – Response framework for the three selected sites

DPSIR	Site 1: Mediterranean (MED) Source: CLICO, (2012)	Site 2: Atlantic Region (ATL) Source: See below	Site 3: Oasean Regions (SOA) Source: See below
Driving force	• Liquid and solid pollution • Over use of ressources • The absence of effective awareness • Biodiversity erosion • Urbanization	• Clearings (Tribak, 2002) • Drought (SIWM, 2001) • Pesticides (Benabid, 2000) • Climate change (Lybbert & Aboudrare, 2011) • Population (Romagny & Boujrouf, 2010) • Overgrazing (Elwahidi *et al*, 2014)	• Salinization of soil and water (BRSMO, 2008) • Decreased fertility and water • Biodiversity, sanitation, treatment, growth and tourist (karmaoui et al., 2015a). • Fragmentation (BRSMO, 2008), • Overeuse of groundwater, climate Aridity, (Karmaoui et al., 2014; Ben Salem, 2014)
Pressure	• Area overgrazing • Illegal cannabis • Over use of water • Agricultural • Hunting and Over fishing • Fires	• Climatic Aridity (Esper *et al*, 2007) • Area deforested and Forest converted into agriculture • Livestock (El Aïch, 2008) • Density of the population (Romagny & Boujrouf, 2010)	BRSMO, (2008): • Farm size, • Crop type • Total irrigated area; • Climatic aridity • Desertification
State	• Area of wetlands • Area affected by dried • Area affected by pollution	• Area degraded / forest • Canopy density % • Eroded area (Aafi, 2007) • Area used for grazing (Elwahidi *et al*, 2014)	• Soil Quality soil salinity (BRSMO, 2008) • Organic matter rates (BRSMO, 2008) • Soil fertility (BRSMO, 2008)
Impact	• Animal biodiversity • Plant biodiversity • Abundance	• Production forests (Karmaoui, 2016) • Biodiversity degradation (El Mrabet, 2014)	• Area productivity of land BRSMO, (2008) • Families abandoning • groundwater level depletion BRSMO, (2008)
Response	• Renewable energy (wind power) (CFCIM, 2015) • Clean industry • Ecotourism • Waste treatment • Reserves & rural tourism • Water management adaptation	• Reforestation (regeneration: El Mrabet, 2014); • Timber extracted (Elwahidi *etal,*2014) • Organization of population: (Lybbert *et al.*2011) • Ecotourism (Karmaoui, 2016) • Migration (Elwahidi *et al*, 2014)	• Area equipped with water-saving technology (BRSMO, 2008) • Area with crop resistant to salinity • Number of collective groups with interests in the water sector (BRSMO, 2008) • Renewable energy (Solar) (MASEN, 2012) • Ecotourism (BRSMO, 2008)

Source: (Authors)

- The decrease in precipitation will be about 20% (from now to 2050) and about 40 in 2080.
- Nationally, the temperature increase will be about 3°C following the A2 scenario until 2080. This will cause an increase of Evapo-transpiration that goes up 40% in the horizon of 2080, and a small increase in the Saharan region (9% in the same period).

It well established that the climate change have considerable impact on net primary production (UNEP, 2012). According to Rochdane *et al.,* (2014), both IPCC SRES A2 and B2, show a projected an increase in demand coupled with a decrease in supply, reporting also that under the first scenario the Moroccan terrestrial ecosystems take more than 1.25 year to produce net primary production used in one year. This pressure is seen also at footprint per-capita. Figure 4 tracks the per-person resource demand Ecological Footprint and bio-capacity in Morocco since 1975. Ecological Footprint is considered a tool of long-term evaluation of an ecosystem (Jorge-Romero et al., 2019).

Figure 4. Per-person resource demand Ecological Footprint and biocapacity in Morocco since 1975; the bio-capacity change following ecosystem degradation and management, agricultural practices, population growth... However, the Footprint varies with consumption and production efficiency. a, Ecological Footprint vs Biocapacity (gha per person); b, Ecological Footprint (Number of Earths).
Source: (Global Footprint Network, 2018)

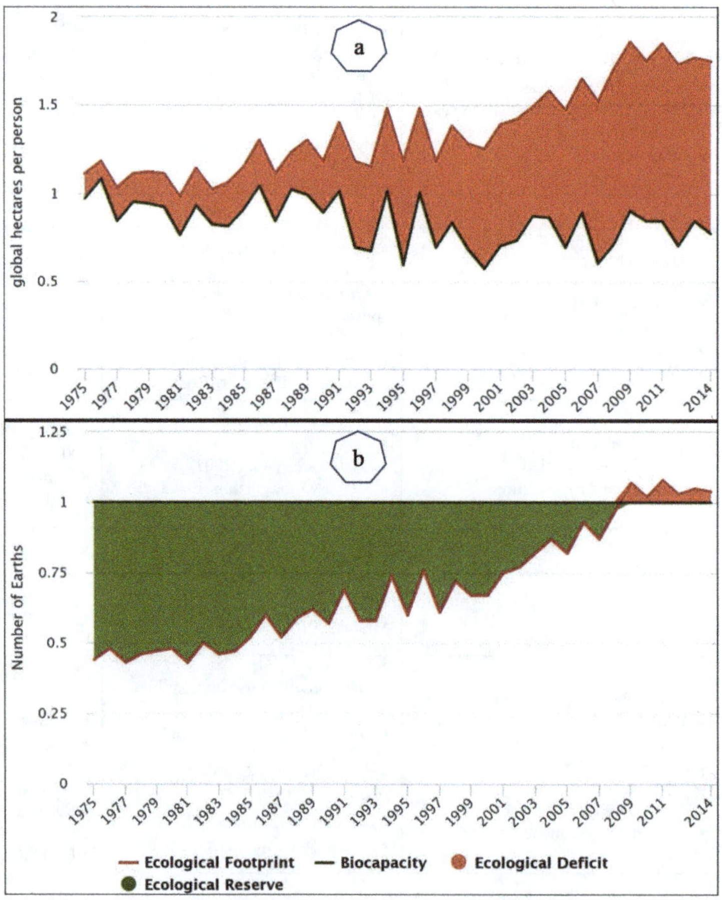

Since 1977, the impact on ecology was significantly changed (Figure 4). In fact, the difference between ecological footprint and bio-capacity is increasing more and more, impacting severely the primary production, which will threaten the future generation. This trend can explain partially the Moroccan environmental vulnerability. The environmental vulnerability profile used in this chapter allows identifying the main vulnerability indicators. The 21 indicators (among the 50 smart indicators) are grouped into four categories (climate, geography, services and resources and anthropogenic). From the results, Morocco is vulnerable to extremely vulnerable for indicators (Wind, Dry, "Hot" and "SST" for the climate category and Relief of geography category; Imbalance "Openness", "Endemics", "Vegetation", "LOS VEG", "Degradation", "Reserves", "MAPs", "Fish Effort", "Water", "Mining", "Sanitation" for the resources & services indicators and "Density", "Growth", "Coastal", "Conflicts" for the anthropogenic indicators. All these indicators can be used as management priorities. We can summarize all these factors in the following areas:

1. Water and soil quality;
2. Biodiversity and reserves;
3. Agriculture and fisheries ;
4. Sea and coastal ;
5. Soil and water overuse and resulted wastes
6. Population density and conflicts

Morocco is exposed to an accelerated desertification, increasing scarcity of water resources, degradation of the marine food chain fishing and changes in the structure of fish populations. The growing installation of population and tourist pressure on coastal ecosystems impact the habitat and then the animal and vegetal biodiversity.

The population concentration in zones at risk causes an increase of natural hazards damages (Leone *et al.,* 2010). As indicated in the major indicators (Scores equal and more than 4), the littoral coastline exposes an important portion of population to natural hazards. Morocco has 3500 kilometers of coastline; over two coastal fronts open to the Mediterranean and the Atlantic. This area attracts more and more anthropogenic and economic activities. In 2010, the Moroccan coastline in 2010 gathered more than 54.7% of the total population of Morocco, and the Atlantic coasts by themselves, concentrate 46.2% of the population of Morocco on only 14% of the national territory (HCP, 2010). The Atlantic coast also occupies a central place in tourist activities. More than 60% of tourist is attracted in this region, in 2008, achieved 54% of tourist nights with a bed capacity of around 51% of the national total (HCP, 2010). The main anthropogenic impact in coastline is the wastewater (Karmaoui, 2018). Another vulnerable zone is the pre-Saharan region (desert oasis region), with an arid and hot climate. Using Environmental vulnerability index at catchment scale, Karmaoui *et al.,* (2014) and Ben Salem, (2014); demonstrate that is also vulnerable to climate change and anthropogenic impacts. Like other models and indices, the used index (Environmental vulnerability index) and the DPSIR model present the challenge of reliable data collection. The use of this last framework (DPSIR) is due to its ability to identify long-term variables and information on the selected sites. It is therefore, used as ILTER whole-system approach. This approach helps understand the interactions of ecosystem variables including socioeconomic factors long-term trends in environmental conditions (Mirtl et al., 2018). Press Pulse Dynamics framework (PPD) is a second important approach suitable for ILTER due to its socio-ecological dimensions. It was operationalized in three Moroccan areas located in the three studied sites, two studies in Saharan site in Oasis Biosphere Reserve (Karmaoui et al., 2014b, 2015b), and a study in the Atlantic site, in Argan Biosphere Reserve (Karmaoui, 2016).

Talking about long-term researches, the data quality affect enormously the exactitude of all scientific long-term researchs because the gathered data are the base of this type of researchs. After Michener et al., (1997), heterogenouse data, missing values, and addition of deletion of parameters influence the ecological researchs. The data (outputs) is clasified in data bank in each site (Sprott, 1998). The local databases (input and output data of the LTER sites) will be networked with a central data management system.

To determine the main sites of the national LTER network, we follow the classification established by Knippertz *et al.,* (2003), based on the Long-term precipitation variability in Morocco and the link to the large-scale circulation in recent and future climates. The determination of the selected sites is based also on the presence of ecological, biological and historical high interest sites. Other criteria were adopted like administrative regionalization, cultural similarity, meteorological directorates, natural resources (water, soil and biodiversity), topography and diversity of ecosystems. These criteria lead us to distinguish three main sites:

Mediterranean Site

The Mediterranean site is bounded on the north by the Strait of Gibraltar and the Mediterranean Sea. In the west is attached by the Atlantic Ocean, south by the Gharb region and to the east by the region of Taza - Al Hoceima – Taounate. According to the website of water Secretarial Department (2009), the Moroccan side of the Intercontinental BR is supposed to be an environmental equilibrium framework for a region now in economic expansion. This part contains many natural ecosystems (forest, coastal and marine) bio-ecological high value, such as the National Park Talassemtane site, Jebel Bouhachem, Jebel Moussa, Smir Lagoon, and Oued Tahaddart.

The Atlantic Site

The Atlantic site can be located in the Argan Biosphere Reserve. This area offers a rich biodiversity that supports a large number of ecosystem services. In fact, at the beginning of the 20th century, the area of the argan forest was estimated at 1.5 million hectares (Boudy, 1952). However, in one century, the forest has lost half of its surface (Tarrier & Benzyane, 2003). The Argan Forest has been reduced both in area and density because of the clearing and removal of floristic cortege (El Mrabet, 2014). Increasing aridity, reported at a regional-scale by Esper, Frank, & Büntgen, (2007) (Esper *et al.,* 2007). The cut of wood is one of the leading contraventions (Elwahidi *et al.,* 2014). According to Aafi, (2007), overgrazing have an impact on soils (damage), causing the erosion.

The Saharan Site

This site is the most vulnerable region of the country. At the border of the hyper-arid deserts, oasis those participate in maintenance of transitional spaces that are buffers. Agriculture is the main economical sector. The diversification of the crops is common and varies with the socio-economical considerations in order to secure the income of the oasis farmers (BROSM, 2008). In ecological terms, these oases are impacted mainly by salinity and desertification; accelerated by the harsh climate, Water shortage, and overgrazing (Laouina, 2006). Environmental vulnerability is much greater in this region, compared to other sites of Morocco. The ecological importance of this system and their vulnerability create a need to establish a site of LTER in oasis area of Morocco.

Establish a LTER Network for Morocco (Mission, Objectives, and Activities of the Moroccan LTER Proposal)

The ILTER approach helps understand the interactions of ecosystem variables including socioeconomic factors long-term trends in environmental conditions (Mirtl et al., 2018). The ILTER approach is based on four conceptual pillars:

- Long-term data
- In-situ data generation
- Process orientation: interactions of ecosystem processes affected by drivers.
- Systems approach: Earth systems, socio-ecological systems,

The mission of LTER Program can gather, control and classify the information and knowledge on the Moroccan ecosystems. The collected and generated data will be used to assess the Moroccan long-term ecological dynamic. This data can be exploited for comparative studies both at national and at global scale.

The objectives are:

- Conduct long-term and multidisciplinary research projects;
- Develop tools for local communities, planers and stakeholders for the sustainable utilization of the local natural resources;
- Provide the scientific foundations to the decision-makers;

The main activities can be as follow:

- Studying the ecosystem process;
- Studying the major natural and human disturbances affecting ecosystem dynamics and ecosystem responses;
- Identifying ecological processes to address environmental changes;
- Forming a metadata by specialty.

The most important lesson in the context of LTER is the multidisciplinary approach to understand the ecological problems (Robertson, 2008). Several countries were associated with the Long-Term Ecological Research (LTER) program, while others are in the process of joining the network. In Morocco, there is an urgent need to be a member of this network because many ecosystems require multidisciplinary long-term research.

Proposed Methods of Work of Moroccan LTER

A Part of the mission will be associated with the generation of methodologies to provide immediate guidance to conservation programs and for the definition of public policies:

- Coordination between Biosphere Reserves, sites of ecological interests, national parks;
- Observe and study specific natural phenomena, including desertification, water and soil salinization…
- Collaborate and participate with other LTER countries;
- Sustainable utilization of natural resources and protection of the environment;
- Standardization of data formats and data quality control;
- Research methodology will support multi-disciplinary research;
- Monitoring and observation, using modern methodologies such as remote sensing, Geographic information Systems (GIS), and mathematical models;
- Formal association in a network is expected to promote standardization of activities and shared opportunities for common research; therefore, networks might begin as a loose aggregation of research sites, but should evolve greater network cohesion as more cooperation and interdependence develop (Johnson *et al.*, 2010).

LTER Generated Data and Its Life Cycle

In an LTER site, the generated data follows a series of stages, starts from data collection, and then its visualization and publication and ends in the emerging of new research questions etc (Figure. 5). The cycle includes crucial data treatments and analysis (Archiving, control, integration, visualization…) and categories (Raw data, verified data, meta-data, archived data…).

Possible Cooperation of the Moroccan LTER With International LTER Network

The Moroccan LTER can develop excellent relationships with other ecological networks and research programs around the world. The following activities are especially important:

Figure 5. The data life cycle, the production of analysis-grade data from scientific question, to measurement, through quality control Analysis, publication, produce new research questions. New research questions drive the integration of existing archival data, which then allows new analyses that result in additional publications and research questions.
Source: (Michener et al., 2011)

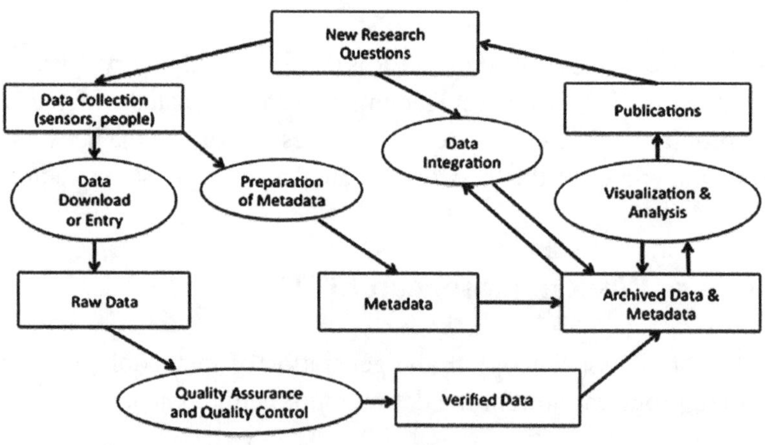

- Developed close relationships with international L TER Network;
- Became an active member of African and European LTER;
- Discussed activities with international scientists.

CONCLUSION

The main challenge of establishing a LTER is to provide concrete data for the national scale, and the state of biodiversity. This data should also be updated regularly. Transform this information in global reference value, taking into account imprecision, local specificities, the long-term use of these services and the relative importance given to the different services. Moroccan environment in general is being subjected to a variety of stresses (Soil and water salinity, groundwater depletion, and desertification) and continuous change in their environmental characteristics. How these changes will alter the biological components of ecosystems. This is a major concern to the pilot sites community because large parts of their economy are based on the use of these ecosystems, forestry and agricultural resources. Together, these effects cause a deterioration of living conditions and poverty especially in the rural population. The environmental phenomena are now very complex to understand. This requires multi-disciplinary long-term researches. We anticipate that the Moroccan Research Network consists in three pilot sites as mentioned above, which include several ecosystems like oases, agriculture, forest, and lakes and other ecosystems. More than 100 scientists and technicians from all socio-ecological institutes through the country can be involved in the project. The Moroccan LTER will be a network at long-term, multidisciplinary research, in order to understand ecosystems changes and sharing knowledge.

REFERENCES

Aafi, A. (2007). Rôle des Cactus dans la Restauration de l'Arganeraie et de ses Composantes Floristique et Faunistique dans la Province de Tiznit. Annales de la recherche forestière au Maroc, 38, 69-76.

Aboutayeb, H. (2014). The biosphere reserve of the argan tree: eco-tourism a new territory South of Morocco. *PASOS: Revista de Turismo y Patrimonio Cultural, 12*(4), 915-922. Retrieved from https://www.cabdirect.org/cabdirect/abstract/20143365231

Agoumi, A. (2003). *Vulnérabilité des pays du Maghreb face aux changements climatiques; Besoin réel et urgent d'une stratégie d'adaptation et de moyens pour sa mise en œuvre*. Available on http://www.cckn.net/pdf/north_africa_fr.pdf

An, G. H., Cha, S. L., & Kim, J. K. (2018). Modification of the long-term deformation models for steel reinforced concrete columns. *Construction & Building Materials, 189*, 245–252. doi:10.1016/j.conbuildmat.2018.08.095

Arrus, R., & Rousset, N. (2007). *L'agriculture du Maghreb au défi du changement climatique: Quelles stratégies d'adaptation face à la raréfaction des ressources hydriques?* Available on https://halshs.archives-ouvertes.fr/halshs-00134115

Babqiqi, A. (2008). *Scenarios de changements climatiques à haute résolution au Maroc*. Casablanca, Maroc: Rapport de la Direction de la Météorologie Nationale.

Baker, K. S., Benson, B. J., Henshaw, D. L., Blodgett, D., Porter, J. H., & Stafford, S. G. (2000). Evolution of a multisite network information system: The LTER information management paradigm. *A.I.B.S. Bulletin*, *50*(11), 963–978. doi:10.1641/0006-3568(2000)050[0963:EOAMNI]2.0.CO;2

Ben Salem, A. (2014). *Vulnérabilité et adaptation aux changements climatiques dans les oasis de la région de Tafilalet- Maroc* (Thèse de doctorat). Département de biologie, Faculté de Semlalia. Université Cadi Ayyad, Marrakesh, Morocco.

Benabid, A., & Fennane, M. (1999). Main forest formations in the great Book of the Forest Moroccan. Mardaga.

Benmohammadi, A., Benmohammadi, L., Ballais, J. L., & Riser, J. (2000). Analysis of anthropic and natural interrelations: their impact on the recrudescence of the phenomena of sand deposition and desertification in south east Morocco (Drâa valley and Ziz valley). *Sécheresse, 11*(4), 297-308. Available on https://www.cabdirect.org/cabdirect/abstract/20013044152

Bretagnolle, V., Berthet, E., Gross, N., Gauffre, B., Plumejeaud, C., Houte, S., ... Gaba, S. (2018). Towards sustainable and multifunctional agriculture in farmland landscapes: Lessons from the integrative approach of a French LTSER platform. *The Science of the Total Environment*, *627*, 822–834. doi:10.1016/j.scitotenv.2018.01.142 PMID:29426207

BROSM. (2008). *Management Framework of the Reserve Biosphere Oasis of southern Morocco*. BRSMO.

Callahan, J. T. (1984). Long-term ecological research. *Bioscience*, *34*(6), 363–367. doi:10.2307/1309727

CFCIM. (2015). *Chambre Française de Commerce et d'Industrie du Maroc (CFCIM)*. L'énergie éolienne au Maroc.

Change, D. P. O. C. (2003). *Vulnerability of North African Countries to Climatic Changes*. Available online: www.iisd.org/cckn/pdf/north_africa.pdf

Christian, R. R., French, C., Gosz, J., & Waide, R. (1999). Perspectives on international long-term ecological research. In *Perspectives in Ecology: A Glance from the VII International Congress of Ecology*. Backhuys.

DEFCS. (1990). *Projet de création du Parc National d'Iriqui. Rapport inédit*. Direction des Eaux et Forêts et de la Conservation des Sols.

DOEC. (1999). *Etat de l'environnement au Maroc, Rapport sur l'état de l'environnement du Maroc (REEM)*. Direction de l'Observation, des Etudes et de la Coordination.

El Aïch, A., Morand-Fehr, P., Bas, P., Araba, A., & Bourbouze, A. (2008). Meat from young goats raised in Argan Tree Forest (Morocco): Emerging product to valorise. *Options Méditerranéennes*, 205-210.

El Fasskaoui, B. (2009). Fonctions, défis et enjeux de la gestion et du développement durables dans la Réserve de Biosphère de l'Arganeraie (Maroc). *Études caribéennes*, (12). Retrieved from https://journals.openedition.org/etudescaribeennes/3711

El Mrabet, S., Ouahmane, L., El Mousadik, A., Msanda, F., & Abbas, Y. (2014). The effectiveness of arbuscular mycorrhizal inoculation and bio-compost addition for enhancing reforestation with Argania spinosa in Morocco. *Open Journal of Forestry*, *4*(1), 14–23. doi:10.4236/ojf.2014.41003

EVI. (2003). *The Demonstration Environmental Vulnerability Index (EVI).* South Pacific Applied Geo-science Commission, SOPAC Technical Report 356.

Fennane, M. & Ibn Tattou, M. (2012). Statistiques et commentaires sur l'inventaire actuel de la flore vasculaire du Maroc. *Bulletin de l'Institut Scientifique, 34*(1), 1-9.

Global Footprint Network. (2018). *Footprint and Biocapacity Atlas of Francophonie Member Nations.* Retrieved from https://www.footprintnetwork.org/content/images/article_uploads/Francophonie_English.pdf

HCEFLCD. (2007). *Etude d'aménagement du parc national d'Iriqui. Rapport de synthèse*. HCEFLCD-DREF SO.

HCP. (2010). *Littoral marocain: un poids économique et démographique important*. Haut Commissariat au Plan du Maroc, présentation du 23/09/2010 à l'IRES lors de la journée d'étude "Littoral marocain et changement climatique".

Hobbies, J. E., Carpenter, S. R., Grimm, N. B., Gosz, J. R., & Seastedt, T. R. (2003). The US long-term ecological research program. *Bioscience*, *53*(1), 21–32. doi:10.1641/0006-3568(2003)053[0021:TULTER]2.0.CO;2

Holmes, R. T. (2011). Avian population and community processes in forest ecosystems: Long-term research in the Hubbard Brook Experimental Forest. *Forest Ecology and Management*, *262*(1), 20–32. doi:10.1016/j.foreco.2010.06.021

Holmes, R. T., & Sherry, T. W. (2001). Thirty-year bird population trends in an unfragmented temperate deciduous forest: Importance of habitat change. *The Auk*, *118*(3), 589–609. doi:10.1642/0004-8038(2001)118[0589:TYBPTI]2.0.CO;2

ILTER. (1998). *International Long-Term Ecological Research Network. US LTER Network.* Albuquerque, NM: Department of Biology, University of New Mexico.

IPCC. (2007). *(Intergovernmental Panel on Climate Change). Climate Change: Synthesis Report. Contributions of Working Groups I, II, and III to the Fourth Assessment Report of the Intergovernmental Panel on Climate Change (AR4). Core writing team* (R. K. Pachauri & A. Reisinger, Eds.). Geneva, Switzerland: IPCC.

Johnson, J. C., Christian, R. R., Brunt, J. W., Hickman, C. R., & Waide, R. B. (2010). Evolution of collaboration within the US long-term ecological research network. *Bioscience*, *60*(11), 931–940. doi:10.1525/bio.2010.60.11.9

Jorge-Romero, G., Lercari, D., Ortega, L., & Defeo, O. (2019). Long-term ecological footprints of a man-made freshwater discharge onto a sandy beach ecosystem. *Ecological Indicators*, *96*, 412–420. doi:10.1016/j.ecolind.2018.09.024

Kaly, U., Pratt, C., & Mitchell, J. (2004). The demonstration environmental vulnerability index (EVI). Pacific Islands Applied Geoscience Commission (SOPAC).

Kaly, U., Pratt, C., & Mitchell, J. (2005). *Building Resilience in SIDS: The Environmental Vulnerability Index. Final Report. SOPAC*. UNEP.

Kaly, U., Pratt, C., Mitchell, J., & Howorth, R. (2003). The demonstration environmental vulnerability index (EVI). *Pacific Islands Applied Geoscience Commission (SOPAC)*. Retrieved from gsd.spc.int/sopac/evi/Files/EVI%202004%20Technical%20Report.pdf

Karasti, H., & Baker, K. S. (2008). Digital data practices and the long term ecological research program growing global. *International Journal of Digital Curation*, *3*(2), 42–58. doi:10.2218/ijdc.v3i2.57

Karmaoui, A., Messouli, M., Yacoubi, Khebiza, M. & Ifaadassan, I. (2014a). Environmental Vulnerability to Climate Change and Anthropogenic Impacts in Dryland (Pilot Study: Middle Draa Valley, South Morocco). *J Earth Sci Clim Change, S11*, 2. doi:10.4172/2157-7617.S11-002

Karmaoui, A. (2016). Ecosystem Services of the Argan Forest, the Current State and Trends. *Advances in Research*, *8*(1), 1–13. doi:10.9734/AIR/2016/21353

Karmaoui, A. (2018). Environmental Vulnerability to Climate Change in Mediterranean Basin: Socio-Ecological Interactions Between North and South. In *Hydrology and Water Resource Management: Breakthroughs in Research and Practice* (pp. 61-96). IGI Global.

Karmaoui, A., Ifaadassan, I., Messouli, M., & Yacoubi Khebiza, M. (2015a). Characterization of Common Environmental Indicators of the Moroccan Oasean Biome, Pilot Study in the Reserve Biosphere of Oases in Southern Morocco. *Advances in Research*. Doi:10.9734/AIR/2015/19369

Karmaoui, A., Messouli, M., Ifaadassan, I., & Khebiza, M. Y. (2014b). A multidisciplinary approach to assess the environmental vulnerability at local scale in context of climate change (Pilot study in upper draa valley, South Morocco). *Glob. J. Technol. Optim*, *6*, 1–11. doi:10.4172/2229-8711.1000167

Karmaoui, A., Messouli, M., & Yacoubi, M. (2015b). Vulnerability of Ecosystem Services to Climate Change and Anthropogenic Impacts in South East of Morocco (Case study: Drying up of Iriki lake). *The International Journal of Climate Change: Impacts and Responses.*, *7*(3), 1835–7156.

Kim, E. S. (2006). Development, potentials, and challenges of the International Long-Term Ecological Research (ILTER) Network. *Ecological Research*, *21*(6), 788–793. doi:10.100711284-006-0049-1

Kingdom of Morocco. (2013). *Green Morocco Plan*. Retrieved from http://www.maroc.ma/en/content/green-morocoo-plan

Knapp, S., Stadler, J., Harpke, A., & Klotz, S. (2016). Dispersal traits as indicators of vegetation dynamics in long-term old-field succession. *Ecological Indicators*, *65*, 44–54. doi:10.1016/j.ecolind.2015.10.003

Knippertz, P., Christoph, M., & Speth, P. (2003). Long-term precipitation variability in Morocco and the link to the large-scale circulation in recent and future climates. *Meteorology and Atmospheric Physics*, *83*(1-2), 67–88. doi:10.100700703-002-0561-y

Kohli, R. K., Batish, D. R., & Singh, H. P. (2005). Need to Establish Long-Term Ecological Research Network in India. Bulletin of the National Institute of Ecology, 15, 271-276.

Kuemmerlen, M., Stoll, S., Sundermann, A., & Haase, P. (2016). Long-term monitoring data meet freshwater species distribution models: Lessons from an LTER-site. *Ecological Indicators*, *65*, 122–132. doi:10.1016/j.ecolind.2015.08.008

Kwon, T. S. (2014). Empirical test of the influence of global warming and forest disturbance on ant fauna at the Gwangneung Forest Long Term Ecological Research site, South Korea. *Journal of Asia-Pacific Biodiversity*, *7*(3), 252–257. doi:10.1016/j.japb.2014.06.004

Laouina, A. (2006). Prospective « Maroc 2030 ». Gestion durable des ressources naturelles et de la biodiversité au Maroc. *Haut commissariat au plan.*

Leone, F., De richemond, N., & Vinet, F. (2010). Aléas naturels et gestion des risques. Édit. PUF, collection L. Géographie.

Lischeid, G., Kalettka, T., Merz, C., & Steidl, J. (2016). Monitoring the phase space of ecosystems: Concept and examples from the Quillow catchment, Uckermark. *Ecological Indicators*, *65*, 55–65. doi:10.1016/j.ecolind.2015.10.067

Lybbert, T. J., Aboudrare, A., Chaloud, D., Magnan, N., & Nash, M. (2011). Booming markets for Moroccan argan oil appear to benefit some rural households while threatening the endemic argan forest. *Proceedings of the National Academy of Sciences of the United States of America*, *108*(34), 13963–13968. doi:10.1073/pnas.1106382108 PMID:21873185

Macpherson, G. L., & Sophocleous, M. (2004). Fast ground-water mixing and basal recharge in an unconfined, alluvial aquifer, Konza LTER Site, Northeastern Kansas. *Journal of Hydrology (Amsterdam)*, *286*(1-4), 271–299. doi:10.1016/j.jhydrol.2003.09.016

Meesenburg, H., Ahrends, B., Fleck, S., Wagner, M., Fortmann, H., Scheler, B., ... Meiwes, K. J. (2016). Long-term changes of ecosystem services at Solling, Germany: Recovery from acidification, but increasing nitrogen saturation? *Ecological Indicators*, *65*, 103–112. doi:10.1016/j.ecolind.2015.12.013

Messouli, M., Bensalem, A., & Rochdane, S. (2013). *Développement d'un concept d'Indice de Vulnérabilité au Changement Climatique et Environnementale au Niveau Régional. Département du développement durable (MNSSD).* Région Moyen-Orient et Afrique du Nord. Banque Mondiale.

Michener, W. K., Brunt, J. W., Helly, J. J., Kirchner, T. B., & Stafford, S. G. (1997). Nongeospatial metadata for the ecological sciences. *Ecological Applications*, *7*(1), 330–342. doi:10.1890/1051-0761(1997)007[0330:NMFTES]2.0.CO;2

Michener, W. K., Porter, J., Servilla, M., & Vanderbilt, K. (2011). Long term ecological research and information management. *Ecological Informatics*, *6*(1), 13–24. doi:10.1016/j.ecoinf.2010.11.005

Mirtl, M., Borer, E. T., Djukic, I., Forsius, M., Haubold, H., Hugo, W., ... Orenstein, D. E. (2018). Genesis, goals and achievements of long-term ecological research at the global scale: A critical review of ILTER and future directions. *The Science of the Total Environment*, *626*, 1439–1462. doi:10.1016/j.scitotenv.2017.12.001 PMID:29898550

Mollenhauer, H., Kasner, M., Haase, P., Peterseil, J., Wohner, C., Frenzel, M., ... Zacharias, S. (2018). Long-term environmental monitoring infrastructures in Europe: Observations, measurements, scales, and socio-ecological representativeness. *The Science of the Total Environment*, *624*, 968–978. doi:10.1016/j.scitotenv.2017.12.095 PMID:29275260

Müller, F., Baessler, C., Frenzel, M., Klotz, S., & Schubert, H. (2010). Long-term ecosystem research between theory and application–an introduction. In *Long-Term Ecological Research* (pp. 3–7). Dordrecht: Springer. doi:10.1007/978-90-481-8782-9_1

Nakano, S. I., & Tetsukazu, Y. (2012). *The Biodiversity observation network in the Asia-Pacific region* (T. Nakashizuka, Ed.). Springer. doi:10.1007/978-4-431-54032-8

PACC. (2011). *Oasis et Zones Arides Résilientes Territoires Engagés*. Retrieved from www.oasisadaptation.com

Panitz, C., Sperl, M. F., Hennig, J., Klucken, T., Hermann, C., & Mueller, E. M. (2018). Fearfulness, neuroticism/anxiety, and COMT Val158Met in long-term fear conditioning and extinction. *Neurobiology of Learning and Memory*, *155*, 7–20. doi:10.1016/j.nlm.2018.06.001 PMID:29883709

PDES. (2004). Le plan de développement économique et social 2000-2004. Région de Tanger-Tétouan, 649-683.

Pereira, H. M., Leadley, P. W., Proença, V., Alkemade, R., Scharlemann, J. P., Fernandez-Manjarrés, J. F., ... Chini, L. (2010). Scenarios for global biodiversity in the 21st century. *Science*, *330*(6010), 1496–1501. doi:10.1126cience.1196624 PMID:20978282

RADDO. (2011). Lutte contre la désertification: les oasis au péril du changement climatique. *Le réseau d'associations oasiennes du Maghreb (RADDO)*. Retrieved from www.raddo.org

Rankou, H., Culham, A., Sghir Taleb, M., Ouhammou, A., Martin, G., & Jury, S. L. (2015). Conservation assessments and Red Listing of the endemic Moroccan flora (monocotyledons). *Botanical Journal of the Linnean Society*, *177*(4), 504–575. doi:10.1111/boj.12258

RBOSM. (2008). *Plan cadre de gestion de la Réserve de Biosphère des Oasis du Sud Marocain (RBOSM)*. Version provisoire.

Rignall, K., & Kusunose, Y. (2018). Governing livelihood and land use transitions: The role of customary tenure in southeastern Morocco. *Land Use Policy*, *78*, 91–103. doi:10.1016/j.landusepol.2018.03.035

Robertson, G. P. (2008). Long-term ecological research: Re-inventing network science. *Frontiers in Ecology and the Environment*, *6*(5), 281–281. doi:10.1890/1540-9295(2008)6[281:LERRNS]2.0.CO;2

Rochdane, S., Bounoua, L., Zhang, P., Imhoff, M. L., Messouli, M., & Yacoubi-Khebiza, M. (2014). Combining satellite data and models to assess vulnerability to climate change and its impact on food security in Morocco. *Sustainability*, *6*(4), 1729–1746. doi:10.3390u6041729

Romagny, B., & Boujrouf, S. (2010). The rush argan oil, chronicles the heritage value of a Moroccan land. *Hesperis Tamuda*, *45*, 151–166.

Rosegrant, M. M. (2009). Changement climatique et sécurité alimentaire au Maroc. Rencontre internationale « Le changement climatique: enjeux et perspectives d'adaptation pour le Maroc », Institut Royal des Etudes Stratégiques (IRES), Rabat, Maroc.

Ruddell, B. L., Zaslavsky, I., Valentine, D., Beran, B., Piasecki, M., Fu, Q., & Kumar, P. (2014). Sustainable long term scientific data publication: Lessons learned from a prototype Observatory Information System for the Illinois River Basin. *Environmental Modelling & Software*, *54*, 73–87. doi:10.1016/j.envsoft.2013.12.015

Schmidt, M. (2003). *Development of a fuzzy expert system for detailed land cover mapping in the Draa catchment (Morocco) using high resolution satellite images*. Retrieved from https://www.researchgate.net/profile/Michael_Schmidt15/publication/34686505_Development_of_a_fuzzy_expert_system_for_detailed_land_cover_mapping_in_the_Dra_catchment_Morocco_using_high_resolution_satellite_images/links/5779f71908aeb9427e2c03cb/Development-of-a-fuzzy-expert-system-for-detailed-land-cover-mapping-in-the-Dra-catchment-Morocco-using-high-resolution-satellite-images.pdf

SECE. (1997). *Etude Nationale sur la biodiversité, Rapport National sur la biodiversité*. Présenté à la quatrième conférence des parties contractantes à la Convention sur la biodiversité.

Secretarial Department. (2009). *Ministry of Agriculture*. Retrieved from http://www.eauxetforets.gov.ma/

Shibata, H. (2016). Impact of winter climate change on nitrogen biogeochemistry in forest ecosystems: A synthesis from Japanese case studies. *Ecological Indicators*, *65*, 4–9. doi:10.1016/j.ecolind.2015.10.063

Simonneaux, V., Cheggour, A., Deschamps, C., Mouillot, F., Cerdan, O., & Le Bissonnais, Y. (2015). Land use and climate change effects on soil erosion in a semi-arid mountainous watershed (High Atlas, Morocco). *Journal of Arid Environments*, *122*, 64–75. doi:10.1016/j.jaridenv.2015.06.002

Soltwedel, T., Bauerfeind, E., Bergmann, M., Bracher, A., Budaeva, N., Busch, K., ... Jacob, M. (2016). Natural variability or anthropogenically-induced variation? Insights from 15 years of multidisciplinary observations at the arctic marine LTER site HAUSGARTEN. *Ecological Indicators*, *65*, 89–102. doi:10.1016/j.ecolind.2015.10.001

Sopac. (n.d.). *Index*. Retrieved from http://www.sopac.org/index

Speth, P., Christoph, M., & Diekkrüger, B. (Eds.). (2010). *Impacts of global change on the hydrological cycle in West and Northwest Africa*. Springer Science & Business Media. doi:10.1007/978-3-642-12957-5

Sprott, P. (1998). *The International Long-Term Ecological Research Network*. US LTER Network Office Albuquerque New Mexico.

Staudinger, M. D., Grimm, N.B., Staud, T. A., Carter, S.L., & Chapin, F.S. (2012). Impacts of climate change on biodiversity, ecosystems, and ecosystem services. United States Global Change Research Program, Washington, DC.

Stotz, O. (2018). The perception of homeownership utility: Short-term and long-term effects. *Journal of Housing Economics*. doi:10.1016/j.jhe.2018.11.003

Tarrier, M., & Benzyane, M. (2003). *Moroccan argan tree is dying: problematic and bio-indication.* Retrieved from http://www.jle.com/fr/revues/sec/e-docs/larganeraie_marocaine_se_meurt_problematique_et_bioindication__262867/article.phtml?cle_doc=000402D3

Tribak, A. (2002). The Environmental Impact Of Management Of Water On The irrigation On Forest Ecosystems: Case Of The argan tree. *H.T.E. Review*, 124.

UICN. (2005). *La France et la biodiversité. Enjeux et responsabilités.* Réalisation. Comité français UICN. Retrieved from www.karibu.fr

UNEP. (2012). *The Fifth Global Environmental Outlook Report.* LAND.

UNESCO. (2010). Changement climatique: Enjeux et perspectives au Maghreb 2010MO0615. UNESCO.

Waide, R. (1998). *The International Long-Term Ecological Research Network.* US LTER Network, National Science Foundation.

Westoby, M. (1991). On long-term ecological research in Australia. In P. G. Risser (Ed.), *Long-Term Ecological Research* (pp. 191–209). New York: John Wiley and Sons.

Willis, C. G., Ruhfel, B., Primack, R. B., Miller-Rushing, A. J., & Davis, C. C. (2008). Phylogenetic patterns of species loss in Thoreau's woods are driven by climate change. *Proceedings of the National Academy of Sciences of the United States of America, 105*(44), 17029–17033. doi:10.1073/pnas.0806446105 PMID:18955707

World Bank. (2011). *Middle-East and Northern Africa water outlook.* Prepared by Future Water.

Xia, S., Liu, Y., Yu, X., & Fu, B. (2018). Challenges in coupling LTER with environmental assessments: An insight from potential and reality of the Chinese Ecological Research Network in servicing environment assessments. *The Science of the Total Environment, 633*, 1302–1313. doi:10.1016/j.scitotenv.2018.03.284 PMID:29758883

Chapter 2
Associations Between Climate, Ecosystems, and Ecosystem Services in the Pre-Sahara:
Case Study of Tafilalet, Morocco

Abdelkrim Ben Salem
Cadi Ayyad University, Morocco

Souad Ben Salem
Cadi Ayyad University, Morocco

Mohammed Yacoubi Khebiza
Cadi Ayyad University, Morocco

Awatif Zine Elabidine
Kennesaw State University, USA

ABSTRACT

This chapter focuses on the richness and wide variety of Sub-Saharan ecosystems particularly in the region of Tafilalet, Morocco. The Tafilalet ecosystems provide an extensive list of goods (see below) to the region, as well as "ecosystem services" for the inhabitants. However, in the last 30 years, these ecosystems are threatened by conjugated impacts of climate change and human overuse of the nature resources. The primary objective of this chapter is to evaluate both the damage inflicted on Tafilalet ecosystems and the associated ecosystem services due to climate change. Adaptation to climate change is of paramount importance to keep the survival of the diversity of Tafilalet ecosystems including agro-ecosystems, forests, and inland waters. Adaptation is a process to reduce the vulnerability of people to prevent displacement and conflict for scarce ecosystem services. The findings of this chapter show that a large number of ecosystem services of the region are limited by climatic conditions that affect the resilience of the whole oasis of Tafilalet.

DOI: 10.4018/978-1-5225-7387-6.ch002

INTRODUCTION

The term ecosystem services (ESs) refers to the benefits that the local population can obtain from ecosystems (MEA, 2005). According to a UN report, ecosystem services include provisioning, regulating, cultural, and supporting services that directly affect human well-being. The trend of climate change (CC) is an important environmental force with significant implication for the state of water resources. The Earth's surface temperature during the last two decades has increased by about 0.5° C and an ongoing rise with similar amplitude is expected up to 2025 (IPCC, 2007). Variability in climatic conditions is already impacting ecosystems and associated ecosystem services. The impact and the link to ecosystem services are obvious and profound. CC is increasing and is affecting some extreme events such as floods and droughts. Several studies explored the impact of climate change on agriculture in general (Xie et al., 2018), on wheat production (Chelleri et al., 2014Hernandez-Ochoa et al., 2018) on forest (Pirovani et al., 2018), oasis desert (Karmaoui et al., 2014), on ecosystem services (Lee et al., 2015), on water and food (Minucci & Karmaoui, 2017; Karmaoui et al., 2018).Besides, several problems are accelerating these changes such as population growth, land-use, overgrazing... Every year CC and the associated events are affecting ecosystem services such as water and soil in both quality and quantity. These problems are affecting many parts of the region called "hot spots", where the frequency of occurrence is greater, the sensitivity is higher, the devastation is more severe, and local communities are more vulnerable. Water resources are a crucial area of vulnerability in Tafilalet, affecting water supply, agriculture, pasture, forest, and others.

Since 1970s, several Moroccan zones have experienced a general rainfall decline (Driouech, 2009; Khomsi, 2015). For the Tafilalt Oases, groundwater and surface water are interconnected resources. This connectivity can have significant implications for both water availability and quality, and presents major challenges for water managers and policy makers (Messouli et al 2009).In this area, a typical gradient can be observed from humid/sub-humid subtropical mountains and sandy desert.In the Oro-Mediterranean (mountainous Mediterranean) zone, between 2600 and 3400 m which coincides, with the most important zone for karstic groundwater creation, a large-scale degradation of vegetation has occurred. Regarding the climatic conditions, Tafilalet is characterized by a low rainfall, high temperatures, and evaporation. Those severe climate conditions influence the availability of water for the vegetation growth.

Regarding to adaptation strategies, the issues related to climate change in the preservation of oases, their preservation and their development "according to a model of resilience to climate change", Morocco have adopted pricing policies and extension services to increase the cropping intensity of existing cultivated land (Sowers, 2010).The oasis rehabilitation project is also part of wider national strategies, such as the 2020 Rural Development Strategy, the National Plan of Action to Combat Desertification and Drought, and the Green Morocco Plan.

This chapter aims to evaluate the ecosystems and the associated ecosystem services under climate change context in Tafilalet zone, which is part of the South East Moroccan oasis biosphere reserve (ReBOSuM) UNESCO 2000. A description of ecosystems, assessment of ecosystem services and the impact of climate change on agro-ecosystems and water resources, as ecosystem services, were addressed through a discussion of available principles, approaches and assessment methods.

STUDY AREA

The current study is carried out in Ziz Valley (Figure 1), located in Errachidia Province in south-oriental part of morocco. According to the 2014 census, the population of this province is 418451 people, with 46,1% in urban area and 53,9% is rural. Population density ranges between 10,8 and 22,4 people per cultivated hectare, 5,1 per uncultivated hectare, and 2 per square kilometer for the rest of the province. Sixty percent of the province's population is engaged in full and part-time farming; the rest is involved in construction, government, commerce, national and international migration (Europe), and services (RGPH, 2015). This region is characterized by the presence of one of the largest oases in the world, Tafilalet Oasis. With an area of about 1370 km^2, it was declared as Biosphere Reserve by UNESCO in 2000. It comprises a series of oases and the reservoir of one of the oldest dams in Morocco, Hassan Adakhil Dam.

The province of Errachidia relief consists of large arid areas encompassing mainly: The High Atlas which crosses all the North-West of the province composed of mountains of difficult access sometimes culminating at 3200m of altitude. The Highlands, vast, rocky and bare forming the east of the province with an altitude that revolves around 1000m, an area of accumulation of alluvial deposits carried by the floods and rivers of Ziz and Gheris is the Plain of Tafilalet, which shelters the towns of Jorf, Erfoud and Rissani, the altitude of this plain varies between 600 and 800m, and oases along the Wadis and at the level of the enclaves of the high mountains.

Rainfall deficit in the Tafilalet region was estimated by using the Mean Deviation Index (Em). The difference at the median is the most used. The difference in mean is the difference between the annual precipitation height (Pi) and the average annual precipitation height (Pm).

$$Em = Pi - Pm$$

The difference is positive for the wet year, and negative for the dry years. The year is in deficit when rainfall is below average and in surplus when the average is exceeded. This index makes it possible to visualize and determine the number of deficit years and their succession.

Calculation of the mean deviation index (Em) for the period (1975 -2008) in the various stations of the Tafilalet region, shows a large water deficit, 20 years out of 33 are moderately dry to extremely dry (Figure 2).

These climatic conditions preclude the practice of rainfed agriculture and necessitate the use of irrigation. Oasis soils are loamy to silt-clay, regenerating from alluvial deposits caused by river floods used for irrigation during the past centuries. The date palm is the basis of life and agricultural activity in the district. Cultivated land is generally located along rivers, in narrow strips of mountains and in larger areas inoases. The main cultivated crop species are: date palm, olive and apple trees (mountainous), cereals, *alfalfa*, vegetables and *henna*.

METHODS

An extensive source of ecological, meteorological, hydrological, hydro-geological data was gathered and used. This data was provided by the regional and local administrations and also extracted from several

Figure 1. Geographic localization and altitude of the study area, Tafilalet
Source: (Authors)

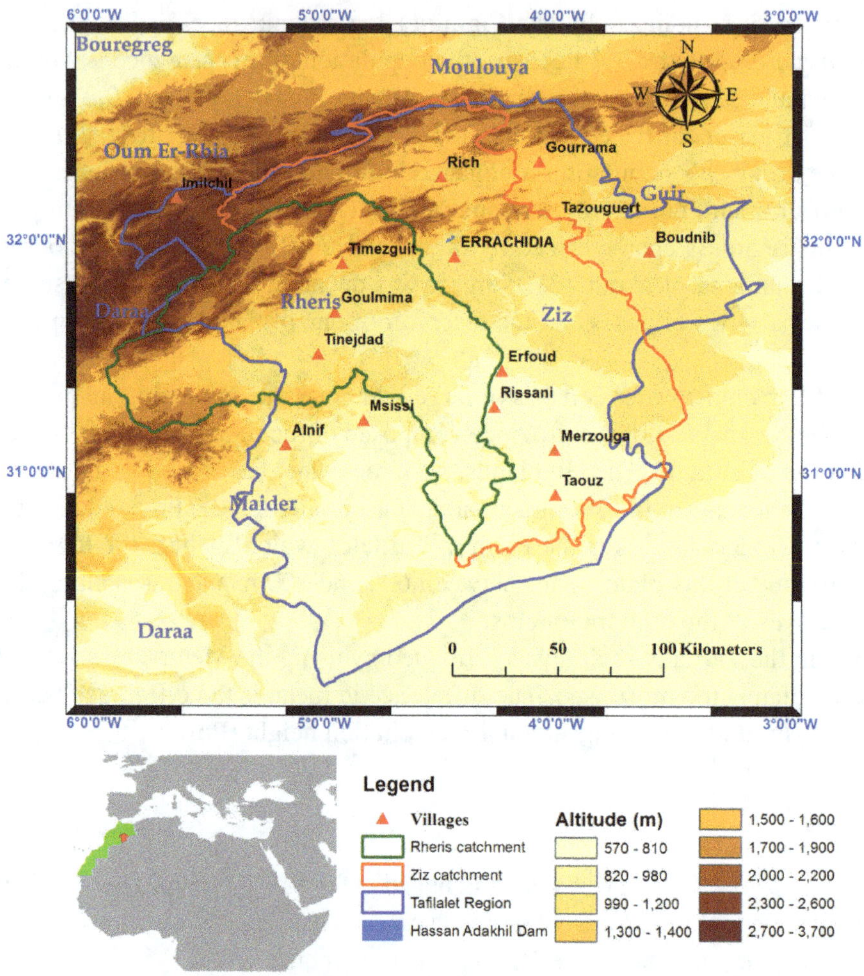

studies (Knippertz et al. 2003; Ilahiane, 2004; Cappy et al. 2004; de Jong et al. 2004; AitBoughrous 2007;). GIS and remote sensing, techniques were used to explore the spatial distribution of climatic, geographic, and ecologic information (land use-land cover).

The Approach Used to Simulate the Water Demand

WEAP software has been used to trace to precipitation projection due to its importance as indicator of ecosystem services productivity under climate change scenarios. This software was developed by the Stockholm Environment Institute (SEI) was selected for the purpose of this study. WEAP system is a demand-, priority-, and preference-driven water planning model (Yates et al., 2005). It facilitates the coupling between biophysical factors influencing the river and socio-economic factors affecting the level of water demand. These factors vary over time, for example due to increased gross domestic product, changes in consumptive attitudes. With its focus on scenario analysis, WEAP supports the planner in

Figure 2. Annual rainfall residuals and average for weather stations. Average Inter-annual Precipitation of all stations is 141.6 mm.
Source:(Ben Salem et al., 2011)

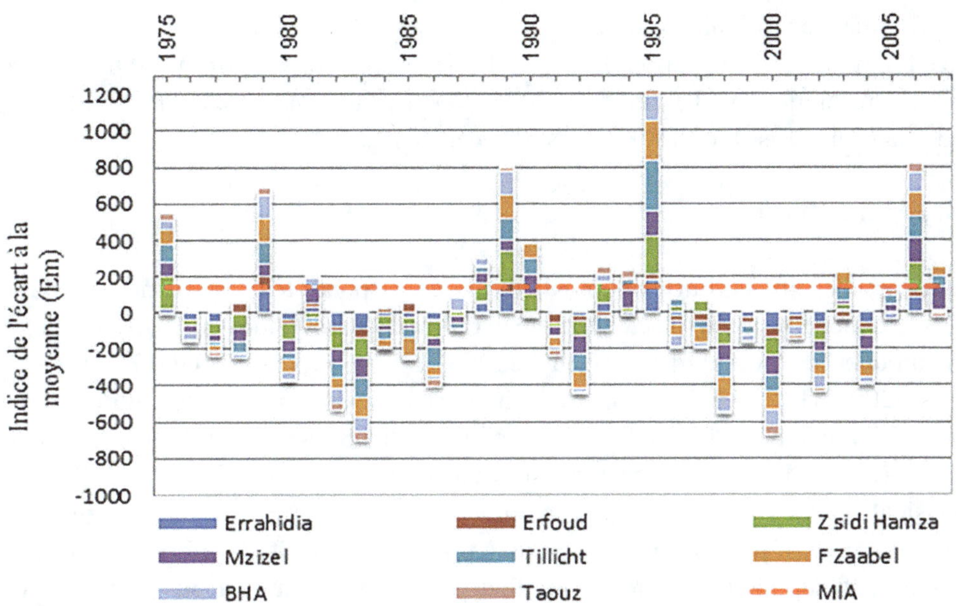

forecasting demand and supply structures under various assumptions and management practices and helps in developing resources management policies to meet future demands and solve allocation problems (Lévite, 2003 ; Yates et al., 2009).

In this study, two climate change scenarios were used: the A2 scenario and the B2 scenario.

A2 (Sc2): This is a pessimistic scenario that describes a world in which the population is growing rapidly, with strong economic growth that relies on polluting technologies in a protectionist world with growing inequalities between North and South. There is a persistent reliance on fossil fuels, and uneven economic growth across regions.

B2 (Sc1): This is an optimistic (less pessimistic) scenario that describes a world where the focus is on local solutions, in a sense of economic, social and environmental sustainability. The world population is growing steadily but at a slower pace than in A2. There are intermediate levels of economic development and technological change is slower and more diverse.

Ecosystem and Vegetation Classification

A biome is a large geographical region having a defining climate to which plants show similar physiological adaptations (Campbell, 1996). Biomes are coarse ecological units based on the phenology, life form, and physiognomy of the dominant vegetation community (Zhou, 2008).

Biome classification is considered as a tool for conservation planning and to assess the sensitivity of vegetation system distribution to climate change. Furthermore, remotely-sensed imagery played a pivotal role in generating information about vegetation cover types and land use changes (Roy, 1993). Among global initiatives, Moderate Resolution Imaging Spectroradiometer-MODIS (Zhang et al., 2003) have been used to generate vegetation cover database on global level. In this study, satellite data was used, from the MODIS at 0.01 degree (MOD1C12) on the NASA Terra and Aqua platforms to describe major vegetal types (biomes) on Tafilalet region.

Normalized Difference Vegetation Index (NDVI) was extracted from MODIS Land Product (MOD13C1), at a resolution of 0.01 degree in 2006. NDVI values was recalculate on ENVI (processing software and image analysis) for NDVI values between -1 and 1.

Net Primary Production

Carbon uptake by canopy plants during photosynthesis, corresponds to gross primary production (GPP). Gross primary production is used to calculate the net primary production (NPP) of the plant product. In this study the product NPP was extracted from the MODIS Land Product (MOD17A3) at a resolution of 5Km for the year 2006. The NPP is influenced by climatic factors especially temperature and precipitation.

Net primary production (NPP) can be calculated using the Standardized Vegetation Index (NDVI). These NDVI data are used to quantify the presence of healthy vegetation, after which NDVI is integrated into a biophysical model with temperature, humidity, precipitation, and type of vegetation cover to estimate the net amount of primary production on a planetary scale. This information has provided the global supply of land-supplied crop production that is available to humans in an average year (Imhoff & Bounoua, 2006).

Eco-Hydrology Approach

Lastly, an eco-hydrology approach was proposed to combine water or hydrology as a basis of all ecosystem services and ecology. This method proposes an integrated scheme taking in account science, engineering and construction, public administration and political decision. The chapter will show how these aspects are liked in Tafilalt Oasis.

The integrated approach to Ziz catchment and river-reservoir management has to address several aspects. Hydropower generation requires stabilization of the hydrological cycle in the catchment by setting spatial planning and land-use regulations. Improving the quality of water requires several parallel actions, including wastewater technology for sewage treatment and eco-hydrological methods. On the basis of eco-hydrology, a simplified model of sustainable groundwater management was proposed.

RESULTS AND DISCUSSION

Bio-Physical Characteristics of the Study Area (Tafilalet)

Climatic Aspects

In this region, climate is continental and arid. The annual rainfall varies from 70 mm in the extreme south, at Erfoud, to 290 mm in the extreme north, at Imilchil. However, the temperature varies from -5 °C to 40° C, with an annual average daytime high of 20° C. Quantifying inter-annual to multi-decadal precipitation variability has a multitude of application in water related research and planning.

In Tafilalet, the presence of the Atlantic barrier culminating in the north at altitudes higher than 3200 m and the arrival of hot winds of Saharan origin are at the origin of the rigor of the climate. The inter-annual averages of precipitation are function of the altitude of each catchment, the highest stations receiving the most abundant precipitation; however, the average stations remain weak. The annual rainfall regime is characterized by two wet seasons, autumn and spring, separated by a brief winter season and a long drought-stricken summer season. During the last thirty years, the average annual precipitation has been marked by a strong irregularity spatial-temporal (Figure 3), from 250 mm on the reliefs of the High Atlas to 130 mm in the area of Errachidia and less than 75 mm in the plain of Tafilalet. This low rainfall in the south is due, on the one hand, to the fact that the atlas chain prevents any oceanic influence, and on the other hand to the presence of hot air masses from the Sahara.

In thermal terms (Figure 4), summers are generally hot with maxima higher than 40° C and winters quite cold with minima lower than 0° C. The average annual air temperatures are generally between 18° C and 21°C, they increase from Foum Zaâbelin the north to Taouz further south. They show a significant trend of about 1.4° C at Errachidia (period 1973-2010).

Ecosystems, agro-ecosystems, and the associated ecosystem services are strongly liked to water resources. Any change in precipitation can impact the local ecosystem services. In order to trace the precipitation projection, WEAP model was applied in Tafilalet region following climate change scenarios.

Figure 3. Spatial distribution of mean annual precipitation in the studied area (period from 1973 to 2010)

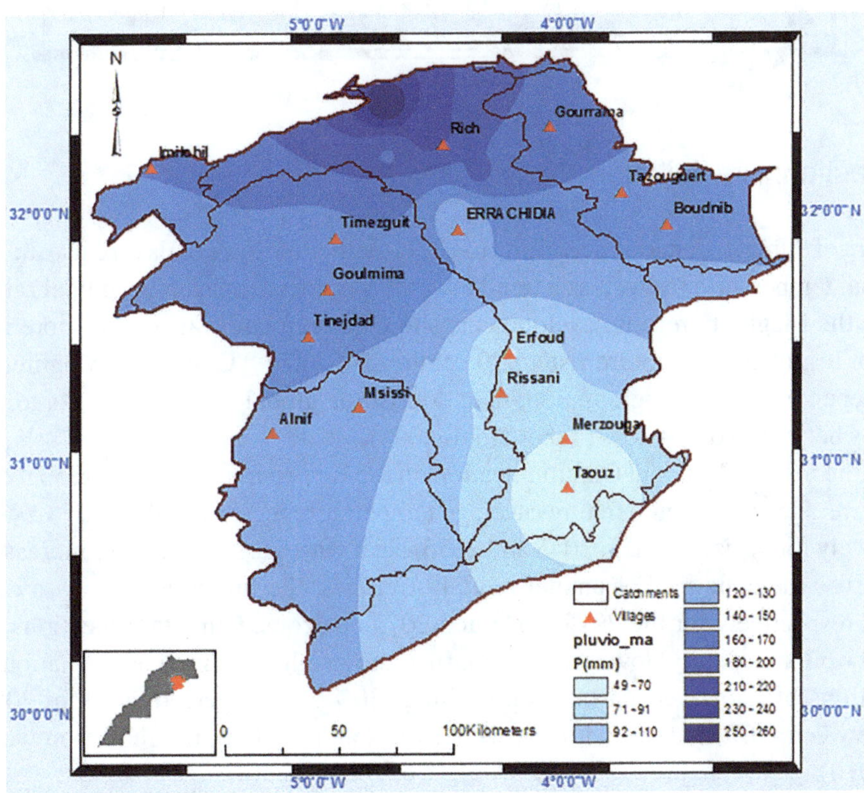

Figure 4. Spatial distribution of mean air temperature in the studied area (period from 1973 to 2010)

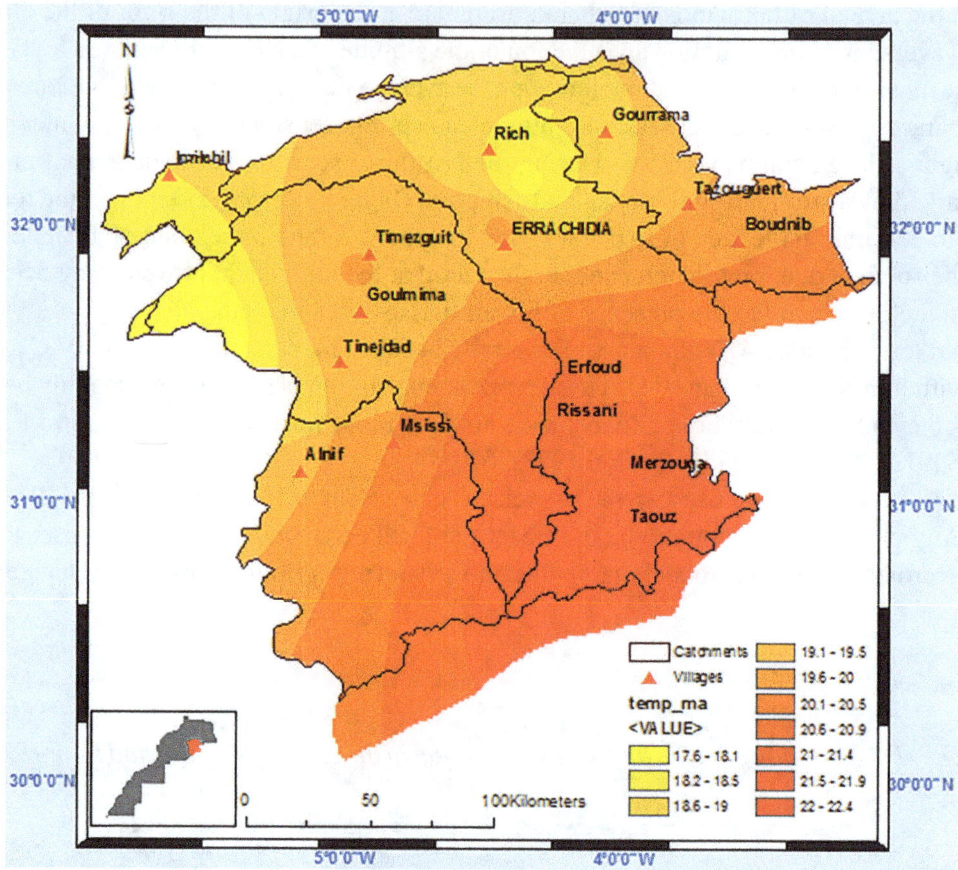

Climate Projection: Precipitation

The development of climate scenarios according to IPCC methodology reveals a trend towards an increase in average annual temperature (as well as a trend towards a decrease in average annual rainfall (Bennani et al., 2001). In the Maghreb, recent developments show that climate warming is more important than average. Indeed, if global temperature rise to 20 century was 0.74 ° C, that the Maghreb was between 1.5 and 2 ° C depending on the region, more than double the global average rise. As for the decline in rainfall, it varies between 10 and 20% (Tabet-Aoul, 2008).

In Morocco,the Second National Communication on the Environment gives as global climate change projections for average annual temperatures average temperature increases of + 0.6 ° C, + 1.8 ° C and + 3.2 ° C respectively for horizons 2015, 2045 and 2075. Heat waves are expected to increase in frequency and severity across the country. For annual rainfall averages, the projections show a decline of -6%, -13% and -19% respectively for the 2015, 2045 and 2075 horizons. Climate projections at the scale of the Souss-Massa pilot basin, by downscaling, confirm the results obtained at the national level above and give reductions of rainy wet seasons approaching -30% for the period 2020; in 2050, the deficit widens to be between -10% and -50% for the winter rains and in 2080, the reduction worsens to reach -80% (DMN 2010).

The analysis of trends observed at the Errachidia station (Figures 5), shows a significant warming, recorded at the level of the region, during the last five decades. On an annual scale, the average temperature increased by 1.4 °C, over the period 1973-2010 in Errachidia. Warming in all seasons more pronounced in summer than in winter; increasing the average temperature from 1 ° C to 3.2 ° C.

In rainfall terms, despite the existence of several drought periods, the cumulative annual rainfall shows an increasing trend at the site. This trend remains, however, weak and mainly due to heavy rains received by the area in recent years. In general, a generalized decreases at the annual scale -20% to -23% and at the seasonal scale from -15% to -59% except for a negligible increase in winter, also a significant lengthening of the maximum period of drought to annual scale. In this chapter, the projection in precipitation at the Tafilalet region scale shows a decreasing trend in precipitation under climate change scenarios 1 and 2 (Figure 5). This trend will impact the ecosystem services related to water resources.

Appling the statistical down-scaling model (SDSM), Ben Salem, (2014) projected an increasing trend in temperature and a decreasing trend in precipitation in Tafilalet. This trend was also found in Draâ Valley by Karmaoui et al., (2016).

Ecosystem, Agro-Ecosystems, and Associated Ecosystem Services

Ecosystems and Landscapes, and Vegetation Classification

Three agro-ecological zones can be distinguished; each one is characterized by a particular production system (Table 1 and Figure 6).

Figure 5. The scenario of observed precipitation at annual time step simulated by WEAP over Ziz basin
Source: (Authors)

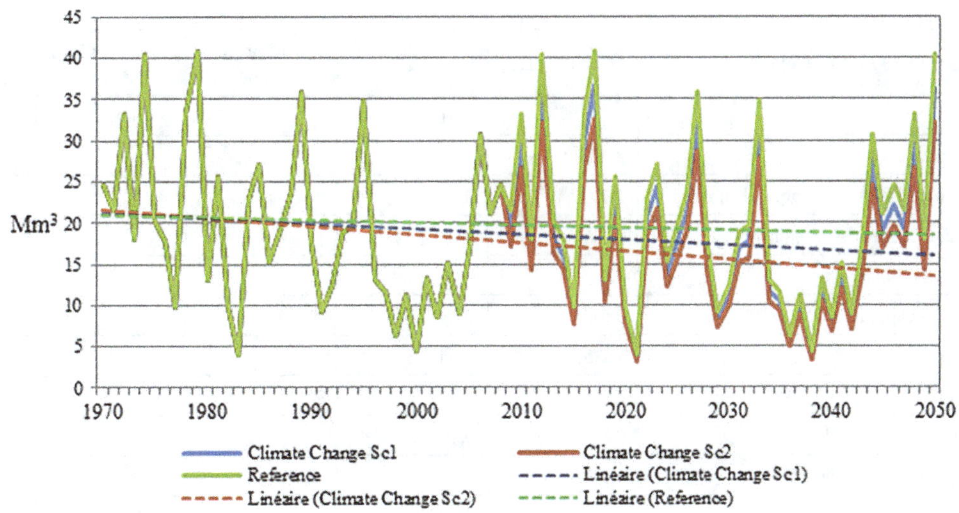

Mountain Zone

In this zone, irrigation derives from perennial waters mobilized using water diversion from dams. This zone includes a large livestock grazing area. Alfalfa and cereals are the main crops cultivated in a two-level crop system in association with olive and rosaceous fruit species.

Intermediate Zone

This zone is located in the hills and valleys at the foot of mountains. It is characterized by an intensive three-level cropping system with date palms as the upper level, olive trees as intermediate one and cereals, forages, legumes and vegetables as the lower level. Hydrologically, river floods and springs are used in irrigation.

Plain Zone

In the plain, water for irrigation comes from river floods, pumping, Khettaras (Old water system irrigation), perennial streams and springs, and Hassan Addakhil Dam. A two-level cropping system is dominant with date palm trees and underneath cereals, alfalfa, and vegetable crops. Animal husbandry is dominated by D'man sheep raised for both milk and meat production.

Land tenure is characterized by small holders (micro-farms). Average farm size is almost 1 ha divided in 3 plots. Ninety percent of the farmers own less than 5 ha. Private ownership represents 95% of land tenure; the remaining 5% belongs to the religious holdings, Habous (Hsain, 1996).

The land cover scheme identifies 13 land cover classes, which includes 10 natural vegetation classes, 1 mosaicked land class, and 2 non-vegetated land classes (Figure 6).

Table 1. Land Use and Land Cover categories and their relationship with the main agro-ecological zones

Land Use / Land Cover	Area (km2)	%	Mountain Zone	Intermediate Zone	Plain Zone
Water bodies	7,1	0,01	*	*	*
Consolidatedbare areas	21003,0	30,25			
Closed to open broadleaved decidous shrubland	4,7	0,01	*		
Mosaic grassland/forest or shrubland	0,2	0,00			
Closed to open shrubland	8,6	0,01	*		
Mosaic forest or shrubland/grassland	580,0	0,84	*		
Sparsegrassland	1835,6	2,64	*		
Sparsevegetation	2800,3	4,03	*		
Rainfedcropland	97,9	0,14		*	*
Rainfed trees and schrub crops	8,1	0,01		*	*
Bare Areas	42862,7	61,73			
Artificial Surfaces and Associated Areas	22,8	0,03	*	*	*
Mosaic vegetation;cropland	145,6	0,21		*	
Mosaic croplandvegetation	62,7	0,09		*	*

Source: (Authors)

The NDVI of Tafilalet is between -0.046800 and 0.277200 for the palm groves and banks of Ziz Wadis and Gheris always have positive values. The oases of Tafilalet show some stability in terms of vegetation throughout the year despite the seasonal climatic variations.

The perimeters of the study area include three classes of the NPP. The first class, from 0 to 52 g (c) / km² / year, in the majority of the territory of Tafilalet (bare and built soil). The second class, from 53 to 170 g (c) / km² / year, in the high mountains of the atlas (upper Ziz) and the plain of Tafilalet (spontaneous plants and forests). The third class, from 180 to 300 g (c) / km² / year, is very limited to a grid (5x5 km) in the upper Ziz where there is a high rainfall and abundance of water resources. However, high-resolution satellite images of PPN can reflect the productivity of the Tafilalet oases. In comparison with Morocco's net primary productivity (NPP) (Rochdane, 2014), which can reach 1900 grams of carbon per square kilometer per year in agricultural areas and evergreen forests, the Tafilalet region and surrounding areas offer a PPN that does not exceed 170 grams of carbon per Km² per year (Figure 7).

Figure 6. Land cover and cover (LULC) in Tafilalet (Modis MOD12C1).
Source: (Authors)

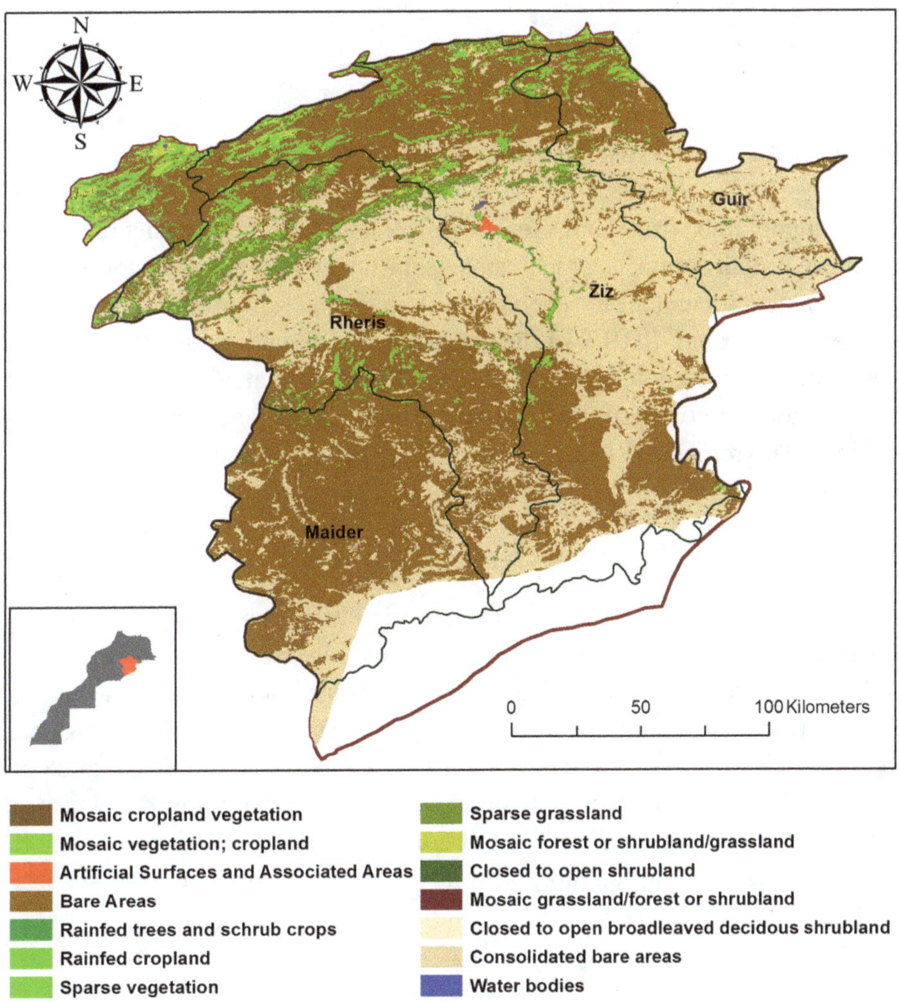

Figure 7. Net Primary Production (NPP) (MOD17A3) at a 5x5km resolution for 2006 Tafilalet and surrounding areas.
Source: (Authors)

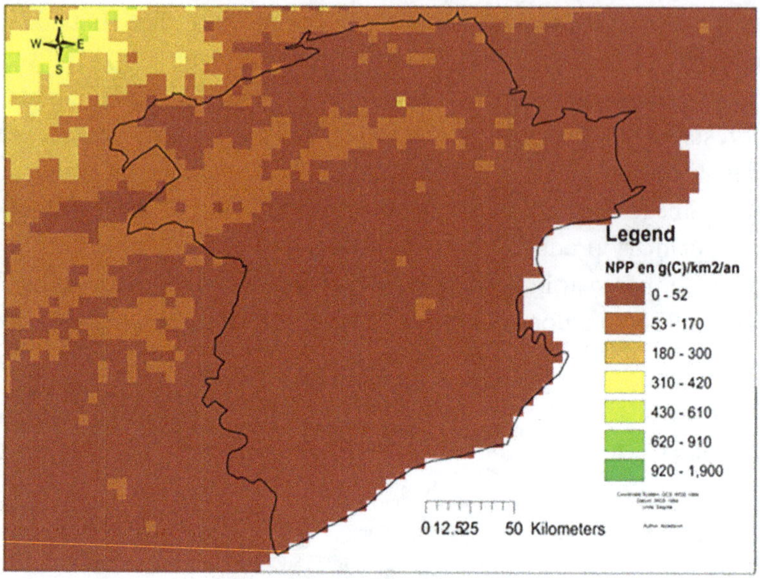

Agro-Ecosystems and Ecosystem Services

According to the result obtained from land use, land cover and NPP the Tafilalet region is characterized by the diversity of ecosystem services, which can be subdivided into four categories (Table 2).

Water is the fundamental element for each ecosystem, in the region of Tafilalet the area occupied by water is estimated at 7, 11 km^2 (0.01% of the total area), to which is added the groundwater which is the main water resource in the region. Vegetation cover is estimated at 5689.33 km^2, or 8.2% of the total area. Water and vegetation are two primary ecosystem services; they act as supply, regulatory, support and cultural services. The remaining land is bare or used for construction, accounting for 92% of the total area (63888.48 km^2).

Agriculture

Agricultural goods and services are mainly cereals, arboriculture such as (olive, apple, and palm trees), and livestock (Figure 8). The area available for the practice of farming is 43,069 hectares and the potential area covers 48,069 hectares making up 45,000 farming household units. On the provincial level, the farming system is partitioned as follows: cereals occupying 70 percent of the farming area (1984-1994) and 54% (1990-2001), *alfalfa* 17 percent, vegetable gardens 4 percent, and *henna* and cumin 1 percent.

Arboriculture, mainly dates and olives, plays a crucial role in the economic development and ecological survival of the area accounting for 60 percent of household income. Livestock is the third important economic aspect of the valley's agriculture, with an average of 5 to 6 head of sheep per household (ORMVAT, 1991).

Table 2. The main categories of Tafilalet ecosystems

Ecosystems	Area (km²)	%
Water	7,11	0,01
Bare area	63865,67	91,97
Vegetation	5689,33	8,19
Artificial Surfaces and Associated Areas	22,82	0,03

Source: (Authors)

In summary, the potential productivity of the region is restricted by climatic conditions affecting the valley's irrigated farms. Water scarcity and its erratic variability over time and space, recurrent droughts, and frequent locust invasions have contributed to the impoverishment of the valley's ecosystems and the associated ecosystem services. The bayoud (*Fusarium oxyisporum*) disease is ravaging the date palm trees. The annual alternating olive production combine to reduce household welfare. All these factors have, in one way or another impeded the optimization of the ecosystems resources.

Irrigation and Water Management

Until the early seventies, irrigation was practiced in Tafilalet oasis along the Ziz River using traditional systems such as Seguia (s) (irrigation canal) and Khettarat (s). The water scarcity caused by the climate

Figure 8. Average cultivated area and number of trees in Tafilalet
Source: (Authors processing using data from ORMVAT, 2004)

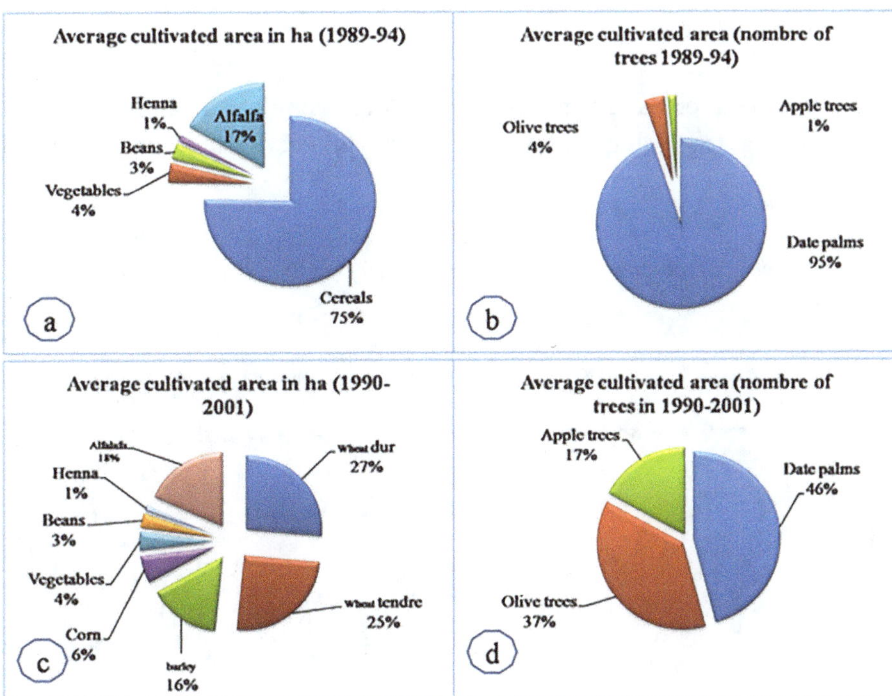

change has constrained farmers to adopt new means to mobilize water for irrigation. Wells were then sinked and equipped with pumping motors for both individual and collective use.

In the prefecture of Tafilalet there are more than 600 private wells, 60 common pumping stations among which 40 are managed by cooperatives and 570 Khettaras. The technology of pumping stations is relatively new for farmers in this region.

Groundwater in Tafilalet

At a national scale, groundwater is a very significant hydraulics heritage. There are 32 layers deep and more than 46 groundwater resources. Groundwater resources which are estimated at four billion m^3 spread unevenly in different parts of the kingdom (Margat& Zebidi,1991). In the Tafilalet, groundwater is estimated at 200 Mm^3 including 120 Mm^3 mobilized annually from moderately deep water (Anonymous, 2000). They are exploited by means of pumping stations, wells and Khettarat (s). From a hydrological viewpoint, there are four levels of surface aquifers to medium deep: The system of groundwater basins Quaternary which is closely related sub-flows, and ground limestone of Turonian, constituting the main aquifer, the intermediate layers composed of Cenomanian horizons of continental waters and red sandstone and deep layers of the Infra-Cenomanian of Ain El Atti.

Groundwater is the only perennial water resources for the people of the region Tafilalet especially during low water. Thus, in response to climate change that knows the area; wet periods are increasingly rare in the region. Thus, the volumes of water extracts of water are on permanent increase that translates into a lower level of the ground (Figure 9), which may influence the water quality of the water by the concentration of chemical elements following the increase the rate of evaporation.

Importance of Ecosystem Services and Their Fragility

List of Ecosystem Services of Tafilalet

Ecosystem services are the goods and benefits that the local population can obtain from ecosystems. These include provisioning, regulating, cultural, and supporting ecosystem services that directly affect human well-being.

Figure 9. Piezometric level variation of aquifers (Fezna-Jorf) in the Tafilalet area.
Source: (Ben Salem et al., 2011)

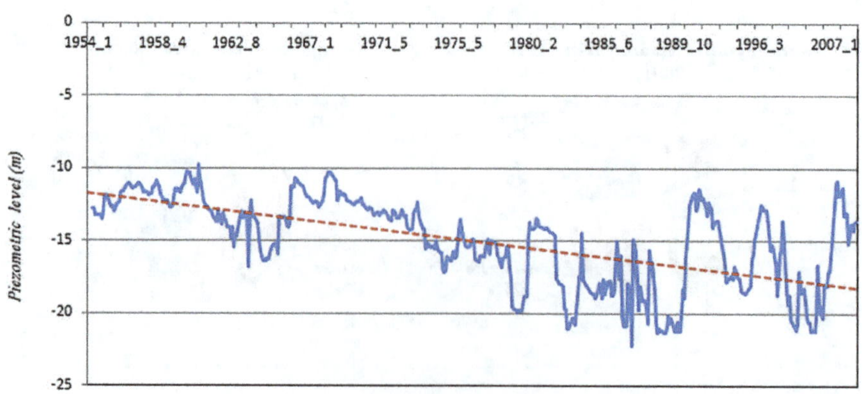

Provisioning Services

This category includes water, food, and fiber. Food products are all goods provided by plants (agricultural products), animals, and microbes. Fiber includes wood and wool provided by the local ecosystems that serve as a source of energy. Natural medicines and pharmaceuticals are biocides, food additives are also derived from agro-ecosystem and natural ecosystems. Freshwater includes groundwater and water surface supplied by local ecosystems, rivers, lakes, reservoirs. Water is also considered as supporting service.

Regulating Services

This type include the benefits obtained from ecosystem processes such as air, water, soil quality regulation. Any change in land use-land cover affects the climatic parameters by favoring or reducing Carbone sequestration. The magnitude of runoff and flooding can is affected by changes in land cover (replacement of forests and urbanization). Additionally, vegetation cover prevents the soil and help in the water purification. Regarding the regulation of diseases, any change in ecosystems can affect the abundance of some pathogens or agent vectors such as the *Leishmania major* that causes cutaneous leishmaniasis.

Cultural Services

This category includes the nonmaterial benefits provided by ecosystems through spiritual enrichment, recreation, aesthetic experiences because ecosystems are key elements that favor the diversity of cultures.

The Tafilalet ecosystems provide a large list of goods and ecosystem services. However, in the last 30 years they are threatened by conjugated impacts of climate change and human pressure. Climate change adaptation will be needed in a variety of ecosystems, including agro-ecosystems (crops, livestock, and grasslands), forests and woodlands, and inland waters.

Adaptation aims to reduce the vulnerability of people (mostly poor) to prevent displacement and conflict for scarce of ecosystem services. Adaptation to climate change in the region of Tafilalet involves the review and if necessary changes in policies, programs, investment and, ultimately, the human behavior toward the nature, in light of accumulated knowledge.

In the studied area, water is at the top of the local population's preoccupations.

The establishment of pilot adaptation projects in collaboration with local people is a key solution. These pilot projects will focus on sectors greatly affected by climate change (water, pasture, agricultural production, biodiversity, and soil fertility).

Agro-Biodiversity

Biodiversity in all its components, genes, species, and ecosystems increase resilience to changing environmental conditions and stresses. Genetically, diverse populations and species-rich ecosystems have greater potential to adapt to climate change. The selection of abiotic-tolerant crops and cultivars (tolerant of high temperature, drought and flood, high soil salt content and pest and disease resistance) allows for the exploitation of genetic variability in new crop varieties..

Soil and Land Management

Climate change adaptation for agricultural cropping systems requires a higher resilience against both excess of water (due to high intensity rainfall) and lack of water (due to extended drought periods). A key element to respond to both problems is soil organic matter, which improves and stabilizes the soil structure so that the soils can absorb higher amounts of water without causing surface run off, which could result in soil erosion and, further downstream, in flooding. Soil organic matter also improves the water absorption capacity of the soil for during extended drought.

Eco-Hydrology as Mitigation Tool

Eco-hydrology is more than just hydrology and ecology combined. It is functional to its best only, if science, engineering and construction, public administration and political decision assist each other in an integrated scheme and on a common scale. The Tafilalt Oasis for example is dependent on all those aspects.

The integrated approach to Ziz catchment and river-reservoir management has to address several aspects. Hydropower generation requires stabilization of the hydrological cycle in the catchment by setting spatial planning and land-use regulations. Improving the quality of water requires several parallel actions, including wastewater technology for sewage treatment and eco-hydrological methods. On the basis of eco-hydrology, a simplified model of sustainable groundwater management was proposed (Figure 10).

Hydraulic modeling will enhance sedimentation in controlled areas in the river mouth. The removal of sediments, which can be used for fertilizer by local farmers, will reduce internal nutrient loading. Particular attention will be given to increasing environmental awareness of local communities.

Application of Eco-Hydrology: Classified Impacts and Adaptations

The integrated approach to watershed management and reservoirs (dams) needs to take into account a number of aspects; hydropower generation requires the stabilization of the hydrological cycle in the watershed by establishing land-use planning and development rules. Improving water quality requires several parallel actions, including new technologies for wastewater treatment and eco-hydrological methods to maintain the balance of ecosystems. Hydraulic modeling will enhance sedimentation in controlled areas in the river mouth. The removal of sediments, which can be used for fertilizer by local farmers, will reduce internal nutrient loading. Particular attention will be given to increasing environmental awareness of local communities. This will result in the use of modern aquaculture technologies, the reduction of degradation of aquatic ecosystems and the creation of agro-economic opportunities.

The oases of Morocco, by their biological, cultural and architectural diversity, offer exceptionally rich and varied landscapes, play many environmental functions and provide a multitude of goods and services of social, ecological or economic nature. They possess assets and potentialities that are at the base of the development of many human activities such as agriculture, rangelands, tourism, crafts and industry. However, these potentialities lie in a constraining context and a fragile environment. The severe climatic conditions, the low capacity of resilience and the scarcity of the water led the man to practice a traditional system of production which allowed to develop a traditional know-how in the mobilization of the water (Khettarat (s), ...), in the practice of agriculture (adoption of an intensive system with

Figure 10. Eco-hydrology approach for decision maker
Source: (Ben Salem 2011)

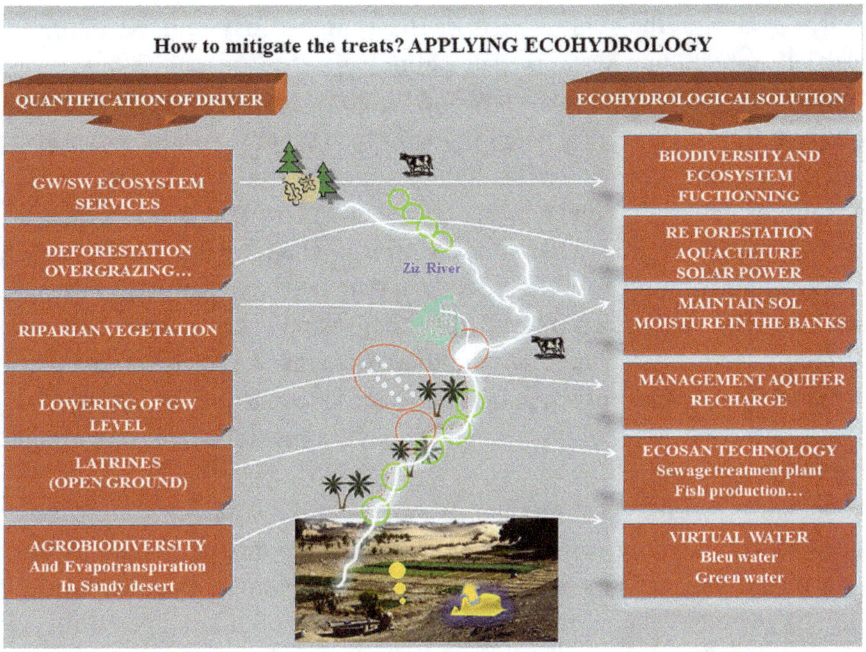

three stages of vegetation associated with livestock and handicrafts) and in the management of natural resources (MEMWE 2009).

The water resources were much more abundant in the past and as a result, the deep aquifers were saturated as evidenced by the extension of the travertine deposits which fossilized the different outlets. Rainfall was more abundant and regular than currently upstream in the High Atlas. Indeed, the results available in neighboring regions (Kabiri, 2004) show that this region of Morocco currently desert in its entirety has known throughout its history and especially for already a few hundred millennia considerable climate and environmental changes. At about 20,000 and 6500 years BP (before Present), for example, there was a mild climate in the Tafilalt. The vegetation was much more diverse and denser, and the groundwater was saturated (Kabiri, 2004).

Tafilalet is experiencing a significant deficit in irrigation water because of prolonged droughts that affect these already fragile regions but also and above all because of human behavior. Indeed, in some oases the motor-pump which replaced the traditional systems of exploitation (Oughrour, Khettarat, Etc.) caused a spectacular reduction of the water table, which is the most solicited. In others it is the silting that is threatening, it is for example Jorf (Kabiri, 2003).

Degradation of natural resources is affected by different factors within existing ecosystems. It is mainly caused by both natural factors and anthropogenic actions that are accentuated with the socio-economic context. Desertification, salinization, soil erosion, Bayoud's disease (Fusarium), silting of cultivated land, and climatic conditions all contributed to the decline of the oases area (Riou, 1990; Belarbi et *al.*, 2004; Vidal, 2005).

CONCLUSION

The most productive system of the study area is the oasis agro-ecosystem. Farming Zone is based on traditional, intensified, poly-culture systems (date palm, fruit trees, olive, alfalfa, cereals & vegetables) developed in response to high population density and limited irrigation. Alfalfa is the main forage used to feed sheep, which are a principal component in the farm economy. Alfalfa is used as green forage in warm months and as dry forage in winter. Though low forage production is related to environmental factors (drought, salinity, and high temperature). The results show that many ecosystem services, particularly water services in the region, are limited by climate change affecting its resilience, as well as the high rate of sediment transport resulting in reduced dam capacity, resulting in insufficient water to meet the downstream population's needs in times of drought and water scarcity.

Adaptation strategies and new technologies are therefore a necessity in various ecosystems, mainly agro-systems, to reduce the vulnerability of people and natural resources.The eco-hydrological approach represents a realistic, effective and socially acceptable environmental strategy, leading to a sustainable management of ecosystem services.

The chapter concludes that the most productive system in the study area are oasis agro-ecosystem and climate change clearly have serious consequences for water resources and associated ecosystem services, but water is insufficient to meet the needs of thc downstream population in times of drought and water scarcity. As a result, adaptation can help to reduce population vulnerability, by applying new technologies, approaches and environmental strategies that are realistic, effective and socially acceptable, leading to sustainable management of ecosystem services.

ACKNOWLEDGMENT

We would like to thank « *Direction régionale de l'Hydraulique des bassins Versants Guir-Rheris-Ziz (DRH GRZ) Errachidia » and "Office Régional de la Mise enValeur Agricole de Tafilalet (ORMVAT) Errachidia.* » for their support in terms of data, our thanks to the "*Africa Climate Leadership Program (AfriCLP)"* and the "*Southern Center for Culture & Sciences (SCCS), Morocco"* for they help to realize this chapter.

REFERENCES

Ait Boughrous, A. (2007). *Biodiversité, écologie et qualité des eaux souterraines dans deux régions arides du Maroc: le Tafilalet et la région de Marrakech* (PhD Thesis). Cadi Ayyad University.

Akter, T., Quevauviller, P., Eisenreich, S. J., & Vaes, G. (2018). Impacts of climate and land use changes on flood risk management for the Schijn River, Belgium. *Environmental Science & Policy, 89*, 163–175. doi:10.1016/j.envsci.2018.07.002

Anonymous. (2000). *Fish collection database of the Gulf Coast Research Laboratory (GCRL). The Gulf Coast Research Laboratory*. Ocean Springs, MI: GCRL.

Belarbi, A., Boayad, A., Diaou, M., Kaassis, N., & Tidjani Miliki, M. (2004). Agrobiodiversité et durabilité des systèmes de production oasiens dans la palmeraie d'Aoufous. ICRA, INRA.

Ben Salem, A. (2014). *Vulnérabilité et adaptation aux changements climatiques dans les oasis du Tafilalet- Maroc* (Doctoral thesis). University Cadi Ayyad.

Ben Salem, A. B., Messouli, M., & Yacoubi-Khebiza, M. (2011). *Developing an oasis-based water management tool: Ecohydrologic approach and weap software for a large arid catchment in Morocco*. Academic Press.

Bennani, A., Buret, J. & Senhaji, F. (2001). *Communication Nationale Initiale a la Convention Cadre des Nations Unies sur les changements climatiques*. Ministère de l'Aménagement du Territoire, de l'Urbanisme de l'Habitat et de l'Environnement.

Campbell, N. A. (1996). Biology (4th ed.). The Benjamin/Cummings Publishing Company.

Cappy, S., Reichert, B., & Thein, J. (2002). Hydrogeological framework of the northern Drâa-catchment:Results of local and regional scale investigations. *Afrikagruppe deutscher Geowissenschaftl er,Jahrestreffen 2002 – Forschungs- und Entwicklungszusammenarbeit in Afrika. Proceedings*.

Chelleri, L., Minucci, G., Ruiz, A., & Karmaoui, A. (2014). Responses to Drought and Desertification in the Moroccan Drâa Valley Region: Resilience at the Expense of Sustainability? *International Journal of Climate Change: Impacts & Responses*, *5*(2).

Driouech, F., Déqué, M., & Mokssit, A. (2009). Numerical simulation of the probability distribution function of precipitation over Morocco. *International Journal Climate Dynamics*, *32*(7-8), 1055–1063. doi:10.100700382-008-0430-6

HCP. (2015). *Caractéristiques Démographiques et Socio-Economiques: REGION DRAA-TAFILALET.* Direction Régionale Drâa-Tafilalet.

Hernandez-Ochoa, I. M., Asseng, S., Kassie, B. T., Xiong, W., Robertson, R., Pequeno, D. N. L., & Hoogenboom, G. (2018). Climate change impact on Mexico wheat production. *Agricultural and Forest Meteorology*, *263*, 373–387. doi:10.1016/j.agrformet.2018.09.008

Ilahiane, H. (1996). Small-Scale Irrigation in a Multiethnic Oasis Environment: The Case of Zaouit AmelkisVillage. *Southeast Morocco Journal of Political Ecology*, *3*, 89. doi:10.2458/v3i1.20459

Imhoff, M. L., & Bounoua, L. (2006). Exploring global patterns of net primary production carbon supply and demand using satellite observations and statistical data. *Journal of Geophysical Research, D, Atmospheres*, *111*(D22), D22S12. doi:10.1029/2006JD007377

IPCC. (2014). *Climate Change 2014: Synthesis Report. Contribution of Working Groups I, II and III to the Fifth Assessment Report of the Intergovernmental Panel on Climate Change. Climate Change 2014: Synthesis Report.* Contribution of Working Groups I, II and III to the Fifth Assessment Report of the Intergovernmental Panel on Climate Change. doi:10.1017/CBO9781107415324

Jong, D. C., Machauer, R., Reichert, B., Cappy, S., Viger, R., & Leavesley, G. (2004). An integrated geomorphological and hydrogeological MMS modeling framework for a semi-arid mountain basin inthe High Atlas, southern Morocco. In *International Congress: Complexity and Integrated Resources Management*. International Environmental Modelling and Software Society.

Kabiri, L. (2003). *Les Ressources en eaux dans les Oasis de Sud Marocain: cas de Tafilalt*. Errachidia, Maroc: Ghèris et de Ferkla.

Kabiri, L. (2004). *Contribution à la connaissance, la préservation et la valorisation des Oasis du Sud marocain: cas de Tafilalt*. Thèse d'habilitation universitaire, Facultés des Sciences et Techniques, Errachidia, Université My Ismaïl.

Karmaoui, A., Ifaadassan, I., Babqiqi, A., Messouli, M., & Khebiza, M. Y. (2016). Analysis of the Water Supply-demand Relationship in the Middle Draa Valley, Morocco, under Climate Change and Socio-economic Scenarios. *Journal of Scientific Research & Reports*, *9*(4), 1–10. doi:10.9734/JSRR/2016/21536

Karmaoui, A., Messouli, M., Khebiza, Y. M., & Ifaadassan, I. (2014). Environmental Vulnerability to climate change and anthropogenic impacts in dryland (Pilot study: Middle Draa Valley, South Morocco). *J Earth Sci Clim Change, S11*, 2. doi:10.4172/2157-7617.S11-002

Karmaoui, A., Minucci, G., Messouli, M., Khebiza, M. Y., Ifaadassan, I., & Babqiqi, A. (2018). Climate Change Impacts on Water Supply System of the Middle Draa Valley in South Morocco. In *Climate Change, Food Security and Natural Resource Management* (pp. 163–178). Cham: Springer.

Khomsi, K., Mahe, G., Tramblay, Y., Sinan, M., & Snoussi, M. (2015). Trends in rainfall and temperature extremes in Morocco. *Natural Hazards and Earth System Sciences*, *3*, 1175–1201. doi:10.5194/nhessd-3-1175-2015

Knippertz, P., Christoph, M., & Speth, P. (2003b). Long-term precipitation variability in Morocco and the linkto the large-scale circulation in recent and future climates. *Meteorology and Atmospheric Physics*, *83*(1-2), 67–88. doi:10.100700703-002-0561-y

Knippertz, P., Fink, A. H., Reiner, A., & Speth, P. (2003a). Three late summer/early autumn cases of tropicalextra-tropical interactions causing precipitation in Northwest Africa. *Monthly Weather Review*, *131*(1), 116–135. doi:10.1175/1520-0493(2003)131<0116:TLSEAC>2.0.CO;2

Lee, C., Schlemme, C., Murray, J., & Unsworth, R. (2015). The cost of climate change: Ecosystem services and wildland fires. *Ecological Economics*, *116*, 261–269. doi:10.1016/j.ecolecon.2015.04.020

Lévite, H., Sally, H., &Cour, J. (2003). Testing water demand management scenarios in a water-stressed basin in South Africa: application of the WEAP model. *Physics and Chemistry of the Earth, 28*(20-27), 779-786.

MEA. (2005). *Dryland Systems. Ecosystems and human well-being: current state and trends.*

Messouli, M., Rochdane, S., El Alami El Filali, A., & Bouarab, L. (2009). Groundwater–surface water interactions and the ecohydrology of arid regions: evidence from the Tafilalt oasis of southern Morocco Groundwater and Climate in Africa. *Proceedings of the Kampala Conference, 334*, 61-67.

Ministry of Energy, Mines, Water and Environment (Secrétariat d'État chargé de l'eau et de l'environpressure). (2009). *Adaptation aux changements climatiques au Maroc: pour des Oasis résilientes*. Author.

Minucci, G., & Karmaoui, A. (2017). Exploring the Water-Food-Energy and Climate Nexus: Insights from the Moroccan Draa Valley. In *Peri-Urban Areas and Food-Energy-Water Nexus* (pp. 89–97). Cham: Springer. doi:10.1007/978-3-319-41022-7_11

Pirovani, D. B., Pezzopane, J. E. M., Xavier, A. C., Pezzopane, J. R. M., de Jesus Júnior, W. C., Machuca, M. A. H., ... dos Santos, A. R. (2018). Climate change impacts on the aptitude area of forest species. *Ecological Indicators*, *95*, 405–416. doi:10.1016/j.ecolind.2018.08.002

Riou, Ch. (1990). Bioclimatologie des oasis. *Option Méditerranéennes, série A*, *11*, 207–213.

Rochdane, S., Bounoua, L., Zhang, P., Imhoff, M. L., Messouli, M., & Yacoubi-Khebiza, M. (2014). Combining Satellite Data and Models to Assess Vulnerability to Climate Change and Its Impact on Food Security in Morocco. *Sustainability*, *6*(4), 1729–1746. doi:10.3390u6041729

Roy, P. S. (1993). *Remote sensing for forest ecosystem analysis and management*. New Delhi, India: Environmental Studies in India. IBH.

Sellers, P. J., Tucker, C. J., Collatz, G. J., Los, S. O., Justice, C. O., Dazlich, D. A., & Randall, D. A. (1996). A revised land surface parameterization (SiB2) for atmospheric GCMs. Part II: The generation of global fields of terrestrial biophysical parameters from satellite data. *Journal of Climate*, *9*(4), 706–737. doi:10.1175/1520-0442(1996)009<0706:ARLSPF>2.0.CO;2

Sowers, J., Vengosh, A., & Weinthal, E. (2010). Climate change, water resources, and the politics of adaptation in the Middle East and North Africa. *Climatic Change*, *104*(3–4), 599–627. doi:10.100710584-010-9835-4

Tabet-Aoul, M. (2008). Impacts du changement climatique sur les agricultures et les ressources hydriques au Maghreb. *Les notes d'alerte du CIHEAM.*

Vidal, M. (2005). Ajustement entre des systèmes irrigués et des systèmes de culture diversifiés. Fonctionnement de deux oasis du Tafilalet Mémoire de Master.

Xie, W., Huang, J., Wang, J., Cui, Q., Robertson, R., & Chen, K. (2018). Climate change impacts on China's agriculture: The responses from market and trade. *China Economic Review*. doi:10.1016/j.chieco.2018.11.007

Yates, D., Purkey, D., Sieber, J., Huber-Lee, A., & Galbraith, H. (2005). WEAP21: A demand, priority, and preference driven water planning model: Part 2, Aiding freshwater ecosystem service evaluation. *Water International*, *30*, 501–512. doi:10.1080/02508060508691894

Yates, D., Purkey, D., Sieber, J., Huber-Lee, A., Galbraith, H., West, J., & Rayej, M. (2009). Climate driven water resources model of the Sacramento Basin, California. *Journal of Water Resources Planning and Management*, *135*(5), 303–313. doi:10.1061/(ASCE)0733-9496(2009)135:5(303)

Zhang, X., Friedl, M. A., Schaaf, C. B., Strahler, A. H., Hodges, J. C., Gao, F., & Huete, A. (2003). Monitoring vegetation phenology using MODIS. *Remote Sensing of Environment*, *84*(3), 471–475. doi:10.1016/S0034-4257(02)00135-9

Zhou, T., Yi, C., Bakwin, P. S., & Zhu, L. (2008). Links between global CO2 variability and climate anomalies of biomes. *Science in China. Series D, Earth Sciences*, *51*(5), 740–747. doi:10.100711430-008-0024-5

KEY TERMS AND DEFINITIONS

Agro-Ecosystem: An agroecosystem is a human-modified ecosystem able to produce vegetal and animal goods and associated services used for human and animal nutrition.

Arid Zone: An arid zone is an area where rainfall is low and inferior to potential evapotranspiration.

Climate Change: Is a change in the average of its characteristics for a long time estimated by decades or more.

Ecosystem: Is a set of living organisms that interact with each other and with the environment in which they live (soil, water, soil, etc.).

Ecosystem Services: Is the set of goods and services provided by an ecosystem.

NDVI: Normalized difference vegetation index is an indicator aimed at assessing the vegetation in the land through the use of remote sensing. Calculated by determining the difference between near-infrared (which vegetation strongly reflects) and red light (which vegetation absorbs).

NPP: Net primary productivity is the amount of organic matter produced by converting solar energy into a living mass.

Oasis: Is a green land in a desert area by dint of a water source or well.

Chapter 3
Ecosystem Services Demand Management Under Climate Change Scenarios:
Use of WEAP Software in Case of Water Demand in Ziz Basin, Morocco

Abdelkrim Ben Salem
Cadi Ayyad University, Morocco

Souad Ben Salem
Cadi Ayyad University, Morocco

Mohammed Khebiza Yacoubi
Cadi Ayyad University, Morocco

Mohammed Messouli
Cadi Ayyad University, Morocco

ABSTRACT

Water ecosystem service is the most important element that supports Tafilalet agro-ecosystems. In this region, drought frequency is increasing, which complicate the management groundwater reserves. The ephemeral flows of the rivers force people to use groundwater to meet the population demand. Consequently, water resource management is of significant importance the sustainability of this area. Water evaluation and planning (WEAP) is useful management software used to evaluate and trace the trend of water demand. This model was applied in case of Ziz basin in order to simulate and analyze the situation of water under different scenarios. The results show an increasing of demand for water irrigation and with introducing modern irrigation scenario. However, a decreasing trend in reservoir storage volume and groundwater storage was projected in Tafilalet.

DOI: 10.4018/978-1-5225-7387-6.ch003

INTRODUCTION

Arid and semi-arid ecosystems provide several goods and services. Water ecosystem service is the most important element that supports the human well-being. Climate change is a real challenge in the dry areas that affect some extreme events such as drought, which increases and complicates the management groundwater reserves. The change in precipitation influences the flows of the rivers, which forces local population to use groundwater to meet the increasing demand. Several studies have shown that it is important to include the effects of climate change in local water planning (Schimmelpfennig, 1996). Risbey (1998),seeking to link present-day planning decisions to uncertain future climate projections. More recently, researchers have used integrated water resource planning models to evaluate the impact of climate perturbations on the performance of current water management systems (Brekke et al., 2004; Vicuna et al., 2007; Zhu et al., 2005; Karmaoui et al., 2018; 2016; 2015).

The trend toward a global warming of the Earth's atmosphere is an important environmental force with significant implication for the state of groundwater ecosystems. The Earth's surface temperature during the last two decades has increased by about 0.5° C and an ongoing rise with similar amplitude is expected up to 2025 (IPCC, 2001; IPCC, 2007). The global warmth in 2001 was unusually high and is considered to be a consequence of anthropogenic greenhouse gases (Hansen et al., 2002).

Since 1970s, Morocco has experienced a general rainfall decline and temperature increase (Driouech, Déqué & Mokssit, 2009) that causes a decreasing trend in water resources. The Ziz-Gheris catchment in the south of Morocco is a concrete example of climate change impact on water resources and groundwater specially.

In this chapter, Water Evaluation and Planning (WEAP) software was used to evaluate and trace the trend of water demand. This model was applied in case of Ziz basin in order to simulate and analyze the situation of water (reservoir and groundwater) under different scenarios.

Study Area

Tafilalet is located in southeastern Atlas Morocco between the South-Atlas latitudes 29°30' and 32°30'. It occupies approximately 8.44% of the surface of Morocco (approximately 60 000 km^2 of which 60 000 ha is irrigated).

Ziz Valley (Figure 1) is located in Errachidia Province, with a density of 2 per square kilometer at provincial scale. The majority of people work in agriculture sector, the main economical source of income. The province host Tafilalet Oasis (1370 km^2), which is an important biosphere reserve. Hydrologically, the region is crossed by Rheris and Ziz Wadis that are fed by Hassan Edakhil Dam.

Tafilalet encompasses three watersheds: Ziz, Gheris, and Guir on the southern slope of the Oriental High Atlas. The high diversity of substrates found within the region, in combination with its large vertical extension and extreme moisture and thermal gradients create diverse ecological systems. This high environmental variability is at the origin of an exceptional biological diversity which makes the High Atlas a center of diversity and endemism at regional scale. The floristic richness does not result only from the superposition of the different regional floras; however, it is characterized by a remarkable individuality.

Figure 1. Localization of Tafilalet region
Source: (Authors)

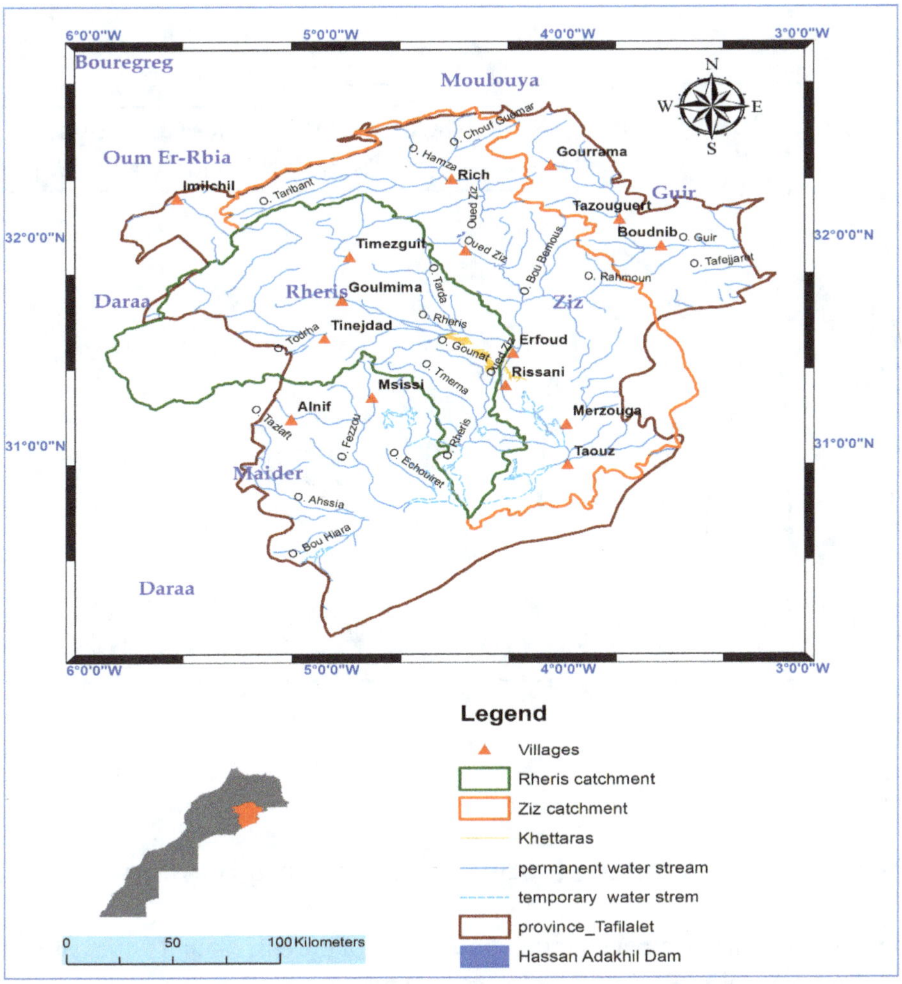

The climate is arid; the annual rainfall varies from 70mm in the extreme south, at Erfoud, to 290 mm in the extreme north, at Imilchil. However, the temperature varies from -5 °C to 40° C, with an annual average daytime high of 20° C. These climatic conditions required the practice of an irrigated agriculture. The date palm is the basis of life and agricultural activity in the district. Cultivated land is generally located along river banks, in narrow strips in the mountains area and in larger tracts in the oases.

Water availability for irrigation is restricted by climatic conditions affecting the whole region. In the Tafilalet, groundwater is estimated at 200 Mm^3 including 120 Mm^3 mobilized annually from moderately deep water (Anonymous, 2000). Groundwater is the only perennial water resources for the people of Tafilalet during dry years. Thus, the volume extracted water is on permanent increase that translates into a lower level of the groundwater level, as shown in figure 2 from 1970 to 2008.

Figure 2. Infiltration of runoff flow from sub catchment (Middle Ziz and plain of Tafilalet) to Ziz Wadi and Tafilalet groundwater simulated using WEAP software
Source: (Authors)

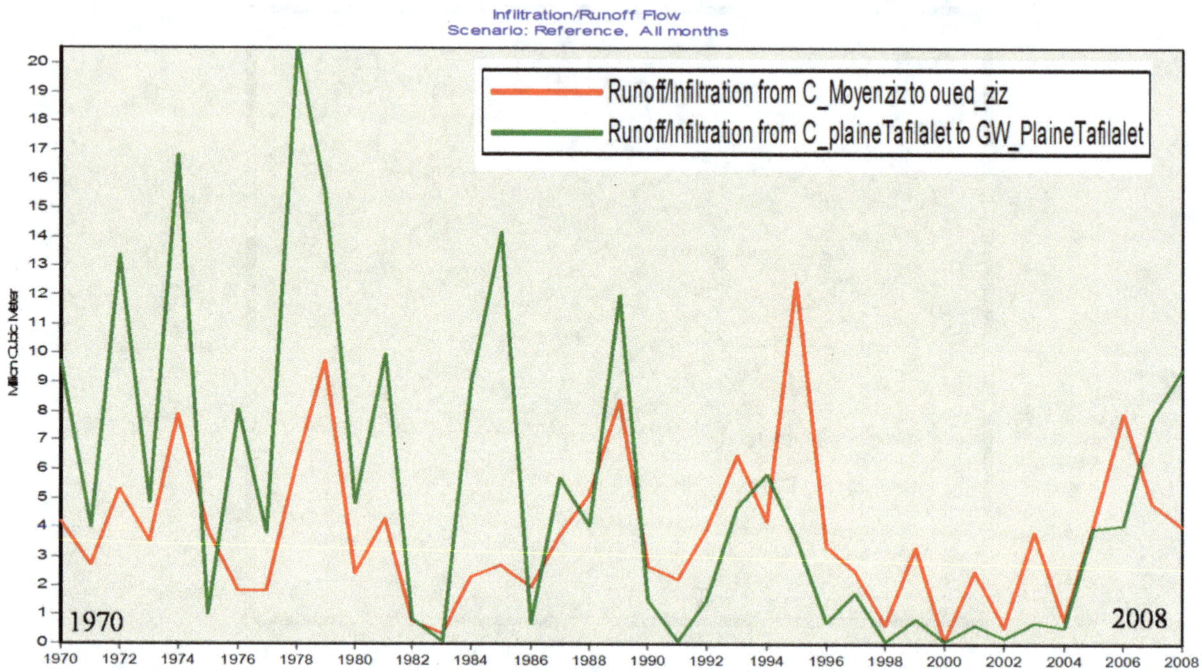

METHODS

Software Description: Water Evaluation and Planning System (WEAP)

Integrated water resources management requires knowledge of the current and future situation of existing resources. For this reason, WEAP 21 was chosen as the approach method for modeling water resources. WEAP is an integrated water management tool that allows the basin assessment using the gathered data. This model was developed by Stockholm Environment Institute (SEI),estimates water demand, supply, infiltration, storage under various hydrological, political, and climate change scenarios. In this chapter, technical irrigation, population growth, economic development, and climate change have been proposed as scenarios and integrated to evaluate water demand and storage.

Data Availability

This work is based on acquired meteorological, hydrological, remote sensing social and economic data. Eight stations were used, see Table 1. Additional information is found in Ait Boughrous, (2007); Cappy et al., (2004); de Jong et al., (2004); Knippertz et al., (2003); and Zeroual, (1995).

Table 1. Details of climatic stations

Station	Latitude	Longitude	Altitude(m)	Mean Annual Rainfall (mm)	Record Period	Number of Years
Z. sidi Hamza	32.43	-4.72	1738	258,13	1970 - 2008	38
M'zizel	32.23	-4.72	1600	176,74	1970 - 2008	38
Tillicht	32.31	-4.55	1627	191,50	1975 - 2008	33
F. Zaâbel	32.16	-4.36	1253	179,0	1970 - 2008	38
BHA	32.00	-4.45	1158	119,8	1973 - 2008	36
Errachidia	31.95	-4.4	1036	119,0	1957 – 2008	51
Radier Erfoud	31.43	-4.23	800	61,91	1957 – 2008	51
Taouz	30.91	-3.98	676	44,73	1970 – 2008	38

Source: (DRH 2008)

WEAP Ziz Basin Model (Tafilalet)

The Upper Ziz hydrology was represented by the WEAP software (Figure 3 and 4). The model has been adapted, calibrated and validated for the Ziz Basin in the framework of the MIOS (Modèle Intégré des Oasis du Sud) and permits the analysis of different hydrological parameters in different climate and policy scenarios. A Current Account of the water system under study is created. Then, based on a variety of demographic, hydrological, and technological trends, a "Reference" scenario projection is established. Other policy scenarios were developed with alternative assumptions about future developments. Only the major water users were described for each major user, the activity level, the annual water demand (net values after accounting for losses), the monthly variation as well as a return flow. Population growth rate is another scenario based on the growth rates between 1993 and 2030 in domestic (1.12-1.2%), livestock (1.2%), and irrigation (1.2%) reported by IRATE, (2006).

The conceptual model showed in Figure 4 represents the water distribution in main parts of the Ziz oasis.

Main Features

Total Ziz Valley Area II = A + B with about 5000 ha dominated and irrigated by water from the Hassan Edakhil Dam (HDD), which passes through three main channels and side channels of modern and traditional channels (Saguia). Supplementary irrigation is provided by pumping stations from wills.

Total Plain Tafilalet = Zone III + IV with approximately 15 400 ha dominated and irrigated by water from the HDD traveling through the Wadi Ziz about 70 km after Douira. They are derived by El Borouj Dam to be sent up perimeters concerned from modern and traditional channels of irrigation. The Ziz and tributaries flood waters diverted from Gheris Wadi contribute the development of the agricultural plain Tafilalet.

Figure 3. Demand and offer sites in Ziz area simulated using WEAP Software
Source: (Ben Salem et al., 2010)

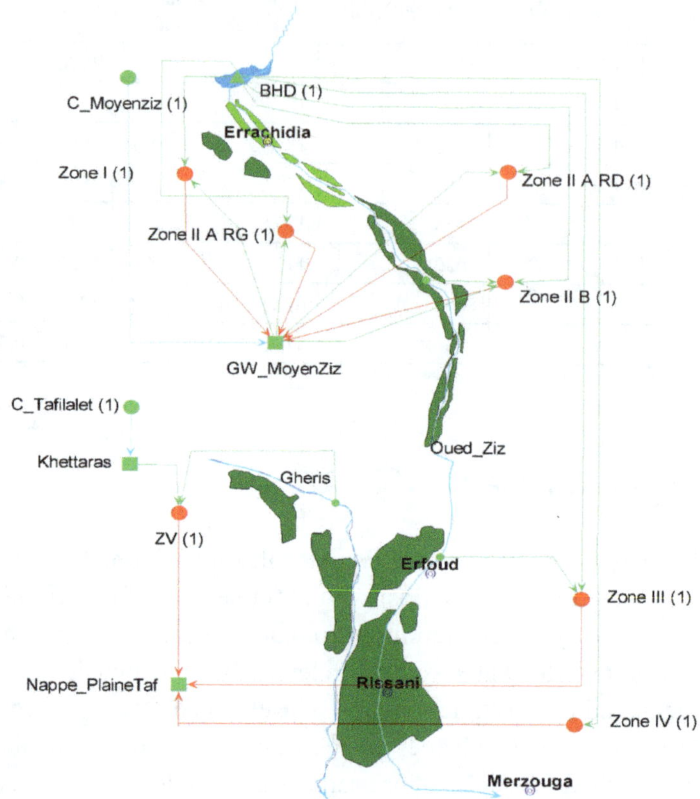

Creating Reference Scenario

The reference scenario is the base scenario that uses the actual data, to help in understanding the best estimates about the studied period. The objective of a reference scenario is to help planner and water resource manager understand what likely could occur if current trend continue and to understand the real situation.

High Level of Growth Population Scenario (1970 – 2050)

Analyzing the living conditions of the population in relation to the use of water and land in the catchment area of Ziz, population dynamics is an important indicator for social change. In order to meet the complexity of demographic processes, quantitative-statistical data and qualitative social-scientific data were collected and evaluated on different spatial levels from regional to local. For scenario analysis and the projected inclusion into information or decision support systems, the findings are integrated in WEAP model. Two scenarios of high level of growth population were used in this study:

Scenario A: Growth rates between 1993 and 2030: domestic 1.12-1.2%, livestock 1.2%, and irrigation 1.2%.

Figure 4. WEAP conceptual Model for water management and irrigated area in Tafilalet.
BHD: Barrage Hassan Edakhil; KH: Khettaras; Z: Zone, GW: Groundwater (See also table 2)
Source: (Ben Salem et al., 2010)

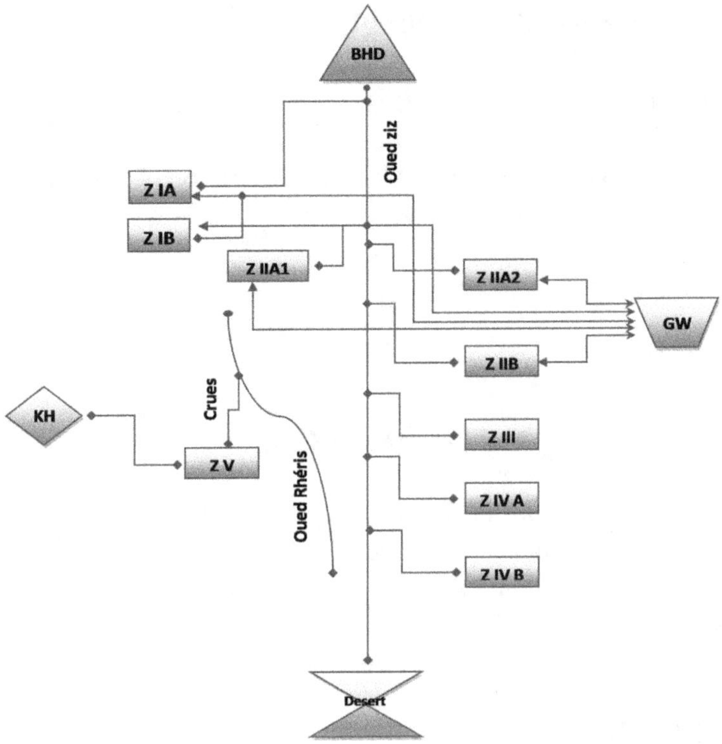

Table 2. Irrigated areas in the Tafilalet region

Irrigated Area	Area (Ha)	Description
ZIZ Valley	5500 Ha	
Zone I	800	1 Principal and secondary supply to sites A and B resettlement
Zone A II	2950	2 Channel principal and secondary supply two stores (= surface area dominated taken into account by the manager of the water releases): Left Bank (about 2600 ha) and right bank (about 350 ha).
Zone II B	2000	Characterized by a modern network exists but not put into operation. Access to water from diversion dam on the Zizoued fueling Seguias and some portions of the modern network. This area is fed by springs.
Plain of Tafilalet	15400 Ha	
Zone III	3300	Sectors of Tizimi and Maadid
Zone IV	12100	Sectors of Erfoud, Rissani and Merzouga,
Zone V	450	Irrigated by Khettaras, Jorf

Source: (Authors using data from archives)

Scenario B: Proposed Scenario of High level of growth population and climate change scenario 2050 were: domestic, livestock and irrigation 3%.

Modern Irrigation Scenario (1970 – 2050)

Climate Change Scenario (1970 – 2050)

Four scenarios of climate change can be adapted for Tafilalet region (Tabet-Aoul, 2008).

- + 1° C P – 10%
- + 1° C P – 20%
- + 2° C P – 10%
- + 2° C P – 20%.

Calibration and Validation of the WEAP Model

The watershed model was calibrated with flow data from HRD. Calibration is performed by comparing the observed and modeled flows of the Hassan Addakhil Dam.

Results are also verified by calculating absolute value of the calculated error between the observed results and the simulation.

RESULTS AND DISCUSSION

In this chapter, the results obtained after modelling and calibration under the WEAP software, are represented in the form of maps and graphs (2 & 3D).

A schematic view of the conceptual WEAP model for the Ziz and Gheris catchment is given in figure 5. it illustrates the agricultural zones studied, surface water and groundwater, and the supply and demand for water.

WEAP conceptual Model for water management and irrigated area in Tafilalet was calibrated. The range of values resulting from the calibration process is illustrated in Figure 6.

Results of calibration corresponding to the period 1978-2006 show an agreement between measured and calculated inputs at the Ziz catchment. This result is also verified by the absolute value of the calculated error between the observed results and the simulation. The error is small (Figure 7), is equivalent to normalized quantities, thus, confirming the relevance of the WEAP model in Ziz area.

The WEAP is programmed to estimate water demand using data acquired. For example, this model can calculate and represent the seasonal water demand of irrigated area in Ziz catchment (Figure 8). In this figure, a comparison of several sites was represented in the same graphic under reference scenario.

WEAP model can also estimate the effect of increased population growth Scenario (A and B) and reference scenario on the water demand. The effect is much more visible on the population, unlike the water demand and unmet demand for site demand purpose, evident to say that irrigation, water demand and unmet water demand percentage can increaseby103.3, 124.1, and 246.8 Mm^3 for August (Figure 9).

Figure 5. Schematic view of the conceptual WEAP model for Ziz-Gheris basin
Source: (Authors)

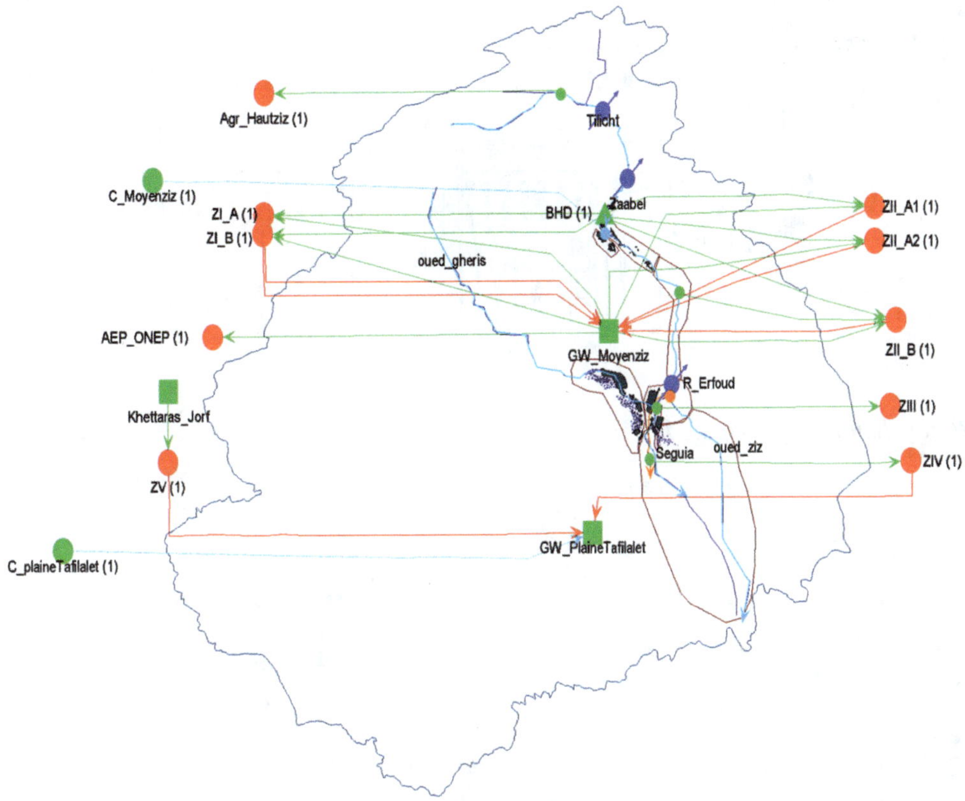

Figure 6. Observed and simulated volumes of the Hassan Addakil Dam (cubic meter) according to the Reference Scenario
Source: (Authors)

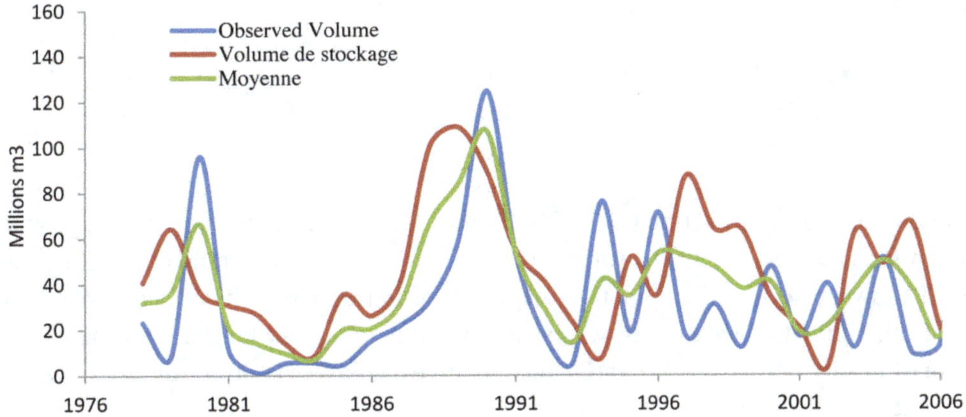

Figure 7. Absolute value of the error for the calibration of the WEAP model
Source: (Authors)

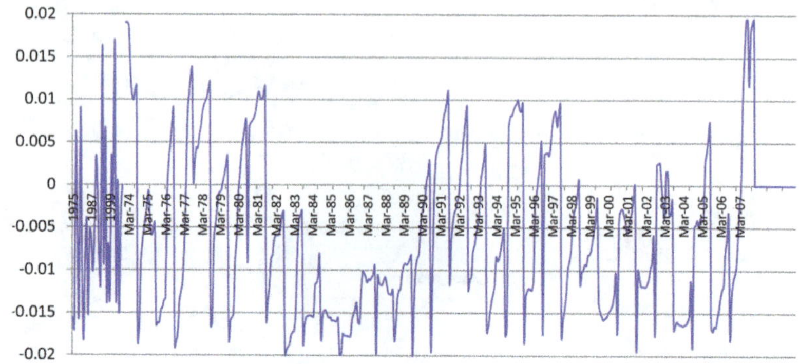

Figure 8. Water demand of irrigated area in Tafilalet
Source: (Authors)

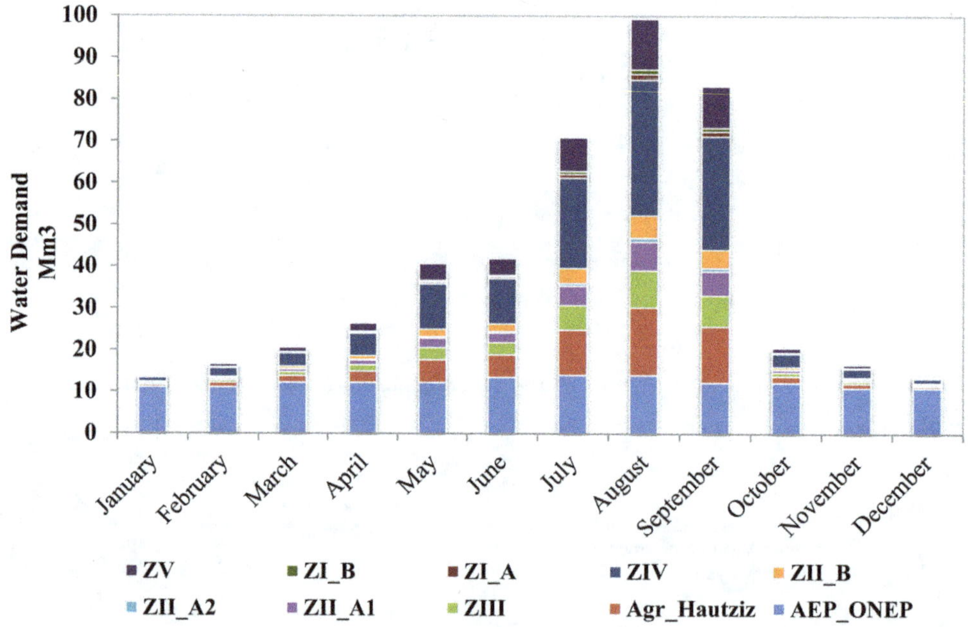

Regarding the long term projection (1970-2050), the effect of increased population growth Scenario (A and B) and reference scenario on the water demand show an increasing trend of water demand (Figure 10). The result shows the effect of population growth on water demand, the effect is much more visible on the population, unlike the demand for water and the unmet demand for demand sites.

Percentage of unsatisfied water demand is increasing especially the domestic use of the town of Er-rachidia and irrigation.

The comparison between two scenarios reference in red and high growth rate of the population in blue, with the normal demographic growth the demands will go from 450Mm3 to 700 Mm3 in 2050 on the other hand with a high growth rate the water demand can reach 1000 Mm3 in 2050.

Figure 9. Example of water demand evolution under population growth and reference scenarios
Source: (Authors)

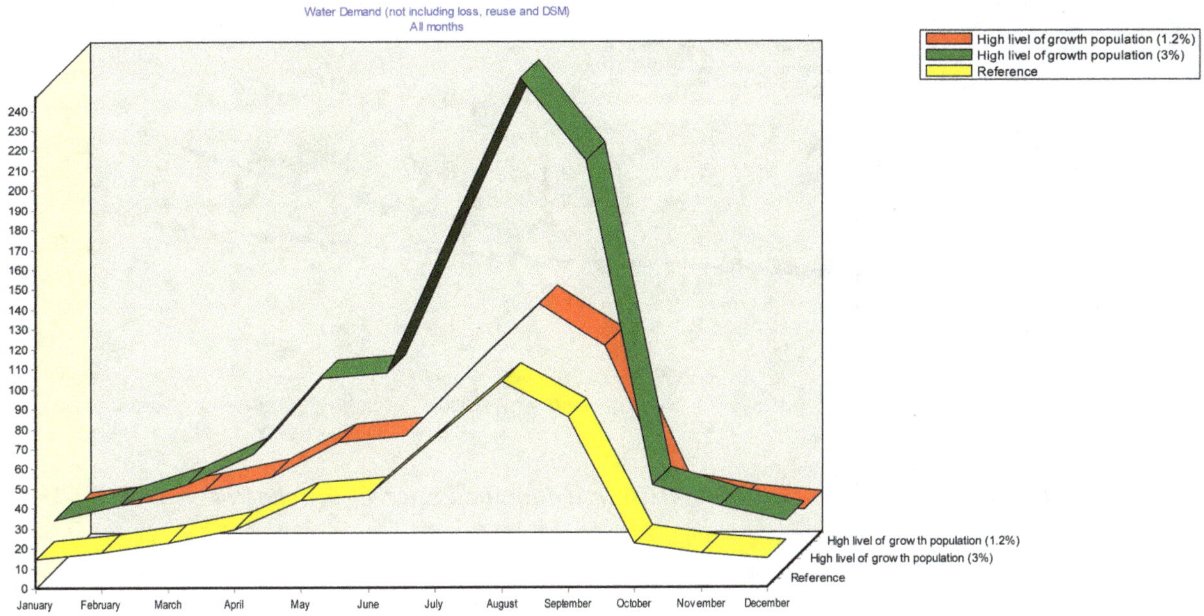

The Tafilalet region with these arid climatic conditions and these scarce water resources will not be satisfied by water.

Changing the scenarios from population growth to the introduction of modern irrigation, a small increase of water demand was found compared to the Reference scenario (Figure 11).

At long term projection, a decrease of water demand was modeled compared to the Reference scenario (Figure 12).

As mentioned above, WEAP model can also simulate the Groundwater volume (Figure 13) and Hassan Edakhil Dam (Figure 14) comparing groundwater inflows and outflows during a year. Long term projection can be modeled as mentioned in figure 15. In this last figure, groundwater storage in Tafilalet was estimated using different scenarios: population growth and the introduction of modern irrigation, and Reference scenario. The results show a decrease of groundwater storage in these scenarios, except for high growth of population (Figure 15) and in reservoir volume as found by Ben Salem et al., (2010) (Figure 16).

Figure 14 shows the inflows and outflows of the main reservoir in Ziz basin. The graph is a balance between inflows and outflows. This graph is very useful to understand the operation of the reservoir throughout the year. The outflow from the reservoir is recorded in the summer time, when there isno rainfall and the "Lachers" are necessary. The "Lachers" bring about the storage volume decreasein the reservoir. Approximately 25% of the reservoir volume is lost by evaporation. At present, some problems of evaporation estimation and modeling are still unsolved and as a result, measures aimed at reducing evaporation from the lake reservoir surface are under discussion.

The reservoir maximum capacity is used yearly, but with declining refill levels no water storage is possible on an annual basis. The combined effects of water planning for the Lachers and of drier conditions compared with the annual average have resulted in a large drop in average water allocation for the

Figure 10. Water demand Scenario for irrigated area in Tafilalet
Source: (Authors)

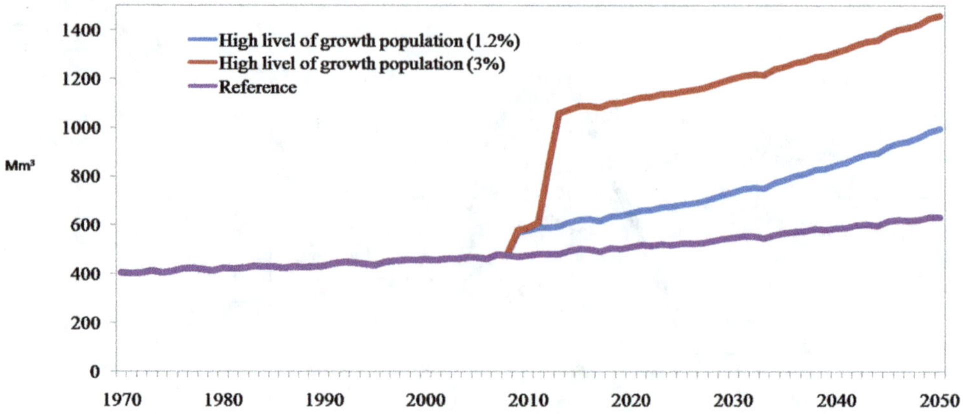

Figure 11. Water demand evolution under modern irrigation and reference scenarios
Source: (Authors)

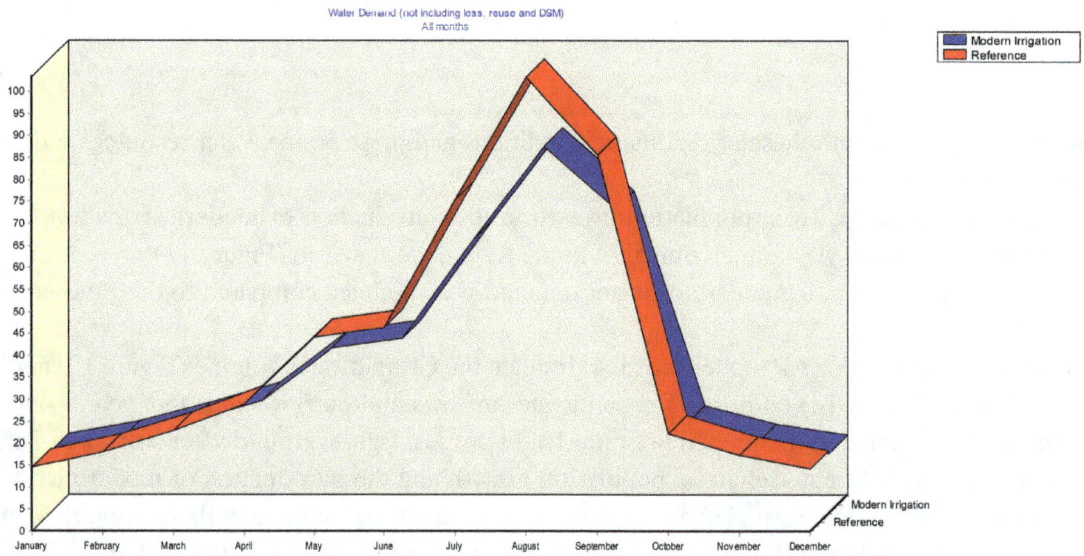

irrigated date palm groves since the 1980s (Messouli et al., 2009). During very dry years the lachers are not used for crop irrigation and are instead exploited to replenish groundwater levels.

Water Demand for Agricultural Sites

In the Middle Ziz

The demand for water from the medium Ziz zones (ZIA, ZIA-1, ZIA-2, ZI-B and ZII-B) (Figure 17-a) is very important during the summer period. The cumulative demand reaches 1.2 Mm^3 in July and does not exceed 0.1 Mm^3 in December in all zones.

Figure 12. Scenario of water demand with introducing modern irrigation in Tafilalet
Source: (Authors)

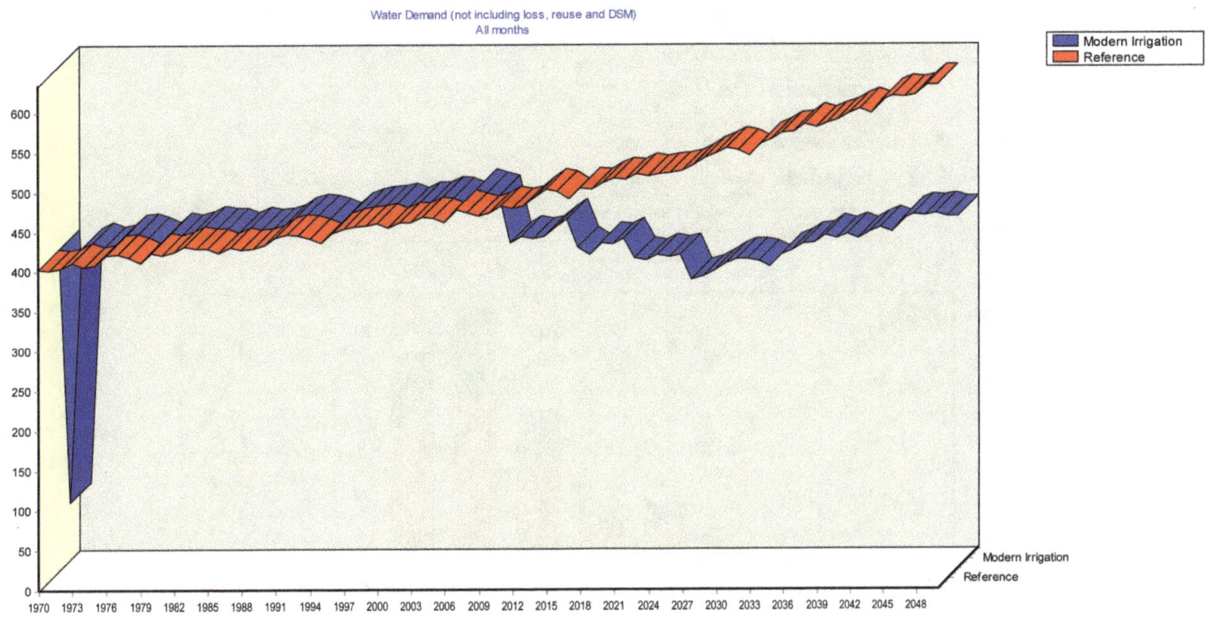

In the Plain of Tafilalet

Demand zones ZIII, ZIV and ZV with larger areas and water-demanding crop types (e.g. Alfalfa) imply a high water demand of up to 3.6 Mm3 in June and July (Figure 17-b).

Total Water Demand for Agricultural Areas of the Ziz Catchment

The demand for water for agriculture and the supply of drinking water in the Ziz catchment area is very important (Ben Salem, 2010). The mobilization of water resources is highly estimated at 480 Mm^3 used by the population of Ziz. The large quantity is mobilized during the summer months (73.1 Mm^3 in July, 99.6 Mm^3 in August and 82.6 Mm^3 in September). Demand is very low during the months (January, February, November and December) (Figure 18).

The water coverage of agricultural areas (ZI-A, ZII-A1, ZII-A2, ZI-B and ZII-B) is 100%. This is explained by the availability of surface water resources (Hassan Addakhil Dam) and the pumping of groundwater. While in the areas of the Tafilalet plain (ZIII, ZIV and ZV), with the high demand for water, combined with the scarcity of surface water resources, the salinity of groundwater and the high evapotranspiration, the agricultural zones suffer from water deficiency during the summer and sometimes throughout the year. Water deficit can reach more than 80 percent during the summer for the zone (Z V) and 70 percent for the zone (Z IV). The situation is critical in mountain areas (High Ziz), in the absence of alternatives to offset the lack of dams, wells and Khettaras the water deficit can reach 94 percent for the area (Agr-Hautziz) (Figure 18).

Figure 13. Groundwater inflows and outflow
Source: (Authors)

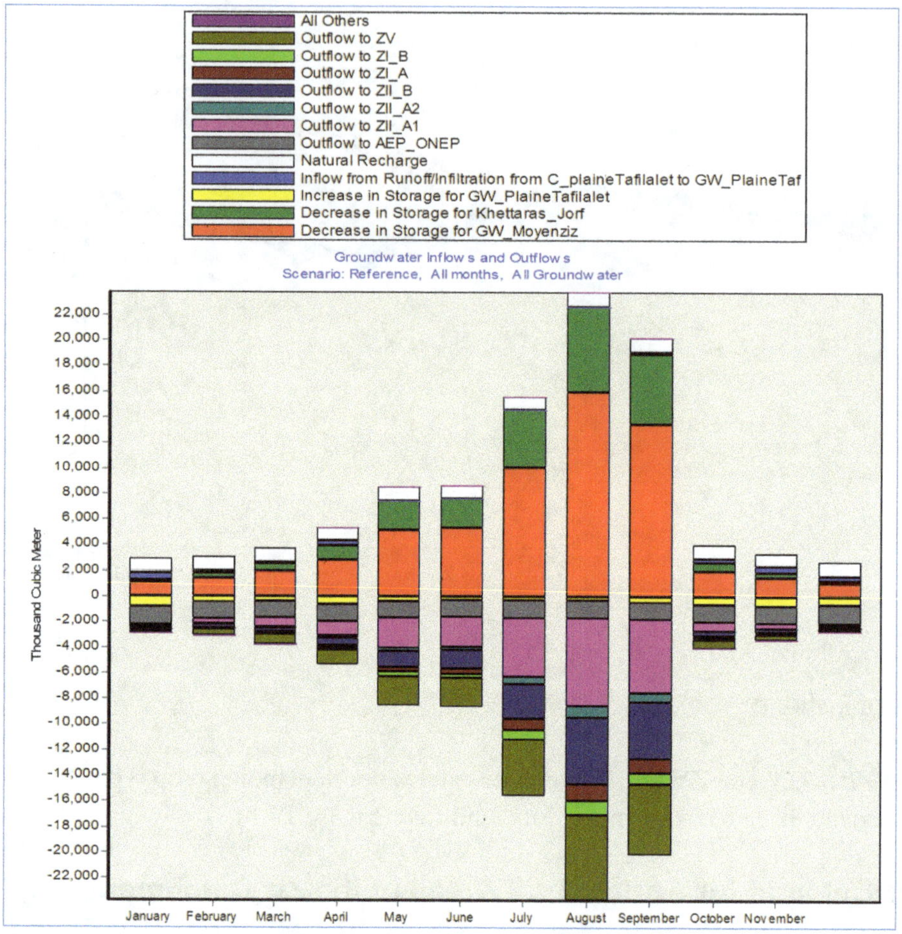

CONCLUSION

Since the 1970s Morocco has attempted to promote sustainable development in oasis areas. Construction of dams, creation of modern infrastructure of irrigation, and adaptation of a set of measures for safeguarding of the environment and natural resource were the important steps of this approach. Despite these efforts, the condition of oases has been considered at risk for several decades (PNUD 2012).

Many studies about current and future water availability in arid and hyper arid mountainous regions have shown that disparities of water supply and water demand are increasing (Alali et al., 2013; Irene et al., 2016). Concomitance of declining renewable water resources in North Africa (Droogers et al., 2012, Messouli et al., 2009) and increasing water demand reveal a strong need for water resource management in northern African arid regions (Helegers et al., 2013, Hamlat et al., 2013). The origins of this situation come from the fragility of the oasis ecosystem and the serious problem related to the availability of water resources and the unsuitable development in the region (Bouaamlat et al., 2016).

This work was carried out in Tafilalet region and the Ziz basin part of the Oasis biosphere reserve. The climate is arid with low rate rainfall. Human actions are affecting the oasis ecosystem services. The

Figure 14. HDD Reservoir inflows and outflows, reference scenario
Source: (Authors)

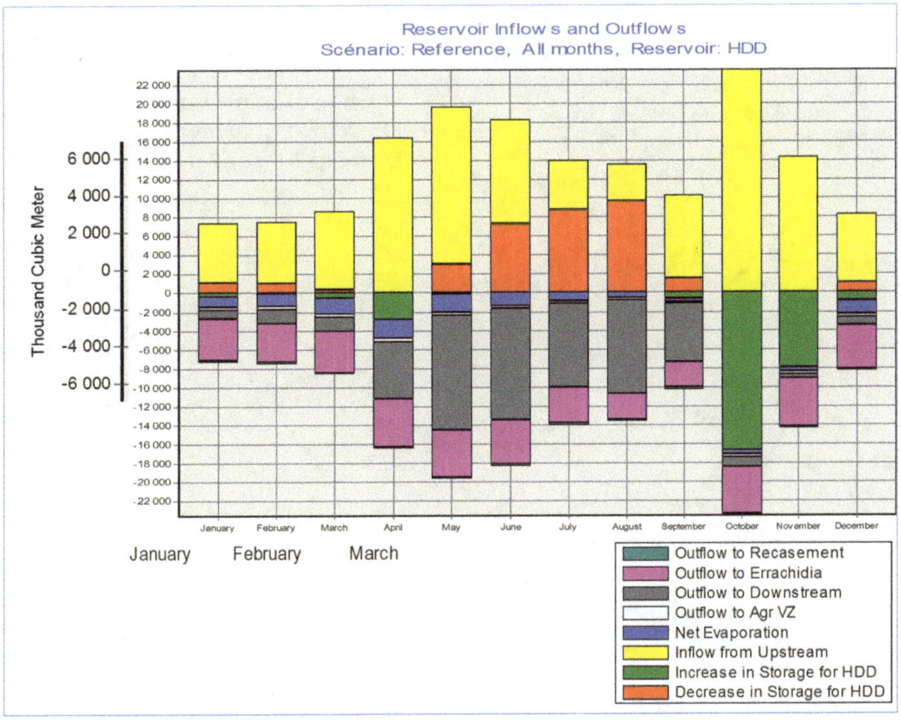

impact of climate change is increasingly felt, with a decrease in river flows, a drop in the piezometric level, a drying up of Khettaras and an advance of sand dunes.

The natural effects linked to climate change are accentuated by human activities, such as the construction of the Hassan Edakhil Dam, which has modified the water regime of Ziz Wadi, subsequently increasing the pressure of demand on downstream groundwater.

Modeling demand and supply, and storage can help to observe and have a wide long-term vision of the problem.

The results show an increasing of water demand for irrigated and following the scenario of introducing modern irrigation in the study area. However, a decreasing trend in Reservoir storage volume and Groundwater storage was projected.

The consequences linked to climate change are increasingly pronounced on groundwater resource, the main water resource of the study area. This trend has also been observed in the decrease of river flow, the fall of the piezometric level, and the drying up of Khettaras (underground canals). The number of functional Khettaras is decreasing, leading to an alteration of the water chemistry, especially due to water salinity.

Hassan Edakhil Dam, located in the center of the Ziz catchment, is likely to be filled rapidly by sediment, due to high rates of sediment transport by Wadi Ziz (Ben Salem, 2010).Currently, the dam's storage capacity is insufficient to meet the needs of downstream farmers, especially during droughts and periods of water scarcity (Figure 69). Once the dam reaches a critical level of siltation (at about half of its capacity) it will lose its regulatory flow (De Jong, 2007). Irrigation will be more subject to irregularity, dominated by floods and flows will depend more heavily on groundwater resources.

Figure 15. Scenarios of Groundwater storage of major groundwater in Tafilalet
Source: (Authors)

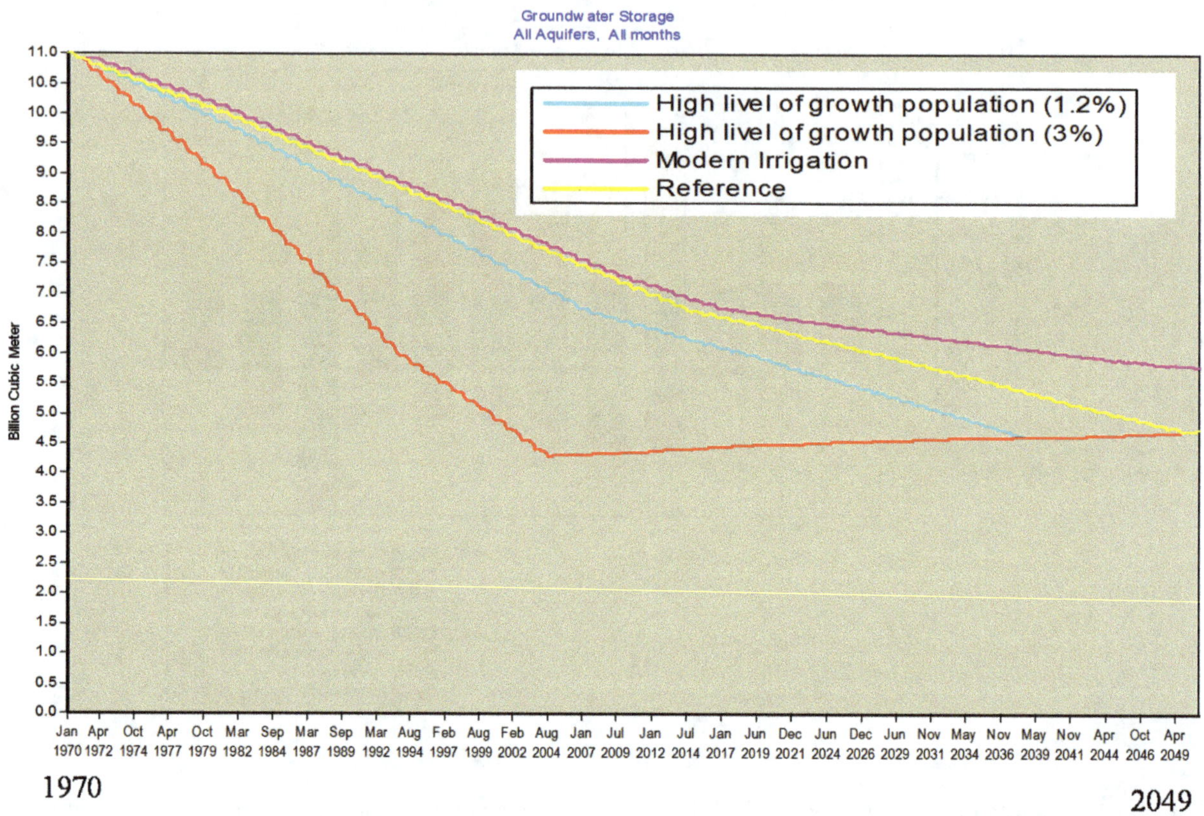

WEAP software is very supportive of establishing a water resources management plan, taking into account vulnerability and climate change in the future. It also allows the establishment of a hydrological balance sheet and a management report (supply and demand). In addition, there are some general adaptation measures for the conservation and protection of water resources, such as the implementation of a realistic water pricing regime to avoid waste, including:

- The forecast of the necessary infrastructures against the risks of flooding.
- Installation and improvement of sanitation systems in villages in the region
- The implementation of an integrated water resources management strategy.

In perspective, more detailed studies in the same basin and or applications of this tool very interesting in other sites of applications and other watersheds (Gheris, Guir) are envisaged.

Adaptation to climate change in the region of Tafilalet involves the review and the change of policies, programs, investment, and ultimately the behavior toward the oasis ecosystems.

Figure 16. Reservoir storage volume under Scenarios of population growth and the introduction of modern irrigation, and the reference
Source: (Ben Salem et al., 2010)

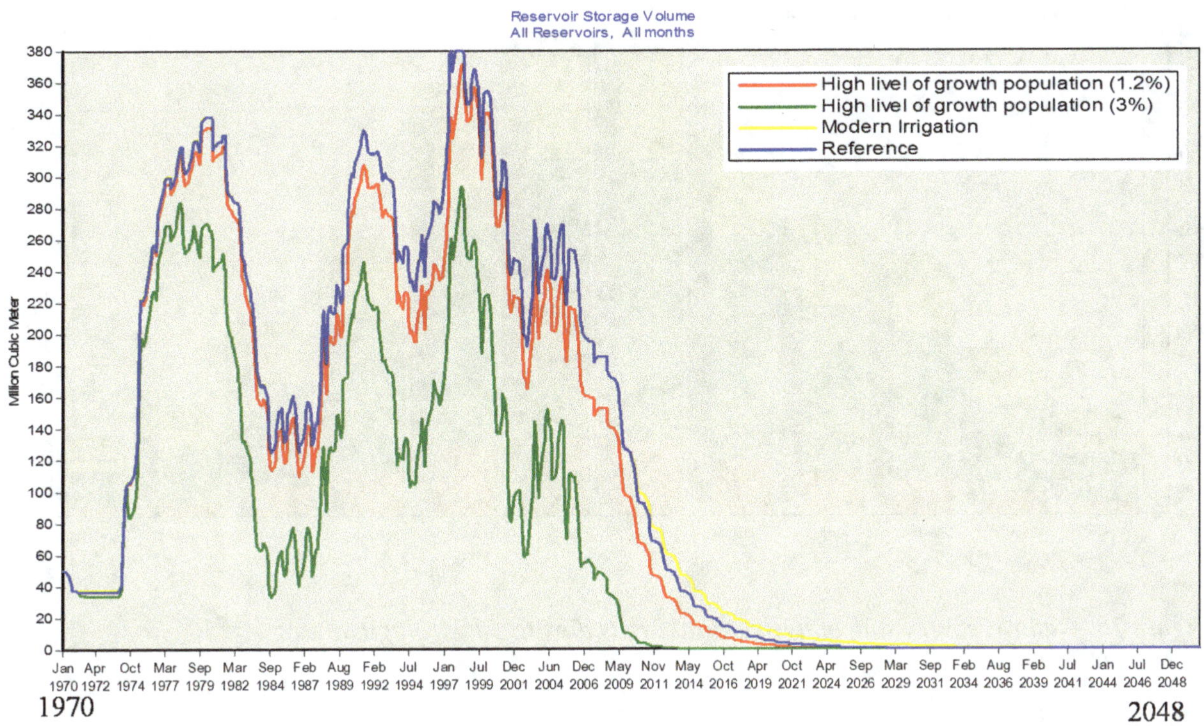

Figure 17a. Water demand for agricultural areas in the Middle Ziz
Source: (Authors)

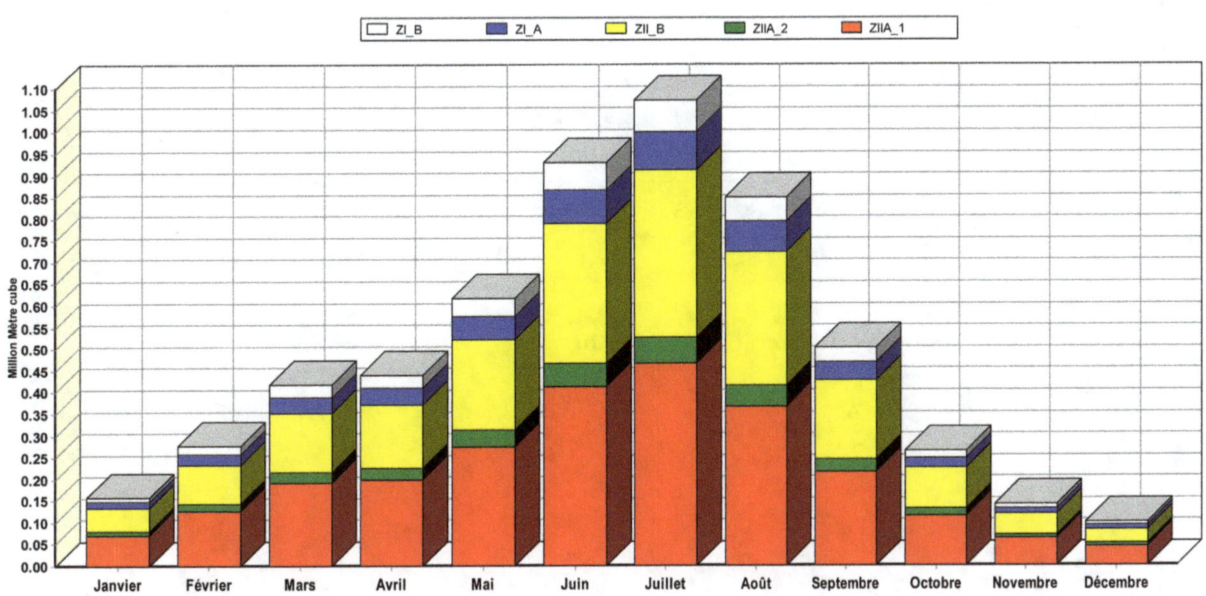

Figure 17b. Water demand for agricultural areas in Tafilalet plain
Source: (Authors)

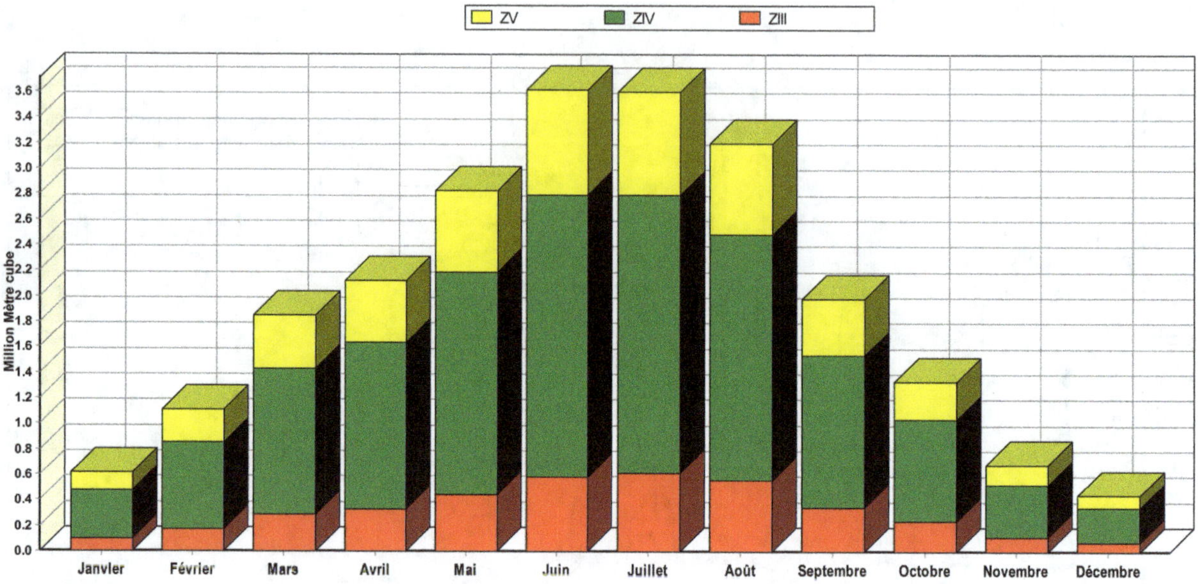

Figure 18. Water recovery rate of agricultural areas during a hydrological year
Source: (Authors)

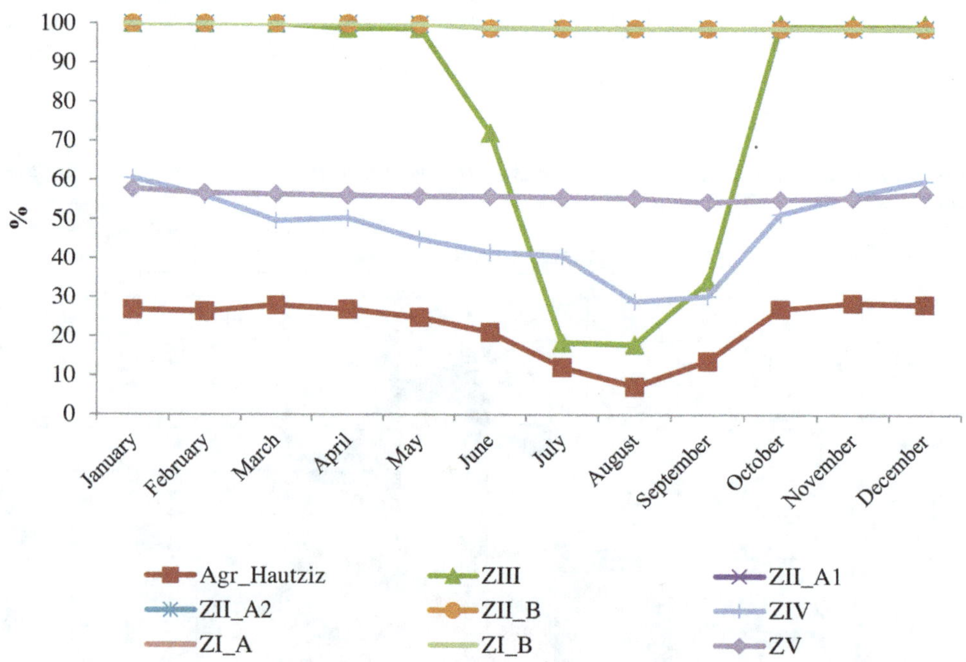

ACKNOWLEDGMENT

We would like to thank Stockholm Environment Institute (SEI) for the software Water Evaluation and Planning system (WEAP), the Direction régionale de l'Hydraulique des bassins Versants Guir-Rheris-Ziz (DRH GRZ) Errachidia, and the Office Régional de la Mise en Valeur Agricole de Tafilalet(ORMVAT) Errachidia for the data. Many thanks go to the Southern Center for Culture and Science (Morocco) for *assistance* and *support*.

REFERENCES

Ait Boughrous, A. (2007). *Biodiversité, écologie et qualité des eaux souterraines de deux regionsarides du Maroc: Le Tafilalet et la région de Marrakech* (PhD Thesis). Cadi Ayyad University, Marrakech, Morocco.

Alali, A. B. (2013). La plaine du Tafilalet (Sud-est, Maroc) face aux problèmes d'environnement. *Rev. Mar. Sci. Agron. Vét., 2*, 68–72.

Anonymous. (2000). *Fish collection database of the Gulf Coast Research Laboratory (GCRL). The Gulf Coast Research Laboratory*. Ocean Springs, MS: GCRL.

Ben Salem, A., Messouli, M., &Yacoubi-Khebiza, M. (2011). *Developing an oasis-based water management tool: Ecohydrologic approach and weap software for a large arid catchment in Morocco*. Academic Press.

Bouaamlat, I., Larbi, A., & Faouzi, M. (2016). Hydrogeological investigation of an oasis-system aquifer in arid southeastern Morocco by development of a groundwater flow model. *Hydrogeology Journal, 24*(6), 1479–1496. doi:10.100710040-016-1409-8

Brekke, L. D., Miller, N. L., Bashford, K. E., Quinn, N. W. T., & Dracup, J. A. (2004). Climate change impacts uncertainty for water resources in the San Joaquin River Basin, California. *Journal of the American Water Resources Association, 40*(1), 149–164. doi:10.1111/j.1752-1688.2004.tb01016.x

De Jong, C. (2007). Impounded rivers: Aridification and geomorphological risks. *IAHS-AISH Publication*.

Driouech, F., Déqué, M., & Mokssit, A. (2009). Numerical simulation of the probability distribution function of precipitation over Morocco. *International Journal. Climate Dynamics, 32*(7-8), 1055–1063. doi:10.100700382-008-0430-6

Droogers, P., Immerzeel, W. W., Terink, W., Hoogeveen, J., Bierkens, M. F. P., van Beek, L. P. H., & Debele, B. (2012). Water Resources Trends in Middle East and North Africa Towards 2050. *Hydrology and Earth System Sciences, 16*(9), 3101–3114. doi:10.5194/hess-16-3101-2012

Hamlat, A., Errih, M., & Guidoum, A. (2003). Simulation of Water Resources Management Scenarios in Western Algeria Watersheds Using WEAP Model. *Arabian Journal of Geosciences, 6*(7), 2225–2236. doi:10.100712517-012-0539-0

Hellegers, P., Immerzeel, W., & Droogers, P. (2003). Economic Concepts to Address Future Water Supply-Demand Imbalances in Iran, Morocco and Saudi Arabia. *Journal of Hydrology (Amsterdam), 502*, 62–67. doi:10.1016/j.jhydrol.2013.08.024

Hsain, I. (1996). Small-Scale Irrigation in a Multiethnic Oasis Environment: The Case of ZaouitAmelkis Village, Southeast Morocco. *Journal of Political Ecology, 3*(1), 89. doi:10.2458/v3i1.20459

IPCC. (2007). *Climate Change 2007: Impacts, Adaptation*. IPCC.

IRATE. (2006). *L'Atlas*. Région Meknes-Tafilalet, Inspection Régionale de l'Aménagement du Territoire et de l'Environnement.

Johannsen, I. M., Hengst, J. C., Goll, A., Höllermann, B., & Diekküger, B. (2016). Future ofWater Supply and Demand in the Middle Drâa Valley, Morocco, under Climate and Land Use Change. *Water (Basel), 8*(8), 313. doi:10.3390/w8080313

Karmaoui, A., Ifaadassan, I., Babqiqi, A., Messouli, M., & Khebiza, M. Y. (2016). Analysis of the Water Supply-demand Relationship in the Middle Draa Valley, Morocco, under Climate Change and Socio-economic Scenarios. *Journal of Scientific Research & Reports, 9*(4), 1–10. doi:10.9734/JSRR/2016/21536

Karmaoui, A., Ifaadassan, I., Messouli, M., & Khebiza, M. Y. (2015). Sustainability of the Moroccan oasean system (Case study: Middle Draa Valley). *Global J TechnolOptim, 6*, 170. doi:10.4172/2229-8711.1000170

Karmaoui, A., Minucci, G., Messouli, M., Khebiza, M. Y., Ifaadassan, I., & Babqiqi, A. (2018). Climate Change Impacts on Water Supply System of the Middle Draa Valley in South Morocco. In *Climate Change, Food Security and Natural Resource Management* (pp. 163–178). Cham: Springer.

Messouli, M., Ben Salem, A., Ghallabi, B., & Yacoubi-Khebiza, M. (2009). Ecohydrology and groundwater resources management under global changes: A pilot study in presaharan basins of southern Morocco. Options Méditerranéennes, 88, 255-264.

PNUD. (2012). *Rapport PNUD Maroc en action*. Programme des Nations Unies pour le développement au Maroc.

RGPH. (2004). *Le Recensement Général de la Population et de l'Habitat 2004*. HCP Maroc.

Risbey, J. S. (1998). Sensitivities of water supply planning decisions to streamflow and climate scenario uncertainties. *Water Policy, 1*(3), 321–340. doi:10.1016/S1366-7017(98)00023-3

Schimmelpfennig, D. (1996). Uncertainty in economic models of climate change impacts. *Climatic Change, 33*(2), 213–234. doi:10.1007/BF00140247

SEI. (2005). *WEAP water evaluation and planning system, Tutorial*. Stockholm Environmental Institute, Boston Center, Tellus Institute.

Vicuna, S. (2006). *Predictions of Climate Change Impacts on California Water Resources Using CALSIM II: A Technical Note*. Berkeley, CA: California Climate Change Center.

Zeroual, E. (1995). *Enregistrements climatiques dans les sédiments du Lac Isli (Haut Atlas du Maroc)* (PhD thesis). University of Neuchatel, Switzerland.

Zhu, T. M. W., Jenkins, M. W., & Lund, J. R. (2005). Estimated impacts of climate warming on California water availability under twelve future climate scenarios. *Journal of the American Water Resources Association*, *14*(5), 1027–1038. doi:10.1111/j.1752-1688.2005.tb03783.x

Section 2

Biodiversity and Ecosystem Services

Chapter 4
Monetary Value Change of Some Provisioning Ecosystem Services of Middle Draa Valley, South of Morocco

Siham Zerouali
Southern Center for Culture and Sciences, Morocco

Mohammed Yacoubi-Khebiza
Cadi Ayyad University, Morocco

Fadoua El Qorchi
Cadi Ayyad University, Morocco

ABSTRACT

The desert oasis is a wetland agro-ecosystem more and more under pressure. This trend threatens the oasis ecosystem services (ES) and then the wellbeing of the local population. In order to quantify this trend in Middle Draa Valley, South of Morocco, the monetary values for a selection of ES have been estimated. In this chapter, the economic values of 28 provisioning ES in 2014 was examined and next compared with the economic values of 2009. These values were standardized to US$ per hectare per year. The total value of the studied provisioning ES in 2014 was found to be 8658.3 US$/ha/year for the arable area (26000ha) and 150 US$/ha/year for the whole Middle Draa Valley (1500000ha). Comparing these findings with the data of 2009, a slight increase of $1245.7 (in 26000ha) and $21.5 US$/ha/year (the whole area) was detected in the period 2009-2014.

DOI: 10.4018/978-1-5225-7387-6.ch004

INTRODUCTION

Wetlands provide important ecological non-replaceable functions: they mitigate pollution, offer habitats, and preserve biodiversity (Sulman et al., 2013). Oases are among the most important wetlands in desert systems (Karmaoui et al., 2015a). The change in the oasis reflects both the ecological status and the changes in water availability (Lu et al., 2015). Water and soil quality are the main factors impacting the oasis system in the south of Morocco influencing the ability to provide ecosystem goods and services. Oasis is like other ecosystem provides a range of services of crucial importance to human well-being, health, livelihoods, and survival (Millennium Ecosystem Assessment (MEA) 2005 and TEEB Foundations, (2010). The ecosystem services (ES) values can express well the environmental change in an oasis. The repeated and prolonged droughts in recent years have severely limited the production of the medium that become too fragile (Chelleri et al., 2014). Ecosystem disturbance leads to the loss of ecosystem services and negative impacts on livelihoods and health. The socio-economic factors affect the ecological system of MDV and the health through the incidence of some neglected disease like cutaneous leishmaniasis (Karmaoui, 2018).

Recently, ES valuation is much studied. The economic valuation is among the most important approaches of ecosystem assessment used globally by several research teams, the most famous are De Groot et al., (2012), Costanza et al., (2014), and (Teoh et al., 2019).

In this chapter, various provisioning ES provided by the oasean system of Middle Draa Valley (MDV) were estimated. The paper priority is to quantify the monetary value change between 2009 and 2014. The outputs can allow tracing the trajectory and evolution of oases in the south east of Morocco.

MATERIAL AND METHODS

Study Area

The study area, located in the south east of Morocco, with an area of about 15 000 km^2 (Figure 1). Hydrologically is the middle part of the Draa Basin. It is subject to an arid climate, the summer temperatures are between 43 and 50 ° C and The mean annual rainfall is about 80 mm.

Figure 1. Localization of the study Area; MDV, Middle Draa Valley

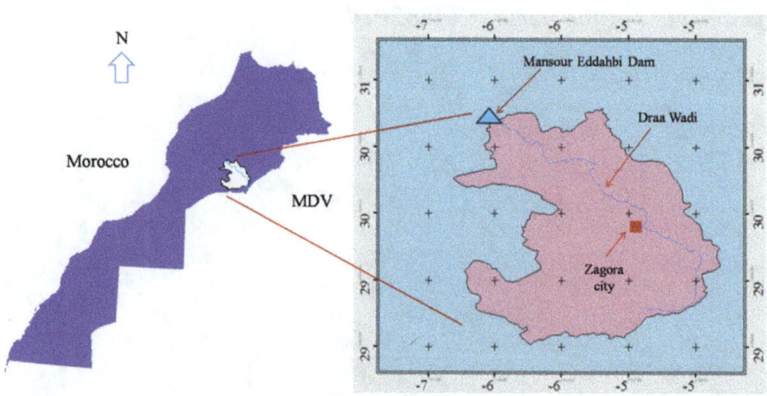

Along the MDV, the greenery oasis agro-ecosystem extends over 26 000 ha in a pre-Saharan environment (Karmaoui et al., 2014a). Its cultivation largely depends on the inflow of water from the outside. Mountains, plains, and desert plateaus are exploited mainly for pastoral purposes. The climate is arid climate and belongs to the Saharan bioclimatic stage. The average annual rainfall is very low and decreases from upstream 108 mm (Agdez) to downstream, 74 mm in Zagora (MPZ, 2014). In addition to agriculture, the tourism is the second economic sector in this area. However, in the period between 2009 and 2014, a continuous decrease in the number of tourists was observed (Figure 2) for unknown reasons.

In this area, the ES with available data and highest importance were identified, gathered, and classified in order to evaluate.

METHODOLOGY

The economic values of these ES have been estimated. The study of Karmaoui et al., (2015b) was used as a baseline to assess changes in provisioning ES values. Table 1 shows the summary of the economic provisioning services as estimated in 2009. In this chapter, the economic values of 28 provisioning ES in 2014 were estimated in order to compare it with the economic values of 2009. The outputs values were standardized to US$ per hectare per year for the arable area (26000ha) and for the whole Middle Draa Valley (1500000ha).

Quantitative analyses were used to estimate the changes in ES of the MDV. The combination values can help understanding the cause of ecosystem disturbance in this arid area. The services provided by oasis are essential for economic and social well-being of local population. The economic value of provisioning ES was estimated using market prices method (prices directly observed in markets). In the available literature, several methods were conceived for estimating the economic value of ecosystem services. In addition to the Market prices approach, there are other methods such as replacement cost (cost of replacing ES), Restoration cost, Travel cost, Contingent valuation (Karmaoui, 2016). The Total Value (V) of ES in $/ha/year for ecosystem type k is V (ES) k (Glaves & Egan, 2013).

Figure 2. Number of overnight stays in Middle Draa Valley
Source: (Authors using data from MPZ, (2014)

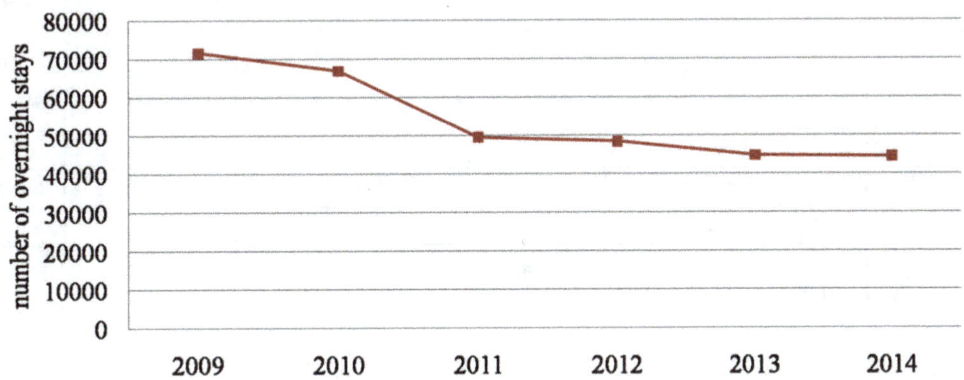

Table 1. Summary of the economic provisioning services of 2009

Ecosystem Services	Area (Ha)/ Number (of Trees of Houses)	Quantity/ Production	Unit Price	Total in Dh	Total in $
Durum Wheat	50 Ha	1300 Qs	4dh/Kg	520000	62400
Common Wheat	16850 Ha	589910 Qs	4dh/Kg	235964000	28315680
Barley	1400 Ha	29300 Qs	4dh/Kg	11720000	1406400
Maize	330 Ha	6675 Qs	4dh/Kg	2670000	320400
Alfalfa	3600 Ha	190000 Qs	2dh/Kg	38000000	4560000
Vegetable Gardening	1445 Ha	260920 Qs	4dh/Kg	104368000	12524160
Watermelon (2013)	1130 Ha	671287 Qs	2.5dh/Kg	167821750	20138610
Almond trees	32800 Trees	164 Ton.	60dh/Kg	9840000	1180800
Olive trees	6850 Trees	158 Ton.	4dh/Kg	632000	75840
Apple trees	31500 Trees	630 Ton.	5dh/Kg	3150000	378000
Palm trees	1421870 Tr.	34125 Ton.	15dh/Kg	511845000	61421400
Apricot trees	40400 Trees	760 Ton.	5dh/Kg	3800000	456000
Bovins	5540 (Nb)	-	4000dh	21130800	2535696
Ovins	83000 (Nb)	-	800dh	62963760	7555651,2
Caprins	59720 (Nb)	-	600dh	38914200	4669704
Camelins	9080 (Nb)	-	6000dh	52218000	6266160
Equins	7795 (Nb)	-	500dh	3868500	464220
Ruches	1370 Ruches	-	1000dh	1128000	135360
Milk	-	1773220 Lit.	4dh	6431840,87	771820,9
Milk Processing	-	Added value	524.000	524000	62880
Honey	-	1772 Kg	200dh	335200	40224
Firewood	240 566	-	1.6Dhs/Kg	140490544	16858865,28
Wood Construction	22000 Houses	-	-	4612500	553500
Water Surface Irrigation	-	225 Mm^3 (85%)	0,18 Dh/M3	40500000	4860000
Groundwater Irrigation	-	40 Mm^3 (15%)	0,75Dh/M3	30000000	3600000
Drinking Water	-	9417363m^3	1,26Dh/M3	2243721,06	269246,53
Water Energy	-	32385,91mw	0,674Dh/Kw	41130105,7	4935612,7
Soil Of Construction	648.8	-	7500Dhs	4278750	513450
Soil Energy	-	-	0,67Dh/Kw	50687,75	6082,53
Soil Transformation (Gold Production)	-	124,485 Kg		41342735,8	4961128,3
Fishing	500 Ton.	-	5dh	2500000	300000
Mining Jobs	91 Pers.	SMIG	2300	2511600	301392
Total general of the selected provisioning ecosystem services				3058633546	367036026

Note: 1 US $ = 8.26; 1MAD = 0.12 US $; www.oanda.com; Sept. 28, 2013. Qs: quintals
Source: (Karmaoui et al., 2015b)

$$V\left(SE\,k\right)=\sum_{i=1}^{n}A\left(LUi\right)\times V\left(SE\,ki\right) \quad (1)$$

where:

A (LUi) = Area of i (Land Use in Ha)
V (ESki) = Annual value of k ES for each i LU (in $/ha/yr)

RESULTS

The used data was collected from different sources (papers and gray literature) and classified according to three ecosystem services categories vegetal, animal, and other products (Table 2).

In the study area, the vegetal products are represented by cereals (Durum wheat and common wheat…), vegetable gardening, and fruit arboriculture (Olive tree, Palm tree…) quantified using quintals (Qs) and tones (T) units. The vegetal products were estimated based on 2009 economic values (Table 3). While the Animal ecosystem services (livestock) are bovine, Ovine, Caprine, Camels… estimated using the number of individuals of each species.

Table 3 shows the number of livestock in the study area, the milk and Honey production in this area. The industrialized milk is about 800 000 and 1000000 liters in 2009 and 1860000 liters in 2014. The total red meat is about 785 tons in 2009 and 806tons in 2014. Concerning jobs, this sector, offers about 3883 jobs (MPZ, 2014).We multiply the minimum guaranteed wage or "SMIG=2800Dh" with the number of jobs created: (3883*2800 Dh = 10872400/month) and 130468800/year (15795254$/year). Regarding the agri-food transformation, the only industrial unit installed in the study area saves an added value of 1038000Dh/year (125665,8$/year).

Table 2. The main ecosystem services categories used in the economic valuation

Ecosystem Services	Vegetal Ecosystem Services	Animal Ecosystem Services	Other Ecosystem Services
Food, water, Fiber and timber ecosystem services as classified by the Millennium ecosystem assessment in 2005	Durum wheat Common wheat Barley Maize Alfalfa Vegetable gardening Watermelon Almond trees Olive trees Apple trees Palm trees Apricot trees	Bovine Sheep Caprine Camels Equine Milk Processing Milk Honey	Firewood Wood construction Water surface irrigation Soil transformation (gold production) Mining jobs Agri-food jobs

Source: (Authors)

The other products of the MDV were the construction wood and the wood used as combustible in rural area. Water surface and soil transformation were also estimated in this paper. The study area houses in 2014, 258137 inhabitants (rural), consuming 1kg/capita/day (1,6dh/1kg), which corresponds to a value of 413019,2/day and 150752008Dh/year (18250848,4 $/year). Concerning the construction wood, the total households in MDV are 38067 (2014); 66% is rural type (25124 households). The construction of 25124 household required (25124*5000): 125620000Dh (15208232,4$). According to Karmaoui et al., (2015b), changing household at the rate of 570,5/year which corresponds to 2,27% of 25124. Therefore 570,5/year × 5000 (price in Dh of Construction wood for a household) = 2852500 Dhs (342300$/year) wood of construction/year. This amount was added to the economic value of the wood used for the construction renewal (renewal of 80% in 50 years) (Karmaoui et al., 2015b). 80% of 25124: 20099, 2/50= 402 households /year; therefore 402×5000=2009920Dhs. As a result 2009920 were added to 2852500 which give 4862420 Dhs/year. Surface water is an important resource of the existence of the MDV oases. This water comes from Draa Wadi fed by Mansour Eddahbi Dam and small lakes reservoirs located in this area. The MDV has more than 250 km store in Mansour Eddahbi Dam in upstream (north-east).

Table 3 gathers the selected provisioning ecosystem services provided by oases agro-ecosystem of MDV. The ecosystem service values illustrated in columns 2, 3, and 4 (2009) were based on the values provided by Karmaoui et al., (2015b) for the provisioning ES of the MDV in 2009. However, columns 5, 6, and 7 (2014) were gathered from provincial Monographic edited in 2015 and/or estimated in this current study.

Source: (Authors)The market prices of cultivated products (Dh/kg) and the water surface irrigation were estimated as reported in Karmaoui et al., (2015b). Table 4 depicts the monetary values of some provisioning ES standardized to US$/hectare/year in 2009 and 2014. The total values of the year 2014 was found to be 8658.3$/ha/year for 26000ha of the oasis area and 150$/ha/year for total area 15000Km^2, with a slight increase of $ 1245,7 (in 26000ha) and $21.5 (the whole area).

DISCUSSION

This paper demonstrates the economic importance of some provisioning ES of MDV. The marked price was used as method to establish economic value change between 2009 and 2014. Several provisioning ES were excluded because of the data lacking. The findings demonstrated, that the total direct economic value of the selected ES in 2014 reached more than $8600 compared to 7412,6 in 2009, with a positive change of $1245,7. This positive trend can be explained based on the estimated environmental vulnerability explored by Karmaoui & Moumane (2016). This later study identified and calculated the indicators of change of environmental vulnerability using the environmental vulnerability index (EVI). Effectively, this Index reflects the environmental state (see also Karmaoui et al., 2014b & Karmaoui et al., 2015c). This Index showed in the period 2009-2014 (Figure 3) a decrease of environmental vulnerability due to high rate of rainfall recorded in 2014. The higher rain intensity in only November was 137 mm, which promoted an increase in water supply and then in vegetation cover.

Finally, this chapter represents just an attempt to give a proximate monetary value change of some ES of MDV. This is due to the fact that the real "price" for the ES is still underestimated (Naruševičius & Matiukas, 2011). In this context, further research is needed. Information on spatial distribution value can help determine strategies to mitigate a loss (Teoh et al., 2019).

Table 3. Economic value of provisioning ecosystem services of Middle Draa Valley. Change of economic valuation between 2009 and 2014

Ecosystem Service	2009 *			2014**			EV Change
Units	**Area/Nb**	**Quant.**	**EV in $**	**Area/Nb**	**Quant.**	**EV in $**	**EV2014-EV2009**
Durum wheat	50 Ha	1300 Qs	62400	50ha	1300Qs	62400	0
Common wheat	16850 Ha	589910 Qs	28315680	16850ha	589910Qs	28315680	0
Barley	1400 Ha	29300 Qs	1406400	1400ha	29300Qs	1406400	0
Maize	330 Ha	6675 Qs	320400	330ha	6675Qs	320400	0
Alfalfa	3600 Ha	190000 Qs	4560000	3600ha	190000Qs	4560000	0
Vegetable gardening	1445 Ha	260920 Qs	12524160	1445ha	260920Qs	12524160	0
Watermelon	150Ha	77650 Qs	20138610	2400ha	1242400Qs	37602905	+17464295
Almond tree	32800 Trees	164 Ton.	1180800	31 850	432 Qs	313801	-866999
Olive tree	6850 Trees	158 Ton.	75840	24400	3030 Qs	146731	+70891
Apple tree	31500 Trees	630 Ton.	378000	28 000	900 Qs	54479	-323521
Palm tree	1421870 Trees	34125 Ton.	61421400	1 498 370	372820 Qs	67703389,8	+6281989,8
Apricot tree	40400 Trees	760 Ton.	456000	38 400	1615 Qs	97760,3	-358239,7
Total Economic Value of the Vegetal Services			**130839690**			***153108106***	**22268416**
Bovine	5540 (Nb)	-	2535696	6600	-	3196125,9	+660429,9
Ovine	83000 (Nb)	-	7555651	93100	-	9016949	+1461298
Caprine	59720 (Nb)	-	4669704	52500	-	3813559	-856145
Camels	9080 (Nb)	-	6266160	9000	-	6537530	+271370
Equine	7795 (Nb)	-	464220	7600	-	460048,4	-4171,6
Beehives	1370 (Nb)	-	135360	1650 (Nb)	-	199757,8	+64397,8
Milk	-	1773220 L	771820,9	-	1860000L	900726,4	+128905,5
Milk Processing	Added value		62880	Added value		62880	0
Honey	-	1,77 T	40224	-	3,65 T	88377,7	+48153,7
Total Economic Value of Animal Services			**22501716**			24275954,2	**+1774238,3**
Firewood	240 566	-	16858865	*See the text*		18250848	+1391983
Wood construction	22000	-	553500	*See the text*		588670	+35170
Water surface irrigation	-	225 Mm^3	4860000	0,18 DH/m3	260 Mm^3	5665859	+805859
Soil transformation (gold production)	-	124,485 Kg	4961128	-	147,583kg	5933876	+972748
Total Economic Value of Other Services			**27233493**			**30439253**	**+3205760**
Mining jobs	91 (Nb)	SMIG (2300)	301392	368	SMIG (2800)	1496949	+1195557
Agrifood jobs	3547 (Nb)	SMIG (2300)	11851961	3883	SMIG (2800)	15795254	+3943293
	Total		**12153353**	**Total**		**17292203**	**+5138850**
	General total						+32387264,4

* data source: cited by Karmaoui et al., (2015b)

** data source: Monography of Zagora and estimated as mentioned in the text.

Note: 1 US $ = 8.26; 1MAD = 0.12 US $; www.oanda.com; Sept. 28, 2013, EV, Economic value; M,Million; T, Tones; L, liters; Nb, Number (units); Qs, quintals

Table 4. Synthesis of economic value of provisioning services of Middle Draa Valley in $/ha/year

Area	2009	2014	Change (2014-2009)
26000ha	7412,6	8658,3	1245,7
1500000 ha	128,5	150	21,5

M (Million); MDV (Middle Draa Valley: area study)
Source: (Authors)

Figure 3. Middle Draa Valley EVI Profiles
Source: (Karmaoui & Moumane, 2016)

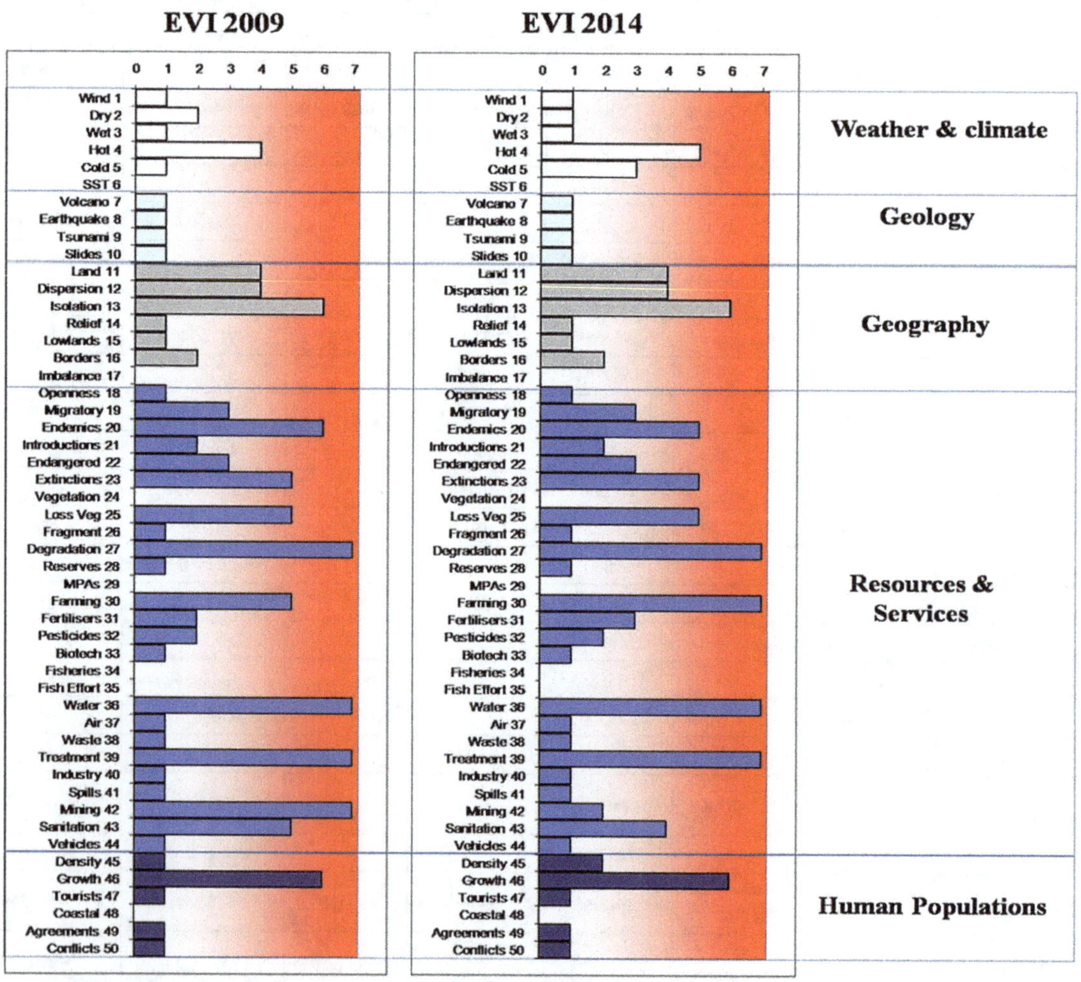

The outputs of this kind of research can support the existing knowledge about ES and their social and economic values for decision makers and local community (Naruševičius & Matiukas, 2011).

Limitations

- To what extent can we consider the selected products as ecosystem services?
- In this paper, we did not evaluate the costs of regulating and cultural ecosystem services due to their interactions and complexity. The most of the economic evaluation studies explore provisioning ES in contrast of other categories of ES.
- Some important ES such as Fishing, soil used in construction, groundwater irrigation, drinking water, water energy have not been covered,
- May be the period of the study is not enough to detect a change
- Due to lack of data on years between 2009 and 2014, is difficult to draw a clear picture of ES trend or change (2010, 2011, 2012, and 2013).

CONCLUSION

MDV contains a vast arid area (15000Km^2) including 26000Ha of oases that provide a multitude of ES. These ecosystem services play an important role in the social and economic well-being of the local population. Due to the ecological, social, and economic importance of these services, a monetary assessment is needed. The economic evaluation is among the most important approaches used globally. In this chapter, the economic value of 28 ES in the years 2009 and 2014 was carried out. The findings reveal that the study area offers a total economic value of *8658.3 US$/ha/year* in 2014 with an increasing change of *$ 1245,7 (in 26000ha)* compared to 2009. This shows that the region has recovered some amount of these ecosystem services due to the high rate of rainfall recorded in 2014. The conjugated climate change and anthropogenic impacts in MDV can cause a loss of ecosystem services estimated to some 1245.7$/ha/year. The findings of this study will support the introduction of ES economic value as concept in ecosystem assessment. These results can be used, when elaborating restoration and protection plans. A dataset on ecosystem services at national scale is suggested to be established in order to support this kind of researches.

ACKNOWLEDGMENT

I would to thank the Southern center for Culture and Sciences of Zagora for the support and assistance.

REFERENCES

Chelleri, L., Minucci, G., Ruiz, A., & Karmaoui, A. (2014). Responses to Drought and Desertification in the Moroccan Drâa Valley Region: Resilience at the Expense of Sustainability? *International Journal of Climate Change: Impacts & Responses*, *5*(2). doi:10.18848/1835-7156/CGP/v05i02/37199

Costanza, R., de Groot, R., Sutton, P., Van der Ploeg, S., Anderson, S. J., Kubiszewski, I., ... Turner, R. K. (2014). Changes in the global value of ecosystem services. *Global Environmental Change*, *26*, 152–158. doi:10.1016/j.gloenvcha.2014.04.002

De Groot, R., Brander, L., Van Der Ploeg, S., Costanza, R., Bernard, F., Braat, L., ... Hussain, S. (2012). Global estimates of the value of ecosystems and their services in monetary units. *Ecosystem Services*, *1*(1), 50–61. doi:10.1016/j.ecoser.2012.07.005

EVICC. (2011). *Evaluation de la vulnérabilité et des impacts du changement climatique dans les oasis du Maroc –EVICCa- et structuration de stratégies territoriales d'adaptation. Mission 1.1: Bilan Diagnostic des vulnérabilités climatiques et des capacités d'adaptation en situation actuelle Bassin oasien de Dadès*. Author.

Glaves, P., & Egan, D. (2013). Some Methods and Approached to Valuing Ecosystem Services. Academic Press.

Karmaoui, A. (2016). Economic Value of Regulating and Cultural Ecosystem Services of Oasean Biome: Case Study in Middle Draa Valley. In Economic Modeling, Analysis, and Policy for Sustainability (pp. 307-326). IGI Global.

Karmaoui, A. (2018). The cutaneous leishmaniasis vulnerability index (CLVI). *Acta Ecologica Sinica*, *38*(4), 288–295. doi:10.1016/j.chnaes.2018.01.001

Karmaoui, A., Ifaadassan, I., Messouli, M., & Khebiza, M. Y. (2015a). Sustainability of the Moroccan oasean system (Case study: Middle Draa Valley). *Global J Technol Optim*, *6*(01), 170. doi:10.4172/2229-8711.1000170

Karmaoui, A., Ifaadassan, I., Messouli, M., & Khebiza, M. Y. (2015c). *Characterization of Common Environmental Indicators of the Moroccan Oasean Biome, Pilot Study in the Reserve Biosphere of Oases in Southern Morocco*. Adv in Res. doi:10.9734/AIR/2015/19369

Karmaoui, A., Ifaadassan, I., Messouli, M., & Yacoubi Khebiza, M. (2015b). Monetarization of Ecosystem Services of Oasean Biome (Case study: Provisioning services of Middle Draa Valley Oases). *British Journal of Applied Science and Technology*, *11*(6), 1–18. doi:10.9734/BJAST/2015/19689

Karmaoui, A., Messouli, M., Ifaadassan, I., & Khebiza, M. Y. (2014b). A Multidisciplinary Approach to Assess the Environmental Vulnerability at Local Scale in Context of Climate Change (Pilot Study in Upper Draa Valley, South Morocco). *Global J Technol Optim*, *6*, 167. doi:10.4172/2229-8711.1000167

Karmaoui, A., Messouli, M., Yacoubi Khebiza, M., & Ifaadassan, I. (2014a). Environmental Vulnerability to Climate Change and Anthropogenic Impacts in Dryland (Pilot Study: Middle Draa Valley, South Morocco). *J Earth Sci Clim Change, S11*, 2. doi:10.4172/2157-7617.S11-002

Karmaoui, A., & Moumane, A. (2016). Changes in the Environmental Vulnerability of Oasean System (desert oasis), Pilot Study in Middle Draa Valley, Morocco. *Expert Opin Environ Biol*, *5*(3), 3. doi:10.4172/2325-9655.1000135

Lu, Z., Wei, Y., Xiao, H., Zou, S., Ren, J., & Lyle, C. (2015). Trade-offs between midstream agricultural production and downstream ecological sustainability in the Heihe River basin in the past half century. *Agricultural Water Management*, *152*, 233–242. doi:10.1016/j.agwat.2015.01.022

Millennium Ecosystem Assessment. (2005). *Ecosystems and Human Well- Being: Synthesis*. Washington, DC: Island Press.

Naruševičius, V., & Matiukas, G. (2011). Introducing the concept of ecosystem services: Inventory and economic valuation on local scale in Lithuania. *Sustainable Development Strategy and Practise*, *1*(5), 120–136.

Sulman, B., Desai, A., & Mladenoff, D. (2013). Modeling soil and biomass carbon responses to declining water table in a wetland-rich landscape. *Ecosystems (New York, N.Y.)*, *16*(3), 491–507. doi:10.100710021-012-9624-1

TEEB Foundations, . (2010). *The Economics of Ecosystems and Biodiversity: Ecological and Economic Foundations*. London: Earthscan.

Teoh, S. H. S., Symes, W. S., Sun, H., Pienkowski, T., & Carrasco, L. R. (2019). A global meta-analysis of the economic values of provisioning and cultural ecosystem services. *The Science of the Total Environment*, *649*, 1293–1298. doi:10.1016/j.scitotenv.2018.08.422 PMID:30308899

KEY TERMS AND DEFINITIONS

Cultural Services: Non-material services provided by an ecosystem.

Economic Valuation: Estimation of the economic value of ecosystem services methods such as market prices, replacement cost, restoration cost, travel cost, contingent valuation.

MEA: Millennium Ecosystem Assessment started by the United Nations in 2005.

Oasis: Agro-ecosystem in a desert zone.

Provisioning Services: Include services like food, water, and fiber.

Regulating Services: Are the services that assure the water, climate, soil regulation.

Supporting Services: Ecosystem services that support other ecosystem services such as the primary production, photosynthesis.

Chapter 5
Potential Role of Endophytes for Sustainable Environment

Zubair A. Dar
Sher-e-Kashmir University of Agricultural Sciences and Technology of Kashmir, India

Bhat Rifat
University of Kashmir, India

Javeed I. A. Bhat
Sher-e-Kashmir University of Agricultural Sciences and Technology of Kashmir, India

Asma Absar Bhatti
Sher-e-Kashmir University of Agricultural Sciences and Technology of Kashmir, India

Shamsul Haq
Sher-e-Kashmir University of Agricultural Sciences and Technology of Kashmir, India

Azra Amin
Sher-e-Kashmir University of Agricultural Sciences and Technology of Kashmir, India

Shakeel Ahmad Dar
Sher-e-Kashmir University of Agricultural Sciences and Technology of Kashmir, India

ABSTRACT

Endophytes are symptomless fungal and bacterial microorganisms found in almost all living plants. They are vital components of plant microbiomes. Endophytes affect plant growth and plant responses to pathogens, herbivores, and environmental change by producing a range of natural products having antifungal, antibacterial, and insecticidal properties. Endophytes have shown particular promise in agriculture particularly as beneficial crop inoculants and are known to enhance abiotic and biotic plant stress tolerance by increasing tolerance to drought and water stress, as well as tolerance to high temperature and high salinity. A better understanding of their plant growth-promoting mechanisms could simplify higher production of energy crops in a more sustainable manner even on marginal land and feed stocks for industrial processes, thus contribute to avoiding conflicts between food and energy production Many endophytes can be exploited to improve the efficiency of phytoremediation as they are found to be resistant to heavy metals and capable of detoxifying organic contaminants.

DOI: 10.4018/978-1-5225-7387-6.ch005

INTRODUCTION

The term endophyte was originally defined by De Bary in 1866 as "Any organism occurring within plant tissues". The word endophyte means in the plant (derived from the Greek-endon = within, phyton=plant) this term can be used for a wide spectrum of potential hosts inhabitants, e.g. bacteria, fungi, etc. (Stone et al., 2000).By definition, an endophytic fungus lives in mycelial form in biological association with living plant at least for some time. Endophytes are very diverse group which include endophytic bacteria and endophytic fungi (Raghukumar, 2008).There are approximately 300,000 plant species on earth and each plant is the host to one or more endophytes, and many of them may colonize certain hosts (Petrini, 1991). Endophytes are ubiquitous in nature as they have been recovered from plants adjusted to a wide range of ecosystems that include hot deserts, temperate, arctic tundra, tropical forests, grasslands and croplands (Arnold, 2007, 2008; Arnold & Lutzoni, 2007). They are present in all the major lineages of land plants (conifers and flowering plants, mosses and other non-vascular plants, ferns and other seedless plants) (Arnold, 2007). It has been estimated that there are at least 1 million endophytic fungal species which signifies the importance of fungal endophytes as vital constituents of fungal biodiversity (Ganley et al., 2004). These microorganisms reside in the intercellular spaces of plants with no antagonistic manifestation and exhibit a relationships with specific plants and produces bioactive substances (plant growth regulatory, antibacterial, antiviral, antifungal, insecticidal, etc.) enhance the growth and competitiveness of the host in nature (Shentu et al., 2014;Hardoim et al., 2015; Card et al., 2016). Some endophytes could be reliable sources of materials of the agricultural and pharmaceutical potential (Nisa et al., 2015). Endophytic microbes initiate the biological degradation of dead or dying host-plant which is necessary for nutrient recycling (Hungria et al., 2010). These microbes inhibit the plant without causing any harm and can ascertain a mutualistic association (Azovedo et al., 2000).

Endophytic bacteria also known as endosymbiotic microorganisms live in internal plant tissues of healthy host plants and inhabit an ecological niche similar to that of phytopathogens, as suitable biocontrol agents (Berg et al., 2005). These biocontrol or biofertilizer agents have replaced harmful chemical fertilizers which have increased the quality of yield without compromising on its quantity. These microorganisms precondition plant internal environment which result into disease suppression and better cope against pathogen in a specific host. Biological control is likely to be less spectacular than most harmful or chemical controls but is generally more stable and environmentally sustainable. Phytoremediation may be repressed because the plant experiences phytotoxic effects from contaminants such as heavy metals, polycyclic aromatic hydrocarbons (PAHs), and halogenated hydrocarbons. Though, some endophytes show plant-growth-promoting potential, but they also possess several traits that can alter contaminants' through the production of iron chelators, siderophores, organic acids and various degrading enzymes (Yousaf et al., 2010; Soleimani et al., 2010). The plant roots offer an impeccable environment for degradation of heavy metal by microbes. These microbes help in production of exudates, enhances the microbial activity, increases the movement of gases and water within soil environment (Weyens et al., 2010).Combined effect of all these influences the contaminant availability for uptake and degradation of pollutants. This type of mechanism is an advantage over bioremediation, where the microbes may not be able to survive at a site or degrade a pollutant because it may not be bioavailable (Newman & Reynolds, 2005). Hence using plant-microbe association for removing heavy metals is a promising green technology. Endophytes are the synthesizers of various bioactive chemicals inside the plant that can be used as potential sources of many pharmaceutical leads. These compounds have been optimized by ecological and environmental factors yielding effective bioactive agents and useful for novel drugs which are anti-bacterial, anti-viral, anti-tumour, anti–malarial, anti-fungal, anti-inflammatory and other compounds (terpenoids, flavonoids, alkaloids, steroids, phenolic compounds) and other structure types (Wagenaar & Clardy, 2001). Thus, endophytes are future producers of an ample and dependable source of bioactive and chemically novel compounds in a wide variety of arenas like; medicine, industry and agriculture. The aim of the present chapter is to address concisely some recent developments in diversity and function of endophytes and their importance in selected areas.

AIM OF THIS CHAPTER

After giving brief introduction about the basic aspects of endophytes, the primary aim of this chapter is to provide insights how endophytes and their secondary metabolites can be efficiently used for improving the environment, plant health and plant growth.

DEFINITIONS

- **Endophytes:** Endophytes may be defined as microorganisms that reside or colonize the internal tissues of plants without causing any harmful effects (Owen & Hundley, 2004).
- **Induced Systemic Resistance:** Induced systemic resistance (ISR) is a defense mechanism induced by plant growth–promoting bacteria and fungi in the rhizosphere against a wide range of pathogens and insect herbivores (Pieterse et al., 2014).
- **Bio Fertilizer:** Biofertilizer is a substance which contains living microorganisms or strains of microorganisms that colonize the rhizosphere or interior of the plants and assist the host plants in uptake of nutrients.

Figure 1. Schematic diagram of the different plant–bacterial endophyte interactions and their applications.
Source: Germaine et al., 2006

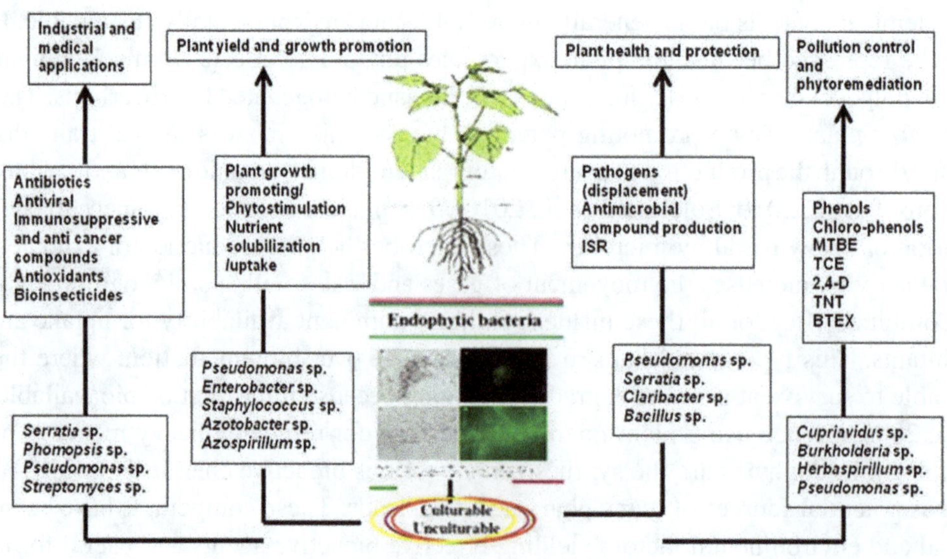

ENDOPHYTES AS BIO-CONTROL OR BIO-FERTILIZERS AGENTS

Why Bio-Control or Bio-Fertilizers

Biological control and bio-fertilizers are better than chemical control because of some important reasons which are as follows:

1. Biological control is self-regulating mechanism and requires no supervision as compared to chemical control. In case of chemical control pest scouting, time and schedule of spray is required before deciding whether to spray or not. Hence, chemical control is time consuming and laborious method.
2. Biological control is economical whereas chemical control methods need some expenses and management.
3. Biological control is environmental friendly, preserve the ecosystem and don't cause any type of pollution while chemical control may cause negative impacts on the ecosystem.
4. Biological control may provide a long term control of harmful organisms.
5. No significant problem of development of resistance against biological control while in chemical control there is a serious problem that results in the failure of chemical control.
6. No side effects of biological control and bio-fertilizers on human health while chemical control has negative effects on the human health and other non-targeting organisms.

Endophytic Nitrogen-Fixing Bacteria as Bio Fertilizer

Endophytes (bacteria or fungi) inhabit the internal tissues of their host plant and form specific relationships with them e.g., symbiotic, commensalistic, mutualistic and trophobiotic (Swarnalatha et al., 2015). Most endophytes originate from the rhizosphere or phyllosphere; however, there are few which may be transmitted through the seed (Ryan et al., 2008). These microbes cause no negative effects on plant growth and can be isolated from the plant after surface disinfection (Gaieroet al., 2013). These bacteria can act as bio-control agents, promote plant growth and yield. Biological nitrogen fixation by endophytes is considered to be the most potential way to provide fixed form of nitrogen to the plants (Dixon& Kahn, 2004). The endophytic diazotrophic (N_2 fixing) bacteria in the free living form or in symbiosis are involved in the fixation process, coverts the atmospheric nitrogen into ammonia with the help of nitrogenase enzyme (Bhattacharjee et al., 2008). Nodulated legumes with endosymbiosis are the most noticeable nitrogen fixing system in agriculture. The first diazotrophic bacterium gene involved in the synthesis and functioning of nitrogenase (nif, N_2 fixation) is *Klebsiellaoxytoca*M5a1. Among the leguminous plants of the Fabaceae family, the soil bacteria of the Rhizobiaceae (rhizobia) family are confined to the root nodules (Willems, 2007). Inside these nodules, rhizobia successfully complete biological nitrogen fixation in the presence of adequate oxygen, an inhibitor of nitrogenase activity.

Endophytes as Plant Growth Microbes

Endophytes have shown positive effects on plant growth and development when applied to the seedling or seed prior to exposure to plant parasitic nematodes infection. These microbes modify themselves to live within plants and have high synergetic impact than the rhizoplanic and rhizospheric counterparts, thereby benefiting their host plant by providing nitrogen through nitrogen fixation (Cocking, 2003). Some

Figure 2. Schematic presentations of plant colonization by endophytic bacteria. Figure in the left panel represents longitudinal section of root showing sites for entry of endophytic bacteria. Figure in the right panel represents transverse section of root showing the distribution and colonization of endophytic bacteria. *Source: (Compant et al., 2010)*

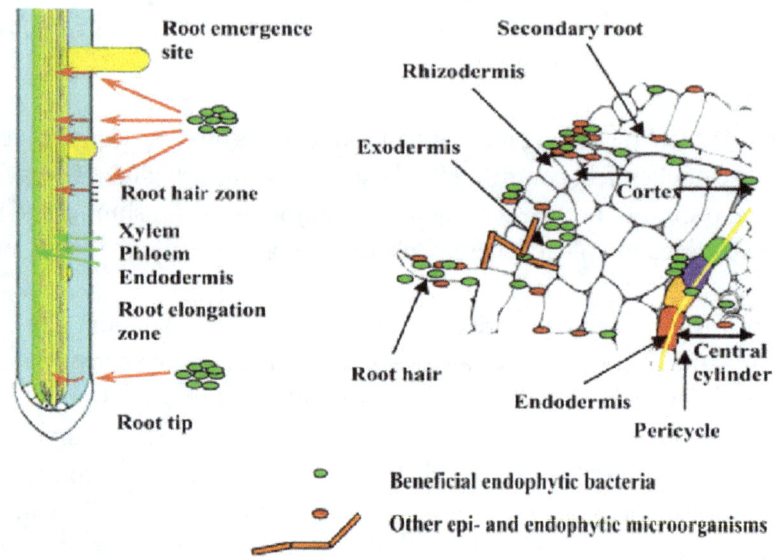

of the important endophytic biofertilizers include the members of *Klebsiella, Burkholderi, Azoarcu, Achromobacte, Gluconoacetobacter, Herbaspirillum, and Serratia* (Rothballer et al., 2008).

Mechanism Employed by Endophytic Plant Growth-Promoting Bacteria

The mechanism employed by microbes to stimulate the plant growth is reasonably well known (Glick, 2012). Endophytes without developing any significant symptoms reside inside the plant tissue. They mostly inhabit the above ground tissues of plants but also have been isolated from roots (Mandyam & Jumpponen, 2005; Leho-Tedersoo et al., 2009). Endophytic bacteria enter and systemically disseminate by colonizing apoplast actively, intracellular spaces and conducting through vessels of most of broad leaf higher plant, facilitate nutrients exchange and enzymes activity. (Khan et al., 2015; (Hallmann et al., 1997).These endophytes have ability of destroying nematodes propagation and benefit other crops in rotation along with the host plants (Sturz & Kimpinski, 2004).

Endophytes or Plant Growth-Promoting Bacteria (PGPB) may affect plant growth either directly or indirectly. Direct promotion of plant growth occurs by:

1. These microbes in association with host plant obtain important nutrients such as nitrogen, phosphorous and iron directly from the environment.
2. These endophytes regulate various plant hormones such as auxin, cytokinin or ethylene, thereby stimulating the plant growth and development.

Indirect promotion of plant growth by endophytes occurs when a bacterium restricts or prevents the damage to plants producing antibiotics, cell wall-degrading enzymes, lowering plant ethylene levels,

Figure 3. Showing multiple applications of diazotrophic (N_2 fixing) endophytic bacteria in various fields environment.
Source: (Hardoim et al., 2008).

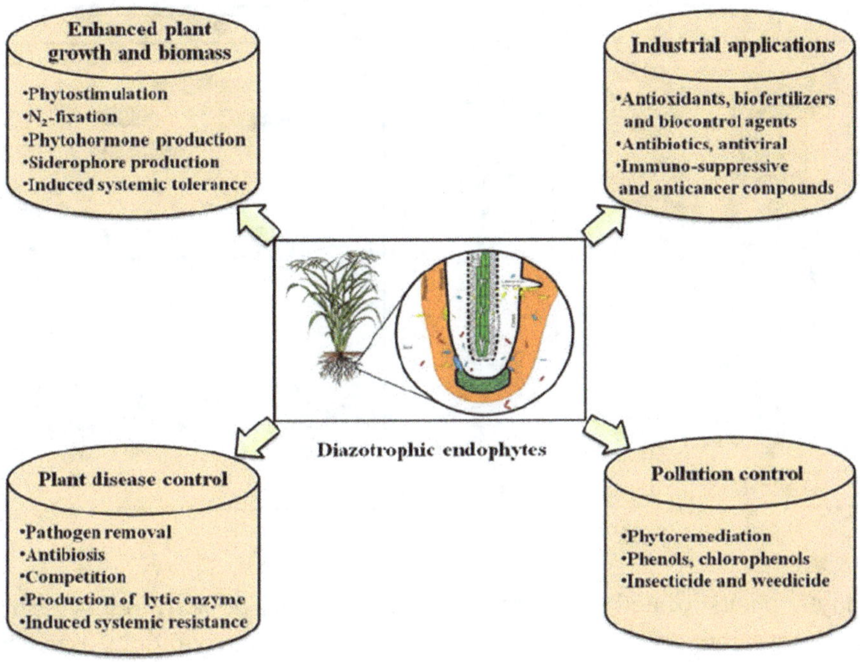

induced systemic resistance, decreasing the amount of iron available to pathogens, and the synthesis of pathogen-inhibiting volatile compounds (Glick, 2015).

ENDOPHYTES FOR CONFERRING ABIOTIC STRESS, INCLUDING IMPACTS OF CLIMATE CHANGE

Mutualistic fungal endophytes help host plants to adapt or survive in stress conditions, thereby increases their tolerance to water stress and drought, as well as tolerance to high salinity and high temperature (Aly et al., 2011). For example, fungal endophytes can increase plant performance in times of water stress (Kannadan & Rudgers, 2008) and heat stress (Rodriguez et al., 2008a), low nutrient availability (Lyons et al., 1990) and intense grazing / herbivore pressure (Hartley & Gange, 2009). Thus, the endophyte symbiosis may play an important role in determining how the plant host responds to climate change. The tropical panic grass, *Dichanthelium lanuginosum,* growing in geothermal soils in Yellowstone National Park forms a mutualistic symbiotic interaction with endophyte, *Curvularia protuberata* which confers heat tolerance (Redman et al., 2002). An experiment was carried out in which the root zones were heated up to 65°C, symbiotic plants infested with the endophyte tolerated and survived the heat but separately, neither the endophyte nor the plant alone were capable of growing at soil temperatures beyond 38 °C (Marquez et al., 2007). These endophytes help host plant to confer stress tolerance and thus provide a novel approach for mitigating the impacts of global climate change on agricultural plant communities (Rodriguez et al., 2008b). *Piriformospora indica* endophyte was originally isolated from the rhizosphere

Figure 4. A) Various types of interaction between soil bacteria and plant roots B) After interaction with roots and/or rhizosphere.
Source: (Souza et al., 2015)

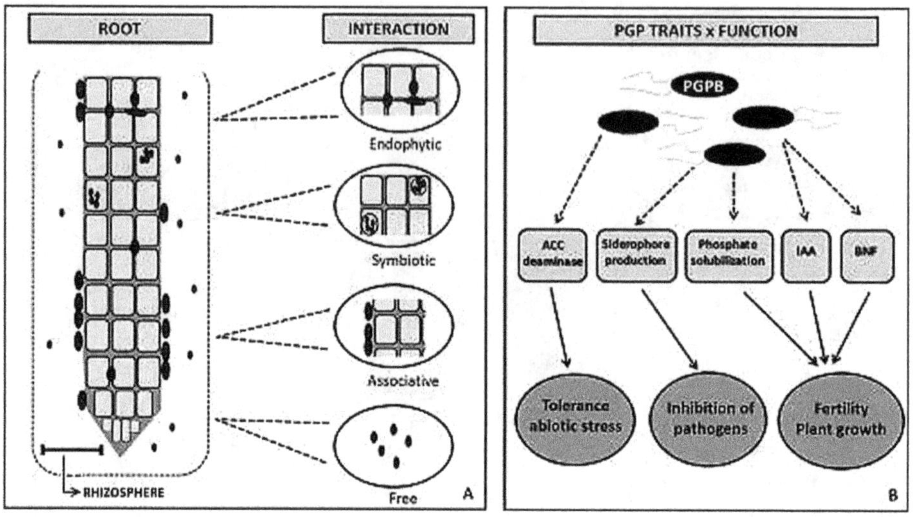

of several xerophytic plants located in the tropical region of Thar dessert, India (Verma et al., 1998). This filamentous fungus confers various host benefits particularly in conditions of nutrient stress, such as development of plant growth and tolerates a wide range of abiotic stresses, as well as biotic stresses. *Piriformospora indica*is well-known for its broad host range, establishing a symbiotic root relationship with both mono- and dicotyledonous plants, including the agriculturally important barley (*Hordeum vulgare*), and the model plants, tobacco *(Nicotiana tabacum)* and *Arabidopsis thaliana* (Johnson et al., 2014).

Ectomycorrhizae respond in different ways to high temperatures, e.g. with reduced colonization like *Cenococcum geophilum* in *Quercus myrsinaefolia* (Kasai et al., 2000) or with reduced respiration as shown for *Cenococcum geophilum, Suillus intermedius* (Malcolm et al., 2008). Colonization of *Lolium perenne* by the endophyte *Neotyphodium lolii* increased tolerance against abiotic stress such as drought (Kane, 2011). *Poaalsodes* showed the same effect when colonized with endophytes in order to make a prediction about the interactions among endophytes and host plants under a global warming scenario (Kannadan & Rudgers, 2008).

NATURAL PRODUCTS FROM ENDOPHYTES

Endophytes one of most important group of microbes produce various beneficial products which have both synergetic as well as antagonistic effects on host plants. Endophytes such as *Pseudomonas, Burkholderia* and *Bacillus* produce varied range of secondary metabolic products including anticancer compounds, insecticidal, volatile organic compounds, immunosuppressant agents, antifungal, antibiotics, and antiviral (Lodewyckx et al., 2002) (Table 1). Many plant associated endophytes produce phytohormones namely auxins, cytokinins, gibberellins, certain volatiles and the co-factor pyro-quinolone quinine which improves lateral root growth formation and thus, nutrient uptake and root exudation by plants (Spaepen et al., 2007).

Table 1. Natural products that have been derived or produced from various endophytic bacteria

Organism	Plant Association	Active Agent	Activity
Taxomycesandreanae	*Taxusbrevifolia*	Taxol	Anticancer
Pseudomonas viridiflava	Grass	Ecomycins B and C	Antimicrobial
Streptomyces griseus	*Kandeliacandel*	p-Aminoacetophenonic acids	Antimicrobial
Cytonaema sp.	*Quercus* sp.	Cytonic acids A and D	Antiviral
Streptomyces sp.	*Monstera* sp.	Coronamycin	Antimalarial antifungal

Source: (Strobel et al., 1993; Miller et al., 1998; Guo et al., 2000; Ezra et al., 2004 and Guan et al., 2005)

ENDOPHYTES FOR CONFERRING DISEASE AND PEST

Endophytes are naturally occurring pathogens of harmful insects which have gained importance as biological control agents. Induced resistance is a kind of stress metabolism; which provides protection against large number detrimental factors and persists in plants for a longer period. Systemic resistance in plants can be induced by several ways. The inducer controls the biochemical behavior of plant at all stages of life. By applying right kind of inducer, plant metabolism can be modulated according to the desire. But the type of defenses may differ according to the age of plant. All these metabolic modifications make plant resistant against abiotic and biotic factors.

Systemic Resistance by Endophyte Fungi (Vesicular-Arbuscular Mycorrhizal) in Apple and Pear Plant

The vesicular arbuscular mycorrhizal fungi (VAM) comprise two characteristic structures like arbuscles and vesicles found in the root cortex of plants. Arbuscules are formed within the root cortex cells by continuous dichotomous branching of hyphae. These structures are mainly present intra and inter- cellular and develop thick walls in older roots, thus provide site of communication between the host and the fungus. The main function of VAM fungi during infection is to stimulate host defense by reducing the access sites. These fungi have been found to reduce the incidence of root-knot nematode (Linderman, 1994). Apple replant disease caused by phytotoxic myxomycetes can be suppressed by inoculating of apple-tree seedlings with the VAM fungi *Glomus fasciculatum* and *Glomus macrocarpum* (Catska, 1994). Also, VAM protects the host pear and apple tree against white rot-infecting pathogenic bacteria. The mechanisms involved during these interactions include chemical interactions, physical protection and indirect effects. Similarly, VAM fungi indirectly suppress plant pathogens include enhanced nutrition to plants; changes in chemical composition like antifungal chitinase, isoflavonoids, etc.; morphological changes in the root by augmenting lignification; improvement of abiotic stress and higher concentration of microbial diversity and composition in the mycorrhizosphere.

Control of Fire Blight Disease

Fire blight caused by *Erwinia amylovora*one of the severe diseases of pear, apple, quince, and loquat. Streptomycin helps in 84.38% to 95.24% disease control and vesicular-arbuscular mycorrhizal (*Glomusintraradices*) application helps in reduction of fire blight by between 9.7% and 50.5% (Akey et al., 2006).

Control of Wilt Diseases

The root colonizing facultative endophyte *Piriformospora indica* forms beneficial relationship with crop plants. *Piriformospora indica* has a potential to inhabit tomato rhizosphere, thereby increases the biomass of the leaves by 20% and reduces the disease (verticillium wilt) caused by *Verticillium dahlia* by more than 30%, (Fakhro et al., 2010). *Piriformospora indica* as a biocontrol solution also helps to manage various root diseases in different crop species such as maize (Kumar et al., 2009) and wheat (Rabiey et al., 2015).

PHYTOREMEDIATION BY ENDOPHYTES

Endophytic bacteria have a natural ability for degradation of xenobiotics. Bacteria degrading recalcitrant compounds are more abundant among endophytic populations than in the rhizosphere of the plants in contaminated sites (Siciliano et al., 2001).Remediation approaches presently applicable to polluted environments (soil and ground water) are often expensive and cause mutilation to the ecosystem(Glass, 2000). The common traditional methods used for treatment are pump and treat, excavation, addition of reactants and incineration (McGuinness & Dowling, 2009). This type of high cost remediation method contributes to the abandonment of many polluted commercial sites or brownfields. To substitute the old traditional methods for detoxification of heavy metals. One method is bioremediation which uses microbes to degrade toxins. At present, attention has been focused on the role of endophytic bacteria in phytoremediation (Newman &Reynolds, 2005). As numerous endophytes have been found to be resistant to heavy metals and adept of degrading organic contaminants. Some plant species have adept themselves to grow in contaminated soils. During phytoremediation, the heavy-metal-resistant endophytes decrease metal phytotoxicity, inhibit metal translocation and hence enhance plant growth (Li et al., 2012).

Following are some of the examples showing heavy metals remediation by plant- endophytes interaction:

1. Endophytes isolated from *Nicotiana tobacum* increases the uptake of trace elements such as zinc and iron by plants thereby decreasing heavy metal toxicity. Similarly, lead resistant endophytes isolated from *Brassica napus* enhance phytoremediation potential by promoting the plant growth (Rajkumar et al., 2010).
2. *Bacillus pumilus* plays a key role in metal uptake in plants, thereby increases plant root growth and shoot length uptake (Ma et al., 2015).
3. Endophytes of yellow lupin have genetically modified themselves for nickel resistance and are able to increase the nickel accumulation (Lodewyckx et al., 2001).

Thus, endophytic bacteria in association with plants efficiently promote the degradation of xenobiotic compounds or their metabolites as they are accumulated, or while being taken up, thereby reducing phytotoxic effects and toxic effects on herbivorous fauna residing on or near contaminated sites.

Mechanism Employed by Endophytic for Phytoremediation

Microbes can improve phytoremediation by increasing the plant biomass, phytoextraction or phytostabilization, as well as simplify metal bioaccumulation (Ma et al., 2011, 2013) (Figure 5). The abil-

Table 2. Endophytes used for phytoremediation of Pollutants

Organism	Host Plant/Tissue	Compound
Pseudomonas aeruginos	Wild rye	Mono and dechlorinated benzoic acids
Methylobacteriumpopuli	Poplar tissues	TNT,RDX,HMX
Pseudomonas sp.	PopulasHzendas	MTBE,BTEX,TCE
B. cepacia	Poplar	Toluene
Burkholderiacepacia	Yellow lupine	Volatile organic compounds
Pseudomonas fluorescens, Microbacteriumsp.	Root	Lead
Phoma sp., Alternaria sp., Peyronellaea sp., Steganosporium sp., Aspergillussp,	Stems and leaves	Lead and Zinc
Enterobacter sp.	Stem	Cadmium and Zinc

Source: (Jalgaonwala & Mahajan, 2011)

ity of endophytes to enter and thrive in the plant tissue makes them unique, having multidimensional interactions within the host plant (Kobayashi & Palumbo, 2000). These endophytes adept themselves to polluted environments, thereby improving the phytoremediation, plant-metal tolerance and promote plant growth (Zhang et al., 2011). They also improve phyto-extraction by altering the availability, solubility and transport of heavy metal. Iron is an essential nutrient and co-factor for various enzymatic reactions. In aerobic conditions, iron occurs mainly in the ferric state (Fe^{3+}) and forms highly insoluble oxy-hydroxides and hydroxides that are mainly unavailable to plants and microorganisms. Siderophores, low-molecular-mass iron chelators (Miethke & Marahiel, 2007), act as solubilizing agents for iron to

Figure 5. Schematic overview of mechanisms of plant- microbe- metal interactions
Source: (Ma et al. 2016)

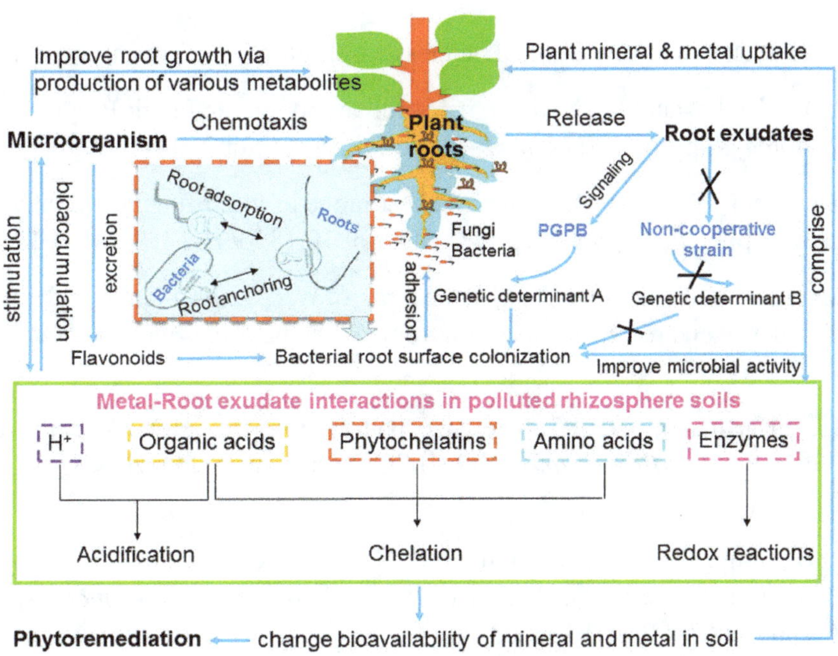

enhance its efficient uptake under conditions of iron limitation. In addition to iron, siderophores have the ability to form complexes with other metals such as aluminum, cadmium, copper and zinc. This binding of the metal to the siderophores increases the soluble metal concentration and thus aids in heavy metal sequestration (Rajkumar et al., 2010).

CONCLUSION

Exploiting the utility of endophytic microorganisms suggests that the bacteria play an important role in balancing plant physiology, plant growth promotion, assist in maintaining the availability of nutrient and phytoremediation. This chapter is a timely collation of state of knowledge developed towards better understanding of endophytes. The present chapter highlights the potential role of endophyte bacteria and fungi in adaptability to biotic or abiotic stresses. Apart from this, endophytic-plant interactions can produce beneficial chemical substances, fixation of nitrogen and can play a significant role in low input sustainable agriculture. In case of phytoremediation endophytes can assist their host plants by detoxifying and degrading the organic containments.

REFERENCES

Akey, A., & Maden, S. (2006). A new approach to fire blight control: *Mycorihiza*. Selcuk University, Faculty of Agriculture, Department of Plant Protection, Campus, Konya. Turkey. *HortScience*, *41*, 1309–1312.

Aly, A. H., Debbab, A., & Proksch, P. (2011). Fungal endophytes: Unique plant inhabitants with great promises. *Applied Microbiology and Biotechnology*, *90*(6), 1829–1845. doi:10.100700253-011-3270-y PMID:21523479

Arnold, A. (2008). Hidden within our botanical richness, a treasure trove of fungal endophytes. Plant Press.

Arnold, A. E. (2007). Understanding the diversity of foliar endophytic fungi: Progress, challenges, and frontiers. *Fungal Biology Reviews*, *21*(2-3), 51–66. doi:10.1016/j.fbr.2007.05.003

Arnold, A. E., & Lutzoni, F. (2007). Diversity and host range of foliar fungal endophytes: Are tropical leaves biodiversity hotspots. *Ecology*, *88*(3), 541–549. doi:10.1890/05-1459 PMID:17503580

Azovedo, J. L., Maccheroni, W., Pereira, J. O., & De Araujo, W. L. (2000). Endophytic microorganisms: A review on insect control and recent advances on tropical plants. *Electronic Journal of Biotechnology*, *3*, 40–65.

Berg, G., Eberl, L., & Hartmann, A. (2005). The rhizosphere as a reservoir for opportunistic human pathogenic bacteria. *Environmental Microbiology*, *7*(11), 1673–1685. doi:10.1111/j.1462-2920.2005.00891.x PMID:16232283

Bhattacharjee, R. B., Singh, A., & Mukhopadhyay, S. N. (2008). Use of nitrogen- fixing bacteria as bio fertilize for non-legumes: Prospects and challenges. *Applied Microbial Biotechnology*, *80*(2), 199–209. doi:10.100700253-008-1567-2 PMID:18600321

Boucher, Y., Douady, C. J., Papke, R. T., Walsh, A., Boudreau, M. E. R., Nesbo, C. L., ... Doolittle, W. F. (2003). Lateral gene transfer and the origins of prokaryotic groups. *Annual Review of Genetics*, *37*(1), 283–328. doi:10.1146/annurev.genet.37.050503.084247 PMID:14616063

Card, S. D., Johnson, L. J., & Teasdale, S. (2016). Deciphering microbial behaviour—the link between endophyte biology and efficacious biological control agents. *FEMS Microbiology Ecology,* 92.

Catska, V. (1994). Interrelationship between *vesicular-arbuscular mycorrhiza* and rhizospheremicroflora in apple replant disease. *Biologia Plantarum*, *36*(1), 99–104. doi:10.1007/BF02921276

Cocking, E. C. (2003). Endophytic colonization of plant roots by nitrogen- fixing bacteria. *Plant and Soil*, *252*(1), 169–175. doi:10.1023/A:1024106605806

Compant, S., Clément, C., & Sessitsch, A. (2010). Plant growth-promoting bacteria in the rhizo and endosphere of plants: Their role, colonization, mechanisms involved and prospects for utilization. *Soil Biology & Biochemistry*, *42*(5), 669–678. doi:10.1016/j.soilbio.2009.11.024

Dixon, R., & Kahn, D. (2004). Genetic regulation of biological nitrogen fixation. *Nature Reviews. Microbiology*, *2*(8), 621–631. doi:10.1038/nrmicro954 PMID:15263897

Ezra, D., Castillo, U. F., & Strobel, G. A. (2004). Coronamycins, peptide antibiotics produced by a verticillate Streptomyces sp. (MSU-2110) endophytic on *Monstera sp. Microbiology*, *150*(4), 785–793. doi:10.1099/mic.0.26645-0 PMID:15073289

Fakhro, A., Andrade-Linares, D. R., Von Bargen, S., Bandte, M., Büttner, C., Grosch, R., ... Franken, P. (2010). Impact of *Piriformospora indica*on tomato growth and on interaction with fungal and viral pathogens. *Mycorrhiza*, *20*(3), 191–200. doi:10.100700572-009-0279-5 PMID:19789897

Gaiero, J. R., McCall, C. A., Thompson, K. A., Day, N. J., Best, A. S., & Dunfield, K. E. (2013). Inside the root microbiome: Bacterial root endophytes and plant growth promotion. *American Journal of Botany*, *100*(9), 1738–1750. doi:10.3732/ajb.1200572 PMID:23935113

Germaine, K., Liu, X., Cabellos, G., Hogan, J., Ryan, D., & Dowling, D. N. (2006). Bacterial endophyte-enhanced phyto-remediation of the organochlorine herbicide 2, 4-dichlorophenoxyacetic acid. *FEMS Microbiology Ecology*, *57*(2), 302–310. doi:10.1111/j.1574-6941.2006.00121.x PMID:16867147

Glass, D. (2000). *U.S and International markets for Phytoremediation, D.* Needham, MA: Glass Associates, Inc.

Glick, B. R. (2012). Plant growth-promoting bacteria: mechanisms and applications. *Scientific*, article ID 963401.

Glick, B. R. (2015). *Beneficial Plant-Bacterial Interactions*. Springer.

Guan, S. H., Sattler, I., Lin, W. H., Guo, D. A., & Grabley, S. (2005). p-Aminoacetophenonic acids produced by a mangrove endophyte: Streptomyces griseus subspecies. *Journal of Natural Products*, *68*(8), 1198–1200. doi:10.1021/np0500777 PMID:16124760

Guo, B., Dai, J. R., Ng, S., Huang, Y., Leong, C., Ong, W., & Carte, B. K. (2000). Cytonic acids A and B: Novel tridepside inhibitors of hCMV protease from the endophytic fungus Cytonaema species. *Journal of Natural Products*, *63*(5), 602–604. doi:10.1021/np990467r PMID:10843568

Hallmann, A., Benhamou, N., & Kloepper, J. W. (1997). Bacterial endophytes in cotton:mechanisms of entering the plant. *Canadian Journal of Microbiology*, *43*(6), 577–582. doi:10.1139/m97-081

Hardoim, P. R., Van, O., Verbeek, L. S., & Berg, G. (2015). The hidden world within plants: Ecological and evolutionary considerations for defining functioning of microbial endophytes. *Microbiology and Molecular Biology Reviews*, *79*(3), 293–320. doi:10.1128/MMBR.00050-14 PMID:26136581

Hardoim, P. R., Van, O., Verbeek, L. S., & Van, E. (2008). Properties of bacterial endophytes and their proposed role in plant growth. *Trends in Microbiology*, *16*(10), 463–471. doi:10.1016/j.tim.2008.07.008 PMID:18789693

Hartley, S. E., & Gange, A. C. (2009). Impacts of plant symbiotic fungi on insect herbivores: Mutualism in a multitrophic context. *Annual Review of Entomology*, *54*(1), 323–342. doi:10.1146/annurev.ento.54.110807.090614 PMID:19067635

Hungria, M., Campo, R. J., Souza, E. M., & Pedrosa, F. O. (2010). Inoculation with selected strains of *A. brasilense* and *A. lipoferum* improves yields of maize and wheat in Brazil. *Plant and Soil*, *331*(1-2), 413–425. doi:10.100711104-009-0262-0

Jalgaonwala, R. E., &Mahajan, R. T. (2011). Isolation and characterization of endophytic bacteria from roots of *Pongamia Glabra Vent. International Journal of Pharmaceutical and Biological Science*, *1*, 280-287.

Johnson, J. M., Alex, T., & Oelmu, R. (2014). *Piriformospora indica*: The versatile and multifunctional root endophytic fungus for enhanced yield and tolerance to biotic and abiotic stress in crop plants. *Journal of Tropical Agriculture*, *52*, 103–122.

Kane, K. H. (2011). Effects of endophyte infection on drought stress tolerance of *Lolium perenne* accessions from the Mediterranean region. *Environmental and Experimental Botany*, *71*(3), 337–344.

Kannadan, S., & Rudgers, J. A. (2008). Endophyte symbiosis benefits a rare grass under low water availability. *Functional Ecology*, *22*(4), 706–713. doi:10.1111/j.1365-2435.2008.01395.x

Kasai, K., Usami, T., Lee, J., Ishikawa, S. I., & Oikawa, T. (2000). Responses of Ectomycorrhizal Colonization and Morphotype Assemblage of Quercusmyrsinaefolia Seedlings to Elevated Air Temperature and Elevated Atmospheric CO_2. *Microbes and Environments*, *15*(4), 197–207. doi:10.1264/jsme2.2000.197

Khan, A. L., Waqas, M., Khan, A. R., Hussain, J., Kang, S. M., Gilani, S. A., ... Lee, I.-J. (2013). Fungal endophytePenicilliumjanthinellum LK5 improves growth of ABA-deficient tomato under salinity. *World Journal of Microbiology & Biotechnology*, *29*(11), 2133–2144. doi:10.100711274-013-1378-1 PMID:23842755

Kobayashi, D. Y., & Palumbo, J. D. (2000). Bacterial endophytes and their effectson plants and uses in agriculture. In C. W. Bacon & J. F. White (Eds.), *Microbialendophytes* (pp. 199–236). New York: Marcel Dekker.

Kumar, M., Yadav, V., Tuteja, N., & Johri, A. K. (2009). Antioxidant enzyme activities in maize plants colonized with *Piriformosporaindica. Microbiology, 155*(3), 780–790. doi:10.1099/mic.0.019869-0 PMID:19246749

Li, H. Y., Wei, D. Q., Shen, M., & Zhou, Z.-P. (2012). Endophytes and their role in phytoremediation. *Fungal Diversity, 54*(1), 11–18. doi:10.100713225-012-0165-x

Linderman, R. G. (1994). Role of AM fungi in biocontrol. In F. L. Pfleger & R. G. Linderman (Eds.), *Mycorrhizae and Plant Health* (pp. 1–25). St. Paul, MN: APS Press.

Lodewyckx, C., Taghavi, S., Mergeay, M., Vangronsveld, J., Clijsters, H., & Van der Lelie, D. (2001). The effect of recombinant heavy metal resistant endophytic bacteria in heavy metal uptake by their host plant. *International Journal of Phytoremediation, 3*(2), 173–187. doi:10.1080/15226510108500055

Lodewyckx, C., Vangronsveld, J., Porteous, F., Moore, E. R. B., Taghavi, S., Mezgeay, M., & Van der Lelie, D. (2002). Endophytic bacteria and their potential applications. *Critical Reviews in Plant Sciences, 21*(6), 583–606. doi:10.1080/0735-260291044377

Lyons, P. C., Evans, J. J., & Bacon, C. W. (1990). Effects of fungal endophyte *Acremonium coenophialum* on nitrogen accumulation and metabolism in tall fescue. *Plant Physiology, 92*(3), 726–732. doi:10.1104/pp.92.3.726 PMID:16667341

Ma, Y., Oliveira, R. S., Freitas, H., & Zhang, C. (2016). Biochemical and Molecular Mechanisms of Plant-Microbe-Metal Interactions: Relevance for Phytoremediation. *Frontiers in Plant Science, 7*, 918. doi:10.3389/fpls.2016.00918 PMID:27446148

Ma, Y., Oliveira, R. S., Nai, F., Rajkumar, M., Luo, Y., Rocha, I., & Freitas, H. (2015). The hyperaccumulator Sedum plumbizincicola harbors metal-resistant endophytic bacteria that improve its phytoextraction capacity in multi-metal contaminated soil. *Journal of Environmental Management, 156*, 62–69. doi:10.1016/j.jenvman.2015.03.024 PMID:25796039

Ma, Y., Rajkumar, M., Luo, Y., & Freitas, H. (2013). Phytoextraction of heavy metal polluted soils using *Sedum plum bizin cicola* inoculated with metal mobilizing *Phyllobacterium myrsinacearum* RC6b. *Chemosphere, 93*(7), 1386–1392. doi:10.1016/j.chemosphere.2013.06.077 PMID:23890964

Ma, Y., Rajkumar, M., Luo, Y. M., & Freitas, H. (2011). Inoculation of endophytic bacteria on host and non-host plants – effects on plant growth and Ni uptake. *Journal of Hazardous Materials, 196*, 230–237. doi:10.1016/j.jhazmat.2011.08.034 PMID:21872991

Malcolm, G. M., Lopez, J. C., Koide, R. T., & Eissenstat, D. M. (2008). Acclimation to temperature and temperature sensitivity of metabolism by ectomycorrhizal fungi. *Global Change Biology, 14*(5), 1169–1180. doi:10.1111/j.1365-2486.2008.01555.x

Mandyam, K., & Jumpponen, A. (2005). Abundance and possible functions of the root-colonising dark septate endophytic fungi. Studies Mycology, 53, 173-189.

Marquez, L. M., Redman, R. S., Rodriguez, R. J., & Roossinck, M. J. (2007). A virus in a fungus in a plant: Three-way symbiosis required for thermal tolerance. *Science*, *315*(5811), 513–516. doi:10.1126cience.1136237 PMID:17255511

McGuinness, M., & Dowling, D. (2009). Plant-Associated Bacterial Degradation of Toxic Organic Compounds in Soil. *International Journal of Environmental Research and Public Health*, *62*, 226–2247. PMID:19742157

Miethke, M., & Marahiel, M. A. (2007). Siderophore-based iron acquisition and pathogen control. *Microbiology and Molecular Biology Reviews*, *71*(3), 413–445. doi:10.1128/MMBR.00012-07 PMID:17804665

Miller, C. M., Miller, R. V., Garton-Kenny, D., Redgrave, B., Sears, J., Condron, M. M., ... Strobel, G. A. (1998). Ecomycins unique antimycotics from *Pseudomonas viridi flava. Journal of Applied Microbiology*, *84*(6), 937–944. doi:10.1046/j.1365-2672.1998.00415.x PMID:9717277

Newman, L. A., & Reynolds, C. M. (2005). Bacteria and phytoremediation: New uses for endophytic bacteria in plants. *Trends in Biotechnology*, *23*(1), 6–8. doi:10.1016/j.tibtech.2004.11.010 PMID:15629849

Nisa, H., Kamili, A. N., Nawchoo, I. A., Shafi, S., Shameem, N., & Bandh, S. A. (2015). Fungal endophytes as prolific source of phytochemicals and other bioactive natural products: A review. *Microbial Pathogenesis*, *82*, 50–59. doi:10.1016/j.micpath.2015.04.001 PMID:25865953

Owen, N. L., & Hundley, N. (2004). Endophytes–the chemical synthesizers inside plants. *Science progress*, *87*(2), 79–99. doi:10.3184/003685004783238553 PMID:15782772

Petrini, O. (1991). Fungalendophytes of tree leaves. In J. H. Andrews & S. S. Hirano (Eds.), *Microbial Ecology of Leaves* (pp. 179–197). Berlin: Springer. doi:10.1007/978-1-4612-3168-4_9

Pieterse, C. M., Zamioudis, C., Berendsen, R. L., Weller, D. M., Van Wees, S. C., & Bakker, P. A. (2014). Induced systemic resistance by beneficial microbes. *Annual Review of Phytopathology*, 52. PMID:24906124

Rabiey, M., Ullah, I., & Shaw, M. W. (2015). The endophytic fungus Piriformosporaindica protects wheat from fusarium crown rot disease in simulated UK autumn conditions. *Plant Pathology*, *64*(5), 1029–1040. doi:10.1111/ppa.12335

Raghukumar, C. (2008). Marine fungal biotechnology: An ecologicalperspective. *Fungal Diversity*, *31*, 5–19.

Rajkumar, M., Ae, N., & Freitas, H. (2010). Endophytic bacteria and their potential to enhance heavy metal phytoextraction. *Chemosphere*, *77*(2), 153–156. doi:10.1016/j.chemosphere.2009.06.047 PMID:19647283

Redman, R. S., Sheehan, K. B., & Stout, R. G. (2002). Thermo-tolerance generated by plant/fungal symbiosis. *Science*, *298*(5598), 1581. doi:10.1126cience.1072191 PMID:12446900

Rodriguez, R. J., Henson, J., Van Volkenburgh, E., Hoy, M., Wright, L., Beckwith, F., ... Redman, R. S. (2008a). Stress tolerance in plants via habitat-adapted symbiosis. *The ISME Journal*, *2*(4), 404–416. doi:10.1038/ismej.2007.106 PMID:18256707

Rodriguez, R. J., Henson, J., Van Volkenburgh, E., Hoy, M., Wright, L., Beckwith, F., ... Redman, R. S. (2008b). Stress tolerance in plants via habitat-adapted symbiosis. *The ISME Journal*, *2*(4), 404–416. doi:10.1038/ismej.2007.106 PMID:18256707

Rothballer, M., Eckert, B., Schmid, M., Fekete, A., Schloter, M., Lehner, A., ... Hartmann, A. (2008). Endophytic root colonization of gramineous plants by Herbaspirillumfrisingense. *FEMS Microbiology Ecology*, *66*(1), 85–95. doi:10.1111/j.1574-6941.2008.00582.x PMID:18761671

Ryan, R. P., Germaine, K., Franks, A., Ryan, D. J., & Dowling, D. N. (2008). Bacterial endophytes: Recent developments and applications. *FEMS Microbiology Letters*, *278*(1), 1–9. doi:10.1111/j.1574-6968.2007.00918.x PMID:18034833

Shentu, X., Zhan, X., Ma, Z., Yu, X., & Zhang, C. (2014). Antifungal activity of metabolites of the endophytic fungus *Trichoderma brevicompactum* from garlic. *Brazilian Journal of Microbiology*, *45*(1), 248–254. doi:10.1590/S1517-83822014005000036 PMID:24948941

Siciliano, S., Fortin, N., & Himoc, N. (2001). Selection of specific endophytic bacterial genotypes by plants in response to soil contamination. *Applied and Environmental Microbiology*, *67*(6), 2469–2475. doi:10.1128/AEM.67.6.2469-2475.2001 PMID:11375152

Soleimani, M., Afyuni, M., Hajabbasi, M. A., Nourbakhsh, F., Sabzalian, M. R., & Christensen, J. H. (2010). Phytoremediation ofan aged petro-leum contaminated soil using endophyt infected and non-infected grasses. *Chemosphere*, *81*(9), 1084–1090. doi:10.1016/j.chemosphere.2010.09.034 PMID:20961596

Souza, R., Ambrosini, A., Luciane, M. P., & Passaglia, T. (2015). Plant growth-promoting bacteria as inoculants in agricultural soils. *Genetics and Molecular Biology*, *38*(4), 401–419. doi:10.1590/S1415-475738420150053 PMID:26537605

Spaepen, S., Vanderleyden, J., & Remans, R. (2007). Indole-3-acetic acid in microbial and microorganismplant signaling. *FEMS Microbiology Reviews*, *31*(4), 425–448. doi:10.1111/j.1574-6976.2007.00072.x PMID:17509086

Stępniewska, Z., & Kuźniar, A. (2013). Endophytic microorganisms—Promising applications in bioremediation of greenhouse gases. *Applied Microbiology and Biotechnology*, *97*(22), 9589–9596. doi:10.100700253-013-5235-9 PMID:24048641

Stone, J. K., Bacon, C. W., & White, J. F. (2000). An overview of endophytic microbes: Endophytism defined. In Microbial endophytes. New York: Dekker.

Strobel, G. A., & Long, D. M. (1998). Endophytic microbes embody pharmaceutical potential. *ASM News*, *64*, 263–268.

Sturz, A., & Kimpinski, J. (2004). Endo root bacteria derived from marigolds (*Tagetes spp.*) can decrease soil population densities of root-lesion nematodes in the potato root zone. *Plant and Soil*, *262*(1/2), 241–249. doi:10.1023/B:PLSO.0000037046.86670.a3

Swarnalatha, Y., Saha, B., & Choudary, L. (2015). Bioactive compound analysis and antioxidant activity of endophytic bacterial extract from *Adhatho dabeddomei*. *Asian Journal of Pharmaceutical and Clinical Research*, *8*(1), 70–72.

Tedersoo, L., Pärtel, K., Jairus, T., Gates, G., Põldmaa, K., & Tamm, H. (2009). Ascomycetes associated with ectomycorrhizas: Moleculardiversity and ecology with particularreference to the *Helotiales. Environmental Microbiology*, *11*(12), 3166–3178. doi:10.1111/j.1462-2920.2009.02020.x PMID:19671076

Verma, S., Varma, A., Rexer, K. H., Hassel, A., Kost, G., Sarbhoy, A., ... Franken, P. (1998). *Piriformo sporaindica*, gen. et sp. nov., a new root-colonizing fungus. *Mycologia*, *90*(5), 896–903. doi:10.1080/00275514.1998.12026983

Wagenaar, M. M., & Clardy, J. (2001). Dicerandrols, new antibiotics and cytotoxic dimmers produced by the fungus *Phomopsis longicolla* isolated from an endangered mint. *Journal of Natural Products*, *64*(8), 1006–1009. doi:10.1021/np010020u PMID:11520215

Weyens, N., Croes, S., Dupae, J., Newman, L., Van der Lelie, D., Carleer, R., & Vangronsveld, J. (2010). Endophytic bacteria improve phytoremediation of Ni and TCE co-contamination. *Environmental Pollution*, *158*(7), 2422–2427. doi:10.1016/j.envpol.2010.04.004 PMID:20462680

Weyens, N., Van der Lelie, D., Taghavi, S., Newman, L., & Vangronsveld, J. (2009). Exploiting plant-microbe partnerships to improve biomass production and remediation. *Trends in Biotechnology*, *27*(10), 591–598. doi:10.1016/j.tibtech.2009.07.006 PMID:19683353

Willems, A. (2007). The taxonomy of rhizobia: An overview. In *First International Meeting on Microbial Phosphate Solubilization*. Springer. 10.1007/978-1-4020-5765-6_1

Wilson, D. (1995). Endophyte - the evolution of term, a classification of its use and defin-ition. *Oikos*, *73*(2), 274–276. doi:10.2307/3545919

Yousaf, S., Andria, V., Reichenauer, T. G., Smalla, K., & Sessitsch, A. (2010). Phylogenetic and functional diversity of alkane degrading bacteria associate with Italian ryegrass (*Lolium multiflorum*) and Birds foot trefoil (*Lotus corniculatus*) in a petroleum oil-contaminated environment. *Journal of Hazardous Materials*, *184*(1-3), 523–532. doi:10.1016/j.jhazmat.2010.08.067 PMID:20851515

Zhang, Y., He, L., Chen, Z., Zhang, W., Wang, Q., Qian, M., & Sheng, X. (2011). Characterization of lead-resistant and ACC deaminase-producing endophytic bacteria and their potential in promoting lead accumulation of rape. *Journal of Hazardous Materials*, *186*(2-3), 1720–1725. doi:10.1016/j.jhazmat.2010.12.069 PMID:21227577

KEY TERMS AND DEFINITIONS

Abiotic Stress: Abiotic stress may be defined as the antagonistic impact of non-living components on the biotic components in a definite environment (e.g., drought, flooding, high and low temperature).

Bioremediation: Bioremediation is the process of using living organisms to remove, detoxify, breakdown or reduce the environmental contaminants from soil, water, and air.

Endophytes: Endophytes may be defined as microorganisms that reside or colonize the internal tissues of plants without causing any harmful effects.

Heavy Metals: Heavy metals are natural metallic components that have a high atomic weight and a density 5 times greater than water. They are generally toxic to organisms even as low level (e.g., lead, mercury, cadmium, etc.).

Nitrogen Fixation: It is biological process by which atmospheric nitrogen (N_2) is converted into nitrogen components (ammonia and nitrates) which are plant nutrients. This process is carried out by specific bacteria in the root nodules of legumes such as beans.

Phytoremediation: Also known as phytodegradation, it has been derived from ancient Greek φυτό (*phyto*), meaning "plant," and Latin *remedium*, meaning "restoring balance," refers to the process that use green plants and their microorganisms to either remove, breakdown, contain, or reduce toxic environmental contaminants.

Chapter 6
Health Effects of Air Pollution in Urban Environment

Banwari Dandotiya
Jiwaji University Gwalior, India

ABSTRACT

In the last few decades, urban development and growing industrialization has caused air pollution to become a major issue in urban areas of developing and developed countries. Urban area is more susceptible compare to other because of higher exposure time of urban residents, due to unbound interference of air pollutants in indoor environment. Exposure to air pollutants has been associated with increased mortality and hospital admissions due to respiratory and cardiovascular diseases. This chapter focused on sources and health effects of air pollution in urban areas in India. Most of the urban areas of India are suffering from higher concentrations of air pollutants including gaseous and particulates.

INTRODUCTION

In the last few decades, urban development and growing industrialization has caused air pollution to become a major issue in urban areas of developing and developed countries. Mostly urban areas suffer from serious air quality problems due to increasing population levels, combined with changes in land use and increases in vehicular traffic. Inhalation of air pollutants deeply connected with increasing hospital admissions and mortality due to respiratory, cardiovascular and other lungs related diseases. The World Health Organization has identified ambient particulate pollution of urban areas as a public health menace, based on estimates of air pollution related deaths and disability. Global efforts to understand and mitigate the health effects of particulate air pollution have a rich and interesting history. United States reported apparent health effects at unexpectedly low concentrations of ambient particulate pollutants. Daily changes in air pollutants concentration proportional to daily hospital admissions in several cities (Schwartz et al., 1990; Fairley, 1990; Schwartz 1991; Pop et al., 1992; Schwartz and Dockery, 1992; Dockery et al., 1992; Schwartz 1993) and long-term exposure was associated with respiratory illness in children and cardiopulmonary mortality in adults (Dockery et al., 1993; Pop et al., 1995). Air pollution was associated with a wide range of health end points, including respiratory hospitalizations (Pop, 1989,

DOI: 10.4018/978-1-5225-7387-6.ch006

1991) lung function and respiratory symptoms (Pop, 1992, 1991, 1993) school absences (Ransom and Pop, 1992) and mortality (Pop, 1992; Archer, 1990).

Global CO_2 emissions were 41% increase against the 1990 level (Olivier et. a., 2011). Urban areas are critically important to the control of greenhouse gas emissions because more than half of the world's population live in urban areas. Developing, implementing and monitoring effective plans and policies that can be linked to reductions in GHG emissions have proven difficult (Albers & Deppisch, 2013; Andrews, 2008; Ewing et. al., 2008; Foley et al., 2005; Lundqvist, 2016; Wheeler, 2008). Residents of large urban areas are exposed to air pollutants concentration that often exceeds the established air quality standards. Diesel vehicles have been identified in several studies as a major contributor to ambient concentrations of air pollutants. It remains unclear to which levels the air pollutant emissions of diesel cars have to decrease to effectively mitigate urban air pollution across urban areas. Residents of many larger cities are still exposed to gaseous and particulate air pollutants concentrations that often exceed the established air quality standards (EEA, 2015). Most cases were found in urban areas, mainly caused by traffic-related pollutant emissions originating from diesel vehicles. Policy makers expected the air pollutants emissions of diesel cars to decrease considerably when introducing more stringent Euro 5 and Euro 6 emission limits (EC, 2012, 2008a, b).

POLLUTANTS IN URBAN ENVIRONMENT

Approximate 3000 different anthropogenic air polluting compounds have been identified, in which most are organic (WHO, 2008). Vehicular emission, a notorious air pollution source in urban environment contains about 500 different compounds, such anthropogenic compound emissions from various domestic, industrial and automobile sources have increased many times and eventually have led to many global environmental problems (Mohanraj and Azeez, 2005). The most hazardous of the air pollutants are products of fossil fuels, primarily the byproduct of combustion fuel engines in most vehicles. Impacts of these substances in the atmosphere can often be aggravated by water vapor, natural dust and sunlight. Which activate chemical reactions and produce a secondary set of hazardous pollutants, including ozone, secondary organic aerosols, sulfates and nitrates. Most urban air pollution comes from road traffic, and is comprised of a mixture of airborne particulate matter (PM), oxides of sulfur (SOx), oxides of nitrogen (NOx), carbon monoxide (CO), carbon dioxide (CO2), volatile organic compounds (VOCs), polycyclic aromatic hydrocarbons (PAHs), and ozone (Thurston, 2008).

Carbon monoxide is produced by incomplete combustion that occurs in an internal combustion engine, or by a furnace, or any other device that uses fuel. Almost all countries in the world have regulations on the maximum allowable Carbon monoxide concentrations in the exhaust of combustion devices. CO's most evident impact is to reduce blood's ability to circulate oxygen in living body. High concentrations of Carbon monoxide can be fatal, but even at its low concentration in the ambient air it causes headaches and dizziness. Nitrogen oxides (NOx), including NO and NO_2, are also produced by internal combustion engines. The most obvious and common producers of NOx are vehicles that run on diesel and on gasoline. Sulfur oxides (SOx) also produced by combustion are the outcome of sulfuric substances in the fuel; SOx is mostly produced by diesel engines since standard gasoline normally contains lower levels of sulfur. Petrochemical plants are another source for emitting SOx. Another major pollutant at lower elevations is ozone (O_3). In the stratosphere ozone acts as a shield, preventing harmful ultraviolet light reaching the earth. However at or near ground level it can be quite harmful. As a result of complex chemical reactions between NOx and unburned hydrocarbons once exposed to sunlight, ozone is produced and reaches the highest level during summer in the middle of the day. Of all major air pollutants, particulate matter or particulates are the most complex and by far the most hazardous. Anthropogenic sources such as fuel combustion produce much smaller particles and are more harmful. Furthermore, the composition of materials in natural particles is generally inert and non-toxic. Anthropogenic combustion sources such as vehicles also produce particles that contain black carbon, organic matters, metal oxides, sulfate and nitrates, which are all highly toxic substances harmful to human health.

INDOOR

The relationship between human health and indoor air quality is an area of current concern and study. Concern needs to focus not only on ambient air contaminant levels, but also on indoor air quality due to the amount of time people spend in these environments (Maesano et. al., 2013; Mendes et. al., 2013). Considerable attention in urban areas has also been given to health risk evaluations for indoor pollutant sources, such as environmental tobacco smoke. Assessments of human exposure to indoor air pollution are performed regularly using indoor air quality models to simulate airborne pollutants concentrations (Lioy et. al., 1991). Despite the widespread use of the well-mixed assumption, indoor air concentrations are not homogeneous. Indoor air pollutant concentrations are affected by many factors such as inefficient air filtration, worker productivity demands, out-gassed products from building and consumer materials, and increased percentage of time spent indoors by residents, and energy efficient building construction and operation (Meckler, 1991). Recently reported experimental research also shows that under certain conditions, including convection flow (Baughman et. al., 1994) and horizontal plug flow (Yost et. al., 1994), air pollutant concentrations in a room are not uniform. There are many types of pollutant emission sources that are used in the immediate vicinity of humans. Examples include cleaning products, personal hygiene products, and art supplies. This type of materials are used by people in a way that results in emissions being released in very close to the resident's breathing area.

HEALTH EFFECTS

Exposure to air pollutants is in relation to both acute and chronic effects on human health. The short-term endpoints of air pollution on human health were amply demonstrated in the twentieth century, when severe air pollution in Europe and in the USA caused mortality and morbidity in thousands of people (Fattore et. al., 2011; Nemery et. al,. 2001). Air pollutant undergoes a complex sequence of events from initial emission to consequent health effect in an individual. Numerous studies worldwide have confirmed that both short and long term exposure to air pollutants are associated with increases in mortality and a number of health effects (Venners et. al., 2003; Kan et. al., 2008; Dockery, 2009). International Agency for Research on Cancer classified outdoor air pollution as carcinogenic to humans (WHO, 2013). Particulate matter is a major component of outdoor ambient pollutant and carcinogenic to humans. The Ambient Air Pollutants are several and have different characteristics and they can be gaseous and solid, carbon monoxide (CO), nitrose dioxide (NO_2), sulfur dioxide (SO_2), ozone (O_3), polycyclic aromatic hydrocarbons (PAHs), and PM in its various fractions PM10 diameter $<10\ \mu m$, PM2.5 diameter $<2.5\ \mu m$, and Ultra Fine Particles diameter $<0.1\ \mu m$ (Polichetti et. al., 2009) are mostly taken into account.

Urban air pollution is emerging as a risk factor, e.g., cancer and mortality due to cardiovascular diseases, particularly in developing countries including India where air pollutant concentrations are elevated (Daryanoosh et. al., 2016; Geravandi et. al., 2016; Ghozikali et. al., 2016; Neisi et. al., 2016; Nourmoradi et. al., 2015). Among different atmospheric pollutants, particulate matter is a pollutant with the most undesired harmful impacts on the human health (Dominici et. al., 2006; Goudarzi et. al., 2015b; Schwartz et. al., 1993; Weuve et. al., 2012). Particulate matter with an aerodynamic diameter equal or less than 10 µm (PM10) has the most adverse effects on the human health (Wang et. al., 2009). Exposure to ambient particulate matter can cause several health effects such as lung irritation, coughing, immune system reactions, damaged cells, asthma, and hospital admissions due to cardiovascular and respiratory diseases (Jeong, 2013). According to the results of epidemiological studies over the world, approximately 5–7% of the lung cancers among ex-smokers and non-smokers can be associated with exposure to high levels of air pollutants containing NO_2 or vicinity to roads with heavy-traffic (Defense, 1996; Fattore et. al., 2011; Ghiyathoddin, 2006). Epidemiological studies showed that the exposure to O_3 levels ranging between 160 and 360 $\mu g/m^3$ for duration of 1–8 h may cause reduction of different parameters of lung function (Zarandi et. al., 2015). Tropospheric O_3 is known as the second critical pollutant of the atmosphere following the particulate matter that show increasing concentrations in the air, associated to human health impacts (Dehghani et. al., 2014; Goudarzi et. al., 2015a, Sicard et. al., 2011; Yari et. al., 2016). There is an association between exposure to gaseous pollutants and pulmonary function (Andersen et. al., 2011; Sicard et. al., 2010). Around 0.7 million peoples per year die worldwide due to exposure to tropospheric O_3 (Anenberg et. al., 2010). Ozone is a secondary gaseous pollutant and greenhouse gas that has an important role in global warming, climate change, chemical composition of the air and deteriorate air quality in ambient atmosphere (Mohnen et. al., 1993; Sicard et. al., 2016). Ozone destroys the photosynthesis system of the plants and decreases the yields of the vegetation (Dingenen et. al., 2009). Nitrogen oxide is the main precursor of nitrogen compounds in the atmospheric reactions (Katsouyanni et. al., 1995; Liang et. al., 2009). In atmospheric reactions, NO can quickly be oxidized by Ozone and then converted to NO_2 by a photochemical reaction (Defense, 1996; Fattore et. al., 2011; Ghiyathoddin, 2006). Air pollution includes gaseous pollutants and particles, but particles are of supreme substance with respect to health effects (Sicard et. al., 2010; Jakubiak et. al., 2015; Khaefi et. al., 2017).

Among all air pollutants, particulate matter with an aerodynamic diameter of 10 µm (PM10) is particularly important in case of human health because PM10 penetrate into the respiratory tract (Schwartz et. al., 1993; Wang et. al., 2009; Weuve et. al., 2012). The main emission sources of PM10 are anthropogenic, e.g., combustion, road traffic, power plant activities, and industrial processes or natural, e.g., desert dust and sea salt (Gharehchahi et. al., 2013). Exposure to ambient PM10 can cause several hazardous health effects such as asthma exacerbation, chronic bronchitis, lung irritation, cancer, mortality and increased hospital admissions resulting from cardiovascular and respiratory illness (Sicard et. al., 2011; Jeong, 2013; Neisi et. al., 2016). In general, the public are most concerned with the elevated levels of particulate matter and ambient gaseous air pollutants both can cause haze conditions and related health problems. As it is well known, photochemical reaction in atmosphere for converting SOx and NOx gases into aerosol sulfate and nitrate is an important process for secondary PM2.5 formation. The contribution of sulfate and nitrate in PM2.5 depends on locality, season, solar radiation, and particularly episode event (Wang et. al., 2016). In addition to the formation of secondary PM2.5 and O_3, the environmental outcome caused by SO_2 and NO_2 are well known including acidification of lakes, affecting cloud formation, acid rain, reduced visibility, photochemical smog, and, more importantly, human health and ecosystem deterioration (Zhang et. al., 2016).

The previous studies estimated the association between PM2.5 (particulate matter with aerodynamic diameter less than or equal to 2.5 μm) exposure during pregnancy and preterm birth, only considered the hazardous effects of high levels of air pollutant exposure, and low levels of pollutant exposure might also be affect pregnancy outcome and preterm effects (Chenchen et. al., 2017). Air pollution is a high priority in global burden of disease assessment, and World Health Organization has estimated that air pollution is responsible for 6.7% of all deaths and 7.6% of disability adjusted life years (4.5 and 3.1% for household and outdoor air pollution) globally (Lim et. al., 2012; WHO, 2016a; WHO, 2014a, b) and fourth highest ranking risk factor for premature mortality in the world (IHME, 2016). According to a World Health Organization (WHO) assessment of the burden of disease due to air pollution, more than two million premature deaths each year can be attributed to the effects of urban ambient air pollution which includes outdoor air pollution and indoor air pollution (WHO, 2005). Outdoor air pollution was associated with roughly 300000 premature deaths per annum in China (Cohen et. al., 2005). The acute and chronic health effects from short and long-term exposures to particulate matter (PM) are well established in the previous literature (Zanobetti et. al., 2008; Pope et. al., 1995, 2004, 2011; Anenberg et. al., 2011; Cesaroni et. al., 2014; Beelen et. al., 2014; Hamra et. al., 2014; Korek et. al., 2015; Brauer et. al., 2012, 2015; Brunekreef and Holgate, 2002). Epidemiological and cohort researches showed that these health impacts rely on long-term ambient (both household and outdoor) exposure of PM and associated risk factor vary country to country (Pope et. al., 2004; Pope and Dockery, 2006; HEI, 2011; CPCB, 2012; Dholakia et. al., 2014). Previous research data have been also reported fluctuating conditions as regards the role of pregnancy time, in particular considering the exposure during the first, mid, and last trimester or the whole period (Bell et. al., 2007; Bobak, 2000; Wang et. al., 1997).

Previous research suggests an association between maternal exposure to air pollution and adverse health effects in birth outcomes, including preterm birth, low birth weight, intrauterine growth restriction, and post-neonatal infant mortality (Glinianaia et. al., 2004; Lacasana et. al., 2005; Maisonet et. al., 2004; Sapkota et. al., 2010; Woodruff et. al., 2008; Wu et. al., 2009). In addition to more acute infant health effects of urban ambient air pollution associated with reduced birth weight including restriction of fetal growth, respiratory distress, variable heart rate, cerebral palsy and deafness has been associated with delays in motor and social development in children (Jedrychowski et. al., 2004; Miranda et. al., 2009). For instance, higher levels of carbon monoxide (CO) and respirable particulate matter were associated with reduced birth weight in some studies (Bell et. al., 2007b; Dejmek et. al., 1999; Gouveia et. al., 2004; Ha et. al., 2001; Maisonet et. al., 2001; Mannes et. al., 2005; Ritz et. al., 2000; Salam et. al., 2005; Wilhelm and Ritz, 2003; Wilhelm and Ritz, 2005; Xu et. al., 2011; Yang et. al., 2003). Ambient air pollution is of specific concern, and there is growing evidences linking to ambient levels of exposure with negative effects on reproductive health (Sram, 1999; Slama et. al., 2008). Human studies have linked maternal ambient air pollutants exposure with reduced fetal growth and increased frequency of visible birth defects (Wang et. al., 1997; Gilboa et. al., 2005) and male exposure with reduced semen quality and sperm DNA integrity (Jurewicz et. al., 2009). Ambient air pollution at any point during adulthood were caused male mediated effects and the genetic integrity of sperms that are a common cause for couples who lives in polluted areas (Oehninger, 2001).

EXPOSURE AND SURROUNDINGS

Exposure time plays very important role in occurrence of health effects, if the duration of exposure is higher than chances of metabolic disturbance is also higher in exposed person or living beings (Sharma et. al., 2017). Previous studies have shown that long term exposure air pollution may induce toxic effects more frequently on female fetuses compared to male fetuses (Ghosh et. al., 2007). Urban areas surrounded by one or more industrial areas are susceptible to higher concentrations due to meteorological conditions of that region. Wind direction plays a crucial role in elevated concentration of air pollutants in urban areas that have nearby industrial area or industrial corridor. In winter season air pollutants disperse very slow because of stagnant air masses in this case due to wind direction and wind speed pollutants enter in the urban environment remain concentrated causing higher concentration in that area. Studies have measured spatial variations in air pollutants concentrations (Isakov et. al., 2007) as well as quantified risks from measured outdoor concentrations of air pollutants and personal exposure in specific areas (McCarthy et. al., 2007), but there is still a great deal to be learned about spatial patterns in air pollutants due to the influences of local sources.

ROLE OF VEGETATION

Urban vegetation is an important measure to reduce ambient air pollution levels in urban areas (Chen and Jim, 2008), evidences indicating that urban trees, plants and vegetation can offer a range of ecosystem management services for urban residents that includes reduction of air pollution (Brack, 2002; Zheng et. al., 2013). Urban vegetation, an unconventional solution to the problem, has been shown to remove particulate matter from the air (Nowak et. al., 2013). Urban vegetation reduces ambient particulate pollution both directly and indirectly. Leaf size and structure plays a key role in direct reduction (Saebo et. al., 2013). The cooling effect reduces the need for energy using fans and air conditioners, which further lowers emissions from power plants. Also, the rates of photo chemical reactions in the urban atmosphere are slowed by the lowered air temperature resulting in decreased production of secondary air pollutants (Nowak et. al., 2000). Vegetation lowers the ambient air temperature by changing the albedos of urban surfaces and evapotranspiration cooling. The direct removal of air pollutant by trees is affected by environmental factors as well as the biophysical characteristics of trees (Zhao et. al., 2013). Environmental factors such as weather conditions, urban morphology and concentrations of PM have a significant impact on the quantity of air pollutant intercepted by trees (Reinap et. al., 2009). Positive effects of urban vegetation and plants on ambient air quality through filtration of polluted air are often mentioned in several studies. Urban vegetation is also away to abate the effects of climate change, e.g. rising sea level and global warming; many cities are increasingly including urban vegetation in their plans (Andersson et. al., 2015). Suspended and respirable particles and gaseous air pollutant molecules can be deposited when they pass close to a leaf surface of plants. Trees and other vegetation have a large leaf surface area, increase deposition of air pollutant particles and molecules compared with the smooth, manufactured surfaces present in urban areas. For example, 10 to 30 times faster deposition was observed for suspended and respirable particles on synthetic grass compared with glass and cement surfaces (Roupsard et. al., 2013). Vegetational deposition is one dimensional vertical deposition on homogeneous leaf surfaces of vegetation in the form of a forest or field.

Dense vegetation affects both phenomenon in ambient air deposition and dispersion. In case of deposition, the vegetation area is either described as leaf area/ground area, dimensionless or as Leaf Area Density. In dispersion, drag force or pressure drop, porosity is important. Green vegetation in the urban environment has been considered as a potential planning solution for improving ambient air quality and in enhancing the sustainability of urban areas for growing urban populations (Salmond et. al., 2016). Urban vegetation can reduce air pollutants through a dry deposition process and microclimate effects. The high surface area and roughness provided by the branches, twigs, and foliage make vegetation an effective sink for air pollutants (Beckett et. al., 1998). Vegetation also has an indirect effect on pollution reduction by modifying microclimates. Plants lower the indoor air temperature through shading, thus reducing the use of electricity for air conditioning (Heisler, 1986).

Several previous studies have attempted to estimate the economic valued benefits of improving urban ambient air quality, although the effect of vegetation on urban air quality is not yet properly understood (Escobedo et. al., 2011). These green solutions include roadside trees, vegetation barriers on road dividers (including hedges), green (or living) vegetated walls, and green (or living) vegetated roofs. These types of trees and vegetation act as porous bodies which influence local ambient air dispersion patterns (Fantozzi et. al., 2015). The air pollution reduction abilities of various types of vegetated infrastructure, either individually or in combination, in different urban environments have a key role in urban areas (Gallagher et. al., 2015). The majority of studies have focused on pollutants such as the Particulate Matter and gaseous pollutants (Chen et. al., 2016, Beevers et. al., 2012, Bigazzi and Figliozzi, 2015), that have implications for the adverse health effects. Particle deposition on plant surfaces removes pollutants from the urban atmosphere, thus reducing their concentration.

CONCLUSION

Air pollution in the urban environment is a major threat to human health. Numerous studies worldwide have confirmed that both short and long term exposure to air pollutants are associated with increases in mortality and a number of health effects. Urban residents are at higher risk compare to others because urban ambient air concentrations of pollutants are high in both indoor and outdoor. As the global population is becoming more concentrated in urbanized areas, new ideas and approaches are needed to help maintain clean air that is safe for everyone to breathe. Tree planting has been proposed as an unconventional measure to control air pollution. The effect of vegetation on urban air pollution depends on leaf structure, surface area, vegetation type and level of air pollution in the area. This chapter demonstrates that urban vegetation improve air quality and by extension public health safety and thereby a perceived improvement in quality of life in urban settings. Clearly, trees had the largest impact on pollutant removal, but shrubs and grass made important contributions to air quality. The chapter explains the degree to which green roofs and green walls can be used in populated urban areas to supplement existing vegetation and improve air quality when installed in sufficient quantities.

REFERENCES

Albers, M., & Deppisch, S. (2013). Resilience in the light of climate change: Useful approach or empty phrase for spatial planning? *European Planning Studies*, *21*(10), 1598–1610. doi:10.1080/09654313.2012.722961

Andersen, Z., Hvidberg, M., Jensen, S., Ketzel, M., Loft, S., Sørensen, M., ... Raaschou-Nielsen, O. (2011). Chronic obstructive pulmonary disease and long-term exposure to traffic related air pollution: A cohort study. *American Journal of Respiratory and Critical Care Medicine*, *183*(4), 455–461. doi:10.1164/rccm.201006-0937OC PMID:20870755

Andersson-Skold, Y., Thorsson, S., Rayner, D., Lindberg, F., Janhall, S., Jonsson, A., ... Granberg, M. (2015). An integrated method for assessing climate related risks and adaptation alternatives in urban areas. *Climate Risk Management*, *7*, 31–50. doi:10.1016/j.crm.2015.01.003

Andrews, C. (2008). Green house gas emissions along the rural-urban gradient. *Journal of Environmental Planning and Management*, *51*(6), 847–870. doi:10.1080/09640560802423780

Anenberg, S., Horowitz, L., Tong, D., & West, J. (2010). An estimate of the global burden of anthropogenic ozone and fine particulate matter on premature human mortality using atmospheric modeling. *Environmental Health Perspectives*, *118*(9), 1189–1195. doi:10.1289/ehp.0901220 PMID:20382579

Anenberg, S. C., Talgo, K., Arunachalam, S., Dolwick, P., Jang, C., & West, J. J. (2011). Impacts of global, regional and sectoral black carbon emission reductions on surface air quality and human mortality. *Atmospheric Chemistry and Physics*, *11*(14), 7253–7267. doi:10.5194/acp-11-7253-2011

Annesi-Maesano, I., Baiz, N., Banerjee, S., Rudnai, P., & Rive, S. (2013). Indoor air quality and sources in schools and related health effects. *J. Toxicol. Environ. Health B*, *16*(8), 491–550. doi:10.1080/10937404.2013.853609 PMID:24298914

Archer, V. E. (1990). Air Pollution and Fatal Lung Disease in Three Utah Counties. *Archives of Environmental Health*, *45*(6), 325–334. doi:10.1080/00039896.1990.10118751 PMID:2270951

Arden Pope, C. III, & Dockery, D. W. (2006). Health Effects of Fine Particulate Air Pollution: Lines that Connect. *Journal of the Air & Waste Management Association*, *56*(6), 709–742. doi:10.1080/10473289.2006.10464485 PMID:16805397

Baughman, A. V., Gadgil, A. J., & Nazaroff, W. W. (1994). Mixing of a point source pollutant by natural convection flow within a room. *Indoor Air*, *4*(2), 114–122. doi:10.1111/j.1600-0668.1994.t01-2-00006.x

Beckett, K. P., Freer-Smith, P., & Taylor, G. (1998). Urban woodlands: Their role in reducing the effects of particulate pollution. *Environmental Pollution*, *99*(3), 347–360. doi:10.1016/S0269-7491(98)00016-5 PMID:15093299

Beelen, R., Raaschou-Nielsen, O., Stafoggia, M., Andersen, Z. J., Weinmayr, G., Hoffmann, B., ... Hoek, G. (2014). Effects of long-term exposure to air pollution on natural-cause mortality: An analysis of 22 European cohorts within the multicentre ESCAPE project. *Lancet (London, England)*, *383*(9919), 785–795. doi:10.1016/S0140-6736(13)62158-3 PMID:24332274

Beevers, S. D., Westmoreland, E., de Jong, M. C., Williams, M. L., & Carslaw, D. C. (2012). Trends in NOx and NO2 emissions from road traffic in Great Britain. *Atmospheric Environment*, *54*, 107–116. doi:10.1016/j.atmosenv.2012.02.028

Bell, M. L., Ebisu, K., & Belanger, K. (2007). Ambient air pollution and low birth weight in Connecticut and Massachusetts. *Environmental Health Perspectives*, *115*(7), 1118–1124. doi:10.1289/ehp.9759 PMID:17637932

Bell, M. L., Ebisu, K., & Belanger, K. (2007b). Ambient air pollution and low birth weight in Connecticut and Massachusetts. *Environmental Health Perspectives*, *7*(7), 1118–1124. doi:10.1289/ehp.9759 PMID:17637932

Bigazzi, A. Y., & Figliozzi, M. A. (2015). Roadway determinants of bicyclist exposure to volatile organic compounds and carbon monoxide. *Transportation Research Part D, Transport and Environment*, *41*, 13–23. doi:10.1016/j.trd.2015.09.008

Bobak, M. (2000). Outdoor air pollution, low birth weight, and prematurity. *Environmental Health Perspectives*, *108*(2), 173–176. doi:10.1289/ehp.00108173 PMID:10656859

Brack, C. L. (2002). Pollution mitigation and carbon sequestration by an urban forest. *Environ. Pollut*, *116*(Suppl. 1), 195–200. doi:10.1016/S0269-7491(01)00251-2 PMID:11833907

Brauer, M., Amann, M., Burnett, R. T., Cohen, A., Dentener, F., Ezzati, M., ... Thurston, G. D. (2012). Exposure assessment for estimation of the global burden of disease attributable to outdoor air pollution. *Environmental Science & Technology*, *46*(2), 652–660. doi:10.1021/es2025752 PMID:22148428

Brunekreef, B., & Holgate, S. T. (2002). Air pollution and health. *Lancet (London, England)*, *360*(9341), 1233–1242. doi:10.1016/S0140-6736(02)11274-8 PMID:12401268

Cesaroni, G., Forastiere, F., Stafoggia, M., Andersen, Z. J., Badaloni, C., Beelen, R., & .... (2014). Long term exposure to ambient air pollution and incidence of acute coronary events: Prospective cohort study and meta-analysis in 11 European cohorts from the ESCAPE project. *BMJ (Clinical Research Ed.)*, *348*, f7412. doi:10.1136/bmj.f7412 PMID:24452269

Chen, L., Liu, C., Zou, R., Yang, M., & Zhang, Z. (2016). Experimental examination of effectiveness of vegetation as bio-filter of particulate matters in the urban environment. *Environ. Pollut*, *208*, 198–208. doi:10.1016/j.envpol.2015.09.006 PMID:26385643

Chen, W., & Jim, C. Y. (2008). Assessment and valuation of the ecosystem services provided by urban forests. In M. Carreiro, Y. C. Song, & J. Wu (Eds.), *Ecology, Planning, and Management of Urban Forests* (pp. 53–83). New York: Springer. doi:10.1007/978-0-387-71425-7_5

Chenchen, L., Jiantao, S., Yuewei, L., Hui, L., Minsheng, W., Chunhong, W., & Tingming, S. (2017). Different exposure levels of fine particulate matter and preterm birth: A meta-analysis based on cohort studies. *Environmental Science and Pollution Research International*, *24*(22), 17976–17984. doi:10.100711356-017-9363-0 PMID:28616740

Cohen, A. J., Anderson, H. R., Ostro, B., Pandey, K. D., Krzyzanowski, M., Kunzli, N., ... Smith, K. (2005). The global burden of disease due to outdoor air pollution. *Journal of Toxicology and Environmental Health, Part A: Current Issues*, *68*(13-14), 1301–1307. doi:10.1080/15287390590936166 PMID:16024504

CPCB. (2012). *Epidemiological study on effect of air pollution on human health (adults) in Delhi*. Central pollution control board, environmental health management series: EHMS/01/2012.

Daryanoosh, S. M., Goudarzi, G., Khaniabadi, Y. O., Armin, H., Bassiri, H., & Khaniabadi, F. O. (2016). Effect of exposure to PM10 on cardiovascular diseases hospitalizations in Ahvaz, Khorramabad and Ilam, Iran during 2014. *Iranian Journal of Health, Safety and Environment*, *3*, 428–433.

Defense, H. (1996). Health effects of outdoor air pollution. *American Journal of Respiratory and Critical Care Medicine*, 477–498. PMID:8564086

Dehghani, M., Anushiravani, A., Hashemi, H., & Shamsedini, N. (2014). Survey on air pollution and cardiopulmonary mortality in shiraz from 2011 to 2012: An analytical-descriptive study. *International Journal of Preventive Medicine*, *5*, 734–740. PMID:25013693

Dejmek, J., Selevan, S. G., Benes, I., Solanský, I., & Srám, R. J. (1999). Fetal growth and maternal exposure to particulate matter during pregnancy. *Environmental Health Perspectives*, *107*(6), 475–480. doi:10.1289/ehp.99107475 PMID:10339448

Dholakia, H. H., Bhadra, D., & Garg, A. (2014). Short term association between ambient air pollution and mortality and modification by temperature in five Indian cities. *Atmospheric Environment*, *99*, 168–174. doi:10.1016/j.atmosenv.2014.09.071

Dockery, D. W. (2009). Health effects of particulate air pollution. *Annals of Epidemiology*, *19*(4), 257–263. doi:10.1016/j.annepidem.2009.01.018 PMID:19344865

Dockery, D. W., Pope, C. A. III, Xu, X., Spengler, J. D., Ware, J. H., Fay, M. E., ... Speizer, F. A. (1933). An Association between Air Pollution and Mortality in Six U.S. Cities. *The New England Journal of Medicine*, *329*(24), 1753–1759. doi:10.1056/NEJM199312093292401 PMID:8179653

Dockery, D. W., Schwartz, J., & Spengler, J. D. (1992). Air Pollution and Daily Mortality: Associations with Particulates and Acid Aerosols. *Environmental Research*, *59*(2), 362–373. doi:10.1016/S0013-9351(05)80042-8 PMID:1464289

Dockery, D. W., Speizer, F. E., Stram, D. O., Ware, J. H., Spengler, J. D., & Ferris, B. G. Jr. (1989). Effects of Inhalable Particles on Respiratory Health of Children. *The American Review of Respiratory Disease*, *139*(3), 587–594. doi:10.1164/ajrccm/139.3.587 PMID:2923355

Dockery, D. W., & Trichopoulos, D. (1997). Risk of Lung Cancer from Environmental Exposures to Tobacco Smoke. *Cancer Causes & Control*, *8*(3), 333–345. doi:10.1023/A:1018401120233 PMID:9498897

Dominici, F., Peng, R. D., Bell, M. L., Pham, L., McDermott, A., Zeger, S. L., & Samet, J. M. (2006). Fine particulate air pollution and hospital admission for cardiovascular and respiratory disease. *Journal of the American Medical Association*, *295*(10), 1127–1134. doi:10.1001/jama.295.10.1127 PMID:16522832

EEA. (2015). *Air Quality in Europe - 2015 Report* (EEA Report No 5/2015). European Environmental Agency. doi:10.2800/62459

Escobedo, F. J., Kroeger, T., & Wagner, J. E. (2011). Urban forests and pollution mitigation: Analyzing ecosystem services and disservices. *Environ. Pollut., 159*(8-9), 2078–2087. doi:10.1016/j.envpol.2011.01.010 PMID:21316130

European Commission. (2008a). *Commission Regulation 692/2008 of 18 July 2008 Implementing and Amending Regulation (EC) No 715/2007 of the European Parliament and of the Council on Type-approval of Motor Vehicles with Respect to Emissions from Light Passenger and Commercial Vehicles*. Author.

European Commission. (2012). *Commission Regulation (EU) No 459/2012 of 29 May 2012 Amending Regulation (EC) No 715/2007 of the European Parliament and of the Council and Commission Regulation (EC) No 692/2008 as Regards Emissions from Light Passenger and Commercial Vehicles (Euro 6)*. Author.

European Commission. (2008b). *Directive 2008/50/EC. Commission Regulation 692/2008 of 18 July 2008 Implementing and Amending Regulation (EC) No 715/2007 of the European Parliament and of the Council on Type-approval of Motor Vehicles with Respect to Emissions from Light Passenger and Commercial Vehicles*. Author.

Ewing, R., Bartholomew, K., Winkelman, S., Walters, J., & Chen, D. (2008). *Growing cooler: The evidence on urban development and climate change*. Washington, DC: Urban Land Institute.

Fairley, D. (1990). The Relationship of Daily Mortality to Suspended Particulates in Santa Clara County, 1980–1986. *Environmental Health Perspectives, 89*, 159–168. doi:10.1289/ehp.9089159 PMID:2088743

Fantozzi, F., Monaci, F., Blanusa, T., & Bargagli, R. (2015). Spatio-temporal variations of ozone and nitrogen dioxide concentrations under urban trees and in a nearby open area. *Urban Clim, 12*, 119–127. doi:10.1016/j.uclim.2015.02.001

Fattore, E., Paiano, V., Borgini, A., Tittarelli, A., Bertoldi, M., Crosignani, P., & Fanelli, R. (2011). Human health risk in relation to air quality in two municipalities in an industrialized area of Northern Italy. *Environmental Research, 111*(8), 1321–1327. doi:10.1016/j.envres.2011.06.012 PMID:21764052

Foley, J., DeFries, R., Asner, G., Barford, C., Bonan, G., Carpenter, S., & Snyder, P. (2005). Global consequences of land use. *Science, 309*(5734), 570–574. doi:10.1126cience.1111772 PMID:16040698

Gallagher, J., Baldauf, R., Fuller, C. H., Kumar, P., Gill, L. W., & McNabola, A. (2015). Passive methods for improving air quality in the built environment: A review of porous and solid barriers. *Atmospheric Environment, 120*, 61–70. doi:10.1016/j.atmosenv.2015.08.075

Geravandi, S., Goudarzi, G., Soltani, F., Salmanzadeh, S., Ghomeishi, A., Zalaghi, E., ... Mohammadi, M. (2016). The cardiovascular and respiratory deaths attributed to sulfur dioxide in Kermanshah. *Majallah-i Danishgah-i Ulum-i Pizishki-i Kirmanshah, 19*, 319–326.

Gharehchahi, E., Mahvi, A. H., Amini, H., Nabizadeh, R., Akhlaghi, A. A., Shamsipour, M., & Yunesian, M. (2013). Health impact assessment of air pollution in Shiraz, Iran: A two- part study. *Iranian Journal of Environmental Health Sciences & Engineering, 11*, 2–8. PMID:24499576

Ghiyathoddin, M. (2006). *Air pollution—sources, effects and control*. Tehran: Tehran University Press.

Ghosh, R., Rankin, J., Pless-Mulloli, T., & Glinianaia, S. (2007). Does the effect of air pollution on pregnancy outcomes differ by gender? A systematic review. *Environmental Research, 105*(3), 400–408. doi:10.1016/j.envres.2007.03.009 PMID:17493608

Ghozikali, M. G., Heibati, B., Naddafi, K., Kloog, I., Conti, G. O., Polos, R., & Ferrante, M. (2016). Evaluation of chronic obstructive pulmonary disease (COPD) attributed to atmospheric O_3, NO_2, and SO_2 using Air Q model (2011–2012 year). *Environmental Research, 144*, 99–105. doi:10.1016/j.envres.2015.10.030 PMID:26599588

Gilboa, S. M., Mendola, P., Olshan, A. F., Langlois, P. H., Savitz, D. A., Loomis, D., ... Fixler, D. E. (2005). Relation between ambient air quality and selected birth defects, Seven County Study, Texas, 1997–2000. *American Journal of Epidemiology, 162*(3), 238–252. doi:10.1093/aje/kwi189 PMID:15987727

Glinianaia, S. V., Rankin, J., Bell, R., Pless-Mulloli, T., & Howel, D. (2004). Particulate air pollution and fetal health: A systematic review of the epidemiologic evidence. *Epidemiology (Cambridge, Mass.), 15*(1), 36–45. doi:10.1097/01.ede.0000101023.41844.ac PMID:14712145

Goudarzi, G., Geravandi, S., Foruozandeh, H., Babaei, A., Alavi, N., Niri, M., ... Mohammadi, M. (2015a). Cardiovascular and respiratory mortality attributed to ground-level ozone in Ahvaz, Iran. *Environmental Monitoring and Assessment, 187*(8), 1–9. doi:10.100710661-015-4674-4 PMID:26141926

Goudarzi, G., Geravandi, S., Foruozandeh, H., Babaei, A., Alavi, N., Niri, M., ... Mohammadi, M. (2015b). Total number of deaths and respiratory mortality attributed to particulate matter (PM10) in Ahvaz, Iran during 2009. International. *Journal of Environmental Health Engineering, 4*, 1–7.

Gouveia, N., Bremner, S. A., & Novaes, H. M. D. (2004). Association between ambient air pollution and birth weight in Sao Paulo, Brazil. *Journal of Epidemiology and Community Health, 58*(1), 11–17. doi:10.1136/jech.58.1.11 PMID:14684720

Ha, E. H., Hong, Y. C., Lee, B. E., Woo, B. H., Schwartz, J., & Christiani, D. C. (2001). Is air pollution a risk factor for low birth weight in Seoul? *Epidemiology (Cambridge, Mass.), 12*(6), 643–648. doi:10.1097/00001648-200111000-00011 PMID:11679791

Hamra, G. B., Guha, N., Cohen, A., Laden, F., Raaschou-Nielsen, O., Samet, J. M., ... Loomis, D. (2014). Outdoor particulate matter exposure and lung cancer: A systematic review and meta- analysis. *Environmental Health Perspectives, 122*(9), 906–911. doi:10.1289/ehp/1408092 PMID:24911630

HEI. (2011). Coordinated studies of short-term exposure to air pollution and daily mortality in two Indian cities. Public Health and Air Pollution in Asia (PAPA), research report 157. Health Effects Institute.

Heisler, G. M. (1986). Effects of individual trees on the solar radiation climate of small buildings. *Urban Ecology, 9*(3/4), 337–359. doi:10.1016/0304-4009(86)90008-2

IHME. (2016). *Poor air quality kills 5.5 million worldwide annually. Institute for health metrics and evaluation*. Available at: http://www.healthdata.org/news- release/poor-air-quality-kills-55-million-worldwide-annually

Isakov, V., Touma, J. S., & Khlystov, A. (2007). A Method of Assessing Air Toxics Concentrations in Urban Areas Using Mobile Platform Measurements. *Journal of the Air & Waste Management Association, 57*(11), 1286–1295. doi:10.3155/1047-3289.57.11.1286 PMID:18069452

Jakubiak-Lasocka, J., Lasocki, J., & Badyda, A. J. (2015). The influence of particulate matter on respiratorymorbidity andmortality in children and infants. *Advances in Experimental Medicine and Biology, 849*, 39–48. doi:10.1007/5584_2014_93 PMID:25381559

Jedrychowski, W., Bendkowska, I., Flak, E., Penar, A., Jacek, R., Kaim, I., ... Perera, F. P. (2004). Estimated risk for altered fetal growth resulting from exposure to fine particles during pregnancy: An epidemiologic prospective cohort study in Poland. *Environmental Health Perspectives, 112*(14), 1398–1402. doi:10.1289/ehp.7065 PMID:15471732

Jeong, S. (2013). The impact of air pollution on human health in Suwon City. *Asian J Atmos Environ, 7*(4), 227–233. doi:10.5572/ajae.2013.7.4.227

Jurewicz, J., Hanke, W., Radwan, M., & Bonde, P. J. (2009). Environmental factors and semen quality. *International Journal of Occupational Medicine and Environmental Health, 22*(4), 1–25. doi:10.2478/v10001-009-0036-1 PMID:20053623

Kan, H. D., London, S. J., Chen, G. H., Zhang, Y. H., Song, G. X., Zhao, N. Q., ... Chen, B. H. (2008). Season, sex, age, and education as modifiers of the effects of outdoor air pollution on daily mortality in Shanghai, China: The public health and air pollution in Asia (PAPA) study. *Environmental Health Perspectives, 116*(9), 1183–1188. doi:10.1289/ehp.10851 PMID:18795161

Katsouyanni, K., Zmirou, D., Spix, C., Sunyer, J., Schouten, J., Pönkä, A., & ... . (1995). Short-term effects of air pollution on health: a European approach using epidemiological time-series data—the APHEA project: background, objectives, design. *The European Respiratory Journal, 8*, 1030–1038. PMID:7589368

Khaefi, M., Hassani, G., Yari, A. R., Soltani, F., Dobaradaran, S., Moogahi, S., ... Khaniabadi, Y. O. (2017). Association of particulatematter impact on prevalence of chronic obstructive pulmonary disease in Ahvaz, Southwest Iran during 2009–2013. *Aerosol and Air Quality Research, 17*(1), 203–207. doi:10.4209/aaqr.2015.11.0628

Korek, M. J., Bellander, T. D., Lind, T., Bottai, M., Eneroth, K. M., Caracciolo, B., ... Penell, J. C. (2015). Traffic related air pollution exposure and incidence of stroke in four cohorts from Stockholm. *Journal of Exposure Science & Environmental Epidemiology, 25*(5), 517–523. doi:10.1038/jes.2015.22 PMID:25827311

Lacasaña, M., Esplugues, A., & Ballester, F. (2005). Exposure to ambient air pollution and prenatal and early childhood health effects. *European Journal of Epidemiology, 20*(2), 183–199. doi:10.100710654-004-3005-9 PMID:15792286

Liang, W., Wei, H., & Kuo, H. (2009). Association between daily mortality from respiratory and cardiovascular diseases and air pollution in Taiwan. *Environmental Research, 109*(1), 51–58. doi:10.1016/j.envres.2008.10.002 PMID:19027107

Lim, S. S., Vos, T., Flaxman, A. D., Danaei, G., Shibuya, K., Adair-Rohani, H., ... Ezzati, M. (2012). A comparative risk assessment of burden of disease and injury attributable to 67 risk factors and risk factor clusters in 21 regions, 1990–2010: A systematic analysis for the global burden of disease study 2010. *Lancet*, *380*(9859), 2224–2260. doi:10.1016/S0140-6736(12)61766-8 PMID:23245609

Lioy, P. J. (1991). *Human Exposure Assessment for Airborne Pollutants -Advances and Opportunities*. Washington, DC: National Academy of Sciences.

Lundqvist, L. (2016). Planning for climate change adaptation in a multi-level context: The Gothenburg metropolitan area. *European Planning Studies*, *24*(1), 1–20. doi:10.1080/09654313.2015.1056774

Maisonet, M., Bush, T. J., Correa, A., & Jaakkola, J. J. (2001). Relation between ambient air pollution and low birth weight in the northeastern United States. *Environmental Health Perspectives*, *109*(suppl. 3), 351–356. PMID:11427384

Mannes, T., Jalaludin, B., Morgan, G., Lincoln, D., Sheppeard, V., & Corbett, S. (2005). Impact of ambient air pollution on birth weight in Sydney, Australia. *Occupational and Environmental Medicine*, *62*(8), 524–530. doi:10.1136/oem.2004.014282 PMID:16046604

Mayer, H. (1999). Air pollution in cities. *Atmospheric Environment*, *33*(24–25), 4029–4037. doi:10.1016/S1352-2310(99)00144-2

McCarthy, M. C., Hafner, H. R., Chinkin, L. R., & Charrier, J. G. (2007). Temporal Variability of Selected Air Toxics in the United States. *Atmospheric Environment*, *41*(34), 7180–7194. doi:10.1016/j.atmosenv.2007.05.037

Meckler, M. (1991, November). Indoor air quality from commissioning through building operation. *ASHRAE Journal*, 42–48.

Mendes, A., Pereira, C., Mendes, D., Aguiar, L., Neves, P., Silva, S., ... Teixeira, J. P. (2013). Indoor air quality and thermal comfort- Results of a pilot study in elderly care centers in Portugal. *Journal of Toxicology and Environmental Health. Part A.*, *76*(4-5), 333–343. doi:10.1080/15287394.2013.757213 PMID:23514075

Miranda, M. L., Maxson, P., & Edwards, S. (2009). Environmental contributions to disparities in pregnancy outcomes. *Epidemiologic Reviews*, *31*(1), 67–83. doi:10.1093/epirev/mxp011 PMID:19846592

Mohanraj, R., & Azeez, P. A. (2005). Urban development and particulate air pollution in Coimbatore city, India. *The International Journal of Environmental Studies*, *62*(1), 69–78. doi:10.1080/0020723042000261713

Mohnen, V., Goldstein, W., & Wang, W. C. (1993). Tropospheric ozone and climate change. *Air & Waste*, *43*(10), 1332–1334. doi:10.1080/1073161X.1993.10467207

Neisi, A., Goudarzi, G., Babaei, A., Vosoughi, M., Hashemzadeh, H., Naimabadi, A., ... Hashemzadeh, B. (2016). Study of heavy metal levels in indoor dust and their health risk assessment in children of Ahvaz City, Iran. *Toxin Reviews*, *35*(1-2), 16–23. doi:10.1080/15569543.2016.1181656

Nemery, B., Hoet, P., & Nemmar, A. (2001). The Meuse Valley fog of 1930: An air pollution disaster. *Lancet*, *357*(9257), 704–708. doi:10.1016/S0140-6736(00)04135-0 PMID:11247570

Nourmoradi, H., Goudarzi, G., Daryanoosh, S.M., Omidi-Khaniabadi, F., & Jourvand, M., & Omidi-Khaniabadi, Y. (2015). Health impacts of particulate matter in air by Air Q model in Khorramabad City, Iran. *J Bas Res Med Sci.*, *2*, 44–52.

Nowak, D. J., Civerolo, K. L., Rao, S. T., Sistla, G., Luley, C. J., & Crane, D. E. (2000). A modeling study of the impact of urban trees on ozone. *Atmospheric Environment*, *34*, 1601–1613. doi:10.1016/S1352-2310(99)00394-5

Nowak, D. J., Hirabayashi, S., Bodine, A., & Hoehn, R. (2013). Modeled PM2.5 removal by trees in ten U.S. cities and associated health effects. *Environmental Pollution*, *178*, 395–402. doi:10.1016/j.envpol.2013.03.050 PMID:23624337

Oehninger, S. (2001). Strategies for the infertile man. *Seminars in Reproductive Medicine*, *19*(03), 231–237. doi:10.1055-2001-18042 PMID:11679904

Olivier, J. G. J., Janssens-Maenhout, G., Peters, J. A. H. W., & Wilson, J. (2011). *Long- term trend in global CO_2 emissions 2011 report*. The Hague: PBL Netherlands.

Pope, C. A. III. (1989). Respiratory Disease Associated with Community Air Pollution and a Steel Mill, Utah Valley. *American Journal of Public Health*, *79*(5), 623–628. doi:10.2105/AJPH.79.5.623 PMID:2495741

Pope, C. A. III. (1991). Respiratory Hospital Admissions Associated with PM10 Pollution in Utah, Salt Lake, and Cache Valleys. *Archives of Environmental Health*, *46*(2), 90–97. doi:10.1080/00039896.1991.9937434 PMID:2006899

Pope, C. A. III, Burnett, R. T., Thurston, G. D., Thun, M. J., Calle, E. E., Krewski, D., & Godleski, J. J. (2004). Cardiovascular mortality and long-term exposure to particulate air pollution: Epidemiological evidence of general pathophysiological pathways of disease. *Circulation*, *109*(1), 71–77. doi:10.1161/01.CIR.0000108927.80044.7F PMID:14676145

Pope, C. A. III, Burnett, R. T., Turner, M. C., Cohen, A., Krewski, D., Jerrett, M., ... Thun, M. J. (2011). Lung cancer and cardiovascular disease mortality associated with ambient air pollution and cigarette smoke: Shape of the exposure-response relationships. *Environmental Health Perspectives*, *119*(11), 1616–1621. doi:10.1289/ehp.1103639 PMID:21768054

Pope, C. A. III, & Dockery, D. W. (1992). Acute Health Effects of PM10 Pollution on Symptomatic and Asymptomatic Children. *The American Review of Respiratory Disease*, *145*(5), 1123–1128. doi:10.1164/ajrccm/145.5.1123 PMID:1586057

Pope, C. A. III, & Dockery, D. W. (2006). Health effects of fine particulate air pollution: Lines that connect. *Journal of the Air & Waste Management Association*, *56*(6), 709–742. doi:10.1080/10473289.2006.10464485 PMID:16805397

Pope, C. A. III, Dockery, D. W., Spengler, J. D., & Raizenne, M. E. (1991). Respiratory Health and PM10 Pollution: A Daily Time Series Analysis. *The American Review of Respiratory Disease*, *144*(3_pt_1), 668–674. doi:10.1164/ajrccm/144.3_Pt_1.668 PMID:1892309

Pope, C. A. III, & Kanner, R. E. (1993). Acute Effects of PM10 Pollution on Pulmonary Function of Smokers with Mild to Moderate Chronic Obstructive Pulmonary Disease. *The American Review of Respiratory Disease*, *147*(6_pt_1), 1336–1340. doi:10.1164/ajrccm/147.6_Pt_1.1336 PMID:8503541

Pope, C. A. III, Schwartz, J., & Ransom, M. R. (1992). Daily Mortality and PM10 Pollution in Utah Valley. *Archives of Environmental Health*, *47*(3), 211–217. doi:10.1080/00039896.1992.9938351 PMID:1596104

Pope, C. A. III, Thun, M. J., Namboodiri, M. M., Dockery, D. W., Evans, J. S., Speizer, F. E., & Heath, C. W. Jr. (1995). Particulate air pollution as a predictor of mortality in a prospective study of U.S. adults. *American Journal of Respiratory and Critical Care Medicine*, *151*(3_pt_1), 669–674. doi:10.1164/ajrccm/151.3_Pt_1.669 PMID:7881654

Pope, C. A. III, Thun, M. J., Namboodiri, M. M., Dockery, D. W., Evans, J. S., Speizer, F. E., & Heath, J. C. W. Jr. (1995). Particulate Air Pollution as a Predictor of Mortality in a Prospective Study of U.S. Adults. *American Journal of Respiratory and Critical Care Medicine*, *151*(3_pt_1), 669–674. doi:10.1164/ajrccm/151.3_Pt_1.669 PMID:7881654

Ransom, M. R., & Pope, C. A. III. (1992). Elementary School Absences and PM10 Pollution in Utah Valley. *Environmental Research*, *58*(1-2), 204–219. doi:10.1016/S0013-9351(05)80216-6 PMID:1511674

Reinap, A., Wiman, B. L. B., Svenningsson, B., & Gunnarsson, S. (2009). Oak leaves as aerosol collectors: Relationships with wind velocity and particle size distribution. Experimental results and their implications. *Trees (Berlin)*, *23*(6), 1263–1274. doi:10.100700468-009-0366-4

Ritz, B., Yu, F., Chapa, G., & Fruin, S. (2000). Effect of air pollution on preterm birth among children born in southern California between 1989 and 1993. *Epidemiology (Cambridge, Mass.)*, *11*(5), 502–511. doi:10.1097/00001648-200009000-00004 PMID:10955401

Roupsard, P., Amielh, M., Maro, D., Coppalle, A., Branger, H., Connan, O., ... Talbaut, M. (2013). Measurement in a wind tunnel of dry deposition velocities of submicron aerosol with associated turbulence onto rough and smooth urban surfaces. *Journal of Aerosol Science*, *55*, 12–24. doi:10.1016/j.jaerosci.2012.07.006

Saebo, A., Hanslin, H. M., Baraldi, R., Rapparini, F., Gawronska, H., & Gawronski, S. W. (2013). Characterization of urban trees and shrubs for particulate deposition, carbon sequestration and BVOC emissions. *Proceedings of International Symposium on Woody Ornamentals of the Temperate Zone*, 509-517. 10.17660/ActaHortic.2013.990.66

Salam, M. T., Millstein, J., Li, Y. F., Lurmann, F. W., Margolis, H. G., & Gilliland, F. D. (2005). Birth outcomes and prenatal exposure to ozone, carbon monoxide, and particulate matter: Results from the Children's Health Study. *Environmental Health Perspectives*, *113*(11), 1638–1644. doi:10.1289/ehp.8111 PMID:16263524

Sapkota, A., Chelikowski, A. P., Nachman, K. E., Cohen, A. J., & Ritz, B. (2010). Exposure to particulate matter and adverse birth outcomes: A comprehensive review and meta analysis. *Air Quality, Atmosphere & Health*, *5*(4), 359–381.

Schwartz, J. (1991). Particulate Air Pollution and Daily Mortality in Detroit. *Environmental Research*, *56*(2), 204–213. doi:10.1016/S0013-9351(05)80009-X PMID:1769365

Schwartz, J. (1993). Air Pollution and Daily Mortality in Birmingham, Alabama. *American Journal of Epidemiology*, *137*(10), 1136–1147. doi:10.1093/oxfordjournals.aje.a116617 PMID:8317443

Schwartz, J., & Dockery, D. W. (1992). Increased Mortality in Philadelphia Associated with Daily Air Pollution Concentrations. *The American Review of Respiratory Disease*, *145*(3), 600–604. doi:10.1164/ajrccm/145.3.600 PMID:1546841

Schwartz, J., & Marcus, A. (1990). Mortality and Air Pollution in London: A Time Series Analysis. *American Journal of Epidemiology*, *131*(1), 185–194. doi:10.1093/oxfordjournals.aje.a115473 PMID:2403468

Schwartz, J., Slater, D., Larson, T., Pierson, W., & Koenig, J. (1993). Particulate air pollution and hospital emergency room visits for asthma in Seattle. *The American Review of Respiratory Disease*, *174*(4), 826–831. doi:10.1164/ajrccm/147.4.826 PMID:8466116

Sharma, H. K., Dandotiya, B., & Jadon, N. (2017). Exposure of Air Pollution and Its Health Effects in Traffic Police Persons of Gwalior City, India. *Environmental Claims Journal*, *29*(4), 305–315. doi:10.1080/10406026.2017.1390357

Sicard, P., Lesne, O., Alexandre, N., Mangin, A., & Collomp, R. (2011). Air quality trends and potential health effects-development of an aggregate risk index. *Atmospheric Environment*, *45*(5), 1145–1153. doi:10.1016/j.atmosenv.2010.12.052

Sicard, P., Mangin, A., Hebel, P., & Malléa, P. (2010). Detection and estimation trends linked to air quality and mortality on French Riviera over the 1990-2005 period. *The Science of the Total Environment*, *408*(8), 1943–1950. doi:10.1016/j.scitotenv.2010.01.024 PMID:20138648

Sicard, P., Serra, R., & Rossello, P. (2016). Spatiotemporal trends of surface ozone concentrations and metrics in France. *Environmental Research*, *149*, 122–144. doi:10.1016/j.envres.2016.05.014 PMID:27200478

Slama, R., Darrow, L., Parker, J., Woodruff, T. J., Strickland, M., Nieuwenhuijsen, M., ... Ritz, B. (2008). Meeting report: Atmospheric pollution and human reproduction. *Environmental Health Perspectives*, *116*(6), 791–798. doi:10.1289/ehp.11074 PMID:18560536

Sram, R. J. (1999). Impact of air pollution on reproductive health. *Environmental Health Perspectives*, *107*(11), 542–543. doi:10.1289/ehp.99107a542 PMID:10627171

Thurston, G. D. (2008). Outdoor air pollution: sources, atmospheric transport, and human health effects. In H. Kris (Ed.), *International Encyclopedia of Public Health* (pp. 700–712). Oxford, UK: Academic Press. doi:10.1016/B978-012373960-5.00275-6

Van Dingenen, R., Dentener, F., Raes, F., Krol, M., Emberson, L., & Cofala, J. (2009). The global impact of ozone on agricultural crop yields under current and future air quality legislation. *Atmospheric Environment*, *43*(3), 604–618. doi:10.1016/j.atmosenv.2008.10.033

Venners, S. A., Wang, B. Y., Peng, Z. G., Xu, Y., Wang, L. H., & Xu, X. P. (2003). Particulate matter, sulfur dioxide, and daily mortality in Chongqing, China. *Environmental Health Perspectives*, *111*(4), 562–567. doi:10.1289/ehp.5664 PMID:12676616

Wang, D., Zhou, B., Fu, Q., Zhao, Q., Zhang, Q., Chen, J., ... Li, J. (2016). Intense secondary aerosol formation due to strong atmospheric photochemical reactions in summer: Observations at a rural site in eastern Yangtze River Delta of China. *The Science of the Total Environment*, *571*, 1454–1466. doi:10.1016/j.scitotenv.2016.06.212 PMID:27418517

Wang, S., Feng, X., Zeng, X., Ma, Y., & Shang, K. (2009). A study on variations of concentrations of particulate matter with different sizes in Lanzhou, China. *Atmospheric Environment*, *43*(17), 2823–2828. doi:10.1016/j.atmosenv.2009.02.021

Wang, X., Ding, H., Ryan, L., & Xu, X. (1997). Association between air pollution and low birth weight: A community-based study. *Environmental Health Perspectives*, *105*(5), 514–520. doi:10.1289/ehp.97105514 PMID:9222137

Weuve, J., Puett, R., Schwartz, J., Yanosky, J., Laden, F., & Grodstein, F. (2012). Exposure to particulate air pollution and cognitive decline in older women. *Archives of Internal Medicine*, *172*(3), 219–227. doi:10.1001/archinternmed.2011.683 PMID:22332151

Wheeler, S. M. (2008). State and municipal climate change plans: The first generation. *Journal of the American Planning Association*, *74*(4), 481–496. doi:10.1080/01944360802377973

WHO. (2002). *World Health Report*. Geneva: World Health Organization.

WHO. (2005). *Air Quality Guideline Global Update 2005*. Copenhagen: Particulate Matter, Ozone, Nitrogen Dioxide and Sulphur Dioxide.

WHO. (2008). *Outdoor Air Pollution, Children's Health and the Environment*. World health organization.

WHO. (2013). *Press release N. 221*. International Agency for Research on Cancer, World Health Organization. Retrieved from http://www.iarc.fr/en/media- centre/pr/2013/ pdfs/pr221_E.pdf

WHO. (2014a). *Burden of disease from household air pollution for 2012*. Geneva, Switzerland: World Health Organization.

WHO. (2014b). *Media centre: 7 million premature deaths annually linked to air pollution*. Geneva: World Health Organization.

WHO. (2016a). *Global Health Observatory (GHO) data: mortality from ambient air pollution—situation and trend*. World Health Organization.

Wilhelm, M., & Ritz, B. (2003). Residential proximity to traffic and adverse birth outcomes in Los Angeles County, California, 1994–1996. *Environmental Health Perspectives*, *111*(2), 207–216. doi:10.1289/ehp.5688 PMID:12573907

Wilhelm, M., & Ritz, B. (2005). Local variations in CO and particulate air pollution and adverse birth outcomes in Los Angeles County, California, USA. *Environmental Health Perspectives*, *113*(9), 1212–1221. doi:10.1289/ehp.7751 PMID:16140630

Woodruff, T. J., Darrow, L. A., & Parker, J. D. (2008). Air pollution and postneonatal infant mortality in the United States, 1999–2002. *Environmental Health Perspectives*, *116*(1), 110–115. doi:10.1289/ehp.10370 PMID:18197308

Wu, J., Ren, C., Delfino, R. J., Chung, J., Wilhelm, M., & Ritz, B. (2009). Association between local traffic-generated air pollution and preeclampsia and preterm delivery in the South Coast Air Basin of California. *Environmental Health Perspectives*, *11*(11), 1773–1779. doi:10.1289/ehp.0800334 PMID:20049131

Xu, X., Sharman, R. K., Talbott, E. O., Zborowski, J. V., Rager, J., Arena, V. C., & … . (2011). PM10 air pollution exposure during pregnancy and term low birth weight in Allegheny County, PA, 1994–2000. *International Archives of Occupational and Environmental Health*, *84*(3), 251–257. doi:10.100700420-010-0545-z PMID:20496078

Yang, C. Y., Tseng, Y. T., & Chang, C. C. (2003). Effects of air pollution on birth weight among children born between 1995 and 1997 in Kaohsiung, Taiwan. *Journal of Toxicology and Environmental Health. Part A.*, *66*(9), 807–816. doi:10.1080/15287390306385 PMID:12746128

Yari, A. R., Goudarzi, G., Geravandi, S., Dobaradaran, S., Yousefi, F., Idani, E., ... Mohammadi, M. J. (2016). Study of ground-level ozone and its health risk assessment in residents in Ahvaz City, Iran during 2013. *Toxin Reviews*, *35*(3-4), 201–206. doi:10.1080/15569543.2016.1225769

Yost, M. G., Gadgil, A. J., Drescher, A. C., Zhou, Y., Simonds, M. A., Levine, S. P., ... Saisan, P. A. (1994). Imaging indoor tracer-gas concentrations with computed tomography: Experimental results with a remote sensing FTIR system. *American Industrial Hygiene Assoc.*, *55*(5), 395–402. doi:10.1080/15428119491018835 PMID:8209844

Zanobetti, A., Bind, M.-A. C., & Schwartz, J. (2008). Particulate air pollution and survival in a COPD cohort. *Environmental Health*, *10*, 7–48. PMID:18847462

Zarandi, S. M., Alimohammadi, M., Moghaddam, V. K., Hasanvand, M. S., Miranzadeh, M. B., Rabbani, D., ... Tehrani, A. M. (2015). Long-term trends of nitrogen oxides and surface ozone concentrations in Tehran city, 2002–2011. *Journal of Environmental Health Science & Engineering*, *15*, 13–63. PMID:26380092

Zhang, Y., Li, C., Krotkov, N. A., Joiner, J., Fioletov, V., & McLinden, C. (2016). Continuation of long-term global SO_2 pollution monitoring from OMI to OMPS. *Atmos Meas Tech Discuss*, *10*(4), 1495–1509. doi:10.5194/amt-10-1495-2017

Zhao, C. X., Wang, Y. J., Wang, Y. Q., & Zhang, H. L. (2013). Interactions between fine particulate matter (PM2.5) and vegetation: A review. [in Chinese]. *Shengtaixue Zazhi*, *32*, 2203–2210.

Zheng, D., Ducey, M. J., & Heath, L. S. (2013). Assessing net carbon sequestration on urban and community forests of northern New England, USA. *Urban Forestry & Urban Greening*, *12*(1), 61–68. doi:10.1016/j.ufug.2012.10.003

KEY TERMS AND DEFINITIONS

Air Pollution: Air pollution is a phenomenon that causes deviation of air from its natural quality.

Cardiovascular Health Effects: Cardiovascular health effects are defined as the effects of urban or ambient air pollutants on any cardiovascular organ.

Gaseous Pollutants: Gaseous pollutants include all hazardous gaseous pollutants present in ambient air of an urban environment.

Health Effects: Health effects are the effects on human body that caused by any air contaminant or pollutant. Higher concentrations of air pollutants as well as less concentration also exerts health effects.

Particulate Matter: Particulate matter is a common term used for respiratory and non-respiratory particles in ambient air.

Respiratory Health Effects: Respiratory health effects include the health effects of ambient or urban air pollutants on any respiratory organ.

Urban Environment: Urban environment is defined as ecosystem of an urban area in which the urban residents interact with biotic and abiotic factors.

Urban Vegetation: The sum of all vegetation present in an urban ecosystem like trees, shrubs, ornamental plants, small plants, etc. is called urban vegetation.

Section 3

Climate Change Impacts on Biodiversity and Ecosystem Services

Chapter 7
Climate Change Impacts on Biodiversity in Arid and Semi-Arid Areas:
Biodiversity Under Climate Change

Hanane Boutaj
Cadi Ayyad University, Morocco

Aicha Moumni
Cadi Ayyad University, Morocco

Oumayma Nassiri
Cadi Ayyad University, Morocco

Abdelhak Ouled Aitouna
Cadi Ayyad University, Morocco

ABSTRACT

Considerable attention has been paid to climate change and its impacts on biodiversity. The climate change has caused several problems such as continuous ecosystem degradation and a resultant biodiversity decline. In addition, climate warming has a range of indirect effects through changes in vegetation type level and sea that affect physical and biological systems. This has also led to changes in the distribution of species, as well as reductions in the size of populations, or even local extinctions of these populations. Moreover, many species are disappearing with time due to climate change combined with the emergence of disease that develops and increases with time. These problems affect different biodiversity components that are close to collapse. This chapter explored the richness of biodiversity in arid and semi-arid zones. It is also illuminates the effects of climate change on distribution of biodiversity. The authors highlight the responses of biodiversity under climate change, in terms of species extinction, biodiversity loss, and the impacts of climate change to ecological tourism. Finally, the authors show how biodiversity can overcome the effect of climate change, by developing some systems that allow to them to survive and conservation of species and ecosystems.

DOI: 10.4018/978-1-5225-7387-6.ch007

INTRODUCTION

Biodiversity is essential to the function and structure of ecosystems and underpins the broad scale of goods and services, which humans procure from natural ecosystems (Walther et al., 2002). It is an intrinsically complex variable composed of processes and numerous biological entities (Naeem, 2009; Mace et al., 2012), interacting individually and collectively at multiple levels of organisation, with biotic propagule dispersal over space and time, and abiotic exchange, flow of materials and energy (Hartvigsen et al., 1998; Noss, 1990). Climate change impacts is considered as the most important global environmental challenges, affecting biodiversity and natural ecosystems because natural ecosystems and biodiversity are closely interconnected on each other (Bharali and Khan, 2011) not just through changes in biodiversity that affect climate change, but also through climate change effects on biodiversity (Secretariat, 2010). Moreover, biodiversity are projected to increase in pervasiveness and magnitude as CO2 levels and temperatures continue to rise, which could increase photosynthesis and water use efficiency (Solomon et al., 2007). The biodiversity in different ways are affected by multiple components of climate change, from individual to biome levels (Figure 1). They primarily interest various forms of physical decrease that are expressed at different degrees, and have impacts on species, ecological networks and ecosystems. Negative effect on some susceptible life forms could be due to the quickly increasing of average temperature and increasing rainfall. It might have even more severe repercussions for biodiversity than the temperature elevation itself. At the most basic level of biodiversity, the rapid migration and directional selection could have an effect on decrease of genetic diversity of populations, which is caused by climate change and might affect in turn the ecosystem resilience and functioning (Botkin et al., 2007; Meyers and Bull, 2002). When one life form is affected, that in turn will affect other life forms. Even in cases where one life form is affected in a positive way, it can in the next step have negative effects on others. Access to the right quantity of water at the right time of year is essential for all life forms, especially for species in arid and semi-arid environments changing precipitation, which can be lethal. Several studies are interested on impacts of genetic effects and at higher organisational levels of climate change, which have been explored just for a very small number of species. Climate change is considered now as a major driver of biodiversity change, extinction and loss due to these pressures and ecosystem responses, which could affect strongly the Mediterranean over the course of this century. Hooper et al., (2005) reported that decrease or loss of any aspect of biodiversity could have direct and indirect effects persistence, ecosystem function and services. Foundation species or keystone have a major role in ecosystems by either providing structure, moderating the availability of different resources to other species, or through trophic processes. Many such species also provide beneficial services to humans in the form of storm and flood protection (e.g. mangroves), food (e.g. fisheries), and maintenance of water quality (Leadley, 2010). The extension of species geographic range limit either towards the poles or to higher elevations; increasing invasion by opportunistic, competitively mobile species; the extinction of local populations along range boundaries at lower elevations or lower latitudes; increasing decoupling of species interactions owing to mismatched phenology. Richardson and Poloczanska (2008) reported that twenty-eight thousand five hundred eighty-six significant biological changes were observed from terrestrial ecosystems associated with climate change, the most important of it is species extinctions. Some species are important for other species as for example food, pollinators, etc. If they disappear, other species will be affected negatively. Different species, which depend on each other can be also affected differently by climate change. However, in many cases there is limited understanding of the functional or interactive role a species or group plays in a system, which in turn limits our ability to predict how the system will respond to changes in climate and other anthropogenic stressors (Salamin et al., 2010). A range of methods and approaches are being employed to predict the effects of climate change on biodiversity including historical trends and relationships, experiments, and model projections, which lead to an overestimation of the ability of species to track climatic changes (Schloss et al., 2012), and an underestimation of extinction rates (Maclean and Wilson, 2011). Projections of species distributions often come from empirical models that relate observed occurrences to current or historical climate conditions. However, increase temperature and greenhouse gases will change rainfall, evaporation runoff and soil humidity storage and will adversely affect growth and productivity. Although, projection could predict future distributions using projected changes in the geographic distributions of abiotic variables. These models are often referred to as niche, climate envelope, or species distribution models, which have the advantage of being relatively easy to apply to large numbers of species (Iverson et al., 2008). Similar empirical models have been used to project changes in biomes and vegetation types (Rehfeldt et al., 2012).

Higher levels of complication and uncertainty were observed during the planification of climate change adaptation for biodiversity (Drielsma et al., 2015). Some individuals, species and ecosystems are proving some capacity for natural adaptation, but others are already showing negative effects under current degrees of climate change. The complex nature of biodiversity, together with uncertain future climate, make it challenging us to anticipate future impacts and to plan effectively for biodiversity conservation (Rissik et al., 2014). Nonetheless, an ability to anticipate, how a changing climate intersects with our current landscapes, enables us to re-examine our present-day priorities and strategies for conservation action; for instance, by maximising the condition and extent of natural environments that aid natural adaptation, and by maintaining or enhancing connectivity in places that will be critical to facilitating climate-induced migrations and range-shifts. Currently, the managers tend to pass over evolutionary processes when conceive ways to develop criteria for conservation (Mace and Purvis, 2008) and protecting biodiversity under climate change (Gebremedhin et al., 2009). The management plans for habitats and species should aim at developing resilient landscapes where the evolutionary potential of populations and species can be conserved. This aim can be accomplished by evolutionary processes ongoing and in situ and can be supported by explicit consideration of processes and the genetic diversity in planning and biodiversity management (Sgro et al., 2011).

BIODIVERSITY RICHNESS IN ARID AND SEMI-ARID AREAS

Biodiversity is all living things, microorganisms, plants, fungi and animals. They are also the reactions that connect them to each other and to the environment in which they live (Esteban and Finlay, 2010). We humans are part of living things, and we interfere in time and space with other components of biodiversity. It is a much broader concept than the simple collection of animal and plant species that is often reduced: it is the diversity of life at all levels of disposition, from gene to species and ecosystems.

Three levels are describing biodiversity (Heywood and Watson, 1995); Genetic diversity: Every living being has unique genetic specificities. Genetic diversity protects the gene diversity of all living organisms. Genes are supported by DNA. They are transmitted during reproduction. Gene diversity influences the diversity of traits of an individual, population, or species (e. g, eye color or disease resistance). Moreover, genetic diversity of mangoes (*Mangifera indica*) is represented by over 1500 varieties, which referring to the diversity of ecosystems within biogeographical boundary. Species diversity: The classification of living beings relies heavily on the concept of species. There are several

Figure 1. Summary of some of the predicted aspects of climate change and some examples of their likely effects on different levels of biodiversity
(Bellard et al., 2012)

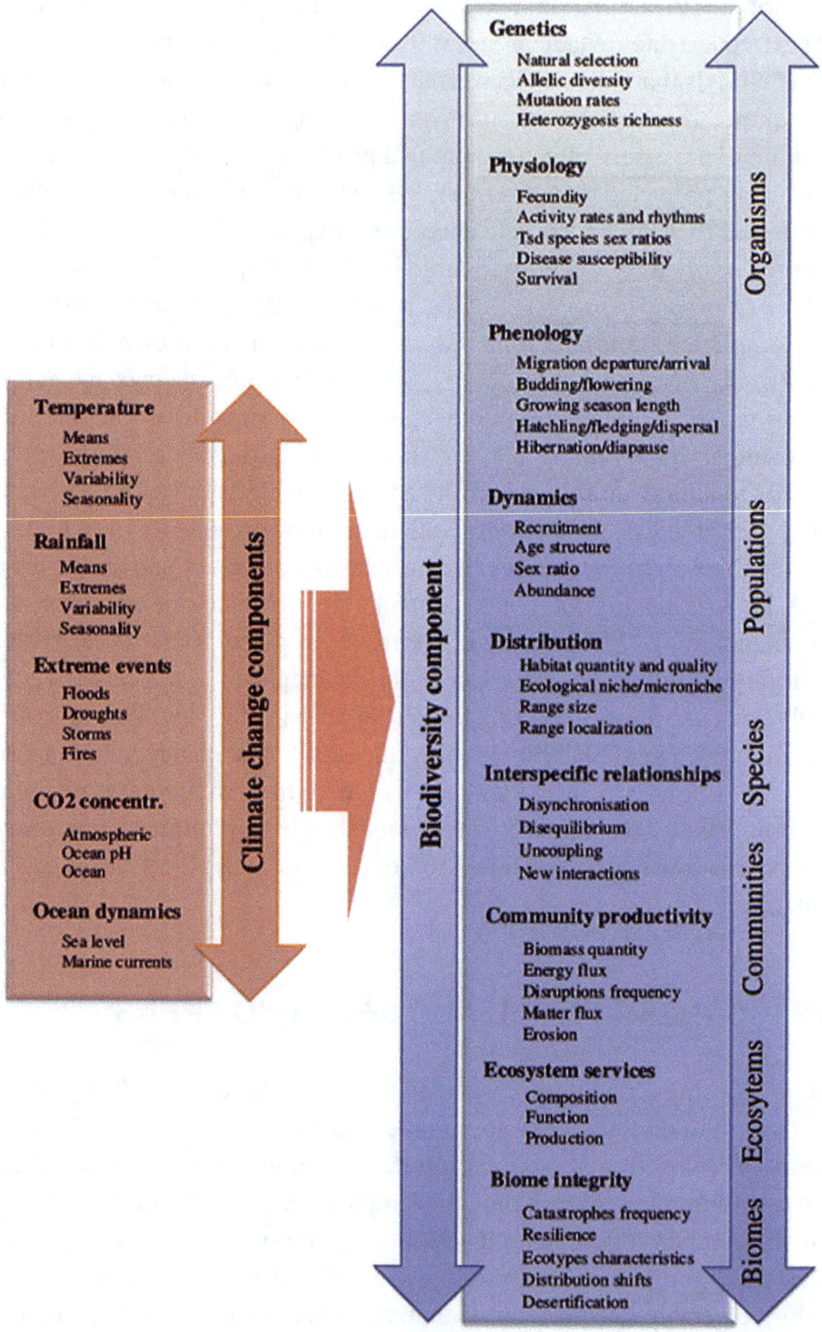

definitions of this concept, but the most classic brings together individuals potentially able to reproduce among themselves and give a viable and fertile descent. Currently no, fewer than two million species have been identified. It is estimated that there are perhaps 5 to 10 times as much to discover. Ecosystem diversity: Sets of populations of different species, forming communities, interact with each other and with their environment (air, land, water, etc.) and thus constitute ecosystems. A forest, a pond, the man and his intestinal flora are for example ecosystems. This diversity includes the richness of relationships and flows of matter and energy between populations of different species, with each other and with their physical and chemical environment. The relationships between species cover all kinds of functions and adaptations; food chain, parasitism, symbiosis, predation, competition, cooperation, etc. The landscapes that surround us express the diversity of ecosystems, fruit of the history of evolution and human influences. The five ecoregions in the world with a Mediterranean climate are characterized by exceptional biodiversity (Cowling et al., 1996) and a high richness in rare plants, mainly concentrated in large plant families (Lozano and Schwartz, 2005). Inherited from a long history, Mediterranean plants represent complex and varied evolutionary models (Raven, 1973; Thompson, 2005). The rapid and high rate of speciation has favored the emergence of rare and restricted-range taxa, particularly on drastic, nutrient-poor substrates (Wisheu et al., 2000). Diversity is also important in drylands; there are, for example, areas of different aridity, temperature and altitude (Smith et al, 2010). Wetlands, forests, mountains and other habitat types provide critical resource areas within arid zones. Species richness is generally lower in arid zones than in tropical forests, but within species, diversity may be higher because of this diversity of habitats and the isolation of populations. However, some arid areas such as Tallgrass prairie in North America are some of the world's most (Packard and Mutel, 1997).

CLIMATE CHANGE EFFECTS ON DISTRIBUTION OF BIODIVERSITY

One of the essential tenets of Earth science is that climate has a major part in the environmental system of our planet. Even minor shifts have impacts that are large and complex. The change of climate influences people and nature, especially biodiversity, in many ways, and it frequently increases existing danger that has already put pressure on the environment. These different effects can operate at many scales individual, population, species, community, ecosystem and biome scales. A considerable effects resulted of climate change is the phenological events, which include changes in leaf out, flowering and blooming in plants, and shifts in the timing of spawning and migrations in animals (Walther et al., 2002). At this level, species have affected the temporal patterns of seasonal migrations and other life cycle events (Doak and Morris, 2010; Dawson et al., 2011). That has to say, the seasonal timing of life events, have been noticed in response to variations in precipitation, temperature, and photoperiod (Parmesan and Yohe, 2003; Parmesan, 2007). For example, higher temperatures have led to earlier flowering in certain plant species (it has advanced by more than 10 days per decade for some species) (Parmesan, 2006), and changes in the dates of egg-laying of birds and Amphibia, among animals, shifts in the timing of springtime bird migrations next to remarkable changes in the dates of egg-laying are the most well recognized biological reactions to warming, and have been documented in the western, mid-western (MacMynowski and Root, 2007) and eastern (VanBuskirk et al., 2009). Table 1 showed the main phenological effects in USA.

Table 1. Examples of observed phenological change across geographical regions of the United States

U.S. Region	Taxonomic Group	Observed Changes in Phenology	Reference
Alaska & The Arctic	Fish	In a 34-year study of an Alaskan creek, fry of pink salmon (*Oncorhynchus gorbuscha*) migrated increasingly earlier over time.	Taylor 2008
Great Plains	Plants	In a study of first flowering dates (FFD) for 178 species of plants from 1910-1961 and 2007-2010 in North Dakota, over 40 percent of plants showed a change in FFD when compared to flowering data for 2007-2010. Most species showed a difference in FFD during the two warmer years (2007 and 2010) of this study.	Dunnell and Travers, (2011)
Midwest & Great Lakes	Birds	Of 44 species of birds monitored in Minnesota over a 40 year period, 36 percent showed significantly earlier arrival dates. Increasing winter temperatures correlated with the earlier arrival of birds, particularly for short-distance migrants.	Swanson and Palmer, (2009)
Northeast	Insects	Native bees have appeared in spring an average of 10 days earlier over the last 130 years (1880-2010). The majority of this advance occurred in the last 40 years in parallel to increasing warming trends. Plants with bee visitation also showed a trend of earlier blooming, helping preserve synchrony in timing between the observed plant species and their insect pollinators.	Bartomeus et al., (2011)
Northwest	Birds	In a 12 year study of Northern Flickers (*Colaptes auratus*), the birds arrived at breeding sites earlier when temperatures along their migration routes were warmer. Temperatures at the breeding site correlated significantly with initiation of egg laying.	Wiebe and Gerstmar, (2010)
Southeast	Reptiles	Two species of autumn-breeding amphibians arrived later, and two winter-breeding species arrived earlier to their breeding area in a 30-year study (1978-2008). Arrival times were up to 76 days later for the autumn-breeding dwarf salamander (*Eurycea quadridigitata*). Overall, rates of change ranged from 5.9 – 37.2 days/decade, and are representative of some of the fastest rates of phenological change observed to date. Increasing overnight temperatures during the breeding season and amount of cumulative rainfall were correlated with these changes.	Todd et al., (2011)
Southwest	Mammals	Yellow-bellied marmots (*Marmota flaviventris*) observed over a 33 year period (1976-2008) in Colorado emerged earlier from hibernation, and gave birth earlier in the season; this gave marmots more time to grow before the end of the season. Consequently, marmots tended to have larger body sizes at the beginning of hibernation, which ultimately led to lower mortality rates and higher population sizes.	Ozgul and others 2010

(Staudinger et al., 2012)

In the southwestern United States, more particularly in the most arid regions, the timing of precipitation has a greater influence on plant phenology than temperature, especially in the form of available soil moisture (Crimmins et al., 2011). In the other hands, a large number of species have already altered their geographical ranges in response to rapid changes in temperature and precipitation regimes. Always populations track temperature gradients by moving poleward, up in elevation, or to increased depths in the oceans. For instance, in Britain, the observed range changes for several butterflies are ascribed to small temperature increases (less than one degree) during this century (Ford, 1982). Furthermore, a study of four decades of Christmas bird counts revealed significant northward movement among 58 percent

(177 of 305) of species tracked in the survey; on average species moved 35 miles northward, and more than 60 species moved in excess of 100 miles north (Heath, 2012). Another latest analyses propose that range shifts treated in 23 taxonomic groups were approximately 2-3 times greater than reported previously (Chen et al., 2011).

Arid zones contain a wide variety of biodiversity, the largest of which is perfectly suited to the ecology of drylands. As a result, in arid areas many species and habitats are not found in wetter areas. Some semi-arid and dry sub-humid areas are part of areas of high biodiversity in the world, including, for example, the Mediterranean Basin, California Chaparral, the Cape Florida Field, southern Madagascar and the Brazilian Cerrado. Flowering plant species are particularly more diversified in dry sub-humid zones than in humid temperate zones. The arid and semi-arid zones are world-famous for their charismatic fauna and flora species, such as the lion (*Panthera leo*), the African elephant (*Loxodonta africana*), the camel of Bactria (*Camelus ferus*), and the bovids like the wild yak (*Bostus mutus*) and the American bison (Bison bison). The arid and semiarid areas are home to groups of emblematic plants such as cacti and succulents, trees such as acacia and baobab, as well as many types of herbs around the world. In sum, there are 10,000 species of mammals, birds and amphibians in the arid zones: 64% of all birds, 55% of all mammals, and 25% of amphibians. By way of comparison, the richest biome in the world (tropical moist forests with large tropical and subtropical leaves) is home to about 70% of the world's terrestrial wildlife (Safriel et al., 2008). Although, Drylands provide habitat for 80% of North American mammal, bird and amphibian species. In Europe, Asia and North Africa, the figure is around 70%, and in sub-Saharan Africa and Latin America it is around 60%. Of the many species found in dryland habitats, 4% of mammals and amphibians and 3% of birds are endemic to drylands and do not exist outside them (Figure 2). Species have adapted to different driving factors in many remarkable ways. The short rainy seasons have favored the selection of species that complete their product cycle over a short period. The periods of prolonged drought have allowed for the selection of a variety of mechanisms at the level of organisms, including the ability to escape, avoid, resist, and endure drought. The pressures exerted by the fires and herbivores have led to the selection of plants that can withstand these impacts or that even require them to reproduce and propagate (SBSTTA, 1999).

Figure 2. Percentage of mammals, birds and amphibians appearing in drylands for different areas (Davies et al., 2012)

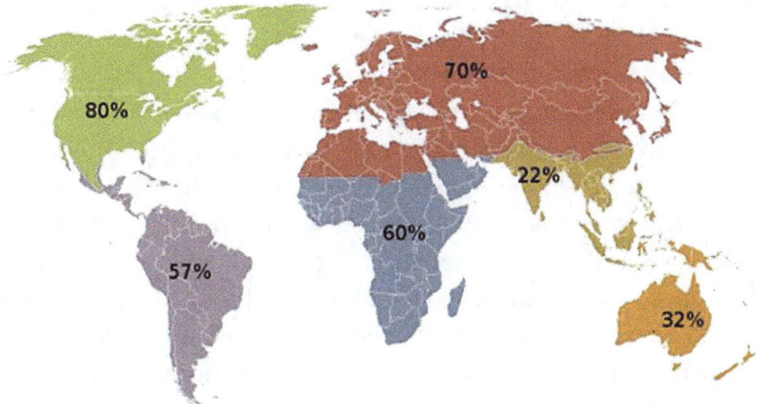

RESPONSES OF BIODIVERSITY TO CLIMATE CHANGE

Notable threats to biodiversity comprise habitat alteration and loss, chemical pollution, over-harvesting, increasing population pressure and invasive species because of climate change. This change may change and increase local anthropogenic disturbances. Concerning responses of biodiversity to climate change, it alters all over species and populations; some of those responses are negative, resulting in localized or widespread declines, while others are positive, guiding to raised range expansions or growth rates (Miller-Rushing et al., 2010; Montoya and Raffaelli, 2010; Geyer et al., 2011). As a result of these shifts, it will probably cause novel inter-specific interactions (Suttle et al., 2007), bring about new species assemblages (Williams and Jackson, 2007), and in worst case scenarios result in some extinctions (Butchart et al., 2010; Barnosky et al., 2011).

Projections of Species Extinction

According to Ephraums (1992), rates of habitat modification are currently very high that virtually all natural terrestrial habitats and protected areas are chosen to become ecological islands in surrounding oceans of habitat much altered. This process of fragmentation and 1387 isolation (Wilson and MacArthur, 1967) is predicted to lead directly and indirectly to accelerated species extinctions at both the local and the global scales, thus reducing the world's biodiversity at all levels (Lawton and May, 1995). Other studies made by Thomas et al. (2004), where they projected extinction rates for a variety of birds, mammals, frogs, reptiles, invertebrates, and plants in areas around the world using a climate envelope modeling approach. Thomas et al. (2004) reported that extinction rates ranged from 11 to 34% for a 0.8°C increase in temperature, and from 33 to 58% for greater than 2°C change in temperature depending on assumptions about species dispersal abilities. Using generalized linear models, Sekercioglu et al. (2008) projected that extinction rates for the majority (87%) of terrestrial bird populations in the Western Hemisphere would range from 1.3% for a 1.1°C to 30% for a 6.4°C increase in temperature. Sinervo et al. (2010) projected 20% loss of global lizard species by 2080 by employing an empirically validated physiological-based modeling approach. As a result of applying International Union for Conservation of Nature (IUCN) Red List Categories and Criteria. Carpenter et al. (2008) reported that 30 percent of reef building corals (845 species) are at risk for extinction in the coming decades. More than that, Malcolm et al. (2006) projected extinction rates among endemic species in global biodiversity hotspots would range from less than1 to 43% using a combined approach of global vegetation models (GVMs) and general circulation models (GCMs). Nonetheless, conclusions among projected studies agree that as the impacts of climate change continue to grow (Kitzes and Harte, 2011). The simplest way for calculating extinction danger is to assume that species go extinct when they no longer have any suitable habitat (Jetz et al., 2007). This may underestimate extinctions because species usually enter an extinction vortex well before they lose their entire habitat. Yet, it could also substantially overestimate extinctions because many species have weak habitat specificity (Malcolm et al., 2006; Willis and Bhagwat, 2009). The species–area relationship (SAR), an empirical relationship between the number of species and the land area of a region, is often used to estimate extinction risk (Thomas et al., 2004). Species extinctions are calculated as a direct function of habitat loss or climate-induced range contraction based on the observation that extinction risk increases with decreasing range and population size. SAR methods may over or underestimate species extinction risk depending on the capacity of species to persist in small populations or adapt to novel environments, and they also fail to provide a timeframe in which extinctions are likely to occur (Chevin et al., 2010) because the extinction debt is not accounted for (Pereira et al., 2010). The IUCN has developed criteria to assess species extinction risk for their red list of the conservation status of plant and animal species (Baillie et al., 2004). These include 90 biological traits identified as enhancing species vulnerability (e.g. habitat specialisation, narrow environmental tolerance, dependence on specific environmental triggers, dependence on inter-specific interactions or poor ability to disperse to or colonise a new range (Foden et al., 2008). The following figure (Figure 3) show the distribution of climate related threats evaluated by the IUCN on 4,161 terrestrial and aquatic species globally. Recent attempts at projecting climate change impacts on biodiversity have used the IUCN status metric to obtain estimates of extinction rates based on projected range shifts (Thuiller et al., 2005), although the use of IUCN status for this purpose has also been questioned (Akçakaya et al., 2006).

Projections of Species Loss

Climate change biology has followed many independent and distinct lines of pattern development. This is fully advantageous, as the convergent anticipations can be considered as a powerful sign of robustness. Essentially, the outcomes show that the extinctions of local species cover an extremely large range, with some regions testing virtually no losses and others facing nearly total loss of current species. In other words, the increased local losses can affect relatively large areas; for example, the estimation of Bakkenes et al. (2002) determined that more than 16% of European landmass would be affected by local species losses exceeding 50% by 2050. However, even these global estimates suggest major losses of biodiversity as a result of global climate change that are generally higher than current rates of loss and far higher than rates of species extinctions registered in the fossil record (Barnosky et al., 2011). For instance, one of the earliest global studies evaluated that by 2050 15 to 37% of species are committed to extinction under intermediate climate warming (Thomas et al., 2004). Birds are projected to be particularly sensitive to climate change. Two studies have calculated losses by 2100 due to climate change: these range from less than 0.3% of the world's 8750 species of land birds would be committed

Figure 3. Distribution of climate related threats evaluated by the IUCN on 4,161 terrestrial and aquatic species globally. Note there may be overlaps in species evaluated among categories
(IUCN, 2010)

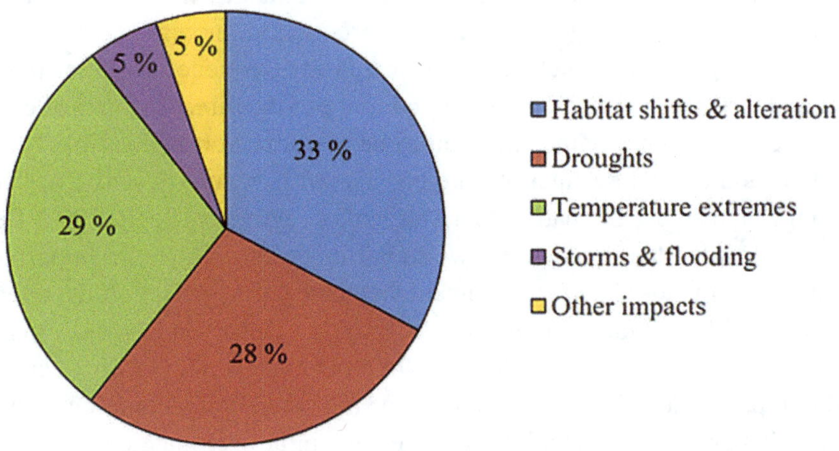

to extinction (Jetz et al., 2007) and up to 30% of the 8400 land bird species in the Western Hemisphere could go extinct (Sekercioglu et al., 2008). Regarding the vulnerability of 25 major biodiversity hotspots. Malcolm and colleagues (2006) suggested that the extinctions of endemic species could reach from 39 to 43% under in worst-case scenarios, showing the great loss of 56 000 endemic plant species and 3700 endemic vertebrate species.

Several studies based on changes in species abundance generally predict an erosion of biodiversity in the same order of magnitude as models of species extinction do, i.e. losses of 11–17% of MSA by the end of the century (Alkemade et al., 2009). Using data from 9856 birds, 6222 amphibians and 799 coral reefs, the IUCN forecasted that c. 35% of the world's birds, 52% of amphibians and 71% of warm-water reef building corals are exceptionally susceptible to climate change (Foden et al., 2008). In fact, the species composition, in some regions, will not alter at all and in others, almost all species will be lost because of climate change. Estimated losses of riverine fish richness at a local level vary remarkably between 0 and 75% according to the river considered (Xenopoulos et al., 2005). In the same way, in Europe, calculated local losses of plant diversity range from 2 to 84% of species lost per pixel (Thuiller et al., 2005). The diversity of taxonomic groups studied, of biodiversity loss metrics; of spatial scales and times periods render complex and uncertain comparisons between these studies (Figure 4). Furthermore, the modelling techniques had large array, which further makes the comparisons more complicated. In addition, numerous studies assess the combined effects on future biodiversity of both land use and climate change, keeping this issue difficult to identified the impact of each. Finally, there is increasing concern that the underlying assumptions of all of these patterns could lead to large over-or underestimations of potential species losses (Pereira et al., 2010). In the other hands, current losses in biodiversity by human-induced environmental changes has renewed interest in research on the significance of biodiversity for ecosystem functioning and resilience with respect to stress and disturbance (Chapin et al., 1997). Grime (1997) concludes that there is currently no convincing evidence that ecosystem processes are crucially dependent on higher biodiversity. Rather, the functional characteristics of the prevalent plant species would be important in monitoring ecosystem processes. Nevertheless, it keeps obvious that species losses or functional groups from an ecosystem could be at some point impair its functioning and its capacity to provide services to society (Chapin et al., 1997). Some species or functional groups achieve unique functions and cannot be substituted by others. In addition, the loss of species with similar ecosystem impacts may decrease ecosystem resilience and narrow the options for adjustments regarding to climate change.

A wide range of climate-related environmental changes will have important consequences for tourism at the destination and regional levels (Gossling and Hall, 2006). Changes in water supply, biodiversity loss, damage to landscape aesthetics, damage to agricultural production, increased natural hazards, Erosion and coastal flooding, damage to infrastructure and the increasing frequency of vector-borne diseases will all have an impact on tourism to varying degrees. In contrary to the direct impacts of climate change on tourism, the indirect effects of environmental change due to climate change are likely to be very negative. Mountain, island and coastal destinations are considered particularly sensitive to climate-related environmental change, and so are the nature tourism market segments (Scott, 2006; Becken and Hay, 2007). UNESCO has already identified several world heritage sites, which are important tourist destinations, as vulnerable to climate-related environmental change (eg Venice, in Italy, Great Barrier Reef, in Australia, International Park Waterton Glacier Peace in the United States and Canada, Chan Chan Archaeological Zone, in Peru) (UNESCO, 2007). While our understanding of the impacts of climate change for different types of destinations has improved since the Djerba conference. It is important to note that there are still significant regional knowledge gaps in how climate change will affect climate

Figure 4. Projections of biodiversity loss due to climate change, for different taxonomic, temporal and spatial scales
(Bellard et al., 2012)

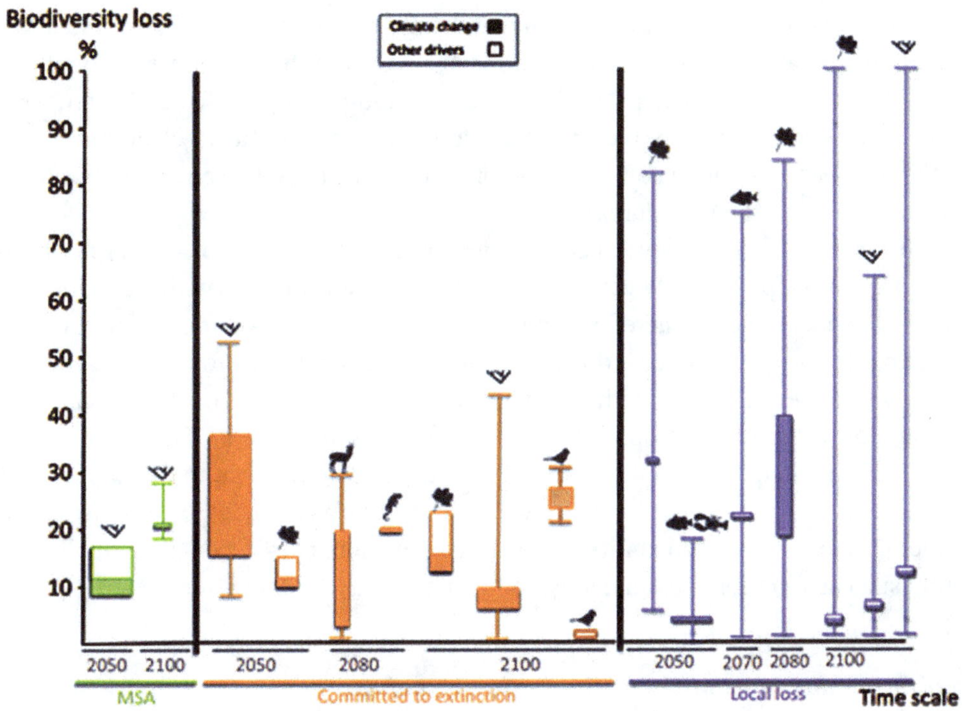

change natural and cultural resources that are essential for tourism in Africa, Caribbean, South America, Middle East and large parts of East Asia (Amelung et al., 2008).

BIODIVERSITY ADAPTATION TO CLIMATE CHANGE

Climate Change and Species Adaptations

Today the impacts of climate change are becoming increasingly harmful, severe and significant on ecosystem biodiversity and ecosystem services. As documented in the other parts of this chapter, the impacts cause disruption of the functioning of the ecosystems, and their compositions (Green et al., 2007). A genetic and demographic change can be probably as affected by this phenomenon of climate change (Suárez et al., 2002). Adaptation in some way is a preventative reaction to a specific action that is climate change. The measures taken to prepare for the anticipated impacts of climate change are often as a proactive or anticipated adaptation, while actions in response to climate-based impacts are often referred to as reactive adaptations (Adger et al., 2005).

The topic remains recent and emerging so the definitions in relation to adaptation remain in intense movement. According to the Intergovernmental Panel on Climate Change (IPCC) adaptation is defined as "*initiatives and measures to reduce the vulnerability of natural and human systems against actual or expected climate change effects and adjustment in natural or human systems in response to actual or expected climatic stimuli or their effects, which moderates harm or exploits beneficial opportunities*" (Meehl et al., 2007).

These climate change adaptations can primarily help the conservation and sustainable use of biodiversity and reduce the impact of extreme climate change on biodiversity. These include preventive practices such as the creation of strong land and marine reserves that are more particularly interconnected, considering the expected changes in the climate (Staudinger et al., 2012). Moreover, adaptation is a structured process that considers the direct, indirect and varied effects of climate change on the resources of the ecosystem, in order to reduce the probable effects on these resources and to bring the appropriate answers to save them (Figure 5).

The adaptive capacity has been developed through a range of natural biological mechanisms (Hoffmann et al., 2011), is more particularly genetic. These mechanisms interact with water, carbon, nitrogen and other major biogeochemical. Therefore, the disturbance or modification of one or more of these biogeochemical parameters can induce a direct change in genetics. In this topic several terms are put in place for adaptation in the context of biodiversity (Anderson et al., 2017; Groffman et al., 2006). These include management, vulnerability, coping skills, resilience and adaptation. The adaptive capacity also known as the "Intrinsic adaptation potential" (Glick et al., 2011), it refers to the capacity of a natural system to take account the impacts of climate change (Smith and Wandel, 2006; Williams et al., 2008). It results from factors such as dispersal ability, genetic diversity, and plasticity at the population or species level, or factors such as functional redundancy and patch size at the ecosystem level.

Figure 5. Implementation framework and adaptation planning
(Glick et al., 2011)

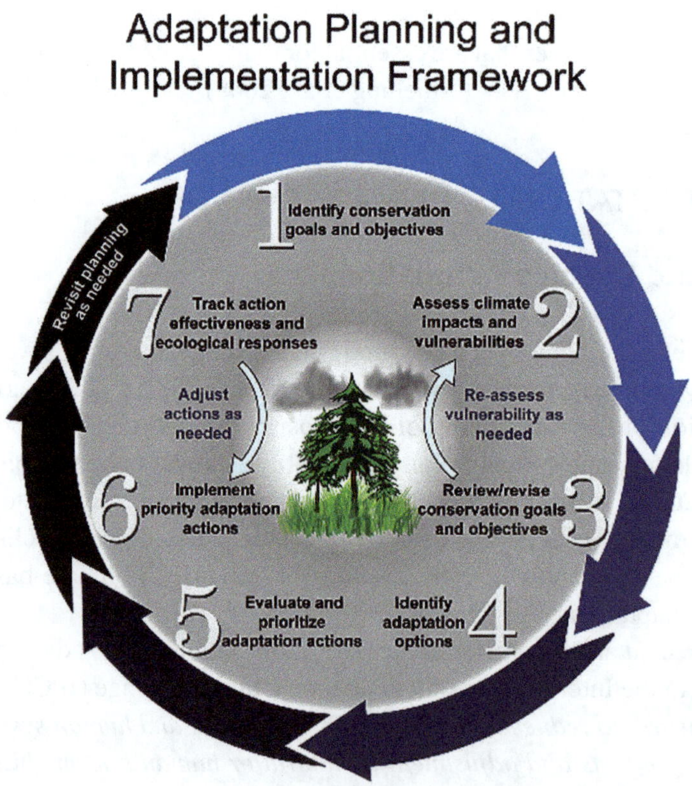

Adaptive management is often practiced in the context of climate adaptation as a means of addressing and responding to the uncertainties associated with predicting human and biological responses to climate change. This concept was applied in order to manage the resources (Williams et al., 2009). Vulnerability is the key of adaptation, it refers to measures in the ecological system or/and individual species, to know the possibility of suffering damage due to climate change (Meehl et al., 2007). Eventually, maladaptation is defined as an adaptation that does not succeed in minimizing vulnerability but instead increases it (McCarthy et al., 2001). The main questions of climate change on ecological effects is whether species can adapt quickly and follow the rapid pace of climate change (Lavergne et al., 2010). For that, two response mechanisms are set up plastic vs. genetic. The response to adaptation is either because of microevolution i.e. genetic adaptation to new condition either by mutations or by the selection of existing genotypes or plasticity, which provides a very short-term means of response individual, live (Charmantier et al., 2008). Empirically speaking, the plastic contribution is more interesting than genetic. However, the plastic contribution indicates intraspecific variation in morphological, physiological, or behavioral characteristics that can occur at small temporal scales within the population spatial extent (Chevin et al., 2010). The case of birds and marmots Several empirical experiments and evidence have shown and confirmed that rapid evolution by mutation and selection could allow species with fast life cycles to adapt very severely with rapid environmental changes (Bell and Gonzalez, 2009; Phillips, 2009).

Along three axes of responses to climate change through phenotypic plasticity or evolutionary reactions (Bellard et al., 2012) (Figure 6); spatial, temporal or self. The first two axes correspond to the parameters of global warming (Parmesan, 2006). The third axe corresponds to physiological and behavioral changes allowing species to adapt to new climatic conditions in a spatial and temporal setting. During rapid climate change, species have three basic alternatives: they can migrate to appropriate environmental conditions; Adapt to new environmental conditions or may disappear as indicated in other part of this chapter.

The resilience of biodiversity to climate change can be reinforced by the reduction of climate change constraints However, the magnitude of these changes is based on the possibility of natural adaptation that will become more and more difficult to increase the adaptive capacity of species and ecosystems in the acceleration of climate change include (Figure 7).

Conservation of Species and Ecosystems

There are highly vulnerable species, so it will be increasingly difficult to achieve conservation goals (Figure 8) (Prober and Dunlop, 2011). Many multidimensional approaches use for understanding the impacts of climate change may have on biodiversity (Dawson et al., 2011) on long-term by prediction models (Beger et al., 2010; Glick et al., 2011), currently, few studies have used conservation modeling (Araújo et al., 2011). It is a global phenomenon provokes several methodological problems, the ecologists must quickly rise the challenge because of the large projected impacts of climate change on biodiversity, with the aim of scientific guidance for the development of conservation strategies (Pressey et al., 2007; Dawson et al., 2011).

The conservation plan is based on reserve networks that protect biodiversity in situ and this to protect the heterogeneity of habitats as well as the genetic diversity within a species to support the ability of a species to adapt (Hansen et al., 2010). The planning of conservation is a matter of spatial planning, and the traditional reserves of the different environmental matrices, it is a statistical conception based on the hypothesis that the threats to biodiversity are also purely static. Today, the proposed areas for biodiversity conservation are geographically isolated (Hannah et al., 2007). The conservation conception

Figure 6. The three direction of responses to climate change
(Bellard et al., 2012)

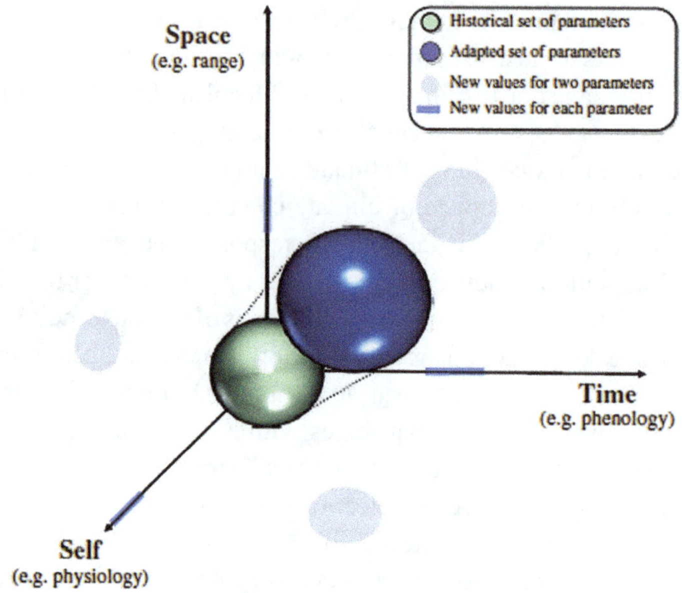

Figure 7. Climate change and biodiversity adaptations approaches
(Ravi Prasad Rao, et al.)

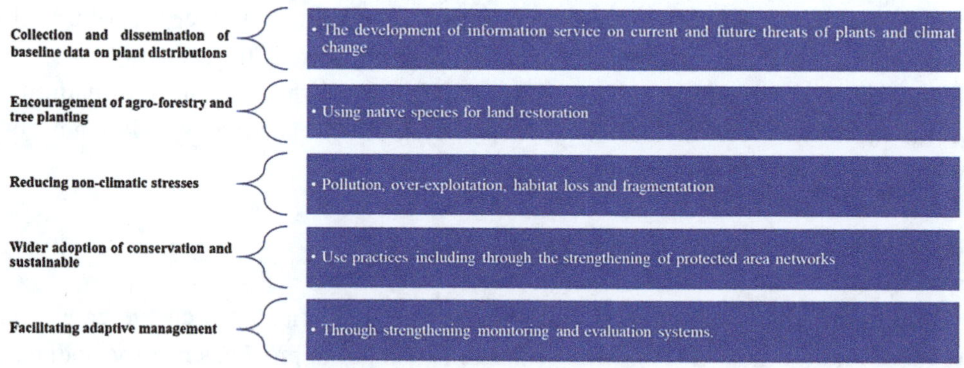

assumes that the climate is a non-variable fixed parameter the same mentioned in review of adaptation policies in Ghana (Mensah et al., 2016) (Table 2), while the climate always changes and it is never stable (Hannah, 2010).

Lastly, Predictive approaches can indicate how, when and where the risks of climate change can influence and affect biodiversity. By classifying species that can be maladapted and negatively affected by climate change and other species that can easily adapt and persist (Hannah et al., 2007). According to (Hansen et al., 2010), global warming will result in extinction of species according to its magnitude. A decrease in these of warming can directly induce the effects that are going to be nonlinear on the conservation of species of extinction and save a significant part of the biodiversity.

Figure 8. The vulnerability of biodiversity to the adaptation options (Pressy et al., 2007).

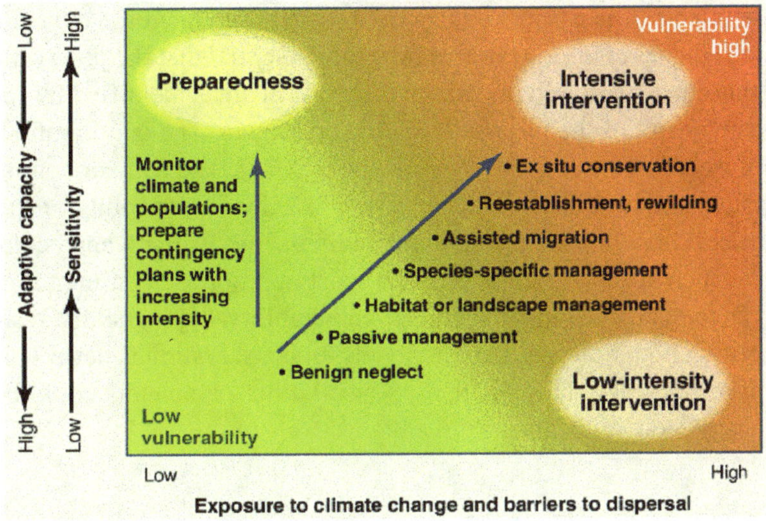

Table 2. Adaptation actions and priorities by sector seen in the NCCAS

Sector	Adaptation Priorities
Agriculture	Build farmer awareness of climate issues Strengthen farmer capacity to increase agricultural productivity Build capacity of extension officers to better support adaptation by farmers Promote alternative livelihood strategies, particularly for vulnerable groups Promote agricultural biodiversity Promote crop and livestock production practices that are climate-resilient Promote post-harvest technologies to minimize losses Identify and document indigenous adaptation strategies
Water Ressources	Water preservation and conservation Increase accessibility and availability of water for domestic, agricultural, industrial and commercial use and energy production, for example through rainwater harvesting. Improve and sustain quality of water resources Build capacity in water resource management. Construct flood protection and water conservation structures. Develop buffer zones along river banks Strengthening and enforcement of land use planning regulations for waterways and flood zones
Fisheries	Promotion of fish farming Design and implementation of programme on fisheries management and disease control. Develop alternative livelihood strategies for fisherfolk
Energy	Increase the use of off-grid alternative energy ressources Expand the use of efficient domestic appliances Develop low-head run of river hydroelectricity schemes Encourage energy conservation on a large scale Increase use natural gas
Health	Raise public awareness on climate change and its impacts on health, livelihoods and environmental sanitation Improve waste management systems and provide new and affordable technologies for environmental sanitation Reduce incidence of water and air-borne diseases Increase capacity and knowledge of health care workers on climate change related health problems Improve and increase existing health facilities and equipment

(Mensah et al., 2016).

It helps to strengthen resilience through systems of connected networks between all tenure and land uses (Drielsma et al., 2009). Finally, the science with all these disciplines seeks to minimize global warming before these effects. It is a priority that could sequentially increase the resilience of biodiversity to climate change (Hughes et al., 2003). It is therefore essential that the policy intervene to assist in the development of this conservation in the arid and semi-arid areas identified by species-level studies that respond to the complexity of the processes to climate change. The conservation of biodiversity in these areas is a major Need. According to (Dawson et al., 2011). The comparison models between the Paleohistory examination and the current observations highlight the variability of species vulnerability. This variability is the best hope for the future to maintain the biodiversity and ecology, by developing effective strategies for their natural adaptation. The most appropriate conservation measures will be based on adaptation factors: Adaptive management, vulnerability, coping skills, resilience and adaptation. This debate is getting hotter, the prediction models for conservation is never going to be global for all species, which makes the case study difficult. Several other parameters could be related with this topic as well as the socio-economic context.

CONCLUSION

Assessing of climate change impact on biodiversity is demanding and difficult, due to the complexity of the problem, the spatial and temporal scale and its interactions with further environmental factors. This greatly complicates the types of research questions of biodiversity responses to climate change, as well as relevant adaptation and conservation strategies for safeguarding these resources in the future. However, little attention has been devoted to researching the interactions between environmental stressors and climate change which, therefore, a nascent understanding of the types of interactions and mechanisms by which climate change could affect species and ecosystems. Taking into account the multivariate context in which ecosystems and biodiversity are highlighted will be very essential for prioritizing and informing strategies for responding to climate change. Nevertheless, food webs will be affected, species distribution will change dramatically, species life cycle timing should also be modified and ecosystem functioning could be seriously compromised, leading in the worst case to extinctions of species. Models of projections that can be used to predict the impacts of climate change on the abundance, distribution and extinction of individuals, populations and species. These models are characterized by their great diversity of structures, with different forecasts depending on the models and species studied. Barnosky et al. (2011) reported that some models have indicated alarming consequences for biodiversity, and the worst-case scenarios that lead to extinction rates may be considered the sixth massive extinction in the earth's history. However, improvements to a new generation of models need to make changes and address gaps to reduce the uncertainties of current models. It is also important to improve understanding of the vulnerability of biodiversity to climate change, to create and develop approaches that can go beyond expectations. Fortunately, the interactive effects of environmental stressors and climate change have been enhanced by such actions already existing examples and by natural resource managers with a toolbox able to tackle these complex challenges.

REFERENCES

Adger, W. N., Arnell, N. W., & Tompkins, E. L. (2005). Successful adaptation to climate change across scales. *Global Environmental Change*, *15*(2), 77–86. doi:10.1016/j.gloenvcha.2004.12.005

Akçakaya, H. R., Butchart, S. H., Mace, G. M., Stuart, S. N., & Hilton-Taylor, C. (2006). Use and misuse of the IUCN Red List Criteria in projecting climate change impacts on biodiversity. *Global Change Biology*, *12*(11), 2037–2043. doi:10.1111/j.1365-2486.2006.01253.x

Alkemade, R., van Oorschot, M., Miles, L., Nellemann, C., Bakkenes, M., & Ten Brink, B. (2009). GLOBIO3: A framework to investigate options for reducing global terrestrial biodiversity loss. *Ecosystems (New York, N.Y.)*, *12*(3), 374–390. doi:10.100710021-009-9229-5

Amelung, B., Moreno, A., & Scott, D. (2008). The place of tourism in the IPCC fourth assessment report: A review. *Tourism Review International*, *12*(1), 5–12. doi:10.3727/154427208785899984

Anderson, E. P., Marengo, J., Villalba, R., Halloy, S., Young, B., Cordero, D., & Martinez, R. (2017). Consequences of climate change for ecosystems and ecosystem services in the tropical. *Andes (Salta)*.

Araújo, M. B., Alagador, D., Cabeza, M., Nogués-Bravo, D., & Thuiller, W. (2011). Climate change threatens European conservation areas. *Ecology Letters*, *14*(5), 484–492. doi:10.1111/j.1461-0248.2011.01610.x PMID:21447141

Baillie, J., Hilton-Taylor, C., & Stuart, S. N. (Eds.). (2004). *IUCN red list of threatened species: a global species assessment*. Iucn.

Bakkenes, M., Alkemade, J. R. M., Ihle, F., Leemans, R., & Latour, J. B. (2002). Assessing effects of forecasted climate change on the diversity and distribution of European higher plants for 2050. *Global Change Biology*, *8*(4), 390–407. doi:10.1046/j.1354-1013.2001.00467.x

Barnosky, A. D., Matzke, N., Tomiya, S., Wogan, G. O., Swartz, B., Quental, T. B., ... Mersey, B. (2011). Has the Earth's sixth mass extinction already arrived? *Nature*, *471*(7336), 51–57. doi:10.1038/nature09678 PMID:21368823

Bartomeus, I., Ascher, J. S., Wagner, D., Danforth, B. N., Colla, S., Kornbluth, S., & Winfree, R. (2011). Climate-associated phenological advances in bee pollinators and bee-pollinated plants. *Proceedings of the National Academy of Sciences of the United States of America*, *108*(51), 20645–20649. doi:10.1073/pnas.1115559108 PMID:22143794

Becken, S., & Hay, J. (2007). *Tourism and climate change – risks and opportunities*. Cleveland: Channel View Publications.

Beger, M., Linke, S., Watts, M., Game, E., Treml, E., Ball, I., & Possingham, H. P. (2010). Incorporating asymmetric connectivity into spatial decision making for conservation. *Conservation Letters*, *3*(5), 359–368. doi:10.1111/j.1755-263X.2010.00123.x

Bell, G., & Gonzalez, A. (2009). Evolutionary rescue can prevent extinction following environmental change. *Ecology Letters*, *12*(9), 942–948. doi:10.1111/j.1461-0248.2009.01350.x PMID:19659574

Bellard, C., Bertelsmeier, C., Leadley, P., Thuiller, W., & Courchamp, F. (2012). Impacts of climate change on the future of biodiversity. *Ecology Letters*, *15*(4), 365–377. doi:10.1111/j.1461-0248.2011.01736.x PMID:22257223

Bharali, S., & Khan, M. L. (2011). Climate change and its impact on biodiversity; some management options for mitigation in Arunachal Pradesh. *Current Science*, 855–860.

Botkin, D. B., Saxe, H., Araujo, M. B., Betts, R., Bradshaw, R. H., Cedhagen, T., & Ferrier, S. (2007). Forecasting the effects of global warming on biodiversity. *A.I.B.S. Bulletin*, *57*(3), 227–236.

Butchart, S. H., Walpole, M., Collen, B., Van Strien, A., Scharlemann, J. P., Almond, R. E., & Carpenter, K. E. (2010). Global biodiversity: Indicators of recent declines. *Science*. PMID:20430971

Carpenter, K. E., Abrar, M., Aeby, G., Aronson, R. B., Banks, S., Bruckner, A., ... Wood, E. (2008). One-third of reef-building corals face elevated extinction risk from climate change and local impacts. *Science*, *321*(5888), 560–563. doi:10.1126cience.1159196 PMID:18653892

Chapin, F. S. III, Walker, B. H., Hobbs, R. J., Hooper, D. U., Lawton, J. H., Sala, O. E., & Tilman, D. (1997). Biotic control over the functioning of ecosystems. *Science*, *277*(5325), 500–504. doi:10.1126cience.277.5325.500

Charmantier, A., McCleery, R. H., Cole, L. R., Perrins, C., Kruuk, L. E., & Sheldon, B. C. (2008). Adaptive phenotypic plasticity in response to climate change in a wild bird population. *Science*, *320*(5877), 800–803. doi:10.1126cience.1157174 PMID:18467590

Chen, I. C., Hill, J. K., Ohlemüller, R., Roy, D. B., & Thomas, C. D. (2011). Rapid range shifts of species associated with high levels of climate warming. *Science*, *333*(6045), 1024–1026. doi:10.1126cience.1206432 PMID:21852500

Chevin, L. M., Lande, R., & Mace, G. M. (2010). Adaptation, plasticity, and extinction in a changing environment: Towards a predictive theory. *PLoS Biology*, *8*(4), e1000357. doi:10.1371/journal.pbio.1000357 PMID:20463950

Cowling, R. M., Rundel, P. W., Lamont, B. B., Arroyo, M. K., & Arianoutsou, M. (1996). Plant diversity in Mediterranean-climate regions. *Trends in Ecology & Evolution*, *11*(9), 362–366. doi:10.1016/0169-5347(96)10044-6 PMID:21237880

Crimmins, S. M., Dobrowski, S. Z., Greenberg, J. A., Abatzoglou, J. T., & Mynsberge, A. R. (2011). Changes in climatic water balance drive downhill shifts in plant species' optimum elevations. *Science*, *331*(6015), 324–327. doi:10.1126cience.1199040 PMID:21252344

Davies, J., Poulsen, L., Schulte-Herbrüggen, B., Mackinnon, K., Crawhall, N., Henwood, W.D., … Gudka, M. (2012). *Conservaton de la biodiversité des zones arides*. Academic Press.

Dawson, T. P., Jackson, S. T., House, J. I., Prentice, I. C., & Mace, G. M. (2011). Beyond predictions: Biodiversity conservation in a changing climate. *Science*, *332*(6025), 53–58. doi:10.1126cience.1200303 PMID:21454781

Doak, D. F., & Morris, W. F. (2010). Demographic compensation and tipping points in climate-induced range shifts. *Nature*, *467*(7318), 959–962. doi:10.1038/nature09439 PMID:20962844

Drielsma, M., & Ferrier, S. (2009). Rapid evaluation of metapopulation persistence in highly variegated landscapes. *Biological Conservation, 142*(3), 529–540. doi:10.1016/j.biocon.2008.11.018

Drielsma, M. J., Manion, G., Love, J., Williams, K. J., Harwood, T., & Saremi, H. (2015). 3C modelling for biodiversity under climate change. NSW Office of Environment and Heritage.

Dunnell, K. L., & Travers, S. E. (2011). Shifts in the flowering phenology of the northern Great Plains: Patterns over 100 years. *American Journal of Botany, 98*(6), 935–945. doi:10.3732/ajb.1000363 PMID:21613073

Ephraums, J. J., & Jenkins, G. J. J. (1992). *Climate change 1992*. Cambridge University Press.

Esteban, G. F., & Finlay, B. J. (2010). Conservation work is incomplete without cryptic biodiversity. *Nature, 463*(7279), 293. doi:10.1038/463293c PMID:20090730

Foden, J., Rogers, S. I., & Jones, A. P. (2008). A critical review of approaches to aquatic environmental assessment. *Marine Pollution Bulletin, 56*(11), 1825–1833. doi:10.1016/j.marpolbul.2008.08.017 PMID:18840385

Ford, M. J. (1982). *The changing climate: responses of the natural fauna and flora*. London: Allen and Unwin.

Gebremedhin, B., Ficetola, G. F., Naderi, S., Rezaei, H. R., Maudet, C., Rioux, D., ... Taberlet, P. (2009). Frontiers in identifying conservation units: From neutral markers to adaptive genetic variation. *Animal Conservation, 12*(2), 107–109. doi:10.1111/j.1469-1795.2009.00255.x

Geyer, J., Kiefer, I., Kreft, S., Chavez, V., Salafsky, N., Jeltsch, F., & Ibisch, P. L. (2011). Classification of climate-change-induced stresses on biological diversity. *Conservation Biology, 25*(4), 708–715. doi:10.1111/j.1523-1739.2011.01676.x PMID:21488958

Glick, P., Stein, B. A., & Edelson, N. A. (2011). *Scanning the conservation horizon: a guide to climate change vulnerability assessment*. Washington, DC: National Wildlife Federation.

Gossling, S., & Hall, C. M. (2006). An introduction to tourism and global environmental change. In *Tourism and Global Environmental Change* (pp. 1–34). Londres: Routledge.

Green, R. E., Harley, M., Miles, L., Scharlemann, J., Watkinson, A., & Watts, O. (2007). Global climate change and biodiversity. *Global Climate Change and Biodiversity*.

Grime, J. P. (1997). Biodiversity and ecosystem function: The debate deepens. *Science, 277*(5330), 1260–1261. doi:10.1126cience.277.5330.1260

Groffman, P. M., Baron, J. S., Blett, T., Gold, A. J., Goodman, I., Gunderson, L. H., & Poff, N. L. (2006). Ecological thresholds: The key to successful environmental management or an important concept with no practical application? *Ecosystems (New York, N.Y.), 9*(1), 1–13. doi:10.100710021-003-0142-z

Hannah, L. (2010). A global conservation system for climate-change adaptation. *Conservation Biology, 24*(1), 70–77. doi:10.1111/j.1523-1739.2009.01405.x PMID:20121843

Hannah, L., Midgley, G., Andelman, S., Araújo, M., Hughes, G., Martinez-Meyer, E., & Williams, P. (2007). Protected area needs in a changing climate. *Frontiers in Ecology and the Environment*, *5*(3), 131–138. doi:10.1890/1540-9295(2007)5[131:PANIAC]2.0.CO;2

Hansen, J., Ruedy, R., Sato, M., & Lo, K. (2010). Global surface temperature change. *Reviews of Geophysics*, *48*(4), RG4004. doi:10.1029/2010RG000345

Hartvigsen, G., Kinzig, A., & Peterson, G. (1998). Complex adaptive systems: Use and analysis of complex adaptive systems in ecosystem science: Overview of special section. *Ecosystems (New York, N.Y.)*, *1*(5), 427–430. doi:10.1007100219900036

Heath, J. A., Steenhof, K., & Foster, M. A. (2012). Shorter migration distances associated with higher winter temperatures suggest a mechanism for advancing nesting phenology of American kestrels Falco sparverius. *Journal of Avian Biology*, *43*(4), 376–384. doi:10.1111/j.1600-048X.2012.05595.x

Heywood, V. H., & Watson, R. T. (1995). *Global biodiversity assessment* (Vol. 1140). Cambridge, UK: Cambridge University Press.

Hoffmann, A. A., & Sgrò, C. M. (2011). Climate change and evolutionary adaptation. *Nature*, *470*(7335), 479–485. doi:10.1038/nature09670 PMID:21350480

Hooper, D. U., Chapin, F. S., Ewel, J. J., Hector, A., Inchausti, P., Lavorel, S., & Schmid, B. (2005). Effects of biodiversity on ecosystem functioning: A consensus of current knowledge. *Ecological Monographs*, *75*(1), 3–35. doi:10.1890/04-0922

Hughes, T. P., Baird, A. H., Bellwood, D. R., Card, M., Connolly, S. R., Folke, C., & Lough, J. M. (2003). Climate change, human impacts, and the resilience of coral reefs. *Science*, *301*(5635), 929–933. doi:10.1126cience.1085046 PMID:12920289

IUCN. (2010). *Building resilience to climate change: Ecosystem-based adaptation and lessons from the field*. Gland, Switzerland: IUCN.

Iverson, L. R., Prasad, A. M., Matthews, S. N., & Peters, M. (2008). Estimating potential habitat for 134 eastern US tree species under six climate scenarios. *Forest Ecology and Management*, *254*(3), 390–406. doi:10.1016/j.foreco.2007.07.023

Jetz, W., Wilcove, D. S., & Dobson, A. P. (2007). Projected impacts of climate and land-use change on the global diversity of birds. *PLoS Biology*, *5*(6), e157. doi:10.1371/journal.pbio.0050157 PMID:17550306

Kitzes, J., & Harte, J. (2014). Beyond the species–area relationship: Improving macroecological extinction estimates. *Methods in Ecology and Evolution*, *5*(1), 1–8. doi:10.1111/2041-210X.12130

Lavergne, S., Mouquet, N., Thuiller, W., & Ronce, O. (2010). Biodiversity and climate change: Integrating evolutionary and ecological responses of species and communities. *Annual Review of Ecology Evolution and Systematics*, *41*(1), 321–350. doi:10.1146/annurev-ecolsys-102209-144628

Leadley, P. (2010). *Biodiversity scenarios: projections of 21st century change in biodiversity, and associated ecosystem services: a technical report for the global biodiversity outlook 3 (No. 50)*. UNEP/Earthprint.

Lozano, F. D., & Schwartz, M. W. (2005). Comparative taxonomic structure of the floras of two Mediterranean-climate regions: Iberia and California. *Diversity & Distributions, 11*(5), 399–408. doi:10.1111/j.1366-9516.2005.00168.x

Mace, G. M., Norris, K., & Fitter, A. H. (2012). Biodiversity and ecosystem services: A multilayered relationship. *Trends in Ecology & Evolution, 27*(1), 19–26. doi:10.1016/j.tree.2011.08.006 PMID:21943703

Mace, G. M., & Purvis, A. (2008). Evolutionary biology and practical conservation: Bridging a widening gap. *Molecular Ecology, 17*(1), 9–19. doi:10.1111/j.1365-294X.2007.03455.x PMID:17696991

Maclean, I. M., & Wilson, R. J. (2011). Recent ecological responses to climate change support predictions of high extinction risk. *Proceedings of the National Academy of Sciences of the United States of America, 108*(30), 12337–12342. doi:10.1073/pnas.1017352108 PMID:21746924

MacMynowski, D. P., & Root, T. L. (2007). Climate and the complexity of migratory phenology: Sexes, migratory distance, and arrival distributions. *International Journal of Biometeorology, 51*(5), 361–373. doi:10.100700484-006-0084-1 PMID:17245563

Malcolm, J. R., Liu, C., Neilson, R. P., Hansen, L., & Hannah, L. E. E. (2006). Global warming and extinctions of endemic species from biodiversity hotspots. *Conservation Biology, 20*(2), 538–548. doi:10.1111/j.1523-1739.2006.00364.x PMID:16903114

McCarthy, J. J., Canziani, O. F., Leary, N. A., Dokken, D. J., & White, K. S. (Eds.). (2001). *Climate change 2001: impacts, adaptation, and vulnerability: contribution of Working Group II to the third assessment report of the Intergovernmental Panel on Climate Change* (Vol. 2). Cambridge University Press.

Meehl, G. A., Stocker, T. F., Collins, W. D., Friedlingstein, P., Gaye, T., Gregory, J. M., & Raper, S. C. (2007). *Global climate projections*. Academic Press.

Mensah, A., Anderson, K., & Nelson, W. (2016). *Review of adaptation related policies in Ghana* (No. 107642). DECCMA working paper, deltas, vulnerability and climate change: migration and adaptation, IDRC project.

Meyers, L. A., & Bull, J. J. (2002). Fighting change with change: Adaptive variation in an uncertain world. *Trends in Ecology & Evolution, 17*(12), 551–557. doi:10.1016/S0169-5347(02)02633-2

Miller-Rushing, A. J., Høye, T. T., Inouye, D. W., & Post, E. (2010). The effects of phenological mismatches on demography. *Philosophical Transactions of the Royal Society of London. Series B, Biological Sciences, 365*(1555), 3177–3186. doi:10.1098/rstb.2010.0148 PMID:20819811

Montoya, J. M., & Raffaelli, D. (2010). *Climate change, biotic interactions and ecosystem services*. Academic Press.

Naeem, S., Bunker, D. E., Hector, A., Loreau, M., & Perrings, C. (Eds.). (2009). *Biodiversity, ecosystem functioning, and human wellbeing: an ecological and economic perspective*. Oxford University Press. doi:10.1093/acprof:oso/9780199547951.001.0001

Noss, R. F. (1990). Indicators for monitoring biodiversity: A hierarchical approach. *Conservation Biology, 4*(4), 355–364. doi:10.1111/j.1523-1739.1990.tb00309.x

Ozgul, A., Childs, D. Z., Oli, M. K., Armitage, K. B., Blumstein, D. T., Olson, L. E., & Coulson, T. (2010). Coupled dynamics of body mass and population growth in response to environmental change. *Nature*, *466*(7305), 482–485. doi:10.1038/nature09210 PMID:20651690

Packard, S., & Mutel, C. F. (Eds.). (1997). *The Tallgrass Restoration Handbook*. Washington, DC: Island.

Parmesan, C. (2006). Ecological and evolutionary responses to recent climate change. *Annual Review of Ecology Evolution and Systematics*, *37*(1), 637–669. doi:10.1146/annurev.ecolsys.37.091305.110100

Parmesan, C. (2007). Influences of species, latitudes and methodologies on estimates of phenological response to global warming. *Global Change Biology*, *13*(9), 1860–1872. doi:10.1111/j.1365-2486.2007.01404.x

Parmesan, C., & Yohe, G. (2003). A globally coherent fingerprint of climate change impacts across natural systems. *Nature*, *421*(6918), 37–42. doi:10.1038/nature01286 PMID:12511946

Pereira, H. M., Leadley, P. W., Proença, V., Alkemade, R., Scharlemann, J. P., Fernandez-Manjarrés, J. F., ... Chini, L. (2010). Scenarios for global biodiversity in the 21st century. *Science*, 1196624. PMID:20978282

Peterson, T. C., & Manton, M. J. (2008). Monitoring changes in climate extremes: A tale of international collaboration. *Bulletin of the American Meteorological Society*, *89*(9), 1266–1271. doi:10.1175/2008BAMS2501.1

Phillips, S. J., Dudík, M., Elith, J., Graham, C. H., Lehmann, A., Leathwick, J., & Ferrier, S. (2009). Sample selection bias and presence-only distribution models: Implications for background and pseudo-absence data. *Ecological Applications*, *19*(1), 181–197. doi:10.1890/07-2153.1 PMID:19323182

Pressey, R. L., Cabeza, M., Watts, M. E., Cowling, R. M., & Wilson, K. A. (2007). Conservation planning in a changing world. *Trends in Ecology & Evolution*, *22*(11), 583–592. doi:10.1016/j.tree.2007.10.001 PMID:17981360

Prober, S. M., & Dunlop, M. (2011). Climate change: A cause for new biodiversity conservation objectives but let's not throw the baby out with the bathwater. *Ecological Management & Restoration*, *12*(1), 2–3. doi:10.1111/j.1442-8903.2011.00563.x

Raven, P. H. (1973). The evolution of Mediterranean floras. In *Mediterranean type ecosystems* (pp. 213–224). Berlin: Springer. doi:10.1007/978-3-642-65520-3_14

Ravi Prasad Rao, B., Salamma, S., & Durga Prasad, K. (2017). From Climate Change and Biodiversity Impacts. *Adaptation and Mitigation-an Overview*.

Rehfeldt, G. E., Crookston, N. L., Sáenz-Romero, C., & Campbell, E. M. (2012). North American vegetation model for land-use planning in a changing climate: A solution to large classification problems. *Ecological Applications*, *22*(1), 119–141. doi:10.1890/11-0495.1 PMID:22471079

Richardson, A. J., & Poloczanska, E. S. (2008). Under-resourced, under threat. *Science-New York Then Washington*, *320*(5881), 1294–1295. doi:10.1126cience.1156129 PMID:18535230

Rissik, D. (2014). *Peer Review Report: Assessment of the Comprehensiveness of the Potential Environmental Impacts, Threats and Risks Identified in the Tamar Lake Inc*. Reports.

Safriel, U., & Adeel, Z. (2008). Development paths of drylands: Thresholds and sustainability. *Sustainability Science, 3*(1), 117–123. doi:10.100711625-007-0038-5

Salamin, N., Wüest, R. O., Lavergne, S., Thuiller, W., & Pearman, P. B. (2010). Assessing rapid evolution in a changing environment. *Trends in Ecology & Evolution, 25*(12), 692–698. doi:10.1016/j.tree.2010.09.009 PMID:20961648

SBSTTA. (1999). Biological diversity of drylands, arid, semi-arid, savannah, grassland and mediterranean ecosystems. Draft recommendations to COP5. Montreal, Canada: Author.

Schloss, C. A., Nuñez, T. A., & Lawler, J. J. (2012). Dispersal will limit ability of mammals to track climate change in the Western Hemisphere. *Proceedings of the National Academy of Sciences of the United States of America, 109*(22), 8606–8611. doi:10.1073/pnas.1116791109 PMID:22586104

Scott, D. (2006). Climate change and sustainable tourism in the 21st century. In Tourism Research: Policy, Planning, and Prospects (pp.175-248). Waterloo: Série Publications du Département de géographie, Université de Waterloo.

Secretariat, C. B. D. (2010). *Global biodiversity outlook 3*. In Secretariat of the Convention on Biological Diversity.

Sekercioglu, C. H., Schneider, S. H., Fay, J. P., & Loarie, S. R. (2008). Climate change, elevational range shifts, and bird extinctions. *Conservation Biology, 22*(1), 140–150. doi:10.1111/j.1523-1739.2007.00852.x PMID:18254859

Sgro, C. M., Lowe, A. J., & Hoffmann, A. A. (2011). Building evolutionary resilience for conserving biodiversity under climate change. *Evolutionary Applications, 4*(2), 326–337. doi:10.1111/j.1752-4571.2010.00157.x PMID:25567976

Sinervo, B., Mendez-De-La-Cruz, F., Miles, D. B., Heulin, B., Bastiaans, E., Villagrán-Santa Cruz, M., ... Gadsden, H. (2010). Erosion of lizard diversity by climate change and altered thermal niches. *Science, 328*(5980), 894–899. doi:10.1126cience.1184695 PMID:20466932

Smith, B., & Wandel, J. (2006). Adaptation, adaptive capacity and vulnerability. *Global Environmental Change, 16*(3), 282–292. doi:10.1016/j.gloenvcha.2006.03.008

Smith, N. (2010). *Uneven development: Nature, capital, and the production of space*. University of Georgia Press.

Solomon, S., Qin, D., Manning, M., Chen, Z., Marquis, M., Averyt, K. B., & Miller, H. L. (2007). IPCC, 2007: Climate change 2007: The physical science basis. Contribution of Working Group I to the fourth assessment report of the Intergovernmental Panel on Climate Change.

Staudinger, M. D., Grimm, N. B., Staudt, A., Carter, S. L., & Chapin, F. S. (2012). *Impacts of climate change on biodiversity, ecosystems, and ecosystem services*. Washington, DC: United States Global Change Research Program.

Suárez, A., Watson, R. T., & Dokken, D. J. (2002). *Climate change and biodiversity*. Academic Press.

Suttle, K. B., Thomsen, M. A., & Power, M. E. (2007). Species interactions reverse grassland responses to changing climate. *Science, 315*(5812), 640–642. doi:10.1126cience.1136401 PMID:17272720

Swanson, D. L., & Palmer, J. S. (2009). Spring migration phenology of birds in the Northern Prairie region is correlated with local climate change. *Journal of Field Ornithology, 80*(4), 351–363. doi:10.1111/j.1557-9263.2009.00241.x

Taylor, S. G. (2008). Climate warming causes phenological shift in Pink Salmon, Oncorhynchus gorbuscha, behavior at Auke Creek, Alaska. *Global Change Biology, 14*(2), 229–235. doi:10.1111/j.1365-2486.2007.01494.x

Thomas, C. D., Cameron, A., Green, R. E., Bakkenes, M., Beaumont, L. J., Collingham, Y. C., & Hughes, L. (2004). Extinction risk from climate change. *Nature, 427*(6970), 145–148. doi:10.1038/nature02121 PMID:14712274

Thompson, J. N. (2005). *The geographic mosaic of coevolution*. University of Chicago Press.

Thuiller, W., Lavorel, S., Araújo, M. B., Sykes, M. T., & Prentice, I. C. (2005). Climate change threats to plant diversity in Europe. *Proceedings of the National Academy of Sciences of the United States of America, 102*(23), 8245–8250. doi:10.1073/pnas.0409902102 PMID:15919825

Todd, B. D., Scott, D. E., Pechmann, J. H., & Gibbons, J. W. (2011). Climate change correlates with rapid delays and advancements in reproductive timing in an amphibian community. *Proceedings of the Royal Society of London B: Biological Sciences, 278*(1715), 2191-2197. 10.1098/rspb.2010.1768

UNESCO. (2007). Études de cas sur le changement climatique et le patrimoine mondial. Paris, France: UNESCO, Centre du patrimoine mondial.

VanBuskirk, K. M., O'Neill, M. T., De La Vega, P., Maier, A. G., Krzych, U., Williams, J., ... Kneteman, N. M. (2009). Preerythrocytic, live-attenuated Plasmodium falciparum vaccine candidates by design. *Proceedings of the National Academy of Sciences of the United States of America, 106*(31), 13004–13009. doi:10.1073/pnas.0906387106 PMID:19625622

Walther, G. R., Post, E., Convey, P., Menzel, A., Parmesan, C., Beebee, T. J., & Bairlein, F. (2002). Ecological responses to recent climate change. *Nature, 416*(6879), 389–395. doi:10.1038/416389a PMID:11919621

Wiebe, K. L., & Gerstmar, H. (2010). Influence of spring temperatures and individual traits on reproductive timing and success in a migratory woodpecker. *The Auk, 127*(4), 917–925. doi:10.1525/auk.2010.10025

Williams, B. K., Szaro, R. C., & Shapiro, C. D. (2009). *Adaptive management: the US Department of the Interior technical guide*. US Department of the Interior.

Williams, J. W., & Jackson, S. T. (2007). Novel climates, no-analog communities, and ecological surprises. *Frontiers in Ecology and the Environment, 5*(9), 475–482. doi:10.1890/070037

Williams, S. E., Shoo, L. P., Isaac, J. L., Hoffmann, A. A., & Langham, G. (2008). Towards an integrated framework for assessing the vulnerability of species to climate change. *PLoS Biology, 6*(12), e325. doi:10.1371/journal.pbio.0060325 PMID:19108608

Willis, K. J., & Bhagwat, S. A. (2009). Biodiversity and climate change. *Science*, *326*(5954), 806–807. doi:10.1126cience.1178838 PMID:19892969

Wilson, E. O., & MacArthur, R. H. (1967). The theory of island biogeography. Princeton, NJ: Academic Press.

Wisheu, I. C., Rosenzweig, M. L., Olsvig-Whittaker, L., & Shmida, A. (2000). What makes nutrient-poor Mediterranean heathlands so rich in plant diversity? *Evolutionary Ecology Research*, *2*(7), 935–955.

Xenopoulos, M. A., Lodge, D. M., Alcamo, J., Märker, M., Schulze, K., & Van Vuuren, D. P. (2005). Scenarios of freshwater fish extinctions from climate change and water withdrawal. *Global Change Biology*, *11*(10), 1557–1564. doi:10.1111/j.1365-2486.2005.001008.x

KEY TERMS AND DEFINITIONS

Adaptation: In biology means the process by which a plant or animal species adapt to the environment in which it lives following a natural selection that acts on a genetic variation.

Biodiversity: Is also known as biological diversity that evokes the lifestyles of one or more individuals, specific diversity (animal or plant), as well as genetic diversity.

Biodiversity Extinction: Known as a crisis of biodiversity, which means the extinction of many animal or vegetable species.

Biodiversity Loss: Designates the disappearance of living beings following its consumption by humans or its maladaptation to climate change.

Biodiversity Richness: (Also known as species richness) is a measure of the ecosystem, which refers to the number of species present in a given location.

Climate Change: Is a disruption of the earth's climate or regional climates, it caused an increase in temperature or cooling, as well as an increase in greenhouse gas emissions caused by human activities.

Conservation: Also known as protection of endangered or vulnerable species plays a vital role as well as the involvement of local people in this process.

Ecosystem: Is the whole of the living beings who live in a given environment, as well as their interaction.

Chapter 8
Impacts of Climate Change on Biodiversity and Ecosystem Services:
Current Trends

Vartika Singh
Amity University Noida, India

ABSTRACT

The chapter illuminates the effect of climate change on biodiversity and ecosystem services. It also explains the present scenario of biological resources with biological diversity. The author describes the known biological resources and their uses with the help of flow chart and also shows the linkage among them. The author clearly explicates the inter-relationship between human activity, climate change, and biological diversity with charts. In the last section, the author talk about the effect on the paleo climate and impact of climate change on different parts of ecosystem services.

INTRODUCTION

In the context of subject matter (Impact of Climate change on Biodiversity and Eco-system services), it is imperative to consider broad definitions for (a) Climate Change (b) Biological diversity including biological resources.

Climate Change

In general, Climate change represent a change in the statistical distribution of weather patterns and that change persists for an extended period (ex. for decades together). Factors that influence climatic variations are termed as climate forcing or forcing mechanisms which can be both internal (and natural to the climatic system itself) and external.

DOI: 10.4018/978-1-5225-7387-6.ch008

Of all the forcing mechanisms, the anthropogenic influences on climate change are considered more critical and relevant in the context of environmental policies at both national and global levels (Rabindranath etal., 1998). Anthropogenic climate changes often referred to as "global warming." The term "global warming" became synonymous with climate change induced by human activity. However, there are intricate differences between global warming and climate change. The former term refers to surface temperature increases while climate change includes global warming and everything else that can be affected by increasing levels of greenhouse gases.

Biological Diversity and Biological Resources

Though these two expressions are used in an inter-mingling fashion, they have visible delineations.

- **Biological Diversity:** Precise and widely acceptable definition of biological diversity is provided by the treaty of Convention on Biological Diversity (CBD) as under:

"Biological diversity" means the variability among living organisms from all sources including, among other things, terrestrial, marine and other aquatic ecosystems and the ecological complexes of which they are part: this includes diversity within species, between species and of ecosystems(Glowka, 1994).

- **Biological Resources:** Any biological entity that has a known or potential application in the economic, environmental, social and cultural well-being of mankind can be defined as "biological resource" in a broader sense. Further, the term biological resource in this broad definition invariably includes the species as a whole or its parts and genetic material. The CBD treaty defined biological resources as under.

"Biological resources' includes genetic resources, organisms or parts thereof, populations, or any other biotic component of ecosystems with the actual or potential use or value for humanity (Cardozo & Comp, 2003).

As a matter of fact, this kind of broader definition seeks to encompass the interactions between the human being and the rest of our nature. Nonetheless, it has been customary to consider the biological resource purely from an economic perspective (existing, potential or perceived) only. The following chart attempts to capture well known economic applications of biological diversity- specifically in the context of botanical species.

Factors Contributing to Climate Change

Regarding human-induced factors leading to climate change, it is important to understand the carbon cycle. An inter-play between oceans, fossil fuel reserves, soil and vegetation and atmosphere comprise of global carbon cycles. To available estimates, the global carbon distribution in nature is as under (Table 1).

As against this distribution, estimates for carbon release volume and absorption patterns since 1850 are as shown in Table 2).

Figure 1. Showing the known uses of biological resources
Source: (Bowker, 2000)

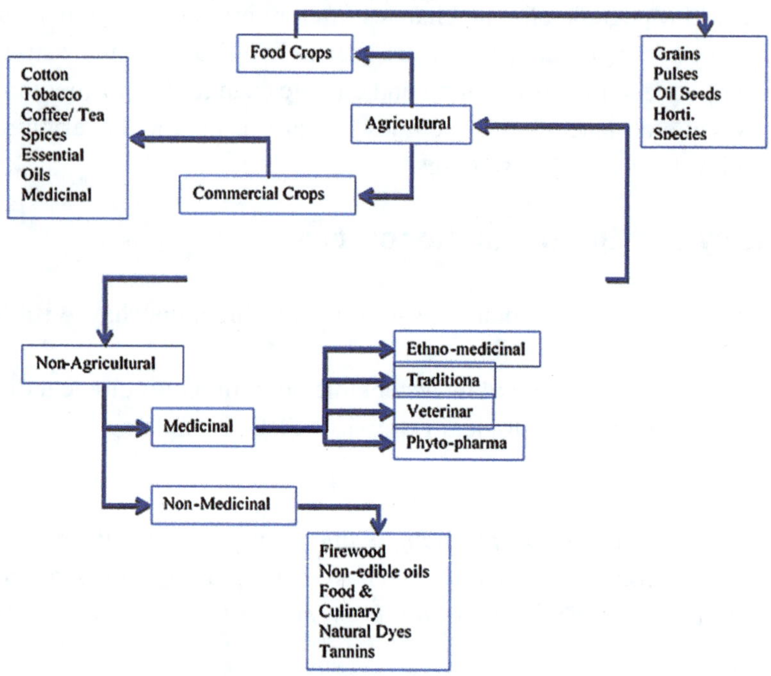

Table 1. Global distribution of natural carbon reserves

Source	Carbon Cont. (Expressed as Giga Tons of Carbon- GtC)	As % of Total Global Carbon Level
Oceans	39000	66.94%
Fossil Fuel Deposits	16000	27.46%
Soils & Vegetation	2500	4.29%
Atmosphere	760	1.34
Total	58260 Giga tons of Carbon	100%

Source: (*Dash etal., 2007*)

Table 2. Carbon release and absorption patterns

Carbon Release		Carbon Absorption	
Source	Volume (GtC)*	Target	Volume (GtC)*& (% Age)
Change in Land use	136	Atmosphere	180 (44.34%)
Fossil Fuel Consumption	270	Growing Vegetation & Soil	110 (27.07%)
Totals of Release:	406 GtC	Oceans	116 (28.57%)
		Totals of absorption	406 GtC

*GtC: Giga tons Carbon
Source: (*Dash etal., 2007*)

It may be observed that human activity is directly contributing to the increasing concentrations of atmospheric carbon as carbon dioxide. It is this concentration leading to persistent and irreversible climate change. Further, it may be noted that the combustion of fossil fuels is the larger contributor to this increasing concentration of atmospheric carbon. This factor may be further strengthened by another observation. Between 1990-1999, it is estimated that 6.3 GtC/ year was released by usage of fossil fuels. Another 1.6 GtC/ year were released due to the burning of forest vegetation- totaling to 7.9 GtC/ year. Growing vegetation and Oceans were estimated each, to absorb 2.3 GtC/ year. Rest of 3.3 GtC/ year accounted for atmospheric carbon - in the form of Carbon-di-oxide(Kumar & Chopra, 2009).

Other Greenhouse Gases and Global Warming

Greenhouse gases occurring in the nature produce greenhouse effect on air-temperature near the earth surface. In general, the natural greenhouse effect is considered to cause 33°C (D. Archer, et al., 2009)., Different constituents of GHG are known to cause different levels of the greenhouse effect. Water vapor is known to cause 36-70%, Carbon dioxide 9-26%, Methane 4-9%, Ozone-3-7% of greenhouse effect. Besides, nitric oxide and CFCs also contribute to this effect (Kumar & Chopra, 2009). However, concentrations of CO2 and methane are calculated to have increased by 36% and 148% respectively since 1750 due to human activities (Burke et al., 2015). Thus, the effects of various constituents of greenhouse gases are also major contributors for persistently increasing levels of global warming. These contributory causes of climate change are fundamental to the ultimate erosion of biological diversity and more importantly, biological resources needed for the very existence of not of human beings alone but for the entire set of the organism of faunal origin.

CLIMATE CHANGE AND ITS EFFECTS ON BIOLOGICAL DIVERSITY AND BIOLOGICAL RESOURCES

Measurements of global mean temperatures though the last 130 years suggest that, the temperatures have risen substantially. However, the rate of this increase during the last few years has been more alarming in relation to the past. In 1996, the Intergovernmental Panel on Climate Change (IPCC) had released its second technical report, which represents a degree of unanimity on various issues of climate. The report estimates that, atmospheric carbon dioxide content will increase from 350 ppmv in 1993 to 525 ppmv by 2050 (Chaturvedi etal., 2011). This estimation is based on the case of IPCC business-as-usual (meaning no deliberate attempt to reduce the carbon dioxide emissions). This will translate into an increase of 0.3° C in global mean temperatures per decade. A potential rise in the global mean surface temperature of 1° to 3.5° was warned based on a recent climate model. Such a change could be 10-50 times faster as compared to a natural change in the parameter (Sathaye et al., 2006). However, projected increases in global temperatures may not be uniform across the globe. The effect will be much more profound on poles (with an increase of 0.8° C per decade) as against equatorial regions (where the increase is projected as 0.1° C per decade) (Thuiller etal., 2005).

- *Eventually, this rise results in increased mean sea levels causing immediate threats to coastal regions as well as for marine biodiversity.*

Substantial scientific evidence gathered through various models suggests that climate change is a potential risk to earth's biodiversity. In comparison to other threats posed by human activities (like pollution, alterations in land use patterns, etc.). Figure 2 depicts an outline for process pathways by which, the climate change can influence the biological diversity. However, it is important to note that:

- The direct effects of recent changes in climate are considered to be slow
- It is difficult to measure/ quantify the inter-relationship between the increases in global means and depletion rate in biodiversity- in a convincing manner
- But these adverse effects are global and are practically irreversible.

As seen from the above figure, Climate change could affect the biological diversity at the genetic level alsoAs per recent estimates, our planet hosts over 13 million species. However, only 1.6 million species have actually been described under different flora and fauna(Larsen et al., 2017). It is also well known that some parts of the globe host highly endemic and exceptionally high concentrations of species. These habitats are also considered to deplete at unusually rapid rates. These areas are called "biodiversity hotspots" and are largely located in tropical forests, coral reefs, and Mediterranean ecosystems.

Loss and alteration of habitats, chemical pollution, over-harvesting, invasive species, increasing population pressures are the well establishes causes of rapid depletion in biological diversity. In the process of depletion, the Biological Resources tend to lose-out first, due to their social and economic relevance leading to over-harvesting.

Climate change may modify and accelerate anthropogenic interventions in local eco-systems. The impacts of climate change on biodiversity/ biological resources need to be assessed under two major timelines

Figure 2. Inter-relationship between human activity- climate change- biological diversity
Source: (Gairola et al., 2010)

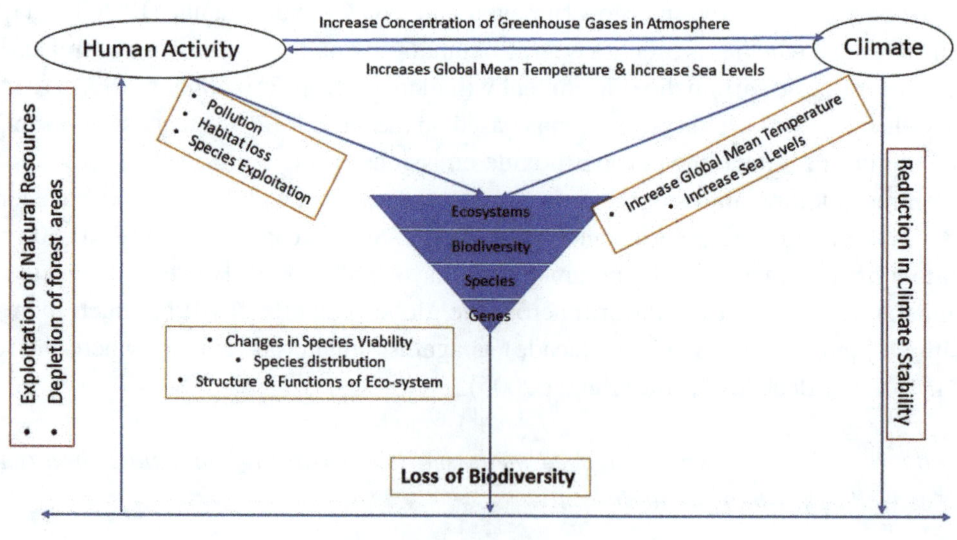

- Paleoclimate changes
- Recent climate changes

Effects of Paleoclimate Changes on Biodiversity

Understanding the effects of past climate changes on plant and animal diversity might help us, in developing appropriate models for prediction of future shifts in species, communities, and ecosystems.

Over the past million years, the most noticeable changes in climate have been the oscillations between colder and warmer (glacial and interglacial) periods- with glacial periods occurring every 1,00,000 years (Sigman & Boyle, 2000),. However, the climatic developments during the last 18000 years (BP- Before Present) including the last glacial- are important in the context of subject matter (Douglass etal. 2005)

The rise in global mean temperature during this period had been five ±1° C due to doubling in the GHG emissions. Itapproximates close to 4.2±1° C, which is the predicted rate for the near future. However, under future global warming conditions- it is predicted that the rates of increase in global mean temperature would be much faster than any of the natural warming rates during the last 18000 years. This particular aspect of global warming phenomenon implies that many species might not be able to move their ranges rapidly enough to catch-up the pace of global warming.

The following trends may be considered to evaluate the impact of paleo climatic changes on biodiversity:

- It was observed that there exists a strong interrelationship between the climate change and floristic dynamics along elevation gradients in tropical mountain regions of Colombian Andes.
    - This particular study relating to Pleistocene climate change (2000,000- 10,000 BP- before present) reported strong altitudinal shifts in the vegetation belts in response to climate change (IPCC, 2007).
- Dendro-chronological studies in Chile and northern hemisphere indicate the sensitivity of tree species to temperature changes over the last few hundred years.
- Data from Africa suggests the impacts of cooler and drier environments on paleotropical rain forests during last ice age (about 70000-12000 years BP). Fossil pollen records from East African mountains show a descent of vegetation zones indicating a 6° C change in temperature.

The above observations point-out that the long-term climate changes modify altitudinal preferences of the species. When global mean temperatures fall, species shift towards lower altitudes as against their behavior to prefer higher altitudes during the global warming.

Impacts of Recent Climate Changes on Biological Diversity

These changes may be classified and described under the following categories:

1. Shifts in major vegetation zones or biomes
2. Shifts in ranges of individual species and compositions of species assemblages
3. Interactions between effects of climate change and habitat fragmentation
4. Changes in ecosystem functioning

Shifts in Major Vegetation Zones or Biomes

In the context of impacts of climate change on biodiversity, vegetation shift is discussed as the foremost influence. This impact is distinctly visible in both altitudinal and latitudinal shifts. As the earth warms, the species tend to shift towards latitudes (poles) and altitudes. For example, timber lines in Finland would rise to about 200 m altitude, where it was inhabited in the past (4500-7500 years ago) during Atlantic thermal periods(Schlyter,2006).Under projected future global warming conditions, tropic alpine vegetation may be fully replaced by montane cloud forests which are presently present in lower altitudes.

Shifts in Ranges of Individual Species and Compositions of Species Assemblages

Long lasting climate changes- tend to affect physiology, phenology and also the interspecific interactions between individual species. As a result, shifts in geographic distributions may occur. For example, range changes were observed for several species of butterflies in Britain are attributed to small but consistent increments in temperature during this century. Similarly, northward expansion was observed in case of Birch towards Swedish tundra. This phenomenon is also attributed to the consistent increase in surface temperatures during the first half- of this century. However, if the species fails to adjust its distribution patterns in response to climate change; it might face the risk of extinction. During the last few decades, 64 species of vascular plants are recorded to be extinct in the Netherlands. Further, some 84 new species have also emerged during the same period. For most of these 84 "neophytes" the northern most limit of the geographic range is in Dutch delta. (Kappelle et al., 1999). Long-term and persisting upward changes in temperature and direct human interventions in the eco-systems are considered to trigger species shifting processes. According to another report, which involved a study on over 1350 species, more than 50% species would become "vulnerable" by 2080 under an assumption of "no-migration."(Garavito, 2015). The study further suggested that at least 22% of these species shall fall under the "critically endangered" category while 2% of them might go extinct. Recent phenological changes in Microlepidoptera species is also an important manifestation of climate change. Between 1975 and 1995 the flight peaks shifted on an average to date 11.6 day earlier (Parmesan,1996). This observation is attributed to the increasing temperatures. During the same period, more than 50% species has undergone significant distribution changes. In this kind of studies aimed to predict the effect of climate change, species which are very sensitive to temperature are used as bio-indicator. Alterations in species assemblages are evaluated through long-term and repeated GIS mapping of selected areas.

Interactions Between Effects of Climate Change and Habitat Fragmentation

During last few decades, depletion in biodiversity at local, regional and global levels is also attributed to destruction and fragmentation of habitats. This particular factor leading to depletion is an additional factor besides the effects of climate change. Both these factors together may set the stage for an even larger wave of extinctions. Habitat fragmentation leads to isolation of species and even sometimes, of the populations within the same species. This leads to a narrowing of genetic diversity within the species. Fragmentation may also restrict the opportunities for migration and dispersions to other suitable areas for the species.

In addition to the effect on migration opportunities of populations, fragmentation of landscapes into smaller units can have a larger impact on climate along edges and inside the remaining fragments. In many forest areas, fragmentation has produced landscapes in which, the forest edges are a dominant feature and thereby restricting opportunities for other species inside the core areas. Studies also show that rain forest remnants surrounded by agricultural fields are exposed to markedly different solar radiation, wind, and water and nutrient regimes- about continuously forested regions (Kappelle et al., 1999). Survival of populations or species in fragmented habitats essentially depends on the opportunities available for migration/ dispersion. It must be underlined that if the migration opportunities enable the populations/ species to adjust their geographic distribution during climate changes. Opportunities for these critical adjustments are influenced by fragment size, quality, and distribution. Thus, it may be stated that climate change and habitat fragmentation together contribute to an enhanced pace for biodiversity depletion.

Changes in Ecosystem Functioning

The functioning of any given ecosystem is the sum of interrelated activities of its organisms, i.e., the ways and rates at which they carry out the ecosystem processes (like respiration, carbon dioxide fixation, nitrification, litter decomposition, etc. Climate change is expected to influence ecosystem functioning by affecting the physiology of the species inhabited within the ecosystem. Climate change directly influences the local temperature patterns and availability of moisture both in the soil and atmosphere. As of now, there is no convincing evidence that ecosystem processes are crucially dependent on wider biodiversity. Rather, the functional characteristics of the dominant plant species would be important in controlling ecosystem processes. However, it is obvious that the losses of species or functional groups from an ecosystem could at some point of time impair its functions and its capacity for rendering ecosystem services. Therefore, it is often agreed that wider biodiversity inhabited in any ecosystem contributes to its resilience and improves adjustments to climate change.

Effects of Climate Change on Plantation Forests

The effects of global warming on natural forests/ ecosystems and the biodiversity thereof are briefly discussed with suitable examples drawn from published research communities and other related literature.

Almost 4% of the global forest area comprises of plantation forests. By and large, these plantation forests are developed to meet the ensuing demand for timber, pulpwood and or such other commercial applications (Barua, 2012). However, the plantation forest areas also render certain other ecosystem services. Since the plantation forest areas initially contain plantation of selected tree species, the area becomes a potential host for diverse herbaceous and ground cover species.

In addition to its direct impact, climate change shall have relevance in the context of plantations management (such as the selection of species, rotation length, etc.). Therefore, it is important to consider the impacts of climate change on the plantation forests. Some of the potential impacts of climate change on plantation forests are summarized in (Table 3)

As seen from the above table, climate change shall have multi-faceted impacts on future of plantation forests. These impacts might often be triggered by regional temperature and moisture balances. Greater understanding of such management impacts covering both biotic and abiotic influences is needed for management of plantation forests. Furthermore, any change in management plans to mitigate the effects

Table 3. Showing the effect of different constituents of climate change

Factor	Potential Impacts	Relevant Examples
Rising Temperature	Adverse effects on temperature sensitive species	• Increase in abundance of thermophilic species • Changes in species richness patterns
Fire frequency	Changing fire frequency in forest areas	Increase in the disturbance in adapted species
Extreme storm events	Greater frequency and magnitude	Large storms across different continents had the devastating effect on plantation forests
Soil abiotic factors	Increase in Soil temperature & altered cycles of wetting and drying cycles	Changes in abundance and composition of soil fauna
	Changing quantity and quality of key inputs for soil carbon like leaf litter etc.	Effect on plant's growth patterns and life cycle
Pests and diseases	Shifts in pest range	The latitudinal and altitudinal spread of the pine processionary moth beyond its previous range in Europe
	Reduced winter mortality and increased number of pest generations per year leading to larger and more persistent pest outbreaks	The outbreak of the mountain pine beetle in pine forests in British Columbia and other partsof western North America
Biocontrol measures	Potential for reduced effectiveness of current biocontrol agents due to phenological or other changes under climate change	

Source: (Yadav et al., 2015)

of climate change shall have long-term effects on biodiversity- where plantations are to be managed on long rotations for timber or carbon.

CLIMATE CHANGE AND BIODIVERSITY IN ARID AND SEMI-ARID ZONES

In fact, the subject of global warming, its effects on the planet's health and the approaches for its mitigation are discussed in a global framework. However, it is imperative to monitor the climate change patterns and their impact on the national, regional and local ecosystems.

At the same time, countries like India face a mammoth task in carrying out these monitoring studies due to multiple complexities. As it well appreciated, the climatic conditions in India exhibit vast spatial and temporal variations. Important weather events in Arid and Semi-Arid regions include floods and drought, monsoon depressions and cyclones, heat waves and cold waves, prolonged fog and snowfalls. With this level of complexities, it would be arduous for scientists with limited resources to bring about a reliable data on climate change patterns across the country.

Basis the available data, it may be stated that:

- Maximum temperature has been increasing during the last century over all the regions of India. However, the magnitude of such an increase is not uniform across the country. West coast shows highest increase in the maximum temperature by about 1.2° C followed by 1° C in north east, 0.9° C in western Himalaya, 0.8° in the north central, 0.6° C both in north-west and east coast (Dash et al., 2007). The lowest level of increase (0.5° C) was seen in an interior peninsula.

- Regarding change in minimum temperatures, it was observed that there is the sharp reduction on this account in north India during 1955-1972 and then an equally sharp increase over the past 3-4 decades in the last century. Such large fluctuation is not seen in Southern parts of India.
- The increase in temperatures in Southern regions goes in tunes with the Sea Surface Temperatures (SST) in the Bay of Bengal and the Arabian Sea.
- Further, available data indicated that, the increase in mean atmospheric surface temperature is about 1° C and 1.1°C during winter and post-monsoon months respectively. During pre-monsoon and monsoon months, the mean temperatures have increased by about 0.3 and 0.4° C respectively (Kumar, 2009).
- Similarly, the decrease in minimum temperatures during summer, monsoon and its increase in post-monsoon months created a large difference (about 0.8° C) in the season temperature anomalies (Syroka& Toumi, 2004). This phenomenon may eventually alter the distribution of heating sources and atmospheric circulation patterns.
- Extreme temperature events have increased in recent past. During heat wave periods, highest daily maximum temperatures in the country were observed in Andhra Pradesh instead of north west and central plains of India. The highest maximum temperature varied between 45-50° C at 4 stations in east coast (Kothawale et al., 2010).
- Rain fall patterns during different seasons indicate small increase during winter months of January-February, Pre-monsoon months of March, April and during the post-monsoon months of October, November, and December(Sinharay & Srivastava, 2000).
- There is an increase in the number of low-pressure areas from June to September. The total no of cyclonic disturbances crossing east coast shows a decreasing trend. At the same time, number of severe cyclonic storms crossing east coast shows an increasing trend.

The above factual statements relating to the climate change trends in India, paints a highly complex and confusing picture. It is practically difficult to predict/ estimate the impacts of these climatic changes on biological diversity.

Biological Diversity of India and Impacts of Climate Change

India is one among the mega-biodiversity zones of the world. To available estimates, the country hosts about 45000 plant species and 65,000 species of fauna. Of the plants, over 15000 species are flowering species (Kapelle et al., 1999). Among the flowering plants, 5000 to 7000 species are endemic. The following table provides a glimpse of faunal diversity hosted in India (Pawson et al., 2013).

The larger part of the above biological diversity is concentrated in Western Ghats and Eastern Himalayanmountains. Thus, these two zones are considered hot-spots for biological diversity. The biodiversity capture in the Himalayan region is shown in Table 5.

In a major attempt to project the impacts of climate change on Indian biological diversity a detailed predictive assessment was carried out two different scenarios. This study provides reliable insight into the trends.

Under Scenario 1 (Atmospheric Concentration of CO_2 reaches 740 ppm by 2085, (Ravindranath et al., 2006)

- On a national average, about 39% forest grids were projected to undergo vegetation type change.

Table 4. Biological Diversity in India about the global scenario

Groups	No. Species World-Wide	No of Species Found in India	As % age Species Reported World-Wide
Mammals	4629	350	7.6%
Birds	9702	1224	12.6%
Reptiles	6552	408	6.2%
Amphibians	4522	197	4.4%
Fishes	21730	2546	11.57%
Flowering Plants	250000	15000	6%

Source: (Khoshoo, 1993.)

Table 5. Endemic species in India

Taxonomic Groups	No. Species	No of Endemic Species	% Age of Endemic Species
Plants	10000	3160	31.6%
Mammals	300	12	4%
Birds	977	15	1.5%
Reptiles	177	48	27.3%
Amphibians	105	42	40%
Fresh water fishes	269	33	12.3%

Source: (Khoshoo, 1993)

- In dominant forest states like Chhattisgarh, Karnataka and Andhra Pradesh, this impact was projected to be more profound- with 73%, 67%, and 62% respectively.
- Net Primary Productivity (NPP) was projected to increase by 68.8%.
- Soil Organic Carbon (SOC) was projected to increase by 37.5%

Under Scenario 2 (Atmospheric Concentration of CO_2 reaches 575 ppm by 2085,(Alonso and Sukumar, 2016)

- By the end of this century, 34% of forest grids are projected to undergo vegetation type change.
- The NPP is projected to increase by 51.2% while SOC was estimated to increase by 30.2%.

Forest Vulnerability Index:

- Under global dynamic vegetation modeling used in this assessment- which is based on data sets of forest density, forest biodiversity as well as model predicted vegetation type changes forest vulnerability index (FVI) was presented by this study.
- The vulnerability index suggests that, Upper Himalayas, northern and central parts of western ghats and parts of central India are most vulnerable to emerging climate changes.
- North eastern forests are projected to be more resilient to cope-up with the climate change.

IMPACT OF CLIMATE CHANGE ON ECOSYSTEM SERVICES

As per Millennium Ecosystem Assessment (MA) 2006 ecosystem services are defined as "benefits people obtain from ecosystems." These services are divided into four distinct groups.

1. **Supporting Services:** Nutrient recycling, primary production and soil formation. These activities of ecosystems are at the base of all other forms of ecosystems services.
2. **Provisioning Services:** provisions required for very existence and prosperity of mankind. These include, food, timber, genetic resources (for crop improvement/ healthcare). Biogenic minerals, medicinal sources (medicinal plants, phytochemicals & microbes), energy (hydro-power & biomass fuels)& ornamental species.
3. **Regulatory Services:** Pollination, carbon sequestration, waste decomposition, purification of water and pest & disease control
4. **Cultural:** Most of the cultural manifestations like music, dance and visual arts were actually took their origin in different ecosystems and practiced by dependent communities.

It might be too complex to predict the effects of climate change on all these kinds of eco-system services. However, ecosystem degradation and alterations are the fundamental impacts of climate change. Once these processes take their rooting in any kind of ecosystem- a snowballing effect on the rest of services can be predicted.

For illustration purposes, the effect of climate change on medicinal plants (provisioning services) and pollination (regulatory services) is described hereunder:

Effects of Climate Change on Medicinal Plants

Ecosystems provide medicinal sources chiefly from higher plants and microbes. Particularly, forest-dependent communities rely upon higher plants for primary healthcare. Available evidence suggests that the impacts of global warming on medicinal plants could be widespread. Some of these changes can be classified into three groups.

1. **Shifts in Phenology:** As suggested above, lifecycle events of medicinal plants are reported to be altering visibly. For example, the flowering times of *Artemisia absinthium* and *Tanacetumparthenium* are reported to be advanced by six days. In case of *Mentha x Piperita*, the flowering season is said to be advanced by ten days.
2. **Shifting of the Ranges:** Certain botanical species are now recorded to be migrating to higher altitudes. Such a phenomenon is correlated to increasing temperatures in their native altitudes of habitat. However, the migratory potential is variable among different species. If the species has a low migratory potential, the process of range shifting would not be in proportion to the pace of climate change.
3. **Effect on Secondary Metabolites**: The medicinal values of any botanical species are largely attributed to the secondary plant metabolites. However, their biosynthesis within the plant is a multi-factorial phenomenon. One of these factors is the levels of biotic or abiotic stress which, the plant is exposed to. Synthesis of alkaloids and terpenoid groups are known to be impacted by the stress exerted on the plant species.

One of the major risk factors for the medicinal plants is their endemicity.

- Some of the medicinal plants by nature are endemic distribution. Their habitats are highly restricted. Any shrinkage in their habitats due to climate change would deplete the species and lead to their extinction.
- Even when the species is known to be pandemic, the quality profiles show high degree variance. The occurrence of specific chemo-type meeting the expected quality profiles might be restricted to limited habitats. In other words, "every medicinal plant is endemic from a quality point of view." In this case, also, shrinkage of habitats due to local climate changes would be detrimental to the resource.

At the same time- it is important to note that- the effects of climate changes on medicinal might not be adverse all the times and on all species. Sometimes, it is possible that these effects might augment medicinal metabolites and lead to the emergence of new chemo-types.

Effects of Climate Change on Pollinators and Pollination Services

Two groups of flowering species depend on insects and bees for pollination (for fruit and seed setting) viz. Agricultural and horticultural crops. The dependence on insect pollination is much wider with almost 80% wild species of plants depend on insect pollination only.

Further, sometimes the insect and plant interactions (mutualism) are unique and show the tendency for exclusivity. In these cases, the survival and sustainability of one species are lost when the other species is depleted leading to mutual extinctions.

Worldwide trends show that there is a visible decline in the diversity of pollinator insects. This would eventually reduce access to pollination services by plant species. A decline in the concerned flora is bound to trigger the further decline in pollinators thereby creating a vicious cycle leading to the extinction of both the stakeholders in the ecosystem.

The global economic value of pollination services is calculated at €153 billion equal to 9.5% of the total economic value of crops involving foods meant for direct consumption (i.e., grains, pulses, and vegetables) (Hegland et al., 2009; Potts et al., 2015).

The decline in the pollinator diversity is precipitated by many factors- such as the indiscriminate use of pesticides and insecticides in agriculture. Climate change is also one among those factors. Therefore, it has been the arduous task to evaluate the impact of climate change on pollinators diversity.

Nonetheless, available scientific evidence suggests that climate change impacts the interactions between the plant species and pollinator insects.

- It is now widely established that climate change induces phenological changes in plant species. Due to this specific phenomenon, the scope for "mismatches" increases substantially.
- Such "temporal mismatch" takes place between the on-set of flowering dates and the first appearance of insect for pollination. Since climate changes lead to delayed or earlier flowering, the insect populations may not be able to match the need of the plant species
- On a long run, there might be opportunities for the insects to undergo adaptive changes. In many cases, the opportunities for adaptive changes are also very narrow.

- Structure of pollination network is observed to be robust when "multi-species plant pollinators" occur as assemblages in the ecosystems.

In a nutshell, it may be stated that- climate change induces a reduction in pollination services by interfering with the natural mutualism among the plants and pollinators.

CONCLUSION

The terms "climate change" and "global warming" have a distinct delineation. Similarly, the terms "biological diversity" and "biological resources" are two different expressions. Climate change might be intrinsic to nature. These changes took place in a very insidious manner and sometime self-corrections might take place. However, the ongoing debate relates to the climate change induced by anthropogenic factors. Further, the debate is also concerned with the pace of climate change leading to increased GHG concentrations and increase in sea levels across the globe. Climate change impacts biological diversity through its gross effects on the ecosystem. By and large, climate change shall have a depleting effect on the existing taxa of flora and fauna. The effects of climate change also impact intra-specific genetic diversity. In addition to the depleting effects climate change may also alter the physiological phenomena of individual species like phenology, altitudinal and latitudinal preferences and certain other adaptive changes. Impacts of climate change on biological diversity most of the times may not be quantifiable precisely. However, these impacts are global and tend to be irreversible. Hence, strategically- appropriate mitigation plans need to be considered. Climate change is also now established to impact the ecosystem services. Particularly, these impacts are profound on provisioning and regulatory services. Through its wide-spread impact on biodiversity and ecosystem services, the climate change has all the potential to deplete the economic opportunities of dependent communities. Further, it is associated with potential risk of degrading the quality of human life.

REFERENCES

Alonso, A., & Raman, S. (2016). *Tropical Conservation Perspectives on Local and Global Priorities.* Oxford University Press.

Archer, D., Eby, M., Brovkin, V., Ridgwell, A., Cao, L., Mikolajewicz, U., ... Tokos, K. (2009). Atmospheric lifetime of fossil fuel carbon dioxide. *Annual Review of Earth and Planetary Sciences*, *37*(1), 117–134. doi:10.1146/annurev.earth.031208.100206. doi:10.1146/annurev.earth.031208.100206

Barua, S. K., Uusivuori, J., & Kuuluvainen, J. (2012). Impacts of carbon-based policy instruments and taxes on tropical deforestation. *Ecological Economics*, *73*, 211–219. doi:10.1016/j.ecolecon.2011.10.029. doi:10.1016/j.ecolecon.2011.10.029

Bowker, G. C. (2000). Biodiversity datadiversity. *Social Studies of Science*, *30*(5), 643–683. doi:10.1177/030631200030005001. doi:10.1177/030631200030005001

Burke, M., Hsiang, S. M., & Miguel, E. (2015). Global non-linear effect of temperature on economic production. *Nature*, *527*(7577), 235–239. doi:10.1038/nature15725 PubMed. doi:10.1038/nature15725 PMID:26503051

Cardozo, J., & Comp, L. (2003). Formulating an International Legal Protection for Genetic Resources, Traditional Knowledge and Folklore: Challenges for the Intellectual Property System. *Weerawit Weerawit*, *11*, 769.

Carle, J., & Holmgren, P. (2008). Wood from planted forests. *Forest Products Journal*, *58*(12), 6.

Chaturvedi, R. K., Gopalakrishnan, R., Jayaraman, M., Bala, G., Joshi, N. V., Sukumar, R., & Ravindranath, N. H. (2011). Impact of climate change on Indian forests: A dynamic vegetation modeling approach. *Mitigation and Adaptation Strategies for Global Change*, *16*(2), 119–142. doi:10.1007/s11027-010-9257-7. doi:10.100711027-010-9257-7

Dash, S. K., Jenamani, R. K., Kalsi, S. R., & Panda, S. K. (2007). Some evidence of climate change in twentieth-century India. *Climatic Change*, *85*(3-4), 299–321. doi:10.1007/s10584-007-9305-9. doi:10.100710584-007-9305-9

Douglass, D. C., Singer, B. S., Kaplan, M. R., Ackert, R. P., Mickelson, D. M., & Caffee, M. W. (2005). Evidence of early Holocene glacial advances in southern South America from cosmogenic surface-exposure dating. *Geology*, *33*(3), 237–240. doi:10.1130/G21144.1. doi:10.1130/G21144.1

Gairola, S., Shariff, N. M., Bhatt, A., & Kala, C. P. (2010). Influence of Climate Change on Secondary Chemicals in High Altitude Himalayas: Issues Need Immediate Attention. *Journal of Medicinal Plants Research*, *4*(18), 1825–1829.

Glowka, L., Burhenne-Guilmin, F., Synge, H., McNeely, J. A., & Gundling, L. (1994). *A guide to the convention on biological diversity* (No. 333.95 G946). IUCN, Bonn (Alemania). Environmental Law Centre IUCN, Gland (Suiza). Biodiversity Programme.

Hegland, S. J., Nielsen, A., Lázaro, A., Bjerknes, A.-L., & Totland, Ø. (2009). How does Climate Warming affect Plant-Pollinator Interactions? *Ecology Letters*, *12*(2), 184–195. doi:10.1111/j.1461-0248.2008.01269.x PubMed. doi:10.1111/j.1461-0248.2008.01269.x PMID:19049509

IPCC. (2007). Synthesis Report. Contribution of Working Groups I, II and III to the Fourth Assessment Report of the Intergovernmental Panel on Climate Change, Geneva, Switzerland.

Kapelle, M., Margret, M. I., Vuiren, V., & Baas, B. (1999). Effects of Climate Change on Biodiversity; A Review & Identification of Key Research Issues. *Biodiversity and Conservation*, *8*, 1393–1394.

Kappelle, M., Van Vuuren, M. M., & Baas, P. (1999). Effects of climate change on biodiversity: A review and identification of key research issues. *Biodiversity and Conservation*, *8*(10), 1383–1397. doi:10.1023/A:1008934324223. doi:10.1023/A:1008934324223

Khoshoo, T. N. (1993). Himalayan biodiversity conservation: an overview. *Himalayanbiodiversity: conservation strategies*, 5-35.

Kothawale, D. R., Munot, A. A., & Kumar, K. K. (2010). Surface air temperature variability over India during 1901–2007, and its association with ENSO. *Climate Research*, *42*(2), 89–104. doi:10.3354/cr00857. doi:10.3354/cr00857

Kumar, K. (2009). Impact of climate change on India's monsoon climate and development of high resolution climate change scenarios for India. *MoEF*.

Kumar, V., & Chopra, A. K. (2009). Impact of Climate Change on Biodiversity in India with Special reference to Himalayan Region- An Overview. *Journal of Applied and Natural Science*, *1*(1), 117–122. doi:10.31018/jans.v1i1.48. doi:10.31018/jans.v1i1.48

Larsen, B. B., Miller, E. C., Rhodes, M. K., & Wiens, J. J. (2017). Inordinate fondness multiplied and redistributed: The number of species on earth and the new pie of life. *The Quarterly Review of Biology*, *92*(3), 229–265. doi:10.1086/693564. doi:10.1086/693564

Parmesan, C. (1996). Climate and species' range. *Nature*, *382*(6594), 765–766. doi:10.1038/382765a0. doi:10.1038/382765a0

Pawson, S. M., Brin, A., Brockerhalf, E. G., Lamb, D., Payn, T. W., Paqquet, A., & Parrotta, J. A. (2013). Plantation Forests, Climate Change & Biodiversity. *Biodiversity and Conservation*, *22*(5), 1203–1227. doi:10.1007/s10531-013-0458-8. doi:10.100710531-013-0458-8

Potts, S.G., Biesmeijer, J.C., Kremen, C., Neumann, P., Schweizer, O., &Kunin, W.E. (2015). Global Pollinator Declines: Trends, Impacts & Drivers. *Trends EcolEvol.*, 345-353.

Ravindranath, Sukumar, & Saxena. (2006). Impact of climate change on forests in India. *Current Science*, *90*(3), 354–361.

Ravindranath, N. H., & Sukumar, R. (1998). Climate change and tropical forests in India. In Potential Impacts of Climate Change on Tropical Forest Ecosystems (pp. 423–441). Dordrecht: Springer. doi:10.1007/978-94-017-2730-3_21 doi:10.1007/978-94-017-2730-3_21

Ray, K. S., & Srivastava, A. K. (2000). Is there any change in extreme events like heavy rainfall? *Current Science*, *79*(2), 155–158.

Sathaye, J., Shukla, P. R., & Ravindranath, N. H. (2006). Climate change, Sustainable development, and India: Global and National Concerns. *Current Science*, *19*(3), 314–325.

Schlyter, P., Stjernquist, I., Bärring, L., Jönsson, A. M., & Nilsson, C. (2006). Assessment of the impacts of climate change and weather extremes on boreal forests in northern Europe, focusing on Norway spruce. *Climate Research*, *31*(1), 75–84. doi:10.3354/cr031075. doi:10.3354/cr031075

Sigman, D. M., & Boyle, E. A. (2000). Glacial/interglacial variations in atmospheric carbon dioxide. *Nature*, *407*(6806), 859–869. doi:10.1038/35038000 PubMed. doi:10.1038/35038000 PMID:11057657

Syroka, J., & Toumi, R. (2004). On the withdrawal of the Indian summer monsoon. *Quarterly Journal of the Royal Meteorological Society, 130*(598), 989-1008.

Thuiller, W., Lavorel, S., Aroujo, M. B., Sykes, M. T., & Prentice, I. C. (2005). Climate Change Threats to Plant Biodiversity in Europe. *Proceedings of the National Academy of Sciences of the United States of America, 102*(23), 8245–8250. doi:10.1073/pnas.0409902102 PubMed. doi:10.1073/pnas.0409902102 PMID:15919825

Yadav, K. K., Gupta, N., & Kumar, V. (2015). *Environmental challenges: Problem and management*. Lap Lambert Academic publishing.

ADDITIONAL READING

Heywood, V. H., & Watson, R. T. (Eds.). (1995). *Global Biodiversity Assessments, UNEP*. Cambridge, UK: Cambridge University Press.

Peters, R. L., & Lovejoy, T. E. (Eds.). (1992). *Global Warming & Biological Diversity*. New Haven, CT: Yale University Press.

Sol-Brig, O. T., Van Emden, H. M., & Oordt, P. G. W. J. (Eds.). (1992). *Biodiversity and Global Change, IUBS Monograph*. Paris: IUBS.

KEY TERMS AND DEFINITIONS

Anthropogenic Factors: Signify the factor of Anthropogenic is leading the origin regulate of species wilts, endangerment and annihilation: land development, in excess of mistreatment, species translocations and introductions, and greenhouse gasses. Leading the Anthropogenic factor to turn out ecological and genetic effects contributing to extinction risk.

Biodiversity: Acclaimed as global importance by the biodiversity, so far the species and habitats carry on under accumulative load from human -made impacts, whether in urban, rural, or wilderness settings.

Biological Diversity: Biological diversity is the variety of life on Earth, the product of millions of years of evolution and thousands of years of cultivation of plants and domestication of animals. It is often referred to in abbreviated form as biodiversity. There are many levels of diversity, from DNA and genes to species, populations, communities and ecosystems. Biological diversity is often considered in terms of the diversity of genes, species and ecosystems – three fundamental and hierarchically related levels of biological organization.

Climate Change: Climate change refers to changes in the earth's climate, especially the gradual rise in temperature caused by high levels of carbon dioxide and other gases.

Ecosystem Services: Ecosystem services are straight or unplanned benefits of natural ecosystem which nourish the human life on the earth and improve the quality of life. Regulation of climate, fresh water, soil biodiversity, genetic resources, and many more are the example of ecosystem services.

Habitat Fragmentation: It is not only cause of habitat loss but it also makes permanent changes in the small habitat. Geological processes, human activities, climate change and many more natural processes are responsible for habitat fragmentation.

Medicinal Plants: Any plant which have medicinal properties or characteristic are called medicinal plant. Some time we use particular part of the plant.

Potential Risk: Potential risk refers to any risk associated with an action that is possible, in certain circumstances. A risk refers to a threat or damage that may occur in operations of any work.

Chapter 9
Impact of Climatic Changes on Earth's Survival:
An Analysis of Changing Environment and Its Consequences at a Global Level

Akash
Gurukula Kangri University, India

Navneet
Gurukula Kangri University, India

ABSTRACT

Every species' survival on earth is dependent on each other for their demand and dependent on the environment and various other sources. These resources include fresh food, clean drinking water, timber for construction, natural gas and coal for industries, fibers for clothing. All the human activity affects the environment severely in different ways. The biggest threats to the environment are climatic changes. Climate is an important factor that affects all survival on earth. The different pollutants, transport, dispersion, chemical transformation, as well as the deposition can be affected by meteorological variable such as humidity, wind, temperature. Climatic changes are expected to worsen the quality of air and water by changing the atmospheric processes and chemistry. Not only human beings but every aspect of the ecosystem is affected due to the changing climate. This chapter will explore the impacts of climatic changes on biodiversity by various activities of humans. Additionally, it will sketch how the impacts can be reduced by plants.

INTRODUCTION

Climate is one of the most important determinants of environment which affects whole components of ecosystem on earth. Atmospheric emission and their dispersion as well as the chemical transformation band deposition can be influenced by different meteorological variables such as temperature, wind, humidity (Kinney, 2008). Change in the climate is expected to worsen the quality of air in densely

DOI: 10.4018/978-1-5225-7387-6.ch009

populated regions of world by changing the atmospheric ventilation and dilution, precipitation band into atmospheric chemistry (Fiore *et al.,* 20015). The changes in environment lead to various problems on earth. Every organism on earth gets affected due to the changes in environment. Reduction in the quality of air due to the air contaminants directly affect ecosystem in that ways which also could affect human as well as the health of other organism on earth by inhaling the air contaminants (Haase *et al.,* 2014 and Who, 2013). Different types of study reveals that climate change has already affected the quality of air. Study from Fang *et al.,* 2013 observed that from 1860 -2000, the global population-weighted fine particle (PM2.5) concentration has been increased by 5% and near-surface ozone concentrations by 2% due to the different changes in environment and climatic changes. The another study from Silva et al., 2013 revels that the change from different industrial revolution resulted in up to 111,000 and 21,400 additional premature fine particle- and ozone-related deaths due to climate change. As the climate changes continues in future these impacts would be more drastic in future on earth. The global warming induced by different types of anthropogenic activity has adverse impact on the earth and its various components (Uysal *et al.,* 2003 and Health and Environment Alliance, 2009). Global greenhouse gas emissions due to the various anthropogenic activity has been growing with an increase of 70% between 1970 and 2004 and would be continue in the next coming years. According to the Intergovernmental Panel on Climate Change 2007 document "rise in globally averaged temperatures on earth since the mid-20th century is very likely due to the recorded increase in anthropogenic greenhouse gas concentrations" (Trenberth *et al.,*2007). The most important green house gas is Carbon dioxide which is continuing increasing on the earth surface. A number of different types of diseases have evolved from climatic changes on earth. Various respiratory tract disease like asthma and other bronchial diseases are results from different types of climatic changes and reduction in the quality of air. A dramatic increase in the prevalence of respiratory diseases such as bronchial asthma, rhino sinusitis has been arises in different countries during the last 3 decades by various anthropogenic activity by human and bad quality of air. The main feature of different respiratory disease is the development of airway inflammation and bronchial hyper responsiveness in the form of a heightened bronchoconstrictor response not only to allergens to which an individual is sensitized, but also to a range of different stimuli, such as cold air and pollutants. On the other way rising trend in allergic respiratory diseases indicates that the higher level of emission comes from vehicles. The adverse effect of air pollution on human health has a quantifiable impact, not only in way of morbidity but also on the mortality of respiratory diseases in the world.

Due to the recent increase in morbidity associated with respiratory diseases is the continuous degradation on the quality of air, as a result of increasing outdoor air pollutants levels by vehicle emissions. Majority of the pollutants are toxic to human health causing serious disease in the human and and at the same time these toxic pollutants also effects the flora and fauna on earth. Sulphur dioxide, Nitrogen dioxide, and particulate matter are toxic air pollutants in the environment. Other are aeroallergens which arises from pollen grains, lead to bronchial obstruction in human. The pollen allergy includes a models that study the relationship between air pollution and respiratory allergic disease (Grinsven *et al.*, 2013; Kuo *et al.,* 2006 and Nawrot *et al.,* 2006).

Different Types of Pollutant

Changes in environment is mainly due to the excessive use for the generation of energy, combustion of fossil fuels as well as the transportation. Different types of air pollutants have been recognized due to the different types of anthropogenic activity. They may exists in different form affects severely to the

Table 1. List of Some of the common atmospheric pollutants and their originating source

Category	Source of Various Pollutants	Emitted Pollutants Into Atmospheric
Quarrying and Mining	Mining of coal	Sulphur dioxide(SO_2), SPM and various oxides of nitrogen
	Natural gas production and Petroleum	Sulphur dioxide (SO_2),
	Non ferrous ore mining	SPM and Lead(Pb)
	Stone quarrying	SPM
	Tobacco and beverages	VOC, Sulphur dioxide (SO_2), carbon mono oxide (CO) and hydrogen sulphite(H_2S)
	Leather industries and textiles	VOC and SPM,
	Products of wood	VOC, SPM,
	Products of paper and printing	SPM, carbon mono oxide, hydrogen sulphide(H_2S) VOC, and Sulphur dioxide(SO_2)
Chemicals manufacturing	Phallic anhydride	Sulphur dioxide(SO_2), carbon mono oxide(CO) and SPM
	Chlor-alkali	Chlorine(Cl)
	Hcl	Hydrochloric acid(Hcl)
	Hydrofluoric acid(Hf)	Hydrogen fluorid(Hf) and SiF_4
	Sulphuric acid (H_2s)	Sulphur dioxide (SO_2) and SO_3
	HNO_2	Oxides of nitrogen(NOx)
	H_3Po_4	SPM and F_2
	Lead oxide and other pigments	SPM and Lead (Pb)
	Ammonia (NH_3)	SPM, Oxides of nitrogen(NOx), ammonia(NH3), VOC and Sulphur dioxide
	$NaHCO_3$	SPM and ammonia
	Calcium carbide	SPM
	Adipic acid	Oxides of nitrogen(NOx), SPM, VOC and carbon mono oxide(CO)
	Alkyl lead	Lead (Pb)
	Maleic anhydride and terephtalic acid	VOC and carbon mono oxide (CO)
	Fertilizers and pesticide production	Ammonia (NH_3) and SPM
	Ammonium nitrate	SPM, Ammonia (NH_3) and nitric acid (HNO_2)
	Ammonium sulphate (NH_4SO_4)	VOC
	Plastic, fibres	VOC and H_2S, SPM, CS_2
	Paints, lacquers and varnishes	VOC and SPM
	Soap	SPM
	Printing ink and Carbon black	NOx, CO, hydrogen sulphide SO_2, VOC and SPM,
	Trinitrotoluene	sulphur dioxide, SPM, Oxides of nitrogen, SO_3 and nitric acid (HNO_2)
Petroleum refineries	Petroleum and coal products	VOC, F2, SPM, sulphur dioxide (SO_2) and carbon dioxide(CO)
	Clay products	
	Plaster cement and lime	SPM, Sulphur dioxide (SO_2), and oxide of nitrogen (NOx)
	Non -ferrous industries	Sulphur dioxide (SO_2), lead (Pb) and SPM,
Transport	filling operation	VOC

continued on following page

Table 1. Continued

Category	Source of Various Pollutants	Emitted Pollutants Into Atmospheric
Music metal industries	Steel and Iron	VOC NOx, sulphur dioxide (SO_2), carbon monoxide(CO) and lead (Pb)
Agriculture	Due to open burning of agricultural wastes	Carbon monoxide(CO) and VOC
Community services	Municipal incinerators	Lead (Pb), Oxides of nitrogen(NOx), SPM, sulphur dioxide (SO_2), carbon monoxide (CO) and VOC,
Wholesale and retail trade	Steam Electricity and Gas	carbon monoxide(CO), Oxides of nitrogen, SO_3, lead and sulphur dioxide (SO_2)

SPM: Suspended particulate matter, VOC**:** Volatile organic compounds
Source: Modified from Economopoulos, 1993

health of human. Plants and animals. Some are gaseous pollutants are CO, NOx, Ozone, $SO_{2,}$ Volatile Organic Compounds whereas some are persistent organic pollutants are dioxins and some are. Mercury, lead and particulate Matter (Birch *et al.,* 2011 and van Grinsven *et al.*, 2013; Tager *et al.,* 2005; Rastogi *et al.,* 1991).

AIR POLLUTION: AN ANALYSIS OF THE CAUSATIVE AGENTS OF AIR POLLUTION AND THEIR SUBSEQUENT EFFECTS

Ozone is a very important greenhouse gas in the environment and which plays significant role in the radiative budget in the stratosphere layer through its interactions with both long wave as well as short wave radiation (Gauss *et al.,* 2003). Ozone is a precursor in oxidizing reactions as it severely influences the effect of greenhouse gases such as hydrofluorocarbons and methane (Gauss *et al.,* 2006). Ozone also affect indirectly to the environment due to the limiting sequestration of carbon dioxide by vegetation (Sitch *et al.,* 2007). The direct radiative effects of ozone provide the positive radiative forcing of the atmospheric gases especially green house gases (Gauss *et al.,* 2003). The ground-level ozone is an important air pollutant that severely affects human and the health of other organism as well as the growth

Table 2. Penetrability of particle according to their size

Size of Particles	Region to Which Penetration Can Occur
> 11 μm	Captured in the nostrils but it do not penetrate the lower respiratory tract
7-11 μm	Nasal passage
4.7-7 μm	Larynx part
3.3-4.7 μm	primary bronchial part and Trachea
2.1-3.3 μm	Secondary part of bronchial section
1.1-2.1 μm	Terminal section of Bronchioles
0.65-1.1 μm	Bronchioles section
0.43-0.65 μm	Alveolar section

Source: Wilson *et al.,* 1996 Ghio *et al.,* 1999.

of forests worldwide (Ainsworth *et al.,* 2012). The increasing level of the emission of the precursor of ozone like NOx, CO have affected different lives on earth (Brasseur *et al.,* 2001). The long-range of trans-Pacific transport of contaminant air masses from Asia is a substantial part of the background ozone in western North America (Vingarzan, 2004). It has been observed that there is significance increasment in the concentration of ozone in the free troposphere during 1984 (0.63-0.34 ppb/ year) in spring season the intercontinental transport across the Pacific is most efficient (Cooper *et al.,* 2010; Law, 2010). The emission from the Asian continent influence, emissions from Canada and Mexico which contribute to background level of ozone concentrations in the eastern region of the United States and in California as well (Wang *et al.,* 2009). In U.S.A, It has also been observed that the highest ozone pollution occurs in the south-western region of United States, especially in California where the fourth highest maximum averages value ranges from 71 to 106 ppb. On the other hand, Moderately elevated O3 levels (51 ppb in Montana to 78 ppb in New Jersey) occurs in the U.S.A (Bytnerowicz *et al.,* 2008).

Reactive nitrogen generally results from different types anthropogenic activity in the environment which severely causes harm to human health as well as the different types of problems. Different forms of the nitrogen as a result of different agricultural activities causes harm in different ways. N_2O, ammonia can affect environment by changing the balance of essentials component of atmosphere. In atmosphere, the oxidative process of ammonia and nitrogen take part in the climatic changes indirectly by altering the production as well as the loss of atmospheric climatic forcers, both greenhouse gases and aerosols scattering incoming solar radiation Further the oxidative products of nitrogen mostly N_2O can harm the atmosphere by increasing the formation of ozone and by removal of CH4 (in cooling effect). So NOx can enhance light-scattering aerosols into the atmosphere (Pinder *et al.,* 2013). It was observed that in some areas of America, the concentration of nitric acid and atmospheric NOx have decreased due to Environmental Protection Agency of the government (Chestut, 2005). Further it was also observed that the deposition of oxides of nitrogen remained stable, mainly due to a lack of regulatory control of various emission source emissions (Davidson *et al.,* 2012).

Different types of phytotoxic effects of the reactive nitrogen has been recorded by Bytnerowicz *et al.,* 1998. In U.S agricultural economy, Synthetic nitrogen fertilizers and cultivation of N-fixing crops are the major parts of it (Smil, 2002 and Houlton *et al.,* 2013). Different types of human activity creates essentials industrial products (Domene and Ayers, 2001) and these creates environmental issues has developed from the use of unintentional leakage of nitrogen. High rate of morbidity and mortality results from climatic changes and harmful pollutants. The water sources also get from contamination by oxides of nitrogen. Hypoxia in freshwater and coastal marine areas as well as the global climatic change results from the ozone-depleting gas oxides of nitrogen and emission of the greenhouse (Davidson *et al.,* 2012).

Sulphur and the deposition of its contents is atmosphere is major problems in the world especially for developing countries. Sulphur Dioxide ($SO_{2)}$ contributes deposition of acid and but its direct phytotoxic effects are limited to local geographic areas (Legge *et al.,* 1998; US EPA, 2008). Study from different workers revelled that wet deposition of sulphur has continuously decreased in recent decades (Greaver *et al.,* 2012; NADP, 2010). Changes in the concentration at small level of sulphur dioxide can cause major changes in environmental chemistry. Oxidation of Sulphur dioxide forms sulphuric acid (H_2SO_4) which has generally low vapour pressure for allowing it to adsorb on tiny dry particles into the atmosphere and after that to increase the size of those particles by molecular condensation of more acid (Ehhalt,1999), finally resulting molecules increases the concentration of aerosol into the atmosphere causes severe harm to livings.

The other property of SO_2 is that it reacts immediately with OH leaving little OH to oxidize many gases like carbon monoxide and methane. Increasing the concentrations of these two gases indicate decreasing oxidizing capacity. The concentration of sulphur dioxide condenses to liquid form below−10 °C. In troposphere layer temperature decreases with increasing altitude at −6.4 °C/km to −55 °C after that increases with altitude in the overlying stratosphere. In most of the cases, the stratosphere layer sulphur dioxide is potentially a liquid. This form of liquid has good soluble ability and can be oxidized readily by hydrogen per oxide in cloud droplets and also in form of precipitation (Ehhalt, 1999). Ehhalt (1999) observed that oxidation of the 70% of the available sulphur dioxide occurs in this way. This form of oxidation process does not lead to nucleation of new particles into the atmosphere (Alexander *et al.,* 2005). Further study of Laj et al.1990, observed that the depletion of gas like hydrogen peroxide in the ice of reached as a gas in Greenland and after that it was oxidized in precipitation. Barrie and Hoff (Barrie *et al.,* 1984) observed that sulphur dioxide in the arctic region originating from burning of fossil fuel in Eurasia.

EFFECTS OF AIR POLLUTION ON SURVIVALS

Air pollutants severely affect the health of human in different ways. The composition of the pollutants, their relative dose and the time of exposure determine the effects on human health. Pollutants in the atmosphere have bad effects on human health which includes breathing problems and various types of cancer like skin cancer, lung cancer. These all are the negative effects of air pollution also includes different birth defects, reduced activity of the immune system as well as developmental delays of children and other diseases. Different epidemiological and animal model data revels that primarily affected systems from air pollutants level are the mainly cardiovascular and respiratory tract. But other study also observed that the function of several other organs can be also influenced by the different level of air pollution (Cohen *et al.,* 2005; Huang and Ghio, 2006; Kunzli and Tager, 2005; Sharma and Agrawal, 2005)..

These pollutants in air also contributes to the contamination of food and water and water which make ingestion in different cases (Thron, 1996). Pollutants enter into the human body with the help of respiratory tract and gastrointestinal path and finally reach to circulation system and deposited to different tissues in the body. The elimination of these tozic substances occurs only to a certain degree by excretion (Madden and Fowler, 2000). Studies from different workers revels that air pollutants affects human in different way in terms of their concentration level. Long- and short-term effects of air pollution have been observed at global level. The long-term exposure of air pollution includes various defects in humnan beings which includes broncho constriction, throat and nose irritation as well as the dyspnoea, especially in those asthmatic human beings which are usually experienced to the exposure of increased concentration of sulphur dioxide (Balmes *et al.,* 1987), nitrogen oxides (Kagawa, 1985). It has been observed that particulate matter penetrates the alveolar epithelium layer (Ghio and Huang, 2004) and different level of ozone level of inflammation (Uysal and Schapira, 2003). Air pollution has severe effects on different organs of human beings. Air pollutants like nitrogen oxides increase the susceptibility to respiratory diseases (Chauhan *et al.,* 1998) whereas certain heavy metals and chronic exposure to ozone reduces functional level of lungs (Rastogi *et al.,* 1991; Tager *et al.,* 2005). The long time and high-level exposure of ozone is also responsible for various types of diseases like asthma, lung cancer emphysema (Kuo *et al.,* 2006; Nawrot *et al.,* 2006). It is also well known that micro exposer of nitrogen dioxide causes serious lesion like Emphysema (Wegmann *et al.,* 2005). The another toxic air pollutant is dioxins, which causes damages to liver cell (Kimbrough *et al.,* 1977), as it indicated by an increase levels of different enzymes in the blood of human (Mandal, 2005).

On the other hand, gases like carbons monoxide binds it to haemoglobin modifying its conformation resulting into the reduction of its capability to transfer oxygen in blood and other organs in body (Badman and Jaffe, 1996). Reduction level of oxygen can immediately affect the function of organ system in human body, resulting in impaired concentration, slow reflexes and confusion. The different concentration of particulate matter (PM) also induces various systemic inflammatory changes affecting the equally blood coagulation (Riediker *et al.,* 2004). It has been also observed that anaemia in human and increased level of blood pressure causes inhibitory effect on haematopoiesis resulting from results of heavy metal pollution like mercury, nickel, arsenic (Huang and Ghio, 2006)).These heavy metals can also induces damages in kidney like m and ercury, nickel tubular dysfunction. The can also risk of stone formation or nephrocalcinosis (Damek-Poprawa and Sawicka-Kapusta, 2003; Jarup, 2003; Loghman-Adham, 1997) and renal cancer (Boffetta *et al.,* 1993; Vamvakas *et al.,* 1993). Further air pollutants causes lung irritation and changes in blood clotting resulting to angina or even to myocardial infarction (Vermylen *et al.,* 2005).

It has been observed that the chemical composition in environment has not always been ideal at every part of world. The increasing of global population causes higher number of intensive livestock production practices. The production of eggs, meat as well as the milk resulted in the generation of large amounts of waste disposal and its accumulation. The other most important affected aspects of air pollutants are animals, which faces different problems from the contaminated environment. As most of the time they

Table 3. Different Potential damage in human organs due to air pollution and as results of climatic changes at global level

Pollutants	Effects	References
Ozone (O3)	Non functioning of lung function	Rastogi *et al.,* 1991 and Tager *et al.,* 2005
	Asthma and Emphysema	Kuo *et al.,* 2006 and Nawrot *et al.,* 2006.
Carbon monoxide (CO_2)	Inflammation in lungs as well as systemic inflammatory changes	Riediker *et al.,* 2004.
Sulphur dioxide (SO_2)	Dyspnoea and Bronchoconstriction	Balmes *et al.,* 1987
Various oxides of Nitrogen (NOx)	Various types of respiratory disease like Asthma, Tuberculosis.	Birch *et al.,* 2011 and van Grinsven *et al.,* 2013
	Declining the ability to see or less visibility—loss of aesthetics	Birch *et al.,* 2011
	Increasing the level of airborne disease	Grinsven *et al* 2013
	Increased level of cancer especially colon cancer	Grinsven *et al* 2013
	Respiratory tract and various cardiovascular problems	Cohen *et al.,* 2005;
	Myocardial infraction	Vermylen *et al.,* 2005.
Reduction in the high level of air pollutants and	Defects in developing foetus	Schell *et al.,* 2006
	Congenital malformations	Bellinger, 2005

live in huge facilities but often got many diseased due to self-made indoor air pollution. These different types of effects from pollution may be poor quality of air which they inhale or ingest in different way or different time and at different concentration. Deposition of particles from different source exhaust on pasture land which may affect adversely to the health of animals resulting in toxic residue in their meat milk and eggs as well as theother products obtained from animals. It has also been seen that high level of dioxin in milk of catttles and zinc-induced arthritis in growing foals are results from industrial air pollutants. Various animals from poultry, Pigs, goats, sheep and cattle are kept in indoor areas. These closed areas and affects the health of these animals in different ways. It was seen observed that air quality of this area is sometime much better in compare to that of the closed facilities for swine and poultry (Wathes *et al.,* 1998). Study from Dunham (1991), revels that maximal concentrations of air pollutants in piggeries are: 2.4 mg dust /m^3; 0.08 mg endotoxin/m^3, 7 ppm ammonia, as 1,540 ppm carbon dioxide, 105 colony-forming units (CFUs) of total microbes/m^3. Scherer & Unshelm, (1995) observed that the different concentrations of bacteria was up to 1.1 x106 CFU/m^3, concentration of ammonia was 27 ppm whereas inhalable dust content of 0.26 mg/m3.

Different types of toxicants are found in manure which containing enteric bacteria and endotoxin resulting from negative effects on the health of animals (Pickrell, 1991).Further generation of harm gasses like ammonia and its greater concentrations in the air affected severely to the health of animals, but also by pig flow management and pig density and volume of the building, (Scherer & Unshelm, 1995). Ammonia is considered as one of the most important toxicants which has adverse effect on the health of animals. Variation in the level levels of ammonia are known to play a role in farrowing units and in different hoses of poultry (Harry, 1978). Study from Dodd & Gross (1980), revels that the 1000 ppm level of ammonia for less than twenty hour causes caused mucosal damage impaired ciliary activity as well as secondary infections in laboratory animals. Further it was also observed that respiratory disease in swine is mostly due to the different combination of primary as well as the opportunistic infectious agents. These primary types of respiratory infectious agents have ability to cause serious infection and finally severe diseases on their own but sometimes uncomplicated infections have also been observed. It was seen that if these primary infections becomes complicated serious respiratory tract disease may occurs in animals by microorganism.

Different types of diseases in animals are porcine reproductive, respiratory syndrome virus and other caused by *Mycoplasma hyopneumoniae, Haemophilus parasuis, Actinobacillus suis, Bordetella bronchiseptica and Streptococcus suis.* Further it was reported from the study of Pickrell (1991), that pig or poultry facilities are exposed to the same increased levels of carbon monoxide, hydrogen sulphide as well as ammonia and the dust particles from feed and manure of the animals. That is why, workers involved in swine production resulting to have greater rates of asthma and various types of respiratory infections.

It has also been observed respiratory distress occurs in Londaon due to various types of smog disaster of prize cattle that were housed in the city for a cattle exhibition (Catcott, 1961). Further high level of sulphur dioxide is responsible for accompanying acute bronchiolitis and emphysema right-sided heart failure.

Manzo et al. (2010) observed that cats with airways inflammatory disease and dogs with chronic bronchitis are at dangers level of exacerbating their conditions if exposed to the prolonged urban air pollutants. Both, outdoor and indoor air pollutants and have negatively affects the health of animals.

The concentration of ozone in terrestrial ecosystems is mostly controlled by epidermal cell of plants called as 'stomata' (Ainsworth *et al.,* 2012). There is a strong relationship between damages of plants by ozone and conductance of stomata (Reich and Amundson, 1985). The stomatal and the non-stomatal

Table 4. Affected birds, mammals, and land invertebrates due to air pollutants

Scientific Name	Effects Due to Poor Quality of Air
Lutra lutra	*Lutra lutra* decline in due to different level of contamination of air which has been reported by Galloway, Scotland (Fry and Cook, 1984), and also for UK (Mason and Macdonald, 1987).
Anas rubripes	Different study reveals that increased duckling mortality due to the high level of atmospheric pollutants (Haramis,1984)
Collembola spp.	*Collembola* spp. decrease in variety and number in forests ecosystem experiencing the bad quality of air (Koposzki,1992).
Aphantopus hyperantus	Declines due to high sulphur dioxide ($SO_{2)}$ levels (Heath *et al.*, 1984).
Lepidoptera spp.	Studies reveals that *Lepidoptera* spp. decline in polluted atmosphere and by other climatic changes (Barbour,1986).
Parnassius apollo	It is decline in polluted areas and also due to the bad air quality (Holm, 1985))

processes of leaf help in reduction of various types of air pollutants (Kurpius and Goldstein, 2003). Sometime the green house gas may also cause visible injury in leaves of plants and act as a barrier in exchange of gases, in photosynthesis, resulting in shortness of foliage, bad health and reduced plant height (Grulke *et al.,* 2009). The different long term chronic exposures to ozone can also lower the rate of photosynthesis in plants as well as the stomatal conductance (Reich and Amundson, 1985). Chronic levels exposure of ozone may limit the opening and closing of the stomata (Grulke *et al.,* 2006), and increasing the water use (Sun *et al.,* 2012). The short and long term exposure reduce the stomatal activity (Sitch *et al.,* 2007). Various study reveals that high level of green house reduces the rate of photosynthetic and stomatal conductance of trees by 11% - 13% (Wittig *et al.,* 2007).

Felzer *et al.* (2004) observed that approximately 2.6–6.8% reduction rate of the annual net primary productivity in U.S.A in response to the levels of ozone since 1950 with the largest decreases in the late 1980s and early in 1990s.On the other hand different levels of nitrogen deposition may have so many negative effects like contamination of water and soils and changes in biodiversity, (Greaver *et al.,* 2012; Pardo *et al.,* 2011; US EPA, 2008). It has been also observed reduction in stomatal activity caused by elevated level of carbon dioxide, methane and ozone may also reduce the ability of plants to sequester carbon dioxide (Sitch *et al.,* 2007). The effects of ozone on hardwood of trees in the north-eastern region of United States from1987–1992 data showed that the declines of annual net primary productivity ranging from 3% to 16% with most pronounced effects at the highest ozone levels (Ollinger *et al.,* 1997).

Phytoremediation of Air Pollutants

Plants are the best remediator of pollutants on earth. Studies from different workers have suggested that plants contribute in reduction of air pollution (Boonen *et al.,* 2015; Angelo *et al.,* 2013). Most of the plant species show substantial effect in controlling of air pollutants (Escobedo *et al.,* 2011). A number of different studies have quantified and identified various ecosystem services in context of air pollution reduction (Nowak, 2000; Nowak *et al.,* 2006). The most essential removal of toxic pollutants by the process of dry deposition, in which the species removes different types of pollutant in troposphere (Cavanagh, 2006; Schlesinger, 1979).

Table 5. Damages to phytodiversity due to various types of air pollutants

Name	Scientific Name	Notes and Source
Bog mosses	*Sphagnum* spp.	*Sphagnum* species are damaged by higher level of sulphur dioxide and also by nitrogen dioxide as well as the deposition of nitrogen.
Liverworts and mosses	*Orthotrichium, Hypnum cupressiforme, Bryum, Grimmia pulvinata*	These species are declines due to the high level of sulphur dioxide and nitrogen dioxide
Fungi	*Cinabrius, Russula* spp, *Lactarius* spp, *Hygrphorus* spp and *Hygrocybe* spp and *Cantharellus*	These fungal species are badly affected by the deposition of nitrogen in acidified forests and other green house gases.
Cyanobacteria	*Anabaena, Scytonema, Nostoc*, etc	Greatly affected due to air high level of air pollutants.
Coniferous trees	*Larix europeaus, Picea,* and *abies.*	It is very sensitive to acute damage by ozone.
Mosses	*Antitrichia curtipendula, Neckera, Orthotrichium,* and *Rhytidiadelphus*	Decline in the due to soil acidification in various areas of worlds like India, Australia, U.S.A.
Herbaceous species	*Vicia sepium, Primula veris, Trifolium medium, Hepatica nutans, Melica nutans* etc	Declined due to soil acidification
Broadleaved trees	*Populus tremulens, Quercus alba, Acer saccarina,* and *Quercus robur*	Reduced due to green house gases like ozone and also by Increased nitrogen deposition.
Aquatic flowering plants	*Littorella, Echinospora, Uniflora, Lobelia dortmanna,* and *Isoetes*	Reduced due to acidification in streams and freshwaters.
Lichens	Many foliar species, ex. *Ramalina, Usnea*	Lichens declined due to the high level of sulphur dioxide.
Woolly fringe moss	*Racomitrium* and *Anuginosum*	Nitrogen deposition is thought be responsible for reduction of this moss in southern part of Scotland.

Dry deposition is mainly dependent on property and nature of air pollutant. Sometimes it is also observed that property of surfaces such as chemical nature, size, roughness, and weather factors such as wind direction and speed as well as temperature is also play important role role in removal of pollutants (Roupsard, 2013). As accumulation of air pollutants on the surface of plant leaf occur, they are removed from different ways. Firstly, they removed through stomata especially through epidermal cell uptake of gaseous pollutants and then by leaf interception of particulate matter (Nowak *et al.,* 2006).

Further the process leads to the diffusion of toxic pollutants into the inner part of plant leaves. The gases pollutants may also be absorbed or react with the surface of plant. The removal may be by re-suspension of intercepted particles from the leaf surfaces through the action of wind. Now it is well known that plant remediate the environmental pollutants by different types of environmental pollutans. Different plant species like *Astragalus sinicus Azolla pinnata, Penninsetum pedicellatum, Triticum aestivum, Hordeum vulgare, Zinnia anguistifolia, Populus deltoids, Quercus robur, Fraxinus excelsior, Aesculus hippocastanum, Zea mays,* removes the environmental pollutants in different way. Table 6 shows various plants species involving removal of atmospheric pollutants.

Table 6. Phytoremediation of environmental pollutants/contamination

Plant Species	Pollutants	References
Kolchia	Metolachlor, Atrazine and trifluralin	Anderson *et al.*,1994
Penninsetum pedicellatum	Fenvalerate, Chlorpyrifos and Cypermethrie	Dubey and Fulekar., 2011
Fragmites australies (Reed)	Fixed Nitrogen	Haleem *et al.*, 2000
Rye Grass	Chlorpyrifos	Korade and Fulekar., 2009
Zinnia anguistifolia	Mefenoxam	Pai *et al.*, 2001
Alpine blue grass and alfalfa	PAHs and Hexadecane	Nichols *et al.*,1997
Triticum vulgare (Wheat)	2,4-D	Kingley *et al.*, 1994
Hordeum vulgare	2,4-D	Jacobsen, 1997
Sugar beet (cv.Rex)	Polychlorinated biphenyls(PCBS)	Brazil *et al.*, 1995
Oryza sativa	Benthiocarb	Sato, 1989
Populus deltoids nigra	1,4-dioxane	Schnoor *et al.*, 1998
Saccharum officinarum	2,4-D	Sandmann and Loose,1984
Astragalus sinicus	Cd	Sriprang *et al.*,2002
Poplar root extract	1,4-dioxane	Kelley *et al.*, 2001
Zea mays (mays)	3-methylbenzoate	Ronchel *et al.*, 2001
Azolla pinnata (Fern)	Diesel fuel	Cohen *et al.*,2002
Zea mays	Atrazine	Seibert *et al.*,1981

Source: Modified from Kala, 2014

RESEARCH AND MANAGEMENT NEEDS

The study deals with sufficient evidence that proved to support that different air pollutants have direct and indirect implications on ecosystem services. Long-term monitoring of ozone and other greenhouse gases and the development of physiologically relevant ozone indices are needed to verify model outputs and understanding of future risk. Different improved techniques are needed to estimate the level of air pollutant exposure. Such techniques would greatly expand our ability to evaluate the future risks caused by various toxic gases and atmospheric deposition of pollutants. Better observation of spatial as well as the temporal distribution of particulate matter and other air pollutants are needed for identification and their sources to develop control strategies. So the study demands more robust studies in the field of air pollution.

CONCLUSION

This chapter presents the adverse effects of different types of pollutants on human health and other survivals. As mention above major impairments of different organs can be observed due to the harmful atmospherics pollutants. The observation also focussed that, in view of increased exposure of different pollutants plants are still providing a protective covering to human as well as the animals from different types of air pollutants. Humans burning fossil fuels are currently emitting high level of sulphur dioxide and various gases like carbon mono oxide, oxides of nitrogen every day. Such high level of these gases causing not only global warming but widespread extinction of many species of plants and animals, the different atmospheric pollutants also causing several harmful diseases in human beings. Both carbon and sulphur dioxide emissions must be reduced significantly to reverse global warming, but reducing of sulphur dioxide should be the highest priority in future.

REFERENCES

Ainsworth, E. E., Yendrek, C. R., Sitch, S., Collins, W. J., & Emberson, L. D. (2012). The effects of tropospheric ozone on net primary productivity and implications for climate change. *Annual Review of Plant Biology*, *63*(1), 637–661. doi:10.1146/annurev-arplant-042110-103829 PMID:22404461

Alexander, B., Park, R. J., Jacob, D. J., Li, Q. B., Yantosca, R. M., Savarino, J., ... Thiemens, M. H. (2005). Sulfate formation in sea-salt aerosols: Constraints from oxygen isotopes. *Journal of Geophysical Research: Atmospheres, 110*(D10).

Anderson, T. A., Kruger, E. L., & Coats, J. R. (1994). Enhanced degradation of a mixture of three herbicides in the rhizosphere of a herbicide-tolerant plant. *Chemosphere*, *28*(8), 1551–1557. doi:10.1016/0045-6535(94)90248-8

Angelo, J., Andrade, L., Madeira, L. M., & Mendes, A. (2013). An overview of photocatalysis phenomena applied to NOx abatement. *Journal of Environmental Management*, *129*, 522–539. doi:10.1016/j.jenvman.2013.08.006 PMID:24018117

Annesi-Maesano, I., Baiz, N., Banerjee, S., Rudnai, P., & Rive, S. (2013). Sinphonie Group. Indoor air quality and sources in schools and related health effects. *Journal of Toxicology and Environmental Health. Part B, Critical Reviews*, *16*(8), 491–550. doi:10.1080/10937404.2013.853609 PMID:24298914

Aunan, K., Pátzay, G., Aaheim, H. A., & Seip, H. M. (1998). Health and environmental benefits from air pollution reductions in Hungary. *The Science of the Total Environment*, *212*(2-3), 245–268. doi:10.1016/S0048-9697(98)00002-3 PMID:9573631

Avery, M. A. (2010). Increasings pringtime ozone mixing ratios in the free troposphere over western North. *American Naturalist*, *463*, 344–348.

Badman, D. G., & Jaffe, E. R. (1996). Blood and air pollution: State of knowledge and research needs. *Otolaryngology - Head and Neck Surgery*, *114*(2), 205–208. doi:10.1016/S0194-5998(96)70166-3 PMID:8637733

Balmes, J. R., Fine, J. M., & Sheppard, D. (1987). Symptomatic bronchoconstriction after short-term inhalation of sulfur dioxide. *The American Review of Respiratory Disease*, *136*(5), 1117–1121. doi:10.1164/ajrccm/136.5.1117 PMID:3674573

Barbour, D. A. (1986). *Why are there so few butterflies in Liverpool? An answer*. London: Royal Entomological Society.

Barrie, L. A., & Hoff, R. M. (1984). The oxidation rate and residence time of sulphur dioxide in the Arctic atmosphere. *Atmospheric Environment (1967), 18*(12), 2711-2722.

Bellinger, D. C. (2005). Teratogen update: Lead and pregnancy. *Birth Defects Research. Part A, Clinical and Molecular Teratology, 73*(6), 409–420. doi:10.1002/bdra.20127 PMID:15880700

Bentayeb M, Simoni M, Norback D, Baldacci S, Maio S, Viegi G.(2013). Indoor air pollution and respiratory health in the elderly. *J Environ Sci Health A Tox Hazard Subst Environ Eng, 48*, 1783-9.

Birch, M. B. L., Gramig, B. M., Moomaw, W. R., Doering, O. C. III, & Reeling, C. J. (2011). Why metrics matter: Evaluating policy choices for reactive nitrogen in the chesapeake bay watershed. *Environmental Science & Technology, 45*(1), 168–174. doi:10.1021/es101472z PMID:20853823

Boffetta, P., Merler, E., & Vainio, H. (1993). Carcinogenicity of mercury and mercury compounds. *Scandinavian Journal of Work, Environment & Health, 19*(1), 1–7. doi:10.5271jweh.1510 PMID:8465166

Boonen, E., Akylas, V., Barmpas, F., Boréave, A., Bottalico, L., Cazaunau, M., ... Beeldens, A. (2015). Construction of a photocatalytic De-polluting field site in the Leopold II tunnel in Brussels. *Journal of Environmental Management, 155*, 136–144. doi:10.1016/j.jenvman.2015.03.001 PMID:25863437

Brasseur, G. P., Muller, J.-F., Tie, X., & Horowitz, L. (2001). Tropospheric ozone and climate: past, present and future. In T. Matsuno & H. Kida (Eds.), *Present and Future of Modeling Global Environmental Change: Toward Integrated Modeling* (pp. 63–75). Tokyo: TERRAPUB.

Brazil, G. M., Kenefick, L., Callanan, M., Haro, A., De Lorenzo, V., Dowling, D. N., & O'gara, F. (1995). Construction of a rhizosphere pseudomonad with potential to degrade polychlorinated biphenyls and detection of bph gene expression in the rhizosphere. *Applied and Environmental Microbiology, 61*(5), 1946–1952. PMID:7646029

Bytnerowicz, A., Arbaugh, M., Schilling, S., Fraczek, W., & Alexander, D. (2008). Ozone distribution and phytotoxic potential in mixed conifer forests of the San Bernardino Mountains, southern California. *Environ. Pollut, 155*(3), 398–408. doi:10.1016/j.envpol.2008.01.046 PMID:18359541

Cardelino, C. A., & Chameides, W. L. (1990). Natural hydrocarbons, urbanization, and urban ozone. *Journal of Geophysical Research, 95*(D9), 13,971–13,979. doi:10.1029/JD095iD09p13971

Carslaw, K. S., Boucher, O., Spracklen, D. V., Mann, G. W., Rae, J. G. L., Woodward, S., & Kulmala, M. (2010). A review of natural aerosol interactions and feedbacks within the Earth system. *Atmospheric Chemistry and Physics, 10*(4), 1701–1737. doi:10.5194/acp-10-1701-2010

Catcott, E. J. (1961). Effects of air pollution on animals. In *World Health Organization Monograph series, no.41* (pp. 221–231). Geneva: WHO.

Cavanagh, J. E. (2006). *Potential of Vegetation to Mitigate Road-generated Air Pollution.* Retrieved April 21st, 2016 from https://www.niwa.co.nz/sites/niwa. co.nz/files/import/attachments/co1x0405 p05.pdf

Cecchi, L. D., Amato, G., & Annesi-Maesano, I. (2013). Climate, urban air pollution, and respiratory allergy. In R. A. Pielke (Ed.), *Climate vulnerability: understanding and addressing threats to essential resources* (pp. 105–113). Waltham, MA: Academic Press. doi:10.1016/B978-0-12-384703-4.00116-7

Chandra, S. V., & Garg, N. (1991). A cross-sectional study of pulmonary function among workers exposed to multimetals in the glass bangle industry. *American Journal of Industrial Medicine, 20*(3), 391–399. doi:10.1002/ajim.4700200311 PMID:1928114

Chanel, O., Henschel, S., Goodman, P. G., Analitis, A., Atkinson, R. W., Le Tertre, A., & Medina, S. (2014). Economic valuation of the mortality benefits of a regulation on SO2 in 20 European cities. *European Journal of Public Health, 24*(4), 631–637. doi:10.1093/eurpub/cku018 PMID:24567289

Chauhan, A. J., Krishna, M. T., Frew, A. J., & Holgate, S. T. (1998). Exposure to nitrogen dioxide (NO2) and respiratory disease risk. *Reviews on Environmental Health, 13*, 73. PMID:9718623

Chestnut, L., & Mills, D. M. (2005). Afresh look at the benefits and costs of the US acid rain program. *Journal of Environmental Management, 77*(3), 252–266. doi:10.1016/j.jenvman.2005.05.014 PMID:16171931

Clancy, L., Goodman, P., Sinclair, H., & Dockery, D. W. (2002). Effect of air-pollution control on death rates in Dublin, Ireland: An intervention study. *Lancet, 360*(9341), 1210–1214. doi:10.1016/S0140-6736(02)11281-5 PMID:12401247

Clifford, A., Lang, L., Chen, R., Anstey, K. J., & Seaton, A. (2016). Exposure to air pollution and cognitive functioning across the life course–A systematic literature review. *Environmental Research, 147*, 383–398. doi:10.1016/j.envres.2016.01.018 PMID:26945620

Cohen, A. J., Ross Anderson, H., Ostro, B., Pandey, K. D., Krzyzanowski, M., Kunzli, N., ... Smith, K. (2005). The global burden of disease due to outdoor air pollution. *Journal of Toxicology and Environmental Health. Part A., 68*(13-14), 1301–1307. doi:10.1080/15287390590936166 PMID:16024504

Cohen, M. F., Williams, J., & Yamasaki, H. (2002). Biodegradation of diesel fuel by an *Azolla*-derived bacterial consortium. *Journal of Environmental Science and Health. Part A, 37*(9), 1593–1606.

Cooper, O. R., Parrish, D. D., Stohl, A., Trainer, M., Nédélec, P., Thouret, V., ... Leblanc, T. (2010). Increasing springtime ozone mixing ratios in the free troposphere over western North America. *Nature, 463*(7279), 344–348. doi:10.1038/nature08708 PMID:20090751

Damek-Poprawa, M., & Sawicka-Kapusta, K. (2003). Damage to the liver, kidney, and testis with reference to burden of heavy metals in yellow-necked mice from areas around steelworks and zinc smelters in Poland. *Toxicology, 186*(1-2), 1–10. doi:10.1016/S0300-483X(02)00595-4 PMID:12604166

Davidson, E. A. (2009). The contribution of manure and fertilizer nitrogen to atmospheric nitrous oxide since 1860. *Nature Geoscience, 2*(9), 659–662. doi:10.1038/ngeo608

Davidson, E. A. (2012). Excess nitrogen in the US environment: Trends, risks, and solutions. *ESA Issues Ecology, 15*, 1–16.

DeAngelo, B. J. (2013). Bounding the role of black carbon in the climate system: A scientific assessment. *Journal of Geophysical Research, D, Atmospheres, 118*(11), 5380–5552. doi:10.1002/jgrd.50171

Dockery, D. W., & Stone, P. H. (2007). Cardiovascular risks from fine particulate air pollution. *The New England Journal of Medicine, 356*(5), 511–513. doi:10.1056/NEJMe068274 PMID:17267912

Dodd, K. T., & Gross, D. R. (1980). Ammonia inhalation toxicity in cats: A study of acute and chronic respiratory dysfunction. *Archives of Environmental Health, 35*(1), 6–14. doi:10.1080/00039896.1980.10667455 PMID:7362273

Domene, L. A. F., & Ayres, R. U. (2001). Nitrogen's role in industrial *systems. Journal of Industrial Ecology*, *5*(1), 77–103. doi:10.1162/108819801753358517

Donham, K. J. (1991). Association of environmental air contaminants with disease and productivity in swine. *American Journal of Veterinary Research*, *52*, 1723–1730. PMID:1767997

Dubey, K. K., & Fulekar, M. H. (2011). Effect of pesticides on the seed germination of *Cenchrus setigerus* and *Pennisetum pedicellatum* as monocropping and co-cropping system: Implications for rhizospheric bioremediation. *Romanian Biotechnological Letters*, *16*(1), 5908–5918.

Economopoulos, A. P. (1993). *Assessment of Sources of Air Water and Land Pollution. A Guide to Rapid Source Inventory Techniques and Their Use in Formulating Environmental Control Strategies. Part One: Rapid Inventory Techniques in Environmental Pollution. Part Two: Approaches for Consideration in Formulating Environmental Control Strategies (Unpublished document WHO/YEP/93.1)*. Geneva: WHO.

Epton, M. J., Dawson, R. D., Brooks, W. M., Kingham, S., Aberkane, T., Cavanagh, J. A., ... Brown, L. (2008). The effect of ambient air pollution on respiratory health of school children: A panel study. *Environmental Health*, *7*(1), 16. doi:10.1186/1476-069X-7-16 PMID:18479529

Escobedo, F., Kroeger, T., & Wagner, J. (2011). Urban forests and pollution mitigation: Analyzing ecosystem services and disservices. *Environ. Pollut*, *159*(8-9), 2078–2087. doi:10.1016/j.envpol.2011.01.010 PMID:21316130

European Commission (EC) Directive. (2008). *50/EC of the European Parliament and of the Council on Ambient Air Quality and Cleaner Air for Europe*. Available online: http://eur-lex.europa.eu/legal-content/

Fang, Y., Naik, V., Horowitz, L. W., & Mauzerall, D. L. (2013). Air pollution and associated human mortality: The role of air pollutant emissions, climate change and methane concentration increases from the preindustrial period to present. *Atmospheric Chemistry and Physics*, *13*(3), 1377–1394. doi:10.5194/acp-13-1377-2013

Felzer, B., Kicklighter, D., Melillo, J., Wang, C., Zhuang, Q., & Prinn, R. (2004). Effects of ozone on net primary production and carbon sequestration in the conterminous United States using a biogeochemistry model. *Tellus*, *56B*(3), 230–248. doi:10.1111/j.1600-0889.2004.00097.x

Fiore, A. M. (2014). Atmospheric chemistry: No equatorial divide for a cleansing radical. *Nature*, *513*(7517), 176–178. doi:10.1038/513176a PMID:25209794

Fiore, A. M., Jacob, D. J., Field, B. D., Streets, D. G., Fernandes, S. D., & Jang, C. (2002). Linking ozone pollution and climate change: The case for controlling methane. *Geophysical Research Letters*, *29*(19), 1919. doi:10.1029/2002GL015601

Fiore, A. M., Naik, V., & Leibensperger, E. M. (2015). Air quality and climate connections. *Journal of the Air & Waste Management Association*, *65*(6), 645–685. doi:10.1080/10962247.2015.1040526 PMID:25976481

Fry, G. (1984). Australia and the South Pacific. *Diplomacy in the marketplace: Australia in world affairs, 90*, 175-77.Filippidou EC, Koukouliata A. Ozone effects on the respiratory system. *Progress in Health Sciences, 2011*(1), 144–155.

Fuzzi, S., Baltensperger, U., Carslaw, K., Decesari, S., Denier Van Der Gon, H., Facchini, M. C., ... Nemitz, E. (2015). Particulate matter, air quality and climate: Lessons learned and future needs. *Atmospheric Chemistry and Physics*, *15*(14), 8217–8299. doi:10.5194/acp-15-8217-2015

Gauderman, W. J., Vora, H., McConnell, R., Berhane, K., Gilliland, F., Thomas, D., ... Peters, J. (2007). Effect of exposure to traffic on lung development from 10 to 18 years of age: A cohort study. *Lancet*, *369*(9561), 571–577. doi:10.1016/S0140-6736(07)60037-3 PMID:17307103

Gauss, M., Myhre, G., Isaksen, I. S. A., Grewe, V., Pitari, G., Wild, O., ... Sanderson, M. G. (2003). Radiative forcing in the 21st century due to ozone changes in the troposphere and the lower stratosphere. *Journal of Geophysical Research*, *108*(D9), 4292. doi:10.1029/2002JD002624

Ghio, A. J., & Huang, Y. C. (2004). Exposure to concentrated ambient particles (CAPs): A review. *Inhalation Toxicology*, *16*, 53.

Ghio, A. J., & Samet, J. (1999). *Air pollution and health: metals and air pollution particles*. Academic Press. doi:10.1016/B978-012352335-8/50104-6

Greaver, T. L., Sullivan, T. J., Herrick, J. D., Barber, M. C., Baron, J. S., Cosby, B. J., ... Novak, K. J. (2012). Ecological effects of nitrogen and sulfur air pollution in the US: What do we know? *Front. Ecol*, *10*(7), 365–372. doi:10.1890/110049

Grulke, N. E., Minnich, R. A., Paine, T. D., Seybold, S. J., Chavez, D. J., Fenn, M. E., . . . Dunn, A. (2009). Air pollution increases forest susceptibility to wildfires In A case study in the San Bernardino Mountains in southern California. Academic Press.

Grulke, N. E., Paoletti, E., & Heath, R. L. (2006). Direct measurements of foliar ozone uptake in crop and tree species. *Environ. Pollut.*, *146*, 640–647. doi:10.1016/j.envpol.2006.04.014 PMID:16777298

Gryparis, A., Forsberg, B., Katsouyanni, K., Analitis, A., Touloumi, G., Schwartz, J., ... Dörtbudak, Z. (2004). al. Acute effects of ozone on mortality from the "air pollution and health: A European approach" project. *American Journal of Respiratory and Critical Care Medicine*, *170*(10), 1080–1087. doi:10.1164/rccm.200403-333OC PMID:15282198

Haase, D., Larondelle, N., Andersson, E., Artmann, M., Borgström, S., Breuste, J., ... Kabisch, N. (2014). A quantitative review of urban ecosystem service assessments: Concepts, models, and implementation. *Ambio*, *43*(4), 413–433. doi:10.100713280-014-0504-0 PMID:24740614

Haleem, D., Von Wintzingerode, F., Moter, A., Moawad, H., & Göbel, U. B. (2000). Phylogenetic analysis of rhizosphere-associated β-subclass proteobacterial ammonia oxidizers in a municipal wastewater treatment plant based on rhizoremediation technology. *Letters in Applied Microbiology*, *31*(1), 34–38. doi:10.1046/j.1472-765x.2000.00760.x PMID:10886611

Hansen, C., Luben, T. J., Sacks, J. D., Olshan, A., Jeffay, S., Strader, L., & Perreault, S. D. (2010). The effect of ambient air pollution on sperm quality. *Environmental Health Perspectives*, *118*(2), 203–209. doi:10.1289/ehp.0901022 PMID:20123611

Haramis, G. M. (1984). Acid alteration of experimental wetlands: Effects on diversity andabundance of macroinvertebrates and growth, behaviours and survival of juvenile black ducks. Draft research summary, US Food and Wildlife Service, Eastern Energy and Land Use Team.

Harry, E. G. (1978). Air pollution in farm buildings and methods of control: A review. *Avian Pathology*, *7*(4), 441–454. doi:10.1080/03079457808418301 PMID:18770401

Health and Environment Alliance. Climate Change. (2009). Retrieved from www.env-health.org/r/93

Heath, Pollard, & Thomas. (1984). *Atlas of Butterflies of Britain and Ireland*. Viking Press.

Houghton. (2001). *Climate Change 2001: The Scientific Basis*. Cambridge University Press.

Huang, Y. C., & Ghio, A. J. (2006). Vascular effects of ambient pollutant particles and metals. *Current Vascular Pharmacology*, *4*(3), 199–203. doi:10.2174/157016106777698351 PMID:16842137

Huang, Y. C., & Ghio, A. J. (2006). Vascular effects of ambient pollutant particles and metals. *Current Vascular Pharmacology*, *4*, 199.

Hulin, M., Simoni, M., Viegi, G., & Annesi-Maesano, I. (2012). Respiratory health and indoor air pollutants based on quantitative exposure assessments. *The European Respiratory Journal*, *40*(4), 1033–1045. doi:10.1183/09031936.00159011 PMID:22790916

Intergovernmental Panel on Climate Change (CH). Climate change. (2007). *The physical science basis*. Geneva: Intergovernmental Panel on Climate Change; 2007 (cited 2007 Jan 01). Available from: http://www.ipcc.ch/pdf/assessment-report/ar4/wg1/ar4_wg1_full_report.pdf

Jacob, D. J., & Winner, D. A. (2009). Effect of climate change on air quality. *Atmospheric Environment*, *43*(1), 51–63. doi:10.1016/j.atmosenv.2008.09.051

Jacobsen, C. S. (1997). Plant protection and rhizosphere colonization of barley by seed inoculated herbicide degrading Burkholderia (*Pseudomonas*) cepacia DBO1 (pRO101) in 2, 4-D contaminated soil. *Plant and Soil*, *189*(1), 139–144. doi:10.1023/A:1004296615446

Jacobson, M. Z. (2007). Effects of ethanol (E85) versus gasoline vehicles on cancer and mortality in the United States. *Environmental Science & Technology*, *41*(11), 4150–4157. doi:10.1021/es062085v PMID:17612204

Jerrett, M., Burnett, R. T., Pope, C. A., Thurston, G., & Krewski, D. (2009). Long-term ozone exposure and mortality. *The New England Journal of Medicine*, *360*(11), 1085–1095. doi:10.1056/NEJMoa0803894 PMID:19279340

Kagawa, J. (1985). Evaluation of biological significance of nitrogen oxides exposure. *The Tokai Journal of Experimental and Clinical Medicine*, *10*, 348.

Kala. (2014). Rhizoremediation: A Promising Rhizosphere Technology. *IOSR-JESTFT, 8*(8), 23-27.

Kelley, S. L., Aitchison, E. W., Deshpande, M., Schnoor, J. L., & Alvarez, P. J. (2001). Biodegradation of 1, 4-dioxane in planted and unplanted soil: Effect of bioaugmentation with *Amycolata* sp. CB1190. *Water Research*, *35*(16), 3791–3800. doi:10.1016/S0043-1354(01)00129-4 PMID:12230161

Kimbrough, R. D., Carter, C. D., Liddle, J. A., Cline, R. E., & Phillips, P. E. (1977). Epidemiology and pathology of a tetrachlorodibenzodioxin poisoning episode. *Archives of Environmental Health*, *32*(2), 77–86. doi:10.1080/00039896.1977.10667259 PMID:557961

Kingsley, M. T., Fredrickson, J. K., Metting, F. B., & Seidler, R. J. (1994). *Environmental restoration using plant-microbe bioaugmentation. In Bioremediation of chlorinated and polyaromatic hydrocarbon compounds* (pp. 287–292). Boca Raton, FL: Lewis Publishers.

Kinney, P. L. (2008). Climate change, air quality, and human health. *American Journal of Preventive Medicine*, *35*(5), 459–467. doi:10.1016/j.amepre.2008.08.025 PMID:18929972

Koposzki, H. (1992). Effects of acid and nitrogen deposition on the mesofauna, especially the collembolan. In *Proceedings of Air Pollution and Interactions between Organisms in Forest Ecosystems, 15th IUFRO International Meeting of Specialists on Air Pollution Effects on Forest Ecosystems* (p. 46). Academic Press.

Korade, D. L., & Fulekar, M. H. (2009). Rhizosphere remediation of chlorpyrifos in mycorrhizospheric soil using ryegrass. *Journal of Hazardous Materials*, *172*(2), 1344–1350. doi:10.1016/j.jhazmat.2009.08.002 PMID:19720454

Kunzli, N., Gutschmidt, K., Pope, A., Romieu, I., Samet, J. M., & Smith, K. (2005). The global burden of disease due to outdoor air pollution. *Journal of Toxicology and Environmental Health*, *68*(13-14), 1301–1307. doi:10.1080/15287390590936166 PMID:16024504

Kuo, C. Y., Wong, R. H., Lin, J. Y., Lai, J. C., & Lee, H. (2006). Accumulation ofchromium and nickel metals in lung tumors from lung cancer patients in Taiwan. *Journal of Toxicology and Environmental Health*, *69*(14), 1337–1344. doi:10.1080/15287390500360398 PMID:16760140

Kurpius, M. R., & Goldstein, A. H. (2003). Gas-phase chemistry dominates O3 loss to a forest, implying a source of aerosols and hydroxyl radicals to the atmosphere. *Geophysical Research Letters*, *30*(7), 1371. doi:10.1029/2002GL016785

Law, K. (2010). More ozone over North America. *Nature*, *463*, 307–308.

Li, S., Williams, G., Jalaludin, B., & Baker, P. (2012). Panel studies of air pollution on children's lung function and respiratory symptoms: A literature review. *The Journal of Asthma*, *49*(9), 895–910. doi:10.3109/02770903.2012.724129 PMID:23016510

Lilleskov, E. A., Liu, L., Lynch, J., Nadelhoffer, K., Perakis, S. S., Robin-Abbott, M. J., ... Dennis, R. L. (2011). Effects of nitrogen deposition and empirical nitrogen critical loads for ecoregions of the United States. *Ecological Applications*, *21*(8), 3049–3082. doi:10.1890/10-2341.1

Lin, J.-T., Patten, K. O., Liang, X.-Z., & Wuebbles, D. J. (2008). Effects of intercontinental transport on surface ozone over the United States at the present and future. *Geophysical Research Letters*, *35*(2), L02805. doi:10.1029/2007GL031415

Liu, Y., Stanturf, J., & Goodrick, S. (2010). Trends in global wildfire potential in a changing climate. *Forest Ecology and Management*, *259*(4), 685–697. doi:10.1016/j.foreco.2009.09.002

Loghman-Adham, M. (1997). Renal effects of environmental and occupational lead exposure. Environ. Health Perspect. 105, 928.Jarup, L., 2003. Hazards of heavy metal contamination. *British Medical Bulletin*, *68*, 167.

Madden, E. F., & Fowler, B. A. (2000). Mechanisms of nephrotoxicity from metal combinations: A review. *Drug and Chemical Toxicology*, *23*(1), 1–12. doi:10.1081/DCT-100100098 PMID:10711385

Mandal, P. K. (2005). Dioxin: A review of its environmental effects and its aryl hydrocarbon receptor biology. *Journal of Comparative Physiology*, *175*(4), 221–230. doi:10.100700360-005-0483-3 PMID:15900503

Manzo, N. D., Slade, R., & Dye, J. A. (2010). Interrelationship of lung inflammation and air pollutant exposure on cellular oxidative stress and epithelial injury. *Proceedings 27th Symposium of the Veterinary Comparative Respiratory Society*.

Mason, C. F., & MacDonald, S. M. (1987). Acidification and otter (Lutra lutra) distribution on a British river. *Mammalia*, *51*(1), 81–87.

McGrath, J. M., Betzelberger, A. M., Wang, S., Shook, E., Zhu, X. G., Long, S. P., & Ainsworth, E. A. (2015). An analysis of ozone damage to historical maize and soybean yields in the United States. *Proceedings of the National Academy of Sciences of the United States of America*, *112*(46), 14390–14395. doi:10.1073/pnas.1509777112 PMID:26578785

Nawrot, T., Plusquin, M., Hogervorst, J., Roels, H. A., Celis, H., Thijs, L., ... Staessen, J. A. (2006). Environmental exposure to cadmium and risk of cancer: A prospective population-based study. *The Lancet Oncology*, *7*(2), 119–126. doi:10.1016/S1470-2045(06)70545-9 PMID:16455475

Neas, L., Williams, R. W., & Devlin, R. B. (2004). Particulate matter exposure in cars is associated with cardiovascular effects in healthy young men. *American Journal of Respiratory and Critical Care Medicine*, *169*(8), 934–940. doi:10.1164/rccm.200310-1463OC PMID:14962820

Nichols, T. D., Wolf, D. C., Rogers, H. B., Beyrouty, C. A., & Reynolds, C. M. (1997). Rhizosphere microbial populations in contaminated soils. *Water, Air, and Soil Pollution*, *95*(1-4), 165–178. doi:10.1007/BF02406163

Nowak, D. J. (2000). Tree species selection, design, and management to improve air quality. *Construction Technology: Annual Meeting Proceedings of the American Society of Landscape Architects*.

Nowak, D. J., Crane, D. E., & Stevens, J. C. (2006). Air pollution removal by urban trees and shrubs in the United States. *Urban Forestry & Urban Greening*, *4*(3-4), 115–123. doi:10.1016/j.ufug.2006.01.007

Ollinger, S. V., Aber, J. D., & Reich, P. B. (1997). Simulating ozone effects on forest productivity: Interactions among leaf-, canopy-, and stand-level processes. *Ecological Applications*, *7*(4), 1237–1251. doi:10.1890/1051-0761(1997)007[1237:SOEOFP]2.0.CO;2

Olsson, D., Mogren, I., & Forsberg, B. (2013). Air pollution exposure in early pregnancy and adverse pregnancy outcomes: A register-based cohort study. *BMJ Open*, *3*(2), e001955. doi:10.1136/bmjopen-2012-001955 PMID:23386578

Pai, S. G., Riley, M. B., & Camper, N. D. (2001). Microbial degradation of mefenoxam in rhizosphere of Zinnia angustifolia. *Chemosphere*, *44*(4), 577–582. doi:10.1016/S0045-6535(00)00368-4 PMID:11482644

Pangtey, B. S., Chandra, S. V., & Garg, N. (1991). A cross-sectional study of pulmonary function among workers exposed to multimetals in the glass bangle industry. *American Journal of Industrial Medicine*, *20*(3), 391–399. doi:10.1002/ajim.4700200311 PMID:1928114

Pardo, L. H., Fenn, M. E., Goodale, C. L., Geiser, L. H., Driscoll, C. T., Allen, E. B., ... Haines, A. (2014). Climate change: Challenges and opportunities for global health. *Journal of the American Medical Association*, *312*(15), 1565–1580. doi:10.1001/jama.2014.13186 PMID:25244362

Peters, A., von Klot, S., Heier, M., Trentinaglia, I., Hörmann, A., Wichmann, H. E., & Löwel, H. (2004). Exposure to traffic and the onset of myocardial infarction. *The New England Journal of Medicine*, *351*(17), 1721–1730. doi:10.1056/NEJMoa040203 PMID:15496621

Pickrell, J. (1991). Hazards in confinement housing - gases and dusts in confined animal houses for swine, poultry, horses and humans. *Veterinary and Human Toxicology*, *33*, 32–39. PMID:2017863

Pinder, R.W., Bettez, N.D., Bonan, G.B., Greaver, T. L., Widier, W.R., Schlesinger,W., & Davidson, E. A. (2013). Impacts of human alteration of the nitrogen cycle in the US on radiative forcing. *Biogeochemistry, 114*, 25–40.

Rastogi, S. K., Gupta, B. N., Husain, T., Chandra, H., Mathur, N., Pangtey, B. S., ... Ravenscroft, J. (2006). Effects of pollution on human growth and development: An introduction. *Journal of Physiological Anthropology*, *25*(1), 103–112. doi:10.2114/jpa2.25.103 PMID:16617215

Reich, P. B., & Amundson, R. G. (Eds.). (n.d.). Ambient levels of ozone reduce net photosynthesis in tree and crop species. Science, 230, 566–570.

Riebau, A. R., & Andersen, C. (1985). Wildland Fires and Air Pollution. *Developments in Environmental Science*, *8*, 365–403.

Riediker, M., Cascio, W. E., Griggs, T. R., Herbst, M. C., Bromberg, P. A., Neas, L., ... Devlin, R. B. (2004). Particulate matter exposure in cars is associated with cardiovascular effects in healthy young men. *American Journal of Respiratory and Critical Care Medicine*, *169*(8), 934–940. doi:10.1164/rccm.200310-1463OC PMID:14962820

Ronchel, M. C., & Ramos, J. L. (2001). Dual system to reinforce biological containment of recombinant bacteria designed for rhizoremediation. *Applied and Environmental Microbiology*, *67*(6), 2649–2656. doi:10.1128/AEM.67.6.2649-2656.2001 PMID:11375176

Roupsard, P. (2013). *Etude Phénoménologique Du dépôt Sec d'aérosols En Milieu Urbain* (Ph.D Thesis). University of Rouen.

Rubin, J. I., Kean, A. J., Harley, R. A., Millet, D. B., & Goldstein, A. H. (2006). Temperature dependence of volatile organic compound evaporative emissions from motor vehicles. *Journal of Geophysical Research*, *111*(D3), D03305. doi:10.1029/2005JD006458

Sandmann, E. R. I. C., & Loos, M. A. (1984). Enumeration of 2, 4-D-degrading microorganisms in soils and crop plant rhizospheres using indicator media; high populations associated with sugarcane (*Saccharum officinarum). Chemosphere, 13*(9), 1073–1084. doi:10.1016/0045-6535(84)90066-3

Scherer, M., & Unshelm, J. (1995). Untersuchungen zur Belastung von Mastschweinen sowie der Umgebung von Mastschweinstallen durch atembaren Feinstaub, stallspezifische Bakterien und Ammoniak. *Zentralblatt fur Hygiene und Umweltmedizin, 196*, 399–415. PMID:7727021

Schlesinger, R. B. (1979). Natural removal mechanisms for chemical pollutants in the environment. *Bioscience, 29*(2), 95–101. doi:10.2307/1307745

Schnoor, J. L., Aitchison, E. W., Kelley, S. L., & Alvarez, P. J. J. (1998). Phytoremediation and bioaugmentation of 1, 4-dioxane. In *Abstr. Adv. Biol. Syst. Rem. Conf* (pp. 87-91). Academic Press.

Schroder, M., Linkert, K. H., Pedersen, S., Takai, H., Johnsen, J. O., Groot Koerkamp, P. W. G., ... Linke, S. (1998). Emissions of Aerial Pollutants in Livestock Buildings in Northern Europe: Overview of a Multinational Project. *Journal of Agricultural Engineering Research, 70*(1), 3–9. doi:10.1006/jaer.1998.0278

Schucht, S., Colette, A., Rao, S., Holland, M., Schöpp, W., Kolp, P., ... Brignon, J. M. (2015). Moving towards ambitious climate policies: Monetised health benefits from improved air quality could offset mitigation costs in Europe. *Environmental Science & Policy, 50*, 252–269. doi:10.1016/j.envsci.2015.03.001

Seibert, K., Fuehr, F., & Cheng, H. H. (1981). *Theory and Practical Use of Soil-Applied Herbicides Symposium*. European Weed Resource Society.

Sharma, R. K., & Agrawal, M. (2005). Biological effects of heavy metals: An overview. *Journal of Environmental Biology, 26*, 301. PMID:16334259

Shine, K. P., Stevenson, D. S., Sudo, K., Szopa, S., & Zeng, G. (2006). Radiative forcing since preindustrial times due to ozone change in the troposphere and the lower stratosphere. *Atmospheric Chemistry and Physics, 6*(3), 575–599. doi:10.5194/acp-6-575-2006

Sillman, S., & Samson, P. J. (1995). Impact of temperature on oxidant photochemistry in urban, polluted rural and remote environments. *Journal of Geophysical Research, D, Atmospheres, 100*(D6), 11497–11508. doi:10.1029/94JD02146

Silva, R. A., West, J. J., Zhang, Y., Anenberg, S. C., Lamarque, J. F., Shindell, D. T., & Horowitz, L. W. (2013). Global premature mortality due to anthropogenic outdoor air pollution and the contribution of past climate change. *Environmental Research Letters, 8*(3), 034005. doi:10.1088/1748-9326/8/3/034005

Sitch, S., Cox, P. M., Collins, W. J., & Huntingford, C. (2007). Indirect radiative forcing of climatechange through ozone effects on the land-carbon sink. *Nature, 448*(7155), 791–794. doi:10.1038/nature06059 PMID:17653194

Smil, V. (2002). Nitrogen and food production: Proteins for human diets. *Ambio, 31*(2), 126–131. doi:10.1579/0044-7447-31.2.126 PMID:12078001

Solomon, S., Qin, D., Manning, M., Marquis, M., Averyt, K., & Tignor, M. l. (2007). Technical summary. In *Climate change. The physical science basis* (pp. 19–940). New York, NY: Cambridge University Press.

Sriprang, R., Hayashi, M., Yamashita, M., Ono, H., Saeki, K., & Murooka, Y. (2002). A novel bioremediation system for heavy metals using the symbiosis between leguminous plant and genetically engineered rhizobia. *Journal of Biotechnology*, *99*(3), 279–293. doi:10.1016/S0168-1656(02)00219-5 PMID:12385715

Sun, G., McLaughlin, S. B., Porter, J. H., Uddling, J., Mulholland, P. J., Adams, M. B., & Pederson, N. (2012). Interactive influences of ozone and climate on streamflow of forested watersheds. *Global Change Biology*, *18*(11), 3395–3409. doi:10.1111/j.1365-2486.2012.02787.x

Tager, I. B., Balmes, J., Lurmann, F., Ngo, L., Alcorn, S., & Kunzli, N. (2005). Chronic exposure to ambient ozone and lung function in young adults. *Epidemiology (Cambridge, Mass.)*, *16*, 751. doi:10.1097/01.ede.0000183166.68809.b0 PMID:16222164

Tai, A. P., Martin, M. V., & Heald, C. L. (2014). Threat to future global food security from climate change and ozone air pollution. *Nature Climate Change*, *4*(9), 817–821. doi:10.1038/nclimate2317

Thiering, E., & Heinrich, J. (2015). Epidemiology of air pollution and diabetes. *Trends in Endocrinology and Metabolism*, *26*(7), 384–394. doi:10.1016/j.tem.2015.05.002 PMID:26068457

Thron, R. W. (1996). Direct and indirect exposure to air pollution. *Otolaryngology - Head and Neck Surgery*, *114*(2), 281–285. doi:10.1016/S0194-5998(96)70184-5 PMID:8637751

Trenberth, K. E., Jones, P. D., Ambenje, P., & Bojariu, R. (2007). Observations: surface and atmospheric climate change. In Climate Change 2007: The Physical Science Basis. Contribution of Working Group I to the Fourth Assessment Report of the Intergovernmental Panel on Climate Change. Cambridge, UK: Cambridge University Press.

Turner, M. C., Jerrett, M., Pope, C. A. III, Krewski, D., Gapstur, S. M., Diver, W. R., ... Burnett, R. T. (2016). Long-term ozone exposure and mortality in a large prospective study. *American Journal of Respiratory and Critical Care Medicine*, *193*(10), 1134–1142. doi:10.1164/rccm.201508-1633OC PMID:26680605

US EPA. (2008). Integrated Science Assessment (ISA) for Oxides of Nitrogen and Sulfur— Environmental Criteria, final report. Washington, DC: Author.

Uysal, N., & Schapira, R. M. (2003). Effects of ozone on lung function and lung diseases. *Current Opinion in Pulmonary Medicine*, *9*, 144.

Vamvakas, S., Bittner, D., & Koster, U. (1993). Enhanced expression of the protooncogenes c-myc and c-fos in normal and malignant renal growth. *Toxicology Letters*, *67*(1-3), 161–172. doi:10.1016/0378-4274(93)90053-Z PMID:8451757

Van Grinsven, H. J. M., Holland, M., Jacobsen, B. H., Klimont, Z., Sutton, M. A., & Willems, W. J. (2013). Costs and benefits of nitrogen for Europe and implications for mitigation. *Environmental Science & Technology*, *47*(8), 3571–3579. doi:10.1021/es303804g PMID:23473305

Vermylen, J., Nemmar, A., Nemery, B., & Hoylaerts, M. F. (2005). Ambient air pollution and acute myocardial infarction. *Journal of Thrombosis and Haemostasis*, *3*(9), 1955–1961. doi:10.1111/j.1538-7836.2005.01471.x PMID:16102102

Vingarzan, R. (2004). A review of surface ozone background levels and trends. *Atmospheric Environment, 38*, 3431–3442.

Wang, H., Jacob, D. J., Le Sager, P., Streets, D. G., Park, R. J., Gilliland, A. B., & van Donkelaar, A. (2009). Surface ozone background in the United States: Canadian and Mexican influences. *Atmospheric Environment, 43*(6), 1310–1319. doi:10.1016/j.atmosenv.2008.11.036

Wathes, C. M., Phillips, V. R., Holden, M. R., Sneath, R. W., Short, J. L., White, R. P. P., & Pedersen, S. (1998). Emissions of aerial pollutants in livestock buildings in Northern Europe: Overview of a multinational project. *Journal of Agricultural Engineering Research, 70*(1), 3–9. doi:10.1006/jaer.1998.0278

Wegmann, M., Fehrenbach, A., Heimann, S., Fehrenbach, H., Renz, H., Garn, H., & Herz, U. (2005). NO2-induced airway inflammation is associated with progressive airflow limitation and development of emphysema-like lesions in C57bl/6 mice. *Experimental and Toxicologic Pathology, 56*(6), 341–350. doi:10.1016/j.etp.2004.12.004 PMID:15945273

WHO Regional Office for Europe (DK). (2013). *Review of evidence on health aspects of air pollution: REVIHAAP Project*. Copenhagen: WHO Regional Office for Europe.

WHO Regional Office for Europe (DK). (2014). *WHO Expert Meeting: methods and tools for asessing the health risks of air pollution at local, national and international level*. Meeting report. Bonn, Germany, 12-13 May 2014. Copenhagen: WHO Regional Office for Europe.

Wilson, R., & Spengler, J. (1996). *Particles in our air*. Harvard University Press.

Wittig, V. E., Ainsworth, E. A., & Long, S. P. (2007). To what extent do current and projected increases in surface ozone affects photosynthesis and stomatal conductance in trees? A metaanalytic review of the last 3 decades of experiments. *Plant, Cell & Environment, 30*(9), 1150–1162. doi:10.1111/j.1365-3040.2007.01717.x PMID:17661752

Woodcock, J., Edwards, P., Tonne, C., Armstrong, B.G., Ashiru, O., Banister, D., & ... Cohen, L. (2009). Public health benefits of strategies to reduce greenhouse-gas emissions: Urban land transport. *Lancet*, 374. PMID:19942277

Wu, S., Mickley, L. J., Jacob, D. J., Rind, D., & Streets, D. G. (2008). Effects of 2000–2050 changes in climate and emissions on global tropospheric ozone and the policy-relevant background ozone in the United States. *Journal of Geophysical Research, 113*(D18), D18312. doi:10.1029/2007JD009639

Xu, X., Ha, S. U., & Basnet, R. (2016). A review of epidemiological research on adverse neurological effects of exposure to ambient air pollution. *Frontiers in Public Health, 4*, 157. doi:10.3389/fpubh.2016.00157 PMID:27547751

Chapter 10
Potential Impacts of Climate Change on Land Degradation and Desertification:
Land Degradation and Climate Change

Dharumarajan S.
National Bureau of Soil Survey and Land Use Planning (ICAR), India

Lalitha M.
National Bureau of Soil Survey and Land Use Planning (ICAR), India

Veeramani S.
National Bureau of Soil Survey and Land Use Planning (ICAR), India

Janani N.
National Bureau of Soil Survey and Land Use Planning (ICAR), India

Kalaiselvi Beeman
National Bureau of Soil Survey and Land Use Planning (ICAR), India

Srinivasan R.
National Bureau of Soil Survey and Land Use Planning (ICAR), India

Rajendra Hegde
National Bureau of Soil Survey and Land Use Planning (ICAR), India

ABSTRACT

Land degradation and desertification have been graded as a major environmental and social dispute in most of the emerging countries. Changes in temperature, wind speed, and precipitation patterns will influence plant biomass production, land use, land cover, soil moisture, infiltration rate, runoff and crop management, and ultimately, land degradation. Close relations between climate change and land degradation processes have been perceived in the past decades. Climate change models and land use models should be combined with hydrologic/erosion models to accurately compute or predict climate change impacts on land degradation. This chapter introduces the advancements in modeling of impact of climate changes in land degradation and need for the critical investigation to better understand and forecast the responses of land degradation processes to a changing climate in the future.

DOI: 10.4018/978-1-5225-7387-6.ch010

INTRODUCTION

Land degradation is a process of deterioration of land caused by extreme weather conditions and human activities that pollute or degrade the quality of land and negatively affect the food production and livelihoods of the people. According to United Nations Convention to Combat Desertification (UNCCD), desertification is the degradation of land in arid, semi-arid and dry sub-humid areas resulting primarily due to human events and climatic variations. Land degradation and desertification is based on time dimensions which vary over a period of years depending on climatic factors such as drought and floods. These studies are conducted at regular interval by mapping the features of large areas which can be assessed effectively using Earth Observation (EO) data's and techniques are used by different researchers and mapped the different degradation /desertification processes (Symeonakis & Higgin bottom, 2014; Dharumarajan et al., 2018a; Dharumarajan et al., 2017c). The indicators of desertification such as soil erosion, vegetal degradation, salinization, water logging etc. can be derived from remote sensing data. Several remote sensors have become available during the past 30 years, with the potential to provide useful information for assessing land degradation (Metternicht et al., 2010). The interpretation of remotely sensed data for monitoring and evaluation of land degradation along with field investigation provides valid datasets for proper planning (Dharumarajan, Lalitha, Vasundhara, & Hegde, 2016; Dharumarajan, Bishop, Hegde, & Singh, 2018b). Different countries are preparing time series land degradation and desertification maps in order to identify the more vulnerable areas for preparation of action plan. Recently India published report cum Atlas on land degradation and desertification mapping of India which shows 29. 32% of total geographical area in India (Table 1 and Figure 1) is affected by desertification and land degradation processes (SAC, 2016).

Land degradation is a global issue affecting both developed and developing nations. Most of the severely affected countries are developing countries and they are already economically disadvantaged and their scope to withstand any downturn in productivity is limited. Global efforts to understand how climate variability has contributed to resource damage and how current and future climate change may further exacerbate the damage are critical for economic and environmental sustainability. There is a clear need for continued research into global and regional climate systems, seasonal climate forecasting, climate trends, risk assessments and communication of those risks to land managers. Decision support tools and policy development based on sound science are prerequisites for the sustainable and productive use of lands.

DIFFERENT PROCESS OF LAND DEGRADATION/DESERTIFICATION

Water Erosion

Water Erosion is detachment and transportation of soil mainly due to rainfall and surface runoff water. Flowing water carries organic and inorganic particles long with soil which deteriorate fertility of land. Severity level of erosion varies with soil type and land covers. The sheet erosion (mostly within agricultural lands) and rills are considered as slight erosion, whereas gullies and ravines are considered as severe stage of erosion (Figure 2).

Figure 1. Desertification and land degradation status map
(source: SAC, 2016)

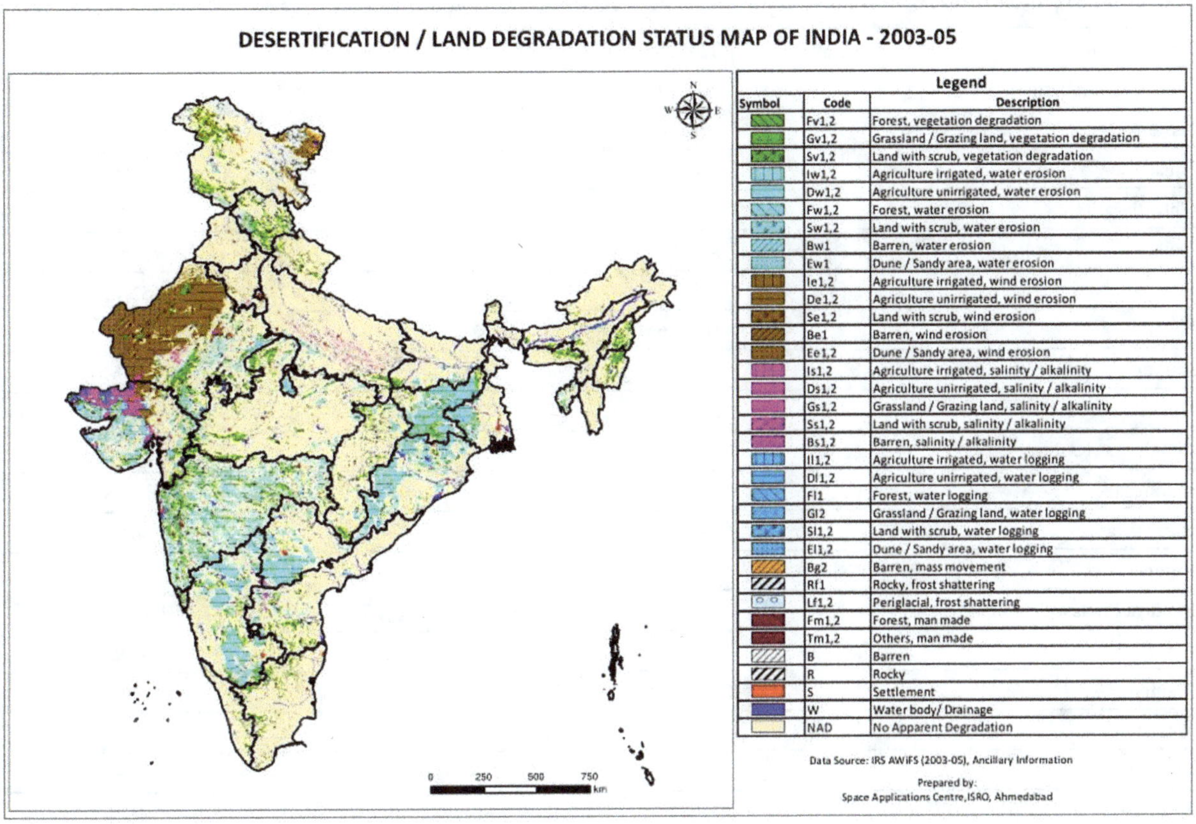

Wind Erosion

Wind erosion is a natural process which moves soil from one place to another due to wind velocity. Wind erosion is most common in arid regions, barren lands, loose soil, sand dunes and desert areas (Figure 3). High wind velocity changes the landscape and condition of land as well as top soil will lose its nutrients available for plant growth.

Vegetal Degradation

Vegetal degradation is a destruction of vegetation by human or natural forces which eventually leads to land degradation (Figure 4). When a land is left barren, the activity of wind and water erodes the top soil leading to decrease in soil aggregation and stability and soil fertility. Vegetal degradation is observed in forest, scrublands and cultivated lands.

Table 1. Desertification and land degradation atlas of India

State Name	Total Area Under Land Degradation and Desertification (ha)	% TGA
Andhra Pradesh	2298758	14.35
Arunachal Pradesh	153933	1.84
Assam	716596	9.14
Bihar	694809	7.38
Chhattisgarh	2211153	16.36
Delhi	89868	60.6
Goa	192973	52.13
Gujarat	10261641	52.29
Haryana	338964	7.67
Himachal Pradesh	2394240	43.01
Jammu and Kashmir	7969607	35.86
Jharkhand	5498726	68.98
Karnataka	6951000	36.24
Kerala	379587	9.77
Madhya Pradesh	3804315	12.34
Maharashtra	13825935	44.93
Manipur	601959	26.96
Meghalaya	494880	22.06
Mizoram	187453	8.89
Nagaland	786678	47.45
Orissa	5304114	34.06
Punjab	144653	2.87
Rajasthan	21526512	62.9
Sikkim	78749	11.1
Tamil Nadu	1543898	11.87
Telangana	3598856	31.34
Tripura	437128	41.69
Uttar Pradesh	1528997	6.35
Uttarakhand	648253	12.12
West Bengal	1733931	19.54
Total	96398166	29.32

Source: SAC (2016)

Figure 2. Water erosion induced land degradation

Figure 3. Wind erosion induced land degradation

Figure 4. Vegetation induced land degradation

Salinization and Alkalization

Salinity or Alkalinity is due to accumulation of water-soluble salts in the soil. It will make the land unsuitable for vegetation and it kills all the microorganisms in the soil, later it transforms fertile land into unfertile or decertified area. Salinity in drylands leads to land degradation. The main causes are excess evapotranspiration, drought, excess irrigation, increase in toxicity, and rise in ground water table. Salinity can be clearly observed through satellite images, but alkalinisation can be inferred based on ground truth and soil sample analysis (Figure 5).

Figure 5. Salinity induced land degradation

Water Logging

Water logging is saturation of water in soil profile for longer duration which clogs the soil pores and decreases the supply of air to the root zone. This may cause due to increased rainfall, excess irrigation and improper drainage planning. If the ground water table in the area is within 2-3 m then that area is considered risk. The severity of water logging is determined based on the period of time the water remains stagnant. Water logging may lead to salinization also. Water logging in agricultural land can be delineated using satellite imagery.

Frost Shattering

Frost shattering is a mechanical disintegration of rock during 0°-degree temperature, where water freezes in rock cracks and exerts forces causing rock to rupture further. During precipitation water enters the cracks of rock which gets frozen at winter increases its volume and exerts tremendous pressure on rocks. This process continues over time, widens the joints/cracks, and causes pieces of rock to shatter from the main rock.

Man Made

Apart from natural processes, land degradation is induced by human beings directly or indirectly. It includes mining, Brick Kiln, Industrial Effluents, City Waste, Urban Agglomeration etc. This manmade process contaminates soil, surface water and ground water which lead to loss of biodiversity and land degradation (Figure 6).

Assessment of Desertification and Land Degradation

A variety of remote sensors have become available to the scientific community during the past 30 year with the potential to provide useful information for assessing land degradation. Such sensors are usually classified according to the source of energy (passive or active sensors), the type of platform (ground, air, or space-borne), the region of the spectrum used to image the earth's surface (optical, infrared, microwave), the platform trajectory (sun-synchronic or geo-stationary satellites), the number and width of spectral bands (panchromatic, multispectral, hyperspectral), the spatial resolution (high, medium also

Figure 6. Manmade activities of land degradation

known as Landsat-like, low), the spatial coverage (point or image view), the temporal resolution (hourly, daily, weekly revisiting frequency), and the radiometric resolution (8, 12, 16,32 bits) (Metternicht, 2007). Landsat series (MSS, TM, ETM+) is nowadays the most commonly used data source (49%), followed by aerial photographs (39%), and microwave sensing (ERS, JERS-1, Radarsat) (27%). About 43% of the works analyzed use multi-scale, multi-sensor, multi-spectral approaches to the cartography of degraded areas, with a combination of visual interpretation and advanced image processing techniques. The use of more expensive hyperspectral and/or very high spatial resolution sensors like AVIRIS, Hyperion (hyperspectral), SPOT-5, and IKONOS is limited to areas of 15 to 150 km^2. Radarsat, JERS, Envisat satellites, and orbital Shuttle SIR-A have been used for mapping degraded areas of 100 to 17,000 km^{-2}spatial resolution sensors (2.5-m spatial resolution or better) to map land degradation over small areas, with a mode around 300 km2.

Assessment of Water Erosion

Shallow mass movements, gullies, rills, sheet erosion, badlands, debris flows, soil slips, and lumps are the indicators most commonly used to assess soil degradation induced by water. Visual interpretation of aerial photographs continues being a popular technique regardless of the size of the surveyed area (Movia, 1980). Optical-IR imagery from moderate resolution sensors (e.g., Landsat TM, CBERS-2 CCD, Terra ASTER) have been used to identify eroded areas using techniques as mathematical morphology (Ishikawa & Silva, 2007), unsupervised ISODATA classification (Mieza, Kovac & Harán, 2004) and spectral indices like the NDVI and Normalized Difference Water Index (NDWI). Other studies to map landslide susceptibility have integrated thematic layers derived from Landsat TM imagery (NDVI, land cover and land use classes) with slope and lithology maps, using a GIS-based fuzzy logic classifier (Schernthanner, 2007) and image segmentation techniques.

Assessment of Wind Erosion

Sand dunes, wind streaks, paleo-aeolian features, desert pavements, sand encroachments, blowouts, and changes in the vegetation cover are indicators commonly applied for mapping wind-induced soil degradation. Land degradation is often assessed from a combination of wind and water erosion processes, a fact that may reflect the one-to-many relationship between indicators and processes. Traditionally, aerial photographs and Landsat imagery are combined to map wind erosion features at local and sub regional levels. Visual interpretation of enhanced imagery, image transformation techniques, digital image classification using neural networks, multi-sensor (e.g., optical, infrared, and microwave), multiscale approaches are applied for combined mapping of water and wind-induced land degradation.

Assessment of Soil Salinization

Surface salinity is a highly dynamic property. This constrains the identification of salt-affected soils and the monitoring of the salinization process, because it influences the spectral, spatial, and temporal behavior of the salt features. The main factors affecting the reflectance of salt-affected soils are quantity and mineralogy of salts, together with soil moisture, color, and surface roughness. Remote sensing-based approaches to map and monitor salinized landscapes should take into consideration (i) the discontinuous way in which salts tend to distribute on the landscape (ii) the mineralogy of salt types, as this controls the occurrence of spectral absorption features in specific regions of the electromagnetic spectrum and influences the appearance of salinity indicators at the soil surface.

Microwave imagery of the Radarsat-1 (C-band, HH polarization, with a spatial resolution of 8 m) and SIR-C (C- and L-bands, multi-polarizationwith a pixel resolution of 25 m) are also used assess the salinisation processes. Variables related to the radar system design (mainly wavelength, number of looks and polarization mode) and to the environmental setting of the area were analyzed to better understand the behavior of the radar backscattering. Four factors were significant when analyzing the variations of the back-scattering coefficients in the environmental setting considered, namely soil type (soil texture), tillage conditions or soil surface aspect, soil moisture or surface water, and the presence of salts. The average backscattering values for all salt-affected soil classes were higher in the L-band than in the C-band of the SIR-C, when the same polarization modes were compared.

Assessment of Vegetation Degradation

Vegetation indices are a measure of greenness that has been successfully related to vegetation types and cover, biological changes of vegetation, leaf-area index, biomass, the fraction of photosynthetically active radiation absorbed by vegetation, crop production, net primary production, and vegetation changes, among others. In areas of closed vegetation canopy, the NDVI can be saturated. In sparsely vegetated areas typical of semiarid environments, retrieving quantitative information on vegetation type, cover, and biomass is more difficult because of the dominance of soil background reflectance and lack of the strong red edge that characterizes vegetation in humid regions The MODIS NPP measures the fraction of the photosynthetically active radiation absorbed by the vegetation from several landscape components. Many symptoms of severe degradation such as gullies and rills rarely cover large areas, thus degradation must be severe indeed to be seen against the signal of surrounding unaffected areas.

Airborne hyperspectral imagery, with advanced image classify cation techniques (e.g., linear mixture modelling), has been used for mapping vegetation degradation and other indicators related to soil degradation over areas in the range of 100 to 1,200 km2. Satellite hyperspectral Hyperion data have been processed using spectral indices (NDVI, SAVI, First Derivative Vegetation Index) and linear spectral mixture analysis integrated with field surveys and ground radiometric measures (Huete, Miura, & Gao, 2003). Furthermore, Monte Carlo spectral mixture analysis of hyperspectral AVIRIS data has been applied to estimate photosynthetic and non-photo synthetic fractional overs.

Figure 7. Illustration of the mechanism of the impact of climate change on soil erosion.
(adopted from Zhiying & Fang, 2016)

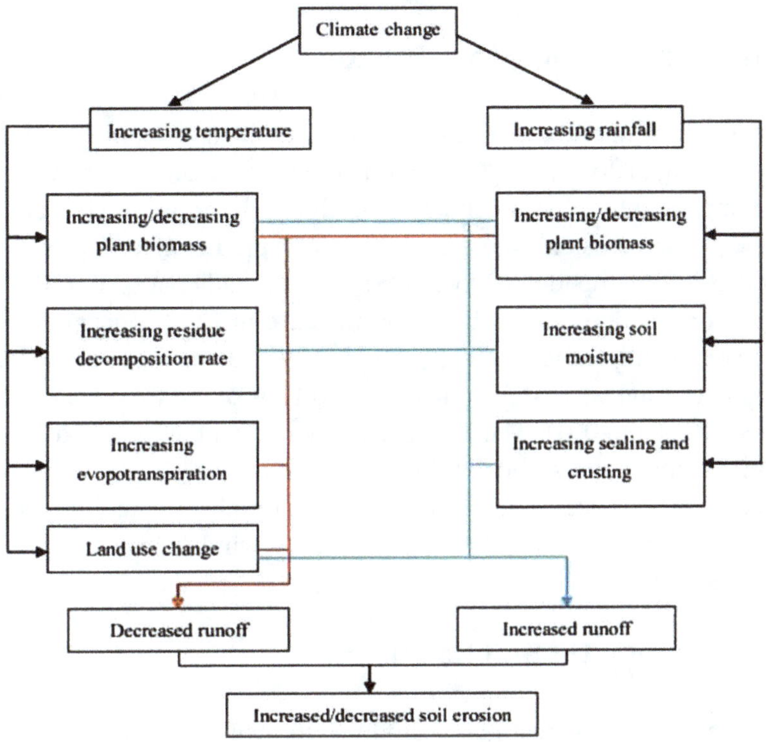

IMPACTS OF CLIMATE CHANGE ON LAND DEGRADATION

Climatic variations are recognized as one of the major issues causes for land degradation. Greater attention must be paid to understand the role of different climatic factors in land degradation. Developing countries are more susceptible to climate change than developed countries, due to the prevalence of economic activities, the scarcity of capital for adaptation measures and their warmer baseline climates.

Impacts of Climate Change on Water Erosion

Impacts of climate change on soil erosion have been reported by many researchers across the world, and many studies suggest that the rainfall is the most direct influencing factor (Figure 7). Lu et al. (2013) reported that every 1% change in precipitation has led to a 2% change in sediment loads and a 1.3% change in water discharge. Leopold (1951) suggested that sheet and gully erosion increased with increasing rainfall amount. The yearly water erosion and nutrient loss under extremely heavy rainfall beats 50% of the total in many environments (Gao et al., 2005). Climate change models' studies estimated that a 7% increase in precipitation will leads to 26% increase in erosion in the United Kingdom (Nearing et al., 2005a). Based on the Erosion Productivity Impact Calculator (EPIC) model, Lee et al. (1996) estimated that 20% increase in precipitation would probably lead to a 37% increase in soil erosion and 40% in surface runoff. For different erosion processes like rill erosion, gully erosion or sheet erosion, separate erosion models are necessary for better prediction. Additionally, more research is needed to explore the impact of crop management practices on water erosion under future changing climate.

Impacts of Climate Change on Wind Erosion

Wind erosion is more problem in arid and semi-arid regions of the developing countries and the severity of wind erosion is increasing in drier years. Changes in temperature and precipitation directly affect soil water content. High temperatures or low precipitation leads to drier soils with low crop residue cover and ultimately more prone to wind erosion. Gao, Ci, & Yu (2002) predicted that wind erosion in grasslands will increase by about 25% if temperatures increase by 2 °C. Liddicoat et al. (2012) predicted greater risk for wind erosion in agricultural lands based on increase in CO2 concentration from 390 to 480 ppm, and 1.5 °C increase in temperature by 2030 compared to the present climate. In Europe, impact of climate change on wind erosion was assessed using Wind Erosion on European Light Soils (WEELS) model (Böhner, Gross, Riksen, 2004). The European Environment Agency (2012) also reported that increased aridity and extreme wind speeds will enhance wind erosion in fine-texture soils. Four-fold increase in wind erosion was predicted across the Corn Belt under a future climate scenario with no change in temperature, precipitation or CO2 and 20% higher wind speeds compared with the present climate (Lee, Phillips, Benson, 1999.).

Impact of Climate Change on Soil Salinisation

Soil salinisation is a major threat for both crop productivity and environment. Global climate change could affect the salt build-up in agricultural lands due to increase in evapotranspiration and temperature. Simulations of the effects of climate change on soil salinisation are vital to identify the vulnerable areas and to suggest proper reclamation measures. Dharumarajan, Bishop, Hegde & Singh (2017a) studied prediction of area under salinisation based on global climate change datasets using IPCC5 (CMIP5) data - HadGEM2-ES model datasets in Anantapur, India. HadGEM2-ES model is a coupled Earth System Model developed by the Met Office Hadley Centre for the CMIP5 centennial simulations. The RCP6 (representative concentration pathways assuming emissions peak around 2080, then decline) of HadGEM2-ES model projected an increase of average maximum temperature from 29.6-34.2 ° C in 2016 to 31.4-35.9 ° C and 32.0 – 36.5° C in 2050 and 2070 respectively in Anantapur district. The spatial distribution of total monthly precipitation projection also increased from 392-673 mm in 2016 to 560-967 mm and 667 -1096 mm in 2050 and 2070 respectively. Random Forest based analysis showed that 14.60 and15.61 percent of Anantapur district is projected to be affected by salinisation process in 2050 and 2070 respectively from the current level of 6.2%. Ashour & Al-Naja (2012) Studied that the increased temperatures by 1°C or 2°C caused an increase of the annual average evapotranspiration by 45 mm to 91mm and leading to increase of irrigation requirements by 3.28% to 6.68%. Change in precipitation and temperature could affect the soil salt build up due to decrease in the salt leaching ability of rainfall because of increase in evapotranspiration. Analysis of effect of climate change on spatial distribution of soil salinity will certainly help the stakeholders to take suitable remedial measures.

CONCLUSION

The actual trend of land degradation and desertification in the world has recently become a serious problem. The consequences are not only related to the loss of soil fertility and land productivity but also to salinization, acidification, surface sealing, desertification and damage to life. Possible future climate change over the region is likely to lead to climate conditions characterized by a decrease in rainfall (drought) especially during the summer and subsequently leading to land degradation and desertification. Presently different models available to assess the different land degradation and desertification processes and these models can be evaluated using various climatic projections to assess the vulnerability.

REFERENCES

Allen, L. H. Jr, Barker, J. T., & Boote, K. J. (1996). The CO2 Fertilization Effect: Higher Carbohydrate Production and Retention as Biomass and Seed Yield. In *Global Climate Change and Agricultural Production. Direct and Indirect Effects of Changing Hydrological, Pedological and Plant Physiological Processes*. Rome, Italy: Food and Agricultural Organization of the United Nations.

Böhner, J., Gross, J., & Riksen, M. (2004). Impact of land use and climate change on wind erosion: prediction of wind erosion activity for various land use and climate scenarios using the WEELS wind erosion model. In D. Goossens & M. Riksen (Eds.), *Wind Erosion and Dust Dynamics: Observations, Simulations, Modelling*. Wageningen, The Netherlands: ESW Publications.

Bot, A., & Benites, J. (2005). *The importance of soil organic matter Key to drought-resistant soil and sustained food and production. In FAO Soils Bulletin 80* (p. 78). Rome: Food and Agriculture Organization of the United Nations.

Butler, R. (2008). 55% Of The Amazon May Be Lost By 2030. *Conservation News*. Retrieved from https://news.mongabay.com/2008/01/55-of-the-amazon-may-be-lost-by-2030/

Chander, S. (2012). Impact of Climate Change on Insects. In H. Pathak & P. K. Aggarwal (Eds.), *Climate Change Impact, Adaptation and Mitigation in Agriculture: Methodology for Assessment and Application* (p. 131). New Delhi, India: Indian Agricultural Research Institute.

Collado, A. D. (2000). *Spatio-temporal dynamics of dune patterns in semiarid Argentina: A neural network analysis. Edaphomatics Bull. no. 20*. Buenos Aires, Argentina: AICET-INTA.

Collado, A. D., Chuvieco, E., & Camarasa, A. (2000). Satellite remote sensing analysis to monitor desertifi cation processes in the crop-rangeland boundary of Argentina. *Journal of Arid Environments*, *52*(1), 121–133. doi:10.1016/S0140-1963(01)90980-2

Davidson, E. A., & Janssens, I. A. (2006). Temperature Sensitivity of Soil Carbon Decomposition and Feedbacks to Climate Change. *Nature*, *440*(7081), 165–173. doi:10.1038/nature04514 PMID:16525463

DEFRA. (2005). *Impact of climate change on soil functions. Final Project Report*. London, UK: Research and Development.

Dharumarajan, Bishop, Hegde, & Singh. (2017a). Prediction of effect of climate change on soil salinity using recent machine language techniques. *Proceedings of XIII Agricultural science congress*, 28.

Dharumarajan, & Lalitha, Hegde, Somashekar, & Singh. (2017c). Biophysical characterisation of Kangeyam grasslands for land degradation vulnerability analysis. *Range Management and Agroforestry*, *38*(1), 10–16.

Dharumarajan, Lalitha, & Hegde, Janani, Rajavat, Sastry, & Singh. (2018a). Status of desertification in South India- Assessment, Mapping and Change detection analysis. *Current Science*, *115*(2), 331–338.

Dharumarajan, S., Bishop, T. F. A., Hegde, R., & Singh, S. K. (2018b). Desertification vulnerability index—an effective approach to assess desertification processes: A case study in Anantapur District, Andhra Pradesh, India. *Land Degradation & Development*, 1–12. doi:10.1002/ldr.2850

Dharumarajan, S., Lalitha, M., Natarajan, A., Naidu, L. G. K., Balasubramanian, R., Hegde, R., ... Singh, S. K. (2017b). Biophysical and Socio Economic Causes for Increasing Fallow Lands in Tamil Nadu. *Soil Use and Management*, *33*(3), 487–498. doi:10.1111um.12361

Dharumarajan, S., Lalitha, M., & Vasundhara, R., & Hegde, R. (2016). The Major Biophysical Indicators of Desertification in Arid and Semi-arid regions of India. *Agropedology (Nagpur)*, *26*, 189–197.

Duru, U. E. (2008). *Impact of climate change on chemical and biological properties in contaminated soils*. Abstract Presented at the Cambridge University Geotechnical Engineering Group.

European Environment Agency. (2012). *Climate Change, Impacts and Vulnerability inEurope 2012: An indicator-based Report*. Report no. 12. Copenhagen, Denmark: Author.

Gao, Q., Ci, L., & Yu, M. (2002). Modeling wind and water erosion in northern China under climate and land use changes. *Journal of Soil and Water Conservation*, *57*, 46–55.

Heikens, A. (2006). *Arsenic contamination of irrigation water, soil and crops in Bangladesh: Risk implications for sustainable agriculture and food safety in Asia. RAP Publication 2006/20*. Bangkok: Food and Agriculture Organization of the United Nations, Regional Office for Asia and the Pacific.

Huete, A. R., Miura, T., & Gao, X. (2003). Land cover conversion and degradation analyses through coupled soil-plant biophysical parameters derived from hyperspectral EO-1 Hyperion. *IEEE Transactions on Geoscience and Remote Sensing*, *41*(6), 1268–1276. doi:10.1109/TGRS.2003.813209

Lee, J. J., Phillips, D. L., & Benson, V. W. (1999). Soil erosion and climate change: Assessingpotential impacts and adaptation practices. *Journal of Soil and Water Conservation*, *54*, 529–536.

Lee, J. J., Phillips, D. L., & Dodson, R. F. (1996). Sensitivity of the US corn belt to climatechangeandelevatedCO2: II. Soil erosion and organic carbon. *Agricultural Systems*, *52*(4), 503–552. doi:10.1016/S0308-521X(96)00015-7

Liddicoat, C., Hayman, P., Alexander, B., Rowland, J., Maschmedt, D., Young, M.-A., . . . Sweeney, S. (2012). Climate change, wheat production and erosion risk in South Australia's cropping zone: linking crop simulation modelling to soil landscape mapping. Government of South Australia, Department of Environment, Water and Natural Resource.

Lu, D., & Weng, Q. (2007). A survey of image classification methods and techniques for improving classification performance. *International Journal of Remote Sensing*, *28*(5), 823–870. doi:10.1080/01431160600746456

Metternicht, G. (2003). Categorical fuzziness: A comparison between crisp and fuzzy class boundary modelling for mapping salt-affected soils using Landsat TM data and a classification based on anion ratios. *Ecological Modelling*, *168*(3), 371–389. doi:10.1016/S0304-3800(03)00147-9

Metternicht, G. (2007). Remote sensing. In K. Kemp (Ed.), Encyclopedia of geographic information science. Sage Publ.

Metternicht, G., Zinck, J. A., Blanco, P. D., & Del Valle, H. F. (2010). Remote Sensing of Land Degradation: Experiences from Latin America and the Caribbean. *Journal of Environmental Quality*, *39*(1), 42–61. doi:10.2134/jeq2009.0127 PMID:20048293

Mieza, M. S., Kovac, F., & Harán, R. (2004). Estimación de superficies inundadasmedianteimágenes. Proc. SimposioLatinoamericano de PercepciónRemota.

Movia, C. (1980). Inventario de la erosiónen la Patagonia Argentina basadoenimágenes Landsat y fotografíaaérea. *Proc. Symposium Arid Land Resource Inventories: Cost-efficient methods.*

Parton, W. J., Scurlock, J. M. O., Ojima, D. S., Schimel, D. S., Hall, D. O., & Members, S. G. (1995). Impact of Climate Change on Grassland Production and Soil Carbon Worldwide. *Global Change Biology*, *1*(1), 13–22. doi:10.1111/j.1365-2486.1995.tb00002.x

Paruelo, J. M., Jobbagy, E., & Sala, O. (2001). Current distribution of ecosystem functional types in temperate South America. *Ecosystems (New York, N.Y.)*, *4*(7), 683–698. doi:10.100710021-001-0037-9

SAC. (2016). Desertification and Land Degradation Atlas of India (Based on IRS AWiFS data of 2011-13 and 2003-05). Space Applications Centre, ISRO.

Schernthanner, H. (2007). Application of the fuzzy logic method for landslide susceptibility mapping, as feasible-ready to use approach, in the context of development, Rio Blanco, Nicaragua. In P. Zeil & S. Kienberger (Eds.), *Geoinformation for development: Bridging the divide through partnerships* (pp. 114–122). Heidelberg, Germany: Hüthig GmbH & Co. KG.

Yue, Y., Li, M., Zhu, A., Ye, X., Mao, R., Wan, J., & Dong, J. (2016). Land Degradation Monitoring in the Ordos Plateau of China Using an Expert Knowledge and BP-ANN-Based Approach. *Sustainability*, *8*(11), 1174. doi:10.3390u8111174

Zhang, X. C. (2005). Spatial downscaling of global climate model output for site-specificassessment of crop production and soil erosion. *Agricultural and Forest Meteorology*, *135*(1-4), 215–229. doi:10.1016/j.agrformet.2005.11.016

Zhiying, L., & Fang, H. (2016). Impacts of climate change on water erosion: A review. *Earth-Science Reviews*, *163*, 94–117. doi:10.1016/j.earscirev.2016.10.004

Chapter 11
Potential Impacts of Climate Change on the Inland Fisheries of Arid and Semi-Arid Regions of Africa:
Impacts of Climate Change on Inland Fisheries

Imefon Udo Udo
University of Uyo, Nigeria

Imekan Isaac Akpan
Akwa Ibom State University, Nigeria

ABSTRACT

Inland fisheries of arid and semi-arid regions of Africa are seriously threatened by negative impacts of climate change. Literature and several models show increase in temperature of 1.1°C in some areas. Sea level rise is projected to increase to 0.8 m by the year 2100. Fish yields have increased almost linearly by around half a million metric tons per decade over the past 60 years, while clear cyclical variations in the residuals of about 20 years' periodicity above and below the trend line have been observed. Although fisher folks, their communities, and local institutions are already constantly adapting to various forms of change, flimsiness in the wider governance and macro-economic environment has weakened the overall adaptive capacity of these regions and fishers are vulnerable to projected climate change. For significant benefits of inland fisheries to be accomplished, planned adaptation at scales from the local to the regional is very necessary.

DOI: 10.4018/978-1-5225-7387-6.ch011

INTRODUCTION

Climate change is the fluctuations in the patterns of climate over long periods. According to the United Nations Framework Convention on Climate Change, it is a change of climate which is attributed directly or indirectly to human activities that alters the composition of the global atmosphere (UNFCCC, 2008). The change may be limited to a specific region, or may occur across the whole earth. According to Akpomi and Vipene (2016) harmful effects of climate change include interuption in seasonal cycle. Ecosystem, agriculture, food supply and water need are also adversely affected. Akuru *et al.*, (2013) posited that the two factors that cause Climate change include natural processes (biogeographical) and human activities (anthropogenic). The natural processes are the astronomical factors which include the changes in the earth's orbit. The anthropogenic factors in climate involves human activities that either emit large amount of greenhouse gases (GHGs) into the atmosphere, which deplete the ozone layer, or reduce the amount of carbon absorbed from the atmosphere. Human activities that emit large amount of greenhouse gases (water vapour, carbon dioxide (CO_2), methane (CH_4), nitrous oxide (N_2O), chlorofluorocarbons (CFCs)) into the environment include industrial burning of fossil fuel, agriculture, urbanization and gas flaring whereas deforestation, alteration in land use, water pollution and agricultural practices reduce the amount of carbon sinks.

The relevance of inland fisheries cannot be overemphasized considering its importance to man such as supply of dietary protein and source of revenue to many. The semi-arid regions harbour some of the most important inland fisheries in the world (Kolding et al., 2016a) and according to IPCC (2014), future climate change will impact seriously on the production systems and livelihoods as well as the fisheries in arid and semi-arid areas of Africa. There is now a strong consensus that climate change presents a fundamental challenge to the well-being of all countries, with potential of being the harshest on countries already suffering from water scarcity. According to Ibid. water scarcity is a well-established context for development in arid and semi-arid countries.

Despite the growing understanding of the global climate change, great uncertainties exist in the prediction of responses of arid and semi-arid regions to global and regional, natural and human-induced climate change. An IPCC report (IPCC, 2007) forecasts that rainfall patterns, river flows and sea levels all over the world will be affected by climate change over the next century whereas an expected precipitation decrease over the next century of 20% or more is envisaged in many parts of the arid and semi-arid regions. Even if efforts to reduce greenhouse gas emissions are successful, it is no longer possible to avoid some degree of global warming and climate change. For any meaningful development to be achieved in the arid and semi-arid zones, the challenges posed by climate change to inland fisheries need to be identified, examined and adaptation strategies proffered. This chapter presents the possible causes of climate change, impact of climate change on inland fisheries, aquatic ecosystem and fishing communities of the arid and semi-arid regions with a focus on case-studies in Africa. Potential adaptation measures are also highlighted and mitigation option suggested.

BACKGROUND

Climate Change or Global Warming

The average weather or change in the distribution of weather events around an average which is attributed directly or indirectly to human activities that alter the composition of global atmosphere and hydrosphere (UNFCCC, 2008) is generally known as climate change. Global climate has changed. The most powerful evidence climate scientists have of global warming is that Earth's temperature is rising brought about by rising levels of GHGs in the atmosphere. Besides air around us, the world oceans absorb about 91 percent of the excess heat trapped by greenhouse gas emissions. Their temperature is rising, too. According to recent study by Cheng *et al.*, (2017), the amount of heat stored in the upper 2000 metres of the ocean during 2015 – 2016 reached its highest point in record. This shows that, since 1960 there has been an increase of 30.4×10^{22} joules of energy blending the earth's systems. According to IPCC (2007) the past 50 years has witnessed a rise in the mean Ocean temperatures of about 0.12 degrees Celsius per decade. It is not just the amount of warming that is significant but also the pace. Increase in the ocean heat causes seawater to expand, raising sea level. According to Lioubimtseva and Cole (2006), predictions based on the paleoanalogues, recent climate trends, and General Circulation Models (GCM) scenarios still rather strongly disagree about the direction, frequency, rapidity, and amplitude of the possible regional climate changes. However, it is likely that any changes in the regional circulation and climate regime will affect various landscape components including hydrological regimes.

Impacts

Generally, refer to influence on inland fisheries, livelihoods and ecosystems brought about by the interface between climate changes or dangerous climate episodes happening within a definite time period and the susceptibility of an exposed society or system (IPCC, 2014). Impacts of climate change on inland fisheries in the arid and semi-arid tropics of Africa can be described as those related to projected temperature increases. Changes in precipitation and the sensitivity of different inland waters to the projected changes as well as the livelihood of the fisherfolks are also implicated.

Exposure and Vulnerability

The extent to which climate change may damage or harm a system; it depends not only on a system's sensitivity, but also its ability to adapt to new climatic conditions" (IPCC, 2001). The character and severity of impacts from climate change emerge from risk that depends on exposure (people and assets at risk) to climate-related hazards. Vulnerability (susceptibility to harm) of human and natural systems are also associated. Paleogeographic researches in many arid regions of the world indicates that deserts and semi-deserts are prone to rapid changes. These include geomorphological, hydrological, biological, and biogeochemical changes in response to global climatic change. These regions have often been predicted to be among the most responsive ecosystems to higher carbon dioxide concentrations and global warming (Lioubimtseva and Cole, 2006). The ecosystems in such regions are therefore fragile (Huang *et al.* 2010). According to Youn *et al.* (2014), fisheries and aquaculture are highly vulnerable to changes in climate and the impacts which could be negative or positive, vary from the coastal areas to the drier parts of the world. This problem of climate change can be understood and ultimately managed from a truly international Perspective.

Sensitivity

The degree to which a system will respond to a change in climatic conditions. This could be measured, for example by a proportional change in ecosystem productivity (or household income and/or expenditure) as a result of perturbations in temperature or precipitation (IPCC, 2014). Strong sensitivity of arid and semi-arid climates to changes in the surface parameters that can be triggered by human activities have been shown by modelling studies (e.g. Claussen *et al.*, 2003). However, there is considerable variability and uncertainty in these estimates between different scenarios.

Adaptation

Human intervention to address the effects of climate change, rather than the response of the ecosystem itself to changing climate (Walker *et al.*, 2004).

Mitigation

Reducing the rate of climate change (Folke *et al.*, 2010).

Inland Fisheries

Fishing activities carried out in inland waters such as streams, rivers, lakes, reservoirs etc. (Cooke, 2016).

Arid

Dry hot climatic zone*:* either of two zones of latitude that are between 15° and 30° north and south of the equator, consisting mostly of desert or semi desert.

Semi-Aridor Hot Steppe

This is the middle between desert climates and moist climates in environmental attributes and agricultural capability. Steppe climate is an areas with annual rainfall less than the starting point but more than half of it. A location with a steppe type climate with the appropriate temperature above either a mean annual temperature of 18°C, or a mean temperature of 0°C or −3°C in the coldest month is classified as hot semi-arid. A location with the appropriate temperature below the given isotherm is classified as cold semi-arid (Peel *et al.*, 2007). The Sahara and the Kalahari-Namib deserts of Africa forms the arid zone. The arid and semi-arid zones of Africa is presented in figure 1. In West Africa, it cuts across Benin, Burkina Faso, Ghana, Liberia, Mali, Mauritania, Nigeria, Senegal, Sierra Leone. The semi-arid countries of East and Southern Africa are: Angola, Botswana, Eritrea, Ethiopia, Kenya, Lesotho, Madagascar, Malawi, Mozambique, Namibia, Rwanda, South Africa, South Sudan, Sudan, Swaziland, Tanzania, Uganda, Zambia, Zimbabwe (MEA, 2005).

Figure 1. Arid and semi-arid areas in African countries
Source: (MEA, 2005)

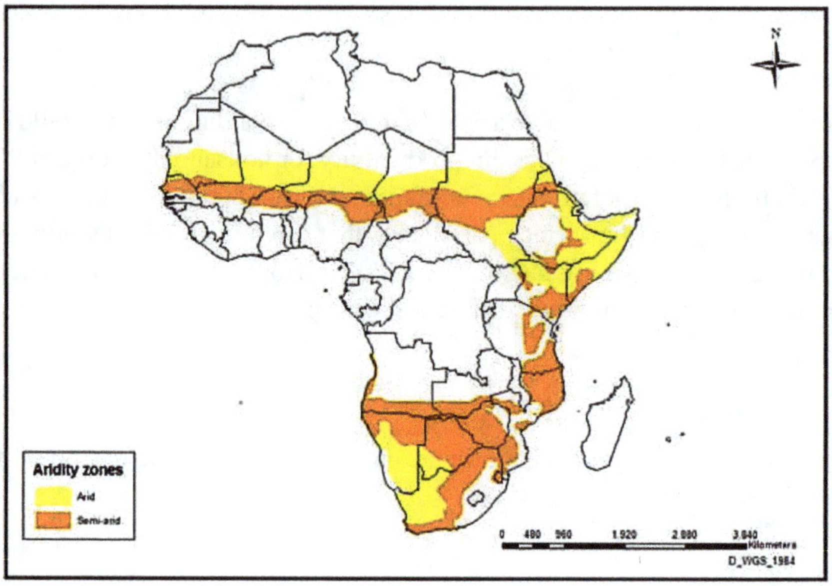

CAUSES OF CLIMATE CHANGE

Causes of climate change can be divided into two categories: those that are due to natural causes and those that are created by man.

Natural Causes

There are a number of natural factors responsible for climate change such as variation in the sun's output etc. Some of the more prominent ones are volcanoes, the earth tilt, and ocean currents (Foster *et al.*, 2007)

Variation in The Sun's Output

Historically, Earth's climate changed during the pre-Industrial Revolution era due to variations in sunlight. The Sun's output varies over time (Lean and Rind, 2008). Tiny wobbles in Earth's orbit altered when and where sunlight falls on Earth's surface. This discrepancy results in corresponding increase and decrease in the amount of solar energy reaching the Earth (Lean *et al.*, 2005). Scientists track these changes using observations of sunspots and (more recently) satellite measurements. Average solar energy has hardly changed in the past 50 years and may even have had a recent cooling effect. The Sun cannot account for the increase in global temperature observed since 1970.

Volcano Eruptions

When volcano erupts, it throws out large volume of sulphur dioxide (SO_2), water vapour, dust and ash into the atmosphere. Although the volcanic activity may last few days, yet the large volumes of gases and ash can influence climate pattern for years. Millions of tons of SO_2 gas can reach the upper levels of the atmosphere (Stratosphere) from a major eruption. This gas combines with dust particles to form aerosols which partially block the incoming rays of the sun, leading to cooling. Sulphur dioxide combines with water to form tiny droplets of sulphuric acid. These droplets are so small that many of them can stay aloft for several years. They are efficient reflector of sunlight, brightening the planet and cooling the climate. Winds in the stratosphere carry the aerosols (suspended particles in gas) rapidly around the globe in either an easterly or westerly direction. Volcanic activity also increases greenhouse gases thereby contributing to global warming (U.S. Geological Survey, 2009).

The Earth's Tilt

The Serbian astrophysicist Milutin Milankovitch developed one of the most important theories relating to Earth motions and long-term climate change. His mathematical theory of climate is based on the seasonal and latitudinal variations of solar radiation received by the Earth. The theory states that "*as the Earth travels through space around the sun, cyclical variations in three elements of Earth-sun geometry combine to produce variations in the amount of solar energy that reaches Earth*":

1. Variations in the Earth's orbital eccentricity—the shape of the orbit around the sun.
2. Changes in obliquity—changes in the angle that Earth's axis makes with the plane of Earth's orbit.
3. Precession—the change in the direction of the Earth's axis of rotation. This means that the axis of rotation behaves like the spin axis of a top that is winding down; hence it traces a circle on the celestial sphere over a period of time.

Together, the periods of these orbital motions have become known as Milankovitch cycles (Figure 2).

Figure 2. Milankovitch cycle
Source: (Hays et al., 1976)

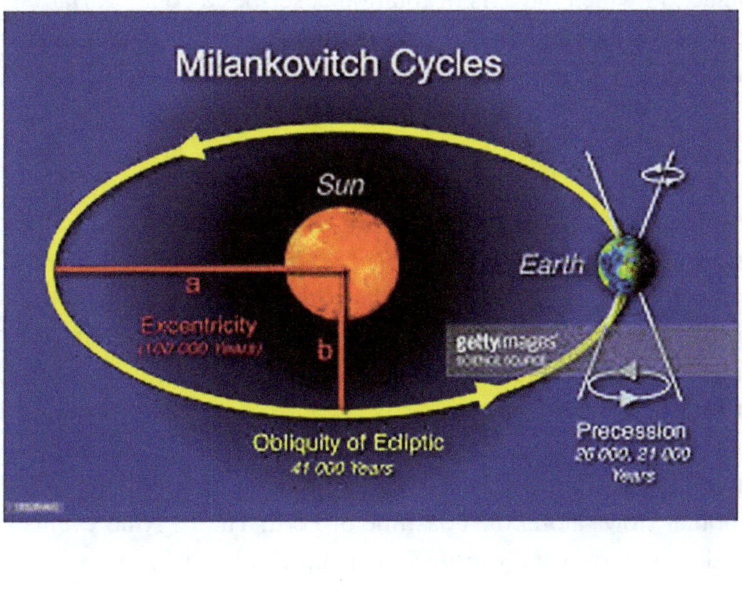

Orbital Variations

According to Hays *et al.*, (1976) Changes in orbital eccentricity affect the Earth-sun distance. At present, a variance of only 3 percent (5 million kilometers) occurs between closest approach (perihelion), which occurs on or about January 3, and furthest departure (aphelion), which occurs on or about July 4. This difference in distance amounts to about a 6 percent increase in incoming solar radiation (insolation) from July to January. The shape of the Earth's orbit changes from being elliptical (high eccentricity) to nearly circular (low eccentricity). This takes place in a cycle that takes between 90,000 and 100,000 years. When the orbit is highly elliptical, the amount of insolation received at perihelion would be on the order of 20 to 30 percent greater than at aphelion. The resultant effect is a substantially different climate from what is experienced today throughout the globe.

Obliquity (Change in Axial Tilt)

The axial tilt increases with the seasonal contrast which results in cooler winters and warmer summers in both hemispheres (NRC, 1982). Today, the Earth's axis is tilted 23.5 degrees from the plane of its orbit around the sun (Jouzel *et al.*, 2007). But this tilt changes. It varies between 22.1 and 24.5 degrees during a mean cycle of about 40,000 years. As a result of this changes in the tilt, the seasons become embellished. This therefore means more tilts more severe seasons—warmer summers and colder winters; less tilt means less severe seasons—cooler summers and milder winters (Hansen *et al.*, 2005). The cool summers ensure year-round accumulation of snow and ice in high latitudes, finally building up into colossal ice sheets. Positive feedbacks are eminent in the climate system as well, because since the Earth is covered with more snow and sheets of ice more of the sun's energy is reflected into space. This reflection results in additional cooling of the Earth (Hansen *et al.*, 2005).

Precession

Precession means conical sweep of axis of rotation. It is the regular motion of a spinning body such as a spinning top or a planet, in which the axis of rotation describes a cone. Variations in axial precession change the dates of perihelion and aphelion. This change brings about increase in the seasonal contrast in one hemisphere and decrease in the seasonal contrast in the other hemisphere (Lutgens *et al.*, 1998).

Ocean Currents

The major surface currents in the world's oceans are caused by prevailing winds. The currents may be cold, as in the instance of the West Wind Drift, or warm, as the Gulf Stream. Currents circulate in paths called gyres, moving in a clockwise direction in the northern hemisphere and a counter clock wise direction in the southern hemisphere. Ocean currents move vast amount of heat across the planet – roughly the same amount as the atmosphere does (Sabine, 2004). But the oceans are surrounded by land masses, so heat transport through the water is through channels (Tol *et al.*, 2009). Wind pushes horizontally against the sea and drive ocean current patterns. Certain parts of the world are influenced by ocean current than others. The coast of Peru and other adjoining regions are directly influenced by the Humboldt Current that flows along the coastline of Peru. The El Ñino events in the Pacific Ocean can affect climate condition all over the world (Soden and Held, 2006).

Ocean currents have been known to change direction or slow down. There has been much of the formation of clouds, which shade the surface and have a net cooling effect. Any of all these phenomena can have impact on climate, as is believed to have happen at the end of the last ice Age, about 14,000 years ago (Manvendra *et al.*, 2010).

These natural causes are still in play today, but their influence is too small or they occur too slowly to explain the rapid warming seen in recent decades. We know this because scientists closely monitor the natural and human activities that influence climate with a fleet of satellites and surface instruments.

Human Causes

Human activities contribute to climate change by causing changes in Earth's atmosphere in the amounts of greenhouse gases, aerosols (small particles), and cloudiness. These activities include livestock and agriculture, land use and deforestation and burning of fossil fuels.

Livestock and Agriculture

According to Stainforth*et al.*, (2005) methane is the second most significant greenhouse gases (GHGs) and cause of climate change, and unit for unit is 21 times more damaging than carbon (iv) oxide (CO_2). A major source of methane comes from livestock and specifically cattle, which is produced by digesting grass and exhale it through their breath. Methane is also a by-product of rice-growing, a considerable impact given that rice is a major staple food for a large part of the world's population. Other GHG emissions resulting from agriculture include nitrous oxide (N_2O) which is released as a by-product of the application of fertilizers. The total global N_2O emission is estimated at 10 to 17.5 million metric tons per year (IPCC, 2007).

Land Use Change and Deforestation

Land use and land use changes can significantly contribute to overall climate change. Vegetation and soils typically act as a carbon sink, storing carbon dioxide that is absorbed through photosynthesis. When the land is disturbed, the stored carbon dioxide along with methane and nitrous oxide are emitted, re-entering the atmosphere. There are two types of land use change: direct anthropogenic (human-caused) changes and indirect changes. Examples of anthropogenic changes include deforestation, reforestation and afforestation, agriculture, and urbanization. Indirect changes include those changes in climate or in carbon dioxide concentrations that force changes in vegetation.

The major changes to the planet's land cover since 1950 have resulted from deforestation in temperate regions where forests and woodlands have been cleared to make room for fields and pastures (Bonan, 2008). More recently, deforestation has been rapidly taking place in tropical regions where economic development has come at the expense of conservation, and much of the world's natural rainforests have given way to palm oil plantations, cattle pastures or the logging industry (MEA, 2005). According to the same findings, deforestation accounts for 20 per cent to 25 per cent of global GHG emission, making it the most significant source of emission in the developing countries.

Burning of Fossil Fuels

The largest known contribution comes from the burning of fossil fuels. This act releases carbon dioxide gas to the atmosphere. Thermal power plants, cars, buses etc. all emit GHGs and aerosols into the atmosphere. These gases affect climate by altering incoming solar radiation and out-going infrared (thermal) radiation that are part of Earth's energy balance (Hansen *et al.*, 2005). Change in the atmospheric abundance or properties of these gases and particles leads to a warming or cooling of the climate system. Since the start of the industrial era (about 1750), the overall effect of human activities on climate has been a warming influence. The human impact on climate during this era greatly exceeds that due to known changes in natural processes, such as solar changes and volcanic eruptions (MEA, 2005).

IMPACTS OF CLIMATE CHANGE ON THE AFRICAN ARID AND SEMI-ARID INLAND FISHERIES

Fisheries is a socio-ecological system and so any significant impact is seen on inland water ecosystem, primary productivity and fish production, and livelihood of the fisherfolks.

Inland Water Ecosystems

The link between inland fisheries and their ecosystems are deeper. It is more significant than those that exist in mainstream agriculture (FAO, 2007). Estuaries, mangroves, corals and sea grass beds are particularly momentous in the provision of ecosystem services, especially as nursery for young fish and are also among those sensitive to climate change.Impacts on ecosystem compriserise in water temperature and acidity, fall in dissolved oxygen (DO), changes in salinity (Roessig*et al.*, 2004).

Rise in Temperature and Acidity

Observed temperatures of the arid and semi-arid regions of Africa have been increasing faster than global warming. The increase varied between 0.2 and 0.8° C since the end of the 1970s and currently up to increase of 1.1°C in some areas. This is obviously higher than the global mean temperature increase of 0.74°Crecorded since 1860 when actual scientific temperature measurement started (IPCC, 2007).

Fish are poikilothermic, their internal temperature varies directly with that of the environment. This makes them more sensitive to temperature of their surrounding environment (Portnar and Knust, 2007). When changes do occur, they move to areas where the external temperature allow them to regain their preferred internal temperature. This behavioural thermoregulation (Roesig *et al.*, 2004) is resulting in rapid migration of fish poleward or into cooler bodies of water (FAO, 2007). Some species also shift from shallow coastal waters and semi-enclosed areas, where temperature will increase fastest, into deeper cooler water. This correspond to the poleward shift of temperature zone. Relatively small temperature changes alter fish metabolism and physiology with consequences for growth, fecundity, feeding behaviour, distribution, migration and abundance. According to ibid, the general effects of increase temperature on parasites include, rapid growth and maturation, increase parasite maturity etc. (Table 1).

Table 1. General effects of increased temperature on parasite life cycles, their hosts and transmission process

Effects of Parasites	Effects on Hosts	Effects on Transmission
Faster rate of development and maturation	Altered behaviour	More generation per year
Faster embryonic development and hatching	Altered feeding	Earlier reproduction in spring
Decreased longevity of larvae and adults	Altered range	Prolonged transmission in the fall
Increased mortality of all stages	• Altered ecology • Reduced host resistance	Potential transmission year round

Source: Marcogliese, (2016)

As the emergence of disease is linked directly to changes in the ecology of the hosts or pathogens, or both, climate change has a profound impact on the spread of parasites and diseases in the inland water ecosystem. climate change affect parasite species directly resulting in the extension of the geographical range of pathogens. In addition, increase temperature causes thermal stress in aquatic animals, leading to reduced growth, sub-optimal behaviour and reduced immune-competence resulting in changes in the distribution and abundance of their hosts (Marcogliese, 2016).

Fall in Dissolved Oxygen (DO)

DO change is inversely proportional to water temperature change and inflow change. Variations in DO can occur seasonally, or even over 24-hour periods, in relation to temperature and biological activity (i.e. photosynthesis and respiration). The biological activity is affected by water temperature change and inflow change. As water temperature rises, the metabolic rates of the most water organisms increase. When inflow rate increases, suspended solids increase. This affects the euphotic zone depth and then algal biomass. Moreover, inflow change affects the phytoplankton stability, biological respiration, including that one related to decomposition processes, reduces DO concentrations. In severe cases of reduced oxygen concentrations (whether natural or man-made), anaerobic conditions can occur (i.e. 0 mg/L of oxygen). In particular, this occurs close to the sediment-water interface as a result of decaying material (Vigil, 2003). According to Ibid. concentrations below 5 mg/L may adversely affect the functioning and survival of biological communities and below 2 mg/L may lead to the death of most fish.

Changes in Salinity

Rise in the water level of the oceans lead to intrusion of saltier water into the river delta areas and inundation of the coastal low-lying areas. This has a profound effect on the distribution of both the freshwater and marine fishes as a result of changes in the physical and chemical properties of the waters.

Fluctuation in the Level of Inland Water Bodies

Owing to the inverse relationship between average annual precipitation and variability, arid and semi-arid water bodies are generally characterized by strong seasonal and interannual fluctuations. The amplitude (variance) increases along the climatic aridity gradient. Rise in seawater level has grave implications for countries whose lands are at elevations similar to sea level. The current global estimate of sea level rise under different scenarios is approximately 0.2 m and it is projected to increase to 0.8 m by the year 2100 (Figure 3).The implication is that this present rise has inundated large portion of the arid and semi-arid regions affecting millions of people.

Primary Productivity and Fish Production

Primary productivity is affected by availability of nutrients in the water, which in turn depends on freshwater runoff, level of light and temperature (FAO, 2007). Reduced precipitation could lead to reduced runoff from lands. According to IPCC (2007), this starves wetlands and mangroves of nutrients and damage local fisheries. In some areas increased precipitation or increased extreme weather events, including flooding, will lead to excessive nutrients levels in rivers, lake and coastal waters as sewage and fertilizer is washed into water bodies causing harmful algal blooms, also known as red tides. With climate change, primary productivity is predicted to decline at lower latitudes where the majority of the world's small-scale fisheries are located. The productivity of the fisheries is thereby reduced. There is increasing evidence that fish production in inland fisheries of the arid and semi-arid regions of Africa is more dependent on the external climatic drivers (amongst which are increased long-term and seasonal variability of surface water bodies, Table 2), than on human exploitation rates and various management interventions (Gownaris *et al*., 2016).

Figure 3. Global sea level rise under different scenarios of climate modelling
Source: (El-Beltagy and Madkour, 2012)

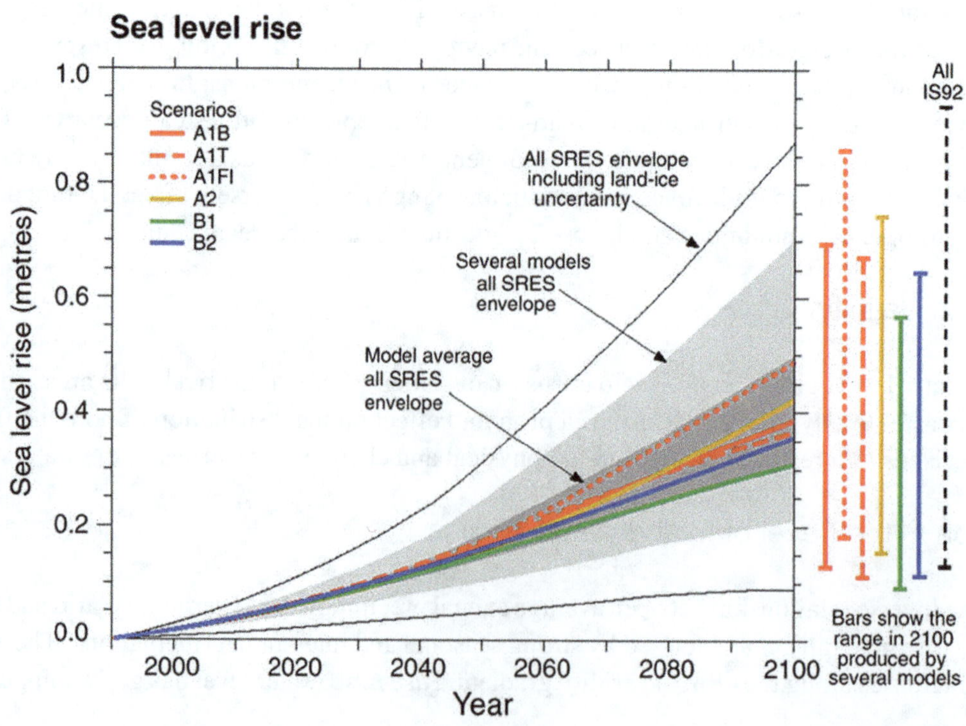

Table 2. Climate impact pathways on inland fisheries

Climatic Variable	Impacts	Potential Outcomes for Fisheries
Surface warming of rivers and lakes	• Warm water species replacing cold water species. • Shifts in distribution of plankton, insect larvae and other key food resources for fish.	• Shifts in distribution of important food fish. • Reduction in biodiversity through loss of sensitive species in inland waters
Changes in precipitation	• Altered river flows, lake levels, flood and drought frequencies. • Rivers affected by snow melt have earlier and faster spring flows	• Changes in fish diversity and production. • Changes in timing and extent of fish production
Changes in cloud cover and wind Speed.	• Changes in thermal stratification, affecting nutrient recycling. • Changes in light penetration and photosynthetic efficiency	• Reduction in ecosystem productivity and fish production where stratification strengthens (reduced wind speed) • Changes in ecosystem structure and function potentially impacting fish production.
Sea level rise	• Backing up of rivers Flooding and salinization of river deltas, coastal lagoons and wetlands. • Loss of coastal fish breeding and nursery habitats	• Increased flooding of riparian settlements. • Reduced production of coastal and brackish-water fisheries and aquaculture. • Declines in biodiversity and provision of ecosystem services
	• Changes in sex ratios • Altered time of spawning • Altered time of migrations • Altered time of peak abundance	Possible impacts on timing and levels of productivity across fresh-water systems.
	Increased invasive species, diseases and algal blooms	Reduced production of target species in fresh water systems
	Changes in fish recruitment success	Abundance of juvenile fish affected and therefore production
Reduced water flows and increased droughts	• Changes in lake water levels • Changes in dry season water flows in rivers. • Reduced water quality (less dilution of pollutants, eutrophication)	• Reduced lake production and productivity. • Reduced river productivity. • Fish kills • Loss of sensitive species. • Reduced production
Reduced surface water availability	• Increased water abstraction for agriculture and other uses • Increased use of built structures (dams, weirs, flood barriers etc.)	• Reduced water availability for aquaculture and fisheries leading to loss of production. • Disruption to life-cycles, reduction in biomass and possible extinction of high-value migratory fish species. • Displacement of fishing communities
Changing levels of precipitation	Where rainfall decreases, reduced opportunities for farming, fishing and aquaculture as part of rural livelihood systems	Reduced diversity of rural livelihoods; greater risks in agriculture; greater reliance on non-farm income.
More droughts or floods	Damage to productive assets (fish ponds, weirs, rice fields etc.) and homes	
Less predictable rain/dry seasons	Decreased ability to plan livelihood activities – e.g. farming and fishing seasonality	

Source: (Modified from Allison et al, 2005and FAO, 2007)

The recorded yields of the arid and semi-arid African inland fisheries have increased almost linearly by around half a million metric tonnes per decade over the past 60 years. However, there are clear cyclical variations in the residuals of about 20 years' periodicity above and below the trend line (Figure 4). Smaller lakes follow the same pattern of climate change. Like most terrestrial farmers, the biggest environmental concern for African fishers is usually uncertainty about annual rainfall. "Fish come with the rain" is the most commonly expressed statement of the African inland fishers when discussing factors controlling fish production (Kolding *et al.*, 2016b).

Examples are the many isolated endorheic (closed) water bodies in African arid and semi-arid regions, such as Lake Ngami (Botswana), Lake Chilwa (Malawi), Lake Mweru Wa' Ntipa (Zambia), Lake Liambezi (Namibia), or Lake Nakuru (Kenya) which periodically dry out completely, becoming muddy swamps or even dusty depressions.

Even large African lakes with considerable buffering capacities are likely to be affected by climate change. For example, the fishery catch of Lake Naivasha in Kenya is strongly related to the water level (Hickley *et al.* 2002), and water level is likely to be affected by anticipated reductions in precipitation. Similar impacts of climate change are also anticipated in Lake Victoria (East Africa). There changes in interannual and interseasonal variability in rainfall and temperature could affect the survival of aquatic life. This can lead to an increase in the variability of fish catches (Sewagudde 2009).

Climate change impacts add to the already difficult water management challenges in these regions (Li *et al.,* 2018) and are likely to climax into fish species extinction before the end of this century (Tedesco *et al.,* 2013). At the same time, river, dam and stream flows mark increase of between 6 and 18 per year with less than 2 floods have been reported (IPCC, 2007). The impacts of increased flooding of the freshwater bodies are negative through erosion of watershed, destruction of fish feeding and breeding habitats. It also results in decrease in primary productivity and alteration of the normal resilience of the aquatic systems. The impacts are positive in expansion of aquatic habitats for primary and fish productions especially during the dry season.

Figure 4. Total annual recorded yield (tonnes) from inland fisheries in 27 African countries comprising arid and semi-arid region and the catch anomalies (residual yield, tonnes) showing an approximate 20-year periodicity around the trend
Source: (FAO FishStat, 2016)

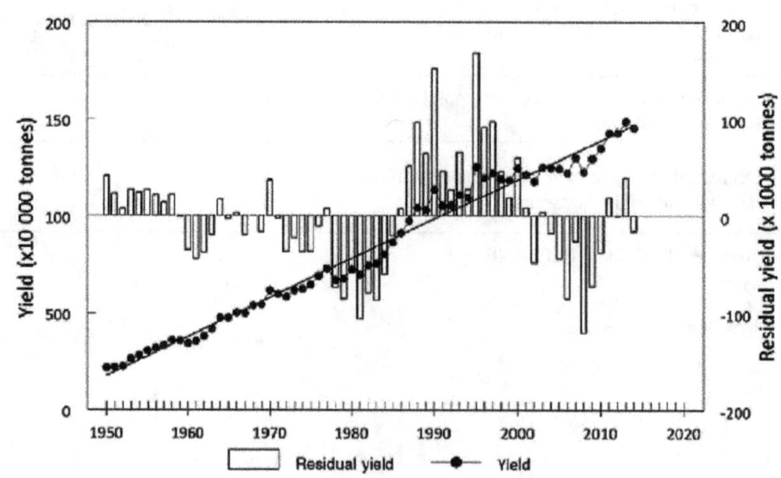

Livelihood of the Fisherfolks

A livelihood can be defined as the capabilities, assets and activities required for means of living (Chambers and Conway,1992). Climate change is one of the major threats to sustained economic growth leading to extended poverty in semi-arid regions of sub Saharan Africa (SSA). With mounting evidence of the impacts of climate change on aquatic ecosystems, the resulting impacts on fisheries livelihoods are likely to be significant. However, this remain a neglected area in climate adaptation policy. According to Badjeck *et al.*, (2009) impacts of climate variability and change on livelihoods assets include: change in natural capital (fish stock) resulting in reduced revenue, damage to physical capital (landing sites, boats, gear, hospitals, schools, sewage system etc.) and reduced financial capital (bank loans and grants), impact on human capital (safety-at-sea and food security) and social capital (unstable settlement, networks etc.).

VULNERABILITY AND ADAPTATION

The Sahara and the Kalahari-Namib deserts of Africa have remained perpetually dry throughout most of geological history. Angola, DR Congo, Mauritania, Mali, Sierra Leone, Mozambique, Niger and Zambia among the semi-arid countries have been identified as highly vulnerable while Malawi, Senegal, and Uganda are mostly vulnerable. Many vulnerable countries are also among the world's least developed. According to Allison *et al.* (2009) the inhabitants of these countries are among the world's poorest. They are twice as reliant on fish which provides 27% of dietary protein compared to 13% in less vulnerable countries. These countries also produce 20% of the world's fish exports.

Although fisher folks and their communities are already constantly adapting to various forms of change (Coulthards, 2009) these existing pressures are expected to intensify in the coming decades. Adaptation at scale in semi-arid regions (ASSAR) aims to transform climate adaptation policy and practice in ways that promote the long-term well-being of the most vulnerable and those with the least agency. Semi-arid regions of Africa where ASSAR is working include Botswana, Namibia, Ghana, Kenya and Uganda (Degefu *et al.*, 2018). In inland fisheries, similar responses can be seen as intensifying fisheries, diversifying species targeted or existing fishing for other livelihoods. According to Adger *et al.* (2005) intensification can be achieved by investing more resources into the fisheries. This can be in terms of increasing fishing efforts (by spending more time at sea), increasing fishing capacity (by number, size, or efficiency of gears or technology) or fishing farther or deeper than previously. However, such adaptation responses obviously have potential negative long-term consequences if overexploited.

Hard (e.g. sea walls) and soft (e.g. wetland rehabilitation or manage retreats) defences are also constructed to adapt to sea level rise, increased storms and surge damage. Improved information systems to integrate knowledge from different coastal sectors, predict and plan for appropriate strategies are also put in place.

As climate change increase environmental variation, more fisheries managers will have to explicitly consider such variations. This will enable them move beyond maximum sustainable yields or similar fixed ideas of the potential productivity of a stock. These are mere static management parameters for a particular stock. Such changes create an additional imperative to implement the ecosystem approach to fisheries (EAF), a holistic, integrated, and participatory approach to obtain sustainable fisheries (FAO, 2007).

Diversified products and markets make fisheries less prone to economic shocks, while information technologies are becoming more available to small-scale fishers which help them navigate international markets and achieve fair prices for their fish.

A technical approach to adaptation cannot underestimate the importance of institution (especially informal) to facilitate or limit adaptation (Badjeck *et al.*, 2009). For instance, many people adapt to declines in fish abundance, by putting pressure on other resources, for example wild plants and other animals. Accordingly, the lack of alternative economic options in these regions exacerbates fishery resource exploitation and declines. This correlation between lack of alternatives and environmental degradation suggests that guiding people to plan and pursue occupations that are not dependent on fish could help people adapt more successfully. This, at the same time, would relieve pressure on fish and could help increase the sustainability and resilience of the resource itself (Coulthard, 2009).

Land ownership and quantity of land owned are highly correlated with success of adaptation, as they help people to be more self-sufficient in obtaining food and can further provide income if rented out.

Access to credit also appears to be important for adaptive capacity. Successful adapters take out fewer loans, but when they borrow money they tend to borrow in larger amounts. Successful adapters take out loans to invest in their livelihood strategies, as opposed to borrowing to meet every day needs. Unsuccessful adapters, who are more likely to borrow for their daily needs or to overcome shocks, have difficulties accessing the larger amounts of credit required to make investment in assets that could help build their adaptive capacity.

Fisherfolks own ideas of what would help them to cope better with environmental change generally focused on starting new supplementary occupations, with business/selling and raising fish or livestock as the most popular ones. These occupations all have a relatively low exposure to environmental change and thus are likely to enhance people's adaptive capacity. Other responses include expansion of current occupations (farming, fishing and business), learning new skills and saving. Finally, improving physical capital (land ownership, boat, fishing gear, motorbike) and access to medicine for livestock can improve the adaptive capacity of the respondents (Schwartz *et al.*, 2016).

The Role of Governance in Adaptation and Mitigation

There is a critical need for informed public policy to address mitigation of greenhouse gas emissions to limit and minimize impact of climate change (FAO, 2007). The safeguarded benefits in the fisheries sector are important factors to be considered. Sound public policy will be required for climate change adaptation in order to reduce ecosystem vulnerability. This will also provide information for planning and simulating adaptation, and ensuring that adaptation actions do not have negative effects on other ecosystem services and the longer-term viability of fisheries (Nicholls *et al.*, 2007).

In addition to good governance principles currently applicable to the sector, aids and management would support climate change mitigation and adaptation in the sector by:

- Building institutional and legal framework that consider and respond to climate change threats and uncertainties along with other pressure such as overfishing, pollution and changing hydrological conditions. This requires effective public, private and NGO partnerships. Integrating research and management across the sectors is necessary. Ensuring that regulation limiting access to resources are appropriate to respond to both threats and benefits of future climate variability is also important (Allan, 2005).

- Moving rapidly towards full implementation of the code of conduct for responsible fisheries (CCRF) which encompasses the ecosystem approach to fisheries.
- Linking disaster risk management with development planning, especially concerning planning coastal or flooding defence and applying soft engineering solutions where possible through conservation of natural storm barriers, floodplains, and erodible shorelines in order to manage costs and damage impacts.
- Increasing cooperation between countries and across the region on climate change challenges and resource requirements.
- Africa lacks both regional and national climate data. There in an urgent need for improved climate modelling and forecasting on the continent, which can provide a basis for informed decision-making and the implementation of adaptation strategies.
- Production of greenhouse gases through deforestation and fuel burning can be reduce by making use of geothermal and solar energy instead of cutting trees and burning wood for fuel (Adejuwon, 2006).
- Climate change education should be taught in formal and informal sector of education.

CONCLUSION

Climate change is predicted to have a wide range of impacts on inland fisheries of arid and semi-arid zones of Africa and those who depend on them. Hot spots include regions, such as Lake Ngami (Botswana), Lake Chilwa (Malawi), Lake Mweru Wa' Ntipa (Zambia), Lake Liambezi (Namibia) and Lake Nakuru (Kenya) which occasionally dry out completely, becoming mud-spattered swamps or even sandy dimples. Future warming across Africa ranging from 0.2°C per decade (low scenario) to more than 0.5°C per decade (high scenario) is anticipated. This warming is greatest over the interior of semi-arid margins of the Sahara and central southern Africa. Biophysical impacts of climate change on aquatic ecosystems, primary productivity and production and the livelihood of the fisherfolks are enormous. Vulnerability of inland fisheries of the arid and semi-arid regions is high. Individuals should avoid burning of fossil fuel and other activities the lead to increase in greenhouse gases. The priority or responsibility of the governments, civil society and international organizations with regards to climate change, is to aggressively pursue reductions in greenhouse gas (GHG) emission. This is because the long-term consequences of climate change are highly complex, unknowable and potentially irreversible. ASSAR should be strengthened and be extended to all the arid and semi-arid regions of Africa.

REFERENCES

Adejuwon, J. O. (2006). *Food security, climate variability and climate change in Sub- Saharan West Africa*. Assessments of impacts and adaptations to climate change. Working paper series. Project No. AF 23. International START Secretariat.

Adger, W. N., Nigel, W., Arnell, N. W., & Tompkins, E. L. (2005). Successful adaptation to climate change across scales. *Global Environmental Change*, *15*(2), 77–86. doi:10.1016/j.gloenvcha.2004.12.005

Akpomi, M. E., & Vipene, J. (2016). Promoting Knowledge of Climate Change (CC) amongst Nigerians: Implications for Education Managers. *Journal of Education and Practice*, *7*(32), 132–138.

Akuru, U. B., Okoro, O. I., & Chikuni, E. (2013, September). Impact of renewable energy deployment on climate change in Nigeria. In AFRICON, 2013 (pp. 1-8). IEEE. doi:10.1109/AFRCON.2013.6757649

Allan, J. D., Abell, R., Hogan, Z. E. B., Revenga, C., Taylor, B. W., Welcomme, R. L., & Winemiller, K. (2005). Overfishing of inland waters. *Bioscience*, *55*(12), 1041–1051. doi:10.1641/0006-3568(2005)055[1041:OOIW]2.0.CO;2

Allison, E. H., Adger, W. N., & Badjeck, M. C. (2005). *Effects of climate change on the sustainability of capture and enhancement fisheries important to the poor: Analysis of the vulnerability and adaptability of fisherfolk living in poverty*. Fisheries Management Science Programme, DfID, UK. No. R4778J. Retrieved from http://www.fmsp.org.uk

Allison, E. H., Perry, A. L., Badjeck, M., Adger, W. N., Brown, K., Conway, D., ... Dulvy, N. K. (2009). Vulnerability of national economies to the impacts of climate change on fisheries. *Fish and Fisheries*, *10*(2), 173–196. doi:10.1111/j.1467-2979.2008.00310.x

Badjeck, M. C., Allison, E. H., Halls, A. H., & Dulvy, N. K. (2009). Impacts of climate variability and change on fishery-based livelihoods. *Marine Policy*. doi:10.1016/j.marpol.2009.08.007

Bonan, G. B. (2008). Forests and climate change: Forcings, feedbacks, and the climate benefits of forests. *Science*, *320*(5882), 1444–1449. doi:10.1126cience.1155121 PMID:18556546

Chambers, R., & Conway, G. R. (1992). *Sustainable rural livelihoods: practical conceptsfor the 21st century*. Discussion paper 296, Institute of Development Studies, Brighton, UK.

Cheng, L., Trenberth, K. E., Fasullo, Boyer, T., Abraham, J., & Zhu, J. (2017). *Improved estimates of ocean heat content from 1960 to 2015*. Retrieved from doi:10.1126ciadv.1601545

Claussen, M., Brovkin, V., Ganopolski, A., Kubatzki, C., & Petoukhov, V. (2003). Climate change in Northern Africa: The past is not the future. *Climatic Change*, *57*(1/2), 99–118. doi:10.1023/A:1022115604225

Cooke, S., Allison, E., Beard, T. D., Arlinghaus, R., Arthington, A., Bartley, D., ... Welcomme, R. (2016). On the sustainability of inland fisheries: Finding a future for the forgotten. *Ambio*, *45*(7), 753–764. doi:10.100713280-016-0787-4 PMID:27312662

Coulthard, S. (2009). Adaptation and conflict within fisheries: insights for living with climate change. In W. N. Adger, I. Lorenzoni, & K. O'Brien (Eds.), *Adapting to climate change: thresholds, values, governance* (pp. 255–268). Cambridge, UK: Cambridge University Press. doi:10.1017/CBO9780511596667.017

Degefu, M. A., Assen, M., & McGahey, D. (2018). *Climate variability and impact in ASSAR's East African region. CARIAA-ASSAR Working Paper*. International Development Research Centre, Ottawa, Canada and UK Aid, London, United Kingdom. Retrieved from www.assar.uct.ac.za

El-Beltagy, A. & Madkour, M. (2012). *Impact of climate change on arid lands agriculture*. doi:10.1186/2048-7010-1-3

FAO. (2007). *The State of world fisheries and aquaculture – 2006*. Rome, Italy: Food and Agricultural Organization.

FAOfishstat. (2016). *The State of world fisheries and aquaculture - 2016. Contributing to food security and nutrition for all*. Rome, Italy: Food and Agricultural Organization.

Folke, C., Carpenter, S. R., Walker, B. H., Scheffer, M., Chapin, T., & Rockström, J. (2010). *Resilience thinking: Integrating resilience, adaptability and transformability*. Retrieved from http://www.ecologyandsociety.org/vol15/iss4/art20/

Forster, P., Ramaswamy, V., Artaxo, P., Berntsen, T., Betts, R., Fahey, D. W., . . . Van Dorland, R. (2007). Changes in atmospheric constituents and in radiative forcing. In S. Solomon, D. Qin, M. Manning, Z. Chen, M. Marquis, K. B. Averyt, M. Tignor, & H.L. Miller (Eds.), Climate change 2007: The physical science basis. Cambridge, UK: Cambridge University Press.

Gownaris, N. J., Pikitch, E., Aller, J. Y., Kaufman, L., Kolding, J., Lwiza, K. M. M., … Rountos, K. J. (2016). *Reconciling fisheries and water level fluctuations in the world's largest desert lake*. Retrieved from doi:10.1002/eco.1769

Hansen, J., Nazarenko, L., Ruedy, R., Sato, M., Willis, J., Del Genio, A., ... Tausnev, N. (2005). Earth's energy imbalance: Confirmation and implications. *Science*, *308*(5727), 1431–1435. doi:10.1126cience.1110252 PMID:15860591

Hays, J. D., Imbrie, J., & Shackleton, N. J. (1976). Variations in the earth's orbit: Pacemaker of the ice ages. *Science*, *194*(4270), 1121–1132. doi:10.1126cience.194.4270.1121 PMID:17790893

Hickley, P., Bailey, R. G., Harper, D. M., Kundu, R., Muchiri, S. M., North, R., & Taylor, A. (2002). The status and future of the Lake Naivasha fishery, Kenya. *Hydrobiologia*, *488*(1/3), 181–190. doi:10.1023/A:1023334715893

Huang, J., Minnis, P., Yan, H., Yi, Y., Chen, B., Zhang, L., & Ayers, J. K. (2010). Dust aerosol effect on semi-arid climate over Northwest China detected from A-Train satellite measurements. *Atmospheric Chemistry and Physics*, *10*(14), 6863–6872. doi:10.5194/acp-10-6863-2010

IPCC. (2001). *Synthesis report 2001 (Contribution of Working Group I, II, and III to the Third Assessment Report of the Intergovernmental Panel on Climate Change. 2001)*. Cambridge, UK: Cambridge University Press.

IPCC. (2007). *Climate change 2007 (Fourth Assessment Report, Intergovernmental Panel on Climate Change)*. Cambridge, UK: Cambridge University Press.

IPCC. (2014). *Climate change 2014: impacts, adaptation, and vulnerability. (Contribution of Working Group II to the Fifth Assessment Report of the Intergovernmental Panel on Climate Change)*. Cambridge, UK: Cambridge University Press.

Jouzel, J., Masson-Delmotte, V., Cattani, O., Dreyfus, G., Falourd, S., Hoffmann, G., ... Wolff, E. W. (2007). Orbital and millennial antarctic climate variability over the past 800,000 years. *Science*, *317*(5839), 793–796. doi:10.1126cience.1141038 PMID:17615306

Kolding, J., Bundy, A., van Zwieten, P. A. M., & Plank, M. J. (2016b). Fisheries, the inverted food pyramid. *ICES Journal of Marine Science*, *73*(6), 1697–1713. doi:10.1093/icesjms/fsv225

Kolding, J., van Zwieten, P. A. M., Marttin, F., & Poulain, F. (2016a). Fisheries in the drylands of Sub-Saharan Africa "Fish come with the Rains". *FAO Fisheries and Aquaculture Circular*, C1118.

Lean, J., Rottman, G., Harder, J., & Kopp, G. (2005). SORCE contributions to new understanding of global change and solar variability. *Solar Physics*, *230*(1-2), 27–53. doi:10.100711207-005-1527-2

Lean, J. L., & Rind, D. H. (2008). *How natural and anthropogenic influences alter global and regional surface temperatures: 1889 to 2006*. doi:10.1029/2008GL034864

Li, R., Wang, C., & Wu, D. (2018). Changes in precipitation recycling over arid regions in the Northern Hemisphere. *Theoretical and Applied Climatology*, *131*(1–2), 489–502. doi:10.100700704-016-1978-4

Lioubimtseva, E., & Cole, R. (2006). Uncertainties of climate change in aridenvironments of Central Asia. *Reviews in Fisheries Science*, *14*(1-2), 29–49. doi:10.1080/10641260500340603

Lutgens, F. K., Edward, J. T., & Tasa, D. G. (1998). *The Atmosphere: An Introduction to Meteorology (Mastering Meteorology Series)* (13th ed.). Prentice-Hall, Incorporated.

Manvendra, K., Dubey, P. C., Zender, C. S., & Folland, C. K. (2010). Global warming and the next ice age. *Bulletin of the American Meteorological Society*, *89*(12), 1905–1909.

Marcogliese, D. J. (2016). The distribution and abundance of parasites in aquatic ecosystems in a changing climate: More than just temperature. *Integrative and Comparative Biology*, *56*(4), 611–619. doi:10.1093/icb/icw036 PMID:27252206

Millennium Ecosystem Assessment. (2005). *Ecosystems and human well-being: synthesis*. Washington, DC: Island Press.

National Research Council. (1982). *Solar variability, weather, and climate*. Washington, DC: National Academy Press.

Nicholls, R. J., Wong, P. P., Burkett, V. R., Codignotto, J. O., Hay, J. E., McLean, R. F., . . . Woodroffe, C. D. (2007). Coastal systems and low-lying areas. In M.L. Parry, O.F. Canziani, J.P. Palutikof, P.J. Linden, & C.E. Hanson (Eds), Climate Change 2007: impacts, adaptation and vulnerability (pp. 315–356). Contribution of working group II to the Fourth Assessment Report of the Intergovernmental Panel on Climate Change. Cambridge, UK: Cambridge University Press.

Peel, M. C., Finlayson, B. L., & McMahon, T. A. (2007). Updated world map of the Koppen-Geiger climate classification. *Hydrology and Earth System Sciences*, *11*(5), 1633–1644. doi:10.5194/hess-11-1633-2007

Pörtner, H. O., & Knust, R. (2007). Climate change affects marine fishes through the oxygen limitation of thermal tolerance. *Science*, *315*(5808), 95–97. doi:10.1126cience.1135471 PMID:17204649

Roessig, J. M., Woodle, C. M., Cech, J. J., & Hansen, L. J. (2004). Effects of global climate change on marine and estuarine fish and fisheries. *Reviews in Fish Biology and Fisheries*, *14*(2), 251–275. doi:10.100711160-004-6749-0

Sabine, C. L. (2004). The oceanic sink for anthropogenic CO_2. *Science*, *305*(5682), 367–371. doi:10.1126cience.1097403 PMID:15256665

Schwartz, N., Gätke, P., & Baran, E. (2016). Adaptation to environmental change among fishing-dependent households in Cambodia. Phnom Penh, Cambodia: Inland Fisheries Research and Development Institute (Fisheries Administration) and WorldFish.

Sewagudde, S. M. (2009). Lake Victoria's water budget and the potential effects of climate change in the 21st century. *African Journal of Tropical Hydrobiology and Fisheries*, *12*, 22–30.

Soden, B. J., & Held, I. M. (2006). An assessment of climate feedbacks in coupled ocean-atmosphere models. *Journal of Climate*, *19*(14), 3354–3360. doi:10.1175/JCLI3799.1

Stainforth, D. A., Aina, T., Christensen, C., Collins, M., Faull, N., Frame, D. J., ... Allen, M. R. (2005). Uncertainty in predictions of the climate response to rising levels of greenhouse gases. *Nature*, *433*(7024), 403–406. doi:10.1038/nature03301 PMID:15674288

Tedesco, M., Willis, I. C., Hoffman, M. J., Banwell, A. F., Alexander, P. & Arnold, N. S. (2013). *Ice dynamic response to two modes of surface lake drainage on the Greenland ice sheet*. doi:10.1088/1748-9326/8/3/034007

Tol, R. S. J., Downing, T. E., Kulik, O. J., & Smith, J. B. (2009). Distributional aspects of climate change impacts. *Global Environmental Change*, *14*(3), 259–272. doi:10.1016/j.gloenvcha.2004.04.007

UNFCCC. (2008). *Report of the Conference of the Parties on its thirteenth session, held in Bali from 3 to 15 December 2007* (Document UNFCCC/CP/2007/6/Add1). UNFCCC.

U.S. Geological Survey. (2009). *Volcanic gases and their effects*. Volcano Hazards Program.

Vigil, K. M. (2003). *Clean water: an introduction to water quality and pollution control*. Corvallis, OR: Oregon State University Press.

Walker, B. H., Holling, C. S., Carpenter, S. R., & Kinzig, A. (2004). Resilience, adaptability and transformability in social–ecological systems. *Ecology and Society*, *9*(2), 5. doi:10.5751/ES-00650-090205

Youn, S. J., Taylor, W. W., Lynch, A. J., Cowx, I. G., Beard, T. D., Bartley, D., & Wu, F. (2014). Inland capture fishery contributions to global food security and threats to their future. *Global Food Security*, *3*(3–4), 142–148. doi:10.1016/j.gfs.2014.09.005

KEY TERMS AND DEFINITIONS

Adaptation: The process or state of changing to fit a new environmental condition.

Arid: A region in which annual rainfall is less than 25 cm/10 in.

Exposure: The act of coming into contact with an environmental condition that has a harmful or beneficial effect.

Inland Fisheries: All fishing activities carried out in inland waters (streams, rivers, lakes, reservoirs, etc.).

Mitigation: The state of making the impacts of climate change less serious and therefore bearable.

Semi-Arid or Hot Steppe: A region with little rainfall and scrubby vegetations.

Sensitivity: The degree to which a system will respond to a change in climatic conditions.

Vulnerability: The state of being easily persuadable or liable to yield to the impacts of environmental changes.

Chapter 12
Climate Change and Its Impacts on Oases Ecosystem in Morocco

Mohamed Ait El Mokhtar
https://orcid.org/0000-0001-8096-9927
Cadi Ayyad University, Morocco

Raja Ben Laouane
Cadi Ayyad University, Morocco

Mohamed Anli
Cadi Ayyad University, Morocco

Abderrahim Boutasknit
Cadi Ayyad University, Morocco

Abdessamad Fakhech
Cadi Ayyad University, Morocco

Said Wahbi
Cadi Ayyad University, Morocco

Abdelilah Meddich
Cadi Ayyad University, Morocco

ABSTRACT

Today, the international community agrees that climate change is one of the most important threats to the equilibrium of ecosystems. For its part, Morocco is not immune to this phenomenon because of climatic disturbances that the country has known for several years. Moroccan oases are among the most affected ecosystems by climate change, which has resulted in reduced yields of agricultural production, poverty, and deterioration of natural resources. The reports showed that in Morocco the area of palm groves has been significantly reduced during the last decades resulting from the upsurge of drought phenomena, soil salinity, and the spread of "Bayoud." Therefore, the deterioration of the socio-economic and environmental importance of the date palm is well established. Aware of these issues, many actions have been initiated by the Moroccan government to preserve this ecosystem. The aim of this chapter is to provide an overview of how climate change affects different levels of oases ecosystem in Morocco and to propose sustainable strategies to limit degradation of this vulnerable area.

DOI: 10.4018/978-1-5225-7387-6.ch012

INTRODUCTION

The oasis is an ecosystem known by its complexity and fragility. It has always existed in areas characterized by an arid climate with scarcity of rainfall, elevated temperatures and elevated evaporation. The structure of this ecosystem relies mainly on the structure of desert-oasis-river components (Yang et *al.*, 2010). The palm grove survived despite the harsh environmental conditions due to groundwater and irrigation practice. The presence of date palm symbolizes water in the desert (Grenade, 2013). Therefore, the existence of palm groves depends on the availability of water resources and especially underground water potential. Oases have played, for centuries, an important role in the development of local economies and in ensuring ecological security. Effectively, the oasis secure many ecological functions and afford large ecological goods besides natural, economic and social services (PNUED, 2013). In Morocco, the sustainability of the oasis ecosystems is increasingly threatened. Over the last century, the oasis ecosystem has deteriorated clearly due to multiple natural and anthropogenic constraints: drought, salinity, Bayoud disease and lack of rehabilitation programs (Botes and Zaid, 2002; Meddich et *al.*, 2018). From the 15 million date palm trees which inhabited the south Atlas oases and whose majority was represented by good cultivars, only 4 million trees, much of which are seedlings of variable and low quality, and thus the total area of the oases has been reduced from about 150,000 to 44,000 ha (Saadi, 1992). At the end of the 19th century, Morocco was ranked third worldwide among dates producing countries and possessed an international remarkable portion of dates trade, due in particular to the quality of its harvest (Haddouch, 1996). Almost a century later, Morocco, which was considered as a major exporter of date palm products, has become an importer (Khardi, 2003).

Even if several researches reported that the ecosystems degradation in arid regions was mostly affected much more by irrational human exploitation than climate change (Huo et *al.*, 2008; Jarsjö et *al.*, 2008; Aus der Beek et *al.*, 2011; Dong et *al.*, 2014; Ma et *al.*, 2014), Moroccan oases figure among the most affected area by climate variation (Karmaoui et *al.* 2015). This situation has caused reduced farmlands yields production, poverty and scarcity of essential natural resources (Kuhn et *al.*, 2010; Ben Salem, 2014; Johannsen et *al.*, 2016; Ait Houssa et *al.*, 2017; Lamqadem et *al.*, 2017)

In this chapter, we will examine climate change status in Morocco with a focus on the possible impacts of the projected global climate change on oases with its different components. We will also discuss the ways forward to ensure the sustainability of this ecosystem considering the challenges posed by climate change.

PREVIEW OF CLIMATE CHANGE IN MOROCCO

Recent Climate Data

Located north-west of the African continent, Morocco opens on both the Atlantic Ocean and the Mediterranean Sea, making it a total of about 3,500 km of coastline (Fertah et *al.*, 2017). With a latitudinal extension from the 37th to the 21st parallel. It covers an area of 710,000 km^2, from the Strait of Gibraltar to almost the southern reaches of the great African Sahara (Almeida-Garcia and Chahine 2015). The country is characterized by a well-diversified geography: the plains, high mountain ranges where many massifs exceed 3,000 meters (Rif and Atlas mountain ranges) and the desert (Driouech, 2010). Three main climatic influences in Morocco are often distinguished: Atlantic, Mediterranean and Saharan (Born et *al.*, 2008).

The climate in Morocco is linked to an atmospheric circulation of global scale. Indeed, rainfall is strongly related to the North Atlantic Oscillation (NAO), formed by a pressure difference between the Azores subtropical anticyclone and the subpolar depression of Iceland (Ward et *al.*, 1999; Knippertz et *al.*, 2003a; Knippertz et *al.*, 2003b). The NAO has been recognized as a major atmospheric circulation pattern influencing Moroccan winter precipitation over at least the past 500 years (Pauling et *al.*, 2006). During the positive phase of the NAO (Figure 1), the strengthening of the southern pressure gradient contributes to the intensification of the circulation from west to northeast Atlantic and the European continent. In this configuration, the depressions rail axis is oriented southwesterly - northeasterly and migrates northward, thereby deflecting the disturbances that develop over the Atlantic to the northwestern part of Europe (Hurrell, 1997; Wanner et *al.*, 2001) which benefits from wet and soft conditions. The rainfall is then reduced on Morocco in comparison to the south of Europe. Conversely, the negative phase (Figure 1) corresponds to a weakening of the southern pressure gradient in the NAO, inducing a decrease in the intensity of the westerly winds. The axis of the rail depressions is then clearly zonal, allowing the passage of Atlantic disturbances towards the south of Europe and the Mediterranean basin. The negative phase of the NAO thus favors rainfall in Morocco and northwestern Africa (Ward et *al.*, 1999; Glueck and Stockton, 2001; Herrera et *al.*, 2001; Knippertz et *al.*, 2003a).

Rainfall in Morocco like in many countries in North Africa is characterized by a wet season in winter and dry conditions in summer. The rainy period, which starts in October and lasts until March, records its maximum in the months from December to February (Endlicher, 2000; Xoplaki, 2002; Lionello et *al.*, 2006). The summer months have a low rainfall and stormy in general. Since the climate of Morocco is more influenced by the Atlantic Ocean in the west, by the Mediterranean in the north and the desert (Sahara) in the south and is locally modulated by the orography, the duration of the rainy season has slight spatial differences (especially between north and south and between west and east). According to

Figure 1. Schematic representation of meteorological conditions associated with the positive (left) and negative (right) phases of the North Atlantic Oscillation (NAO).
Source:(Bell & Visbeck, n.d.)

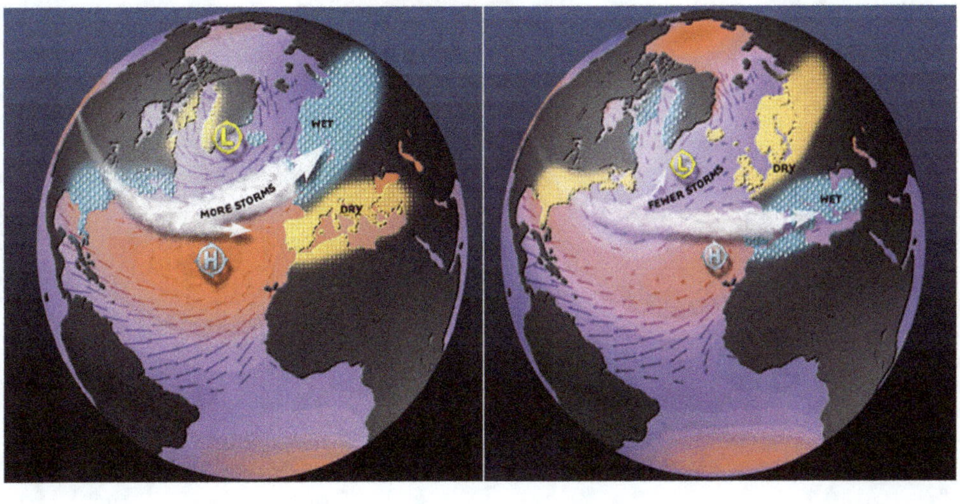

Agoumi and Debbarh (2006), rainfall ranges from 800 mm/year in the North to less than 25 mm/year in the South, and from 600 mm/year in the West to 100 mm/year in the East. The average rainfall varies from 500 to 2,000 mm/year in the most watered zone of the country in the North-West, while it is less than 200 mm in the arid zones of the South of the country (Ministry of Environment of Morocco, 2001). Figure 2 shows the decrease from north to south and from west to east annual rainfall averaged over the period 1971-2000. Only the high mountains and the extreme north-west receive average annual levels exceeding 800 mm. Most of the country does not exceed 400 mm annual average.

Although a higher mean annual rainfall, due to the effect of the Atlas Mountains, the country is characterized by acute inter-annual rainfall levels. The coefficient of variability (ratio of the standard deviation to the mean) varies on Morocco from 30% to 40% in the west to more than 70% in the southern regions (Knippertz et *al.*, 2003a; Driouech et *al.*, 2009). Thereby, long-term mean precipitation reflects averages over many dry years and some relatively humid years. For north-eastern Morocco, many studies showed a below average annual rainfall rates which have taken over since about the mid-1970s (Hertig, 2004; Fink et *al.*, 2010; Meddi et *al.*, 2010). Likewise, the southern region of the Moroccan Atlantic coast as well as the Atlas Mountains showed several periods of below average precipitation in the second half of the 20th century in the winter season (Hertig, 2004). Based on the observed changes, the last decades for

Figure 2. Average annual rainfall in Morocco calculated over the period 1971-2000
(Source: Direction de la Météorologie Nationale, Maroc)

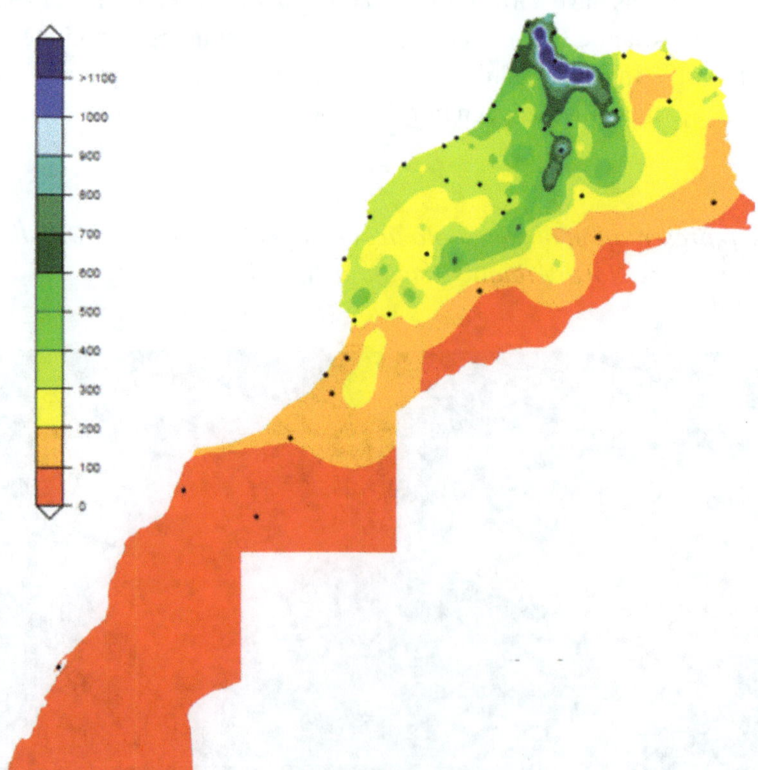

the above mentioned regions are characterized by a general trends toward warmer and drier conditions (Gerstengarbe and Werner, 2007; Born et *al.*, 2008).

Climate Change Trends in Morocco

In Morocco, the observations of the last three decades provide evidences of probable impacts of the expected climate changes: frequency and intensity of drought periods, unusual disastrous floods, reduction of the snow duration of Rif and Atlas peaks, changes in the spatial and temporal distribution of rains (Royaume du Maroc, 2010). Some of these events have already cost Morocco in social, economic and environmental terms. Morocco is experiencing more and more recurring and severe droughts combined with an increasing demand for water (Agoumi and Debbarh, 2006). It has experienced more than forty years of drought of all kinds accompanied by excessive temperature variations that marked the past century. In the twenty years alone (1980-2000), two droughts of 4 years and one of 3 consecutive years have taken place and have severely affected its agricultural productivity (fall of more than 50% of yields) (FAO, 2018).

Morocco has also experienced the last two decades an increasingly high frequency of meteorological phenomena with high flood risk. Torrential rainfall with floods (Figure 3) caused enormous damage in the province of Errachidia in August 2006 (Merzouga region), in the East in May 2007 and in October 2008 again in the province of Errachidia (Gourrama and Boudnib regions) and in the North of the country (Tangier, Tetouan, Nador, ...).

Figure 3. Torrential rainfall with floods in Tazarine from Drâa oases, Morocco (Summer 2018)

Precipitation Trends

For the region of North Africa, climate change researches indicate that annual precipitation is expected to decline during the current century (Gibelin and Déqué, 2003; Giorgi and Bi, 2005; Rowell, 2005). According to the Regional Model REMO, rainfall in Morocco, as other countries in North Africa, is likely to decrease between 10% and 20% by 2050 under SRES A1B scenario conditions (Paeth et *al.*, 2009). An evaluation of rainfall changes within the IMPETUS project shows that Moroccan precipitation might be reduced in the period 2011–2050 between 5 (mountainous areas) and 30% in the southern parts of the country for the SRES A1B scenario and by 5 to 20% for the B1 scenario (Christoph et *al.*, 2010).

Based on rainfall anomaly (scenarios A2 and B2) and compared to the period 1961-1990, the rainfall decline will affect the whole country (Figure 4). According to the least pessimistic scenario B2, the annual rainfall will decrease by 20% by 2050 and by 40% by 2080, with the exception of the Saharan zone where the decrease will be 16% in 2080. The precipitation decrease will be greater according to the pessimistic A2 scenario. It is during the autumn and spring that the rainfall decline will be felt, that is to say during the periods during which we normally record peaks of rainfall.

Statistical downscaling of precipitation, developed from assessments of precipitation changes under the SRES B2 scenario hypothesis, for a region covering the Northern parts of Morocco shows rainfall increases in December/January of up to about 60 mm in the period 2071–2100 compared to the time period 1990–2019 (Figure 5) (Schilling et *al.*, 2012). As shown in Figure 5, there are negative precipitation variations at the beginning of the rainy season in October/November. However, in December/

Figure 4. Rainfall anomaly for 2050 and 2080 horizons in Morocco under the A2 and B2 emission scenarios compared to 1961-1990 (WB/MAPM, FAO, INRA, DMN, 2008)

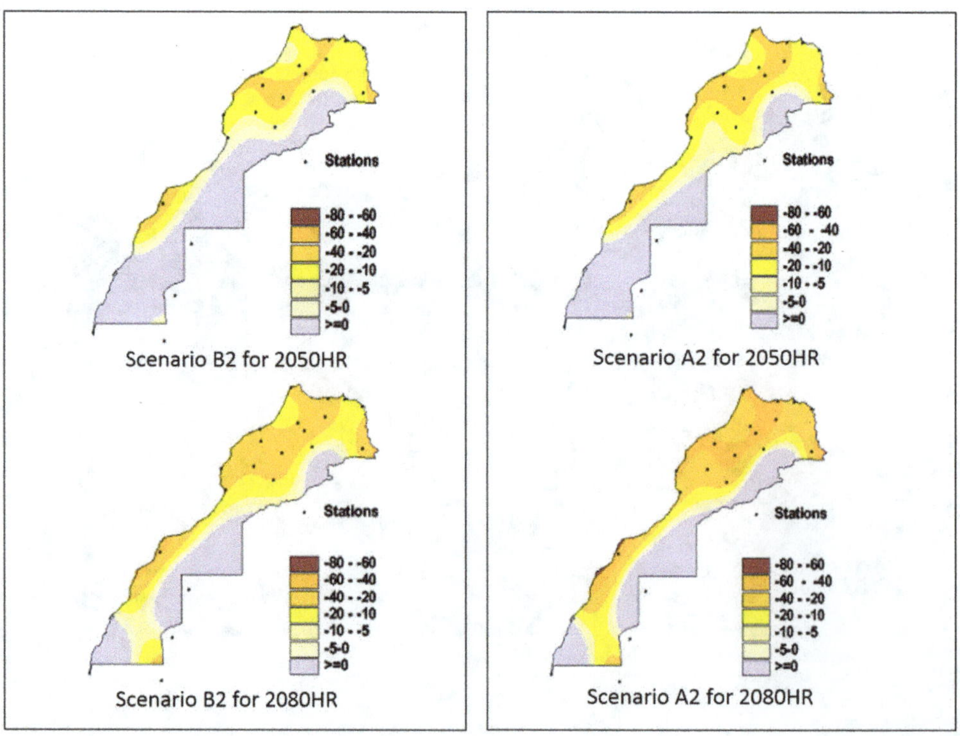

January significant increases are estimated. This could be linked to changing circulation patterns (e.g. to a more southerly path of storm tracks in connection with a pressure rise centered over Central Europe in December and January as simulated by some coupled global general circulation models (GCMs) (Hertig, 2004). In this context, enhanced atmospheric humidity can be regarded as an additional factor for higher rainfall amounts. A simulation with a local model within the IMPETUS project shows more important precipitation events south of the Atlas Mountains as a result of enhanced humidity advection (Christoph et *al*., 2010). Otherwise, it should be noted also that a decline of precipitation in winter are projected for the Northern region of Morocco, for example by Giorgi and Lionello (2008) under the use of an ensemble of GCM simulations and by Paeth et *al*. (2009) by means of dynamical downscaling.

In the statistical assessment of Schilling et *al*., (2012), remarkable decreases could be noticed in the most Northern parts of Morocco in February/March (Figure 5). Thereafter, the whole region is affected by drier conditions in spring (April/May). The signal- to-noise-ratio (the rainfall difference of the two 30-years periods in relation to natural variability) of the precipitation reduction in spring is greater than one, indicating that the climate change signal is greater than the recent natural variability. A study by Palutikof and Wigley (1996) also estimates decreased rainfall south of the Mediterranean Sea during spring. Around this drier region, the north-western parts of northern Africa stand out as a region of most pronounced decreases in precipitation. This was also reported by Jacobeit (1994).

Figure 5. Changes of Mediterranean precipitation for the main rainy season from October to May according to statistical downscaling assessments; Differences of the mean 2-month precipitation between the periods 2071–2100 and 1990–2019 in mm
(Schilling et al., 2012)

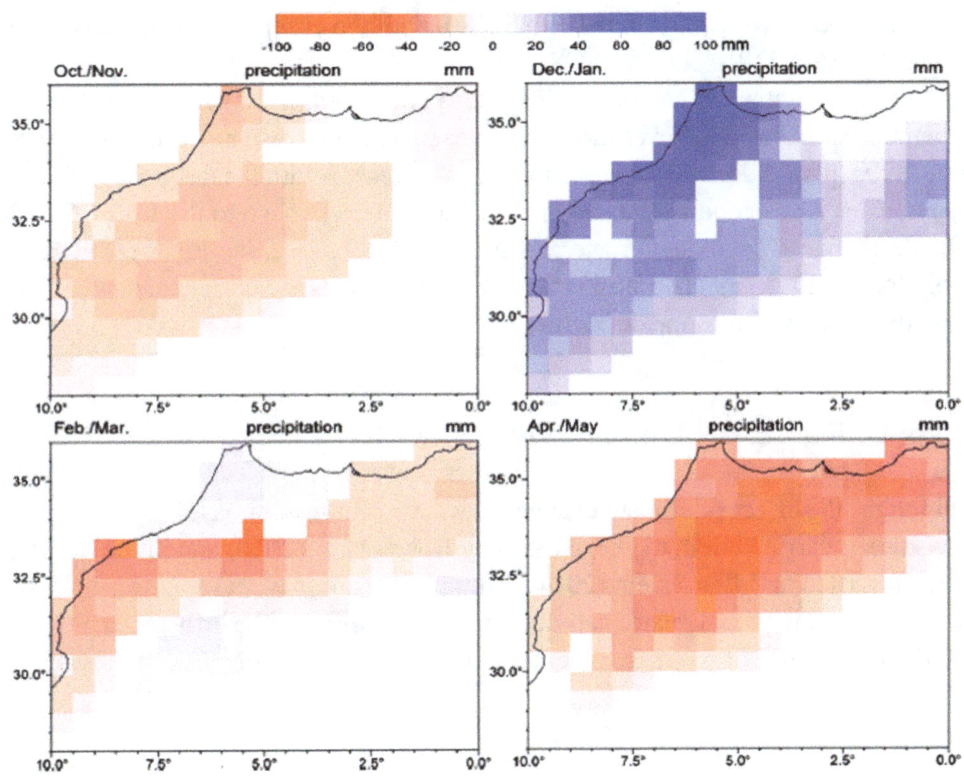

Temperature Trends

Model projections available for North Africa indicate a clear increase in temperature over the next 20 years that is expected to continue throughout the 21st century, probably at a rate higher than the estimated global average (Agoumi, 2003). A new reconstruction of the climate history of the Maghreb concluded that 20th century was the driest in North Africa since severe droughts in the 13th and 16th centuries (Touchan et *al.*, 2011).

The dynamical regional model REMO proposes a temperature increase in North Africa between 2 and 3 °C by 2050 under A1B scenario conditions (Paeth et *al.*, 2009). The temperature enhancement for Morocco is estimated to 1.2 °C in the SRES A1B and 1 °C in the B1 scenario. Both scenarios provide a slightly more intense increase in the mountain region (Christoph et *al.*, 2010). An application of regional and global climate models by Patricola and Cook (2010) indicates a very strong warming of about 6 °C over Morocco in the 21st century compared to the 20th century. Likewise, the results from statistical downscaling by Hertig and Jacobeit (2008) show a noticeable temperature rise for western North Africa. The temperature assumptions indicate increases of mean temperature in all months of the year, with largest warming rates in summer (June/July) and autumn (October/November) of partly more than 4 °C until the end of the 21st century. The lowest warming rate is assessed for the high winter months (December and January) with values of up to about 1 °C. In general, the spatial warming pattern has important values on the mountainous areas of the Atlas Mountains, and weakens toward the coastal areas of the Atlantic Ocean and the Mediterranean Sea.

According to A2 scenario (Figure 6), temperature increases will affect the whole country and warming will be close to 3 °C by 2080 and will reach 5 °C in the Eastern and mountainous zones. This increase in temperature will lead to an increase in evapotranspiration (the sum of soil evaporation and plant transpiration) of around 20% by 2050 and 40% by 2080, with the exception of the Saharan zone (9% in 2080) (WB/MAPM, FAO, INRA, DMN, 2008).

Overall, the air temperature in Morocco, all seasons combined, has increased on average from 0.6 to 1.4 °C depending on the region, over the last 40 years. This increase has been most effective since the 1980s and 1990s (period during which warming has increased with the frequency of droughts). The climate projections of the Second National Communication of Morocco to the UNFCCC (2010) give, as global climate change projections for average annual temperatures, increases of +1.8 °C and +3.2 °C respectively for the 2045 and 2075 horizons. Heat waves are expected to increase in frequency and severity across the country (Royaume du Maroc, 2010).

IMPACTS OF CLIMATE CHANGE IN OASIS ECOSYSTEMS

In order to highlight the effect of climate change on water resources in oases ecosystem, we took the results of two studies, carried out in the Drâa oases by Kuhn et *al.*, (2010) and in the Tafilalet oases by Ben Salem et *al.*, (2011) and Ben Salem (2014). These two examples make a clear assessment of the changes in water requirements and related activities in the oasis that may be influenced by climate change and therefore the importance of water management in the oasis ecosystem toward its sustainability.

Figure 6. Change in average, maximum and minimum temperatures for 2050 and 2080 horizons in Morocco under the A2 emission scenario
(Babqiqi et al., 2011)

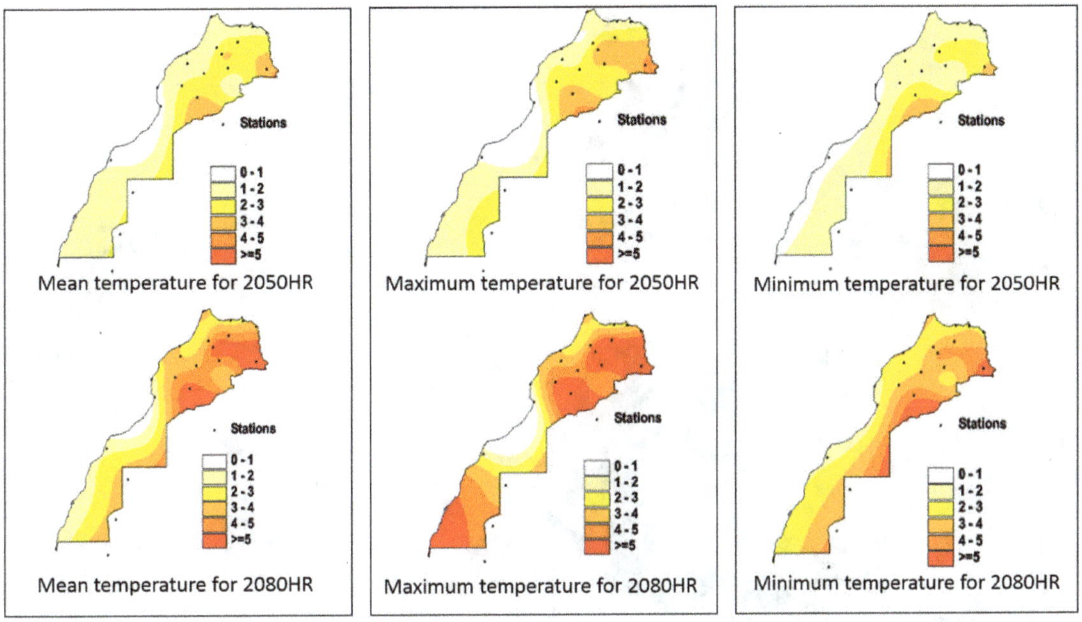

Importance of Water Resources Management for Livelihood Security Under Climate Change in Southern Morocco (Drâa Oases)

Case Study

The Middle Drâa valley south of the city of Ouarzazate is part of the region Souss-Massa-Drâa south east of Morocco comprising an area of about 15,000 km^2 (Figure 7) (Klose, 2009). Climatic conditions are arid to hyper-arid. Rainfall adds up to 200 mm annually in the northwestern parts of the basin to less than 30 mm in its southeast, with an average potential evaporation of up to 3,000 mm (Klose, 2012). A mean annual temperature of 22.4 °C and a mean annual precipitation of 58 mm are observed at the Zagora meteorological station (Schulz, 2006). The actual evapotranspiration is higher in the oases along the Oued Drâa than in the scarcely vegetated surroundings (Huebner et *al.*, 2004), where rates of 60 to 80% of the total annual precipitation are calculated (Weber, 2004).

Typical desert soils are prevailing in the Middle Drâa Valley. The agriculturally used oasis soils are different, as they are formed on loamy loess-like river sediments (Brancic, 1968).

Cultivation is based on irrigation by both surface water and groundwater. Several crops are cultivated in the oases farms, such as wheat, barley, date palms, alfalfa, maize, henna, and some vegetables. Depending on the amount of available water in the dam for the coming season and the crop rotation, farmers adapt the extent of cultivated area (Heidecke, 2009). Furthermore, in years of water scarcity, farmers use groundwater for irrigation, which has led to a decline of the groundwater table and a widespread salinization of soils (Klose, 2009; Klose, 2012). Farms' livestock wealth is made of cattle, sheep, goats and camels provide meat and milk for own consumption and, somewhat, for income (Sraïri et *al.*, 2017).

Figure 7. Middle Drâa valley area
(Karmaoui et al., 2014)

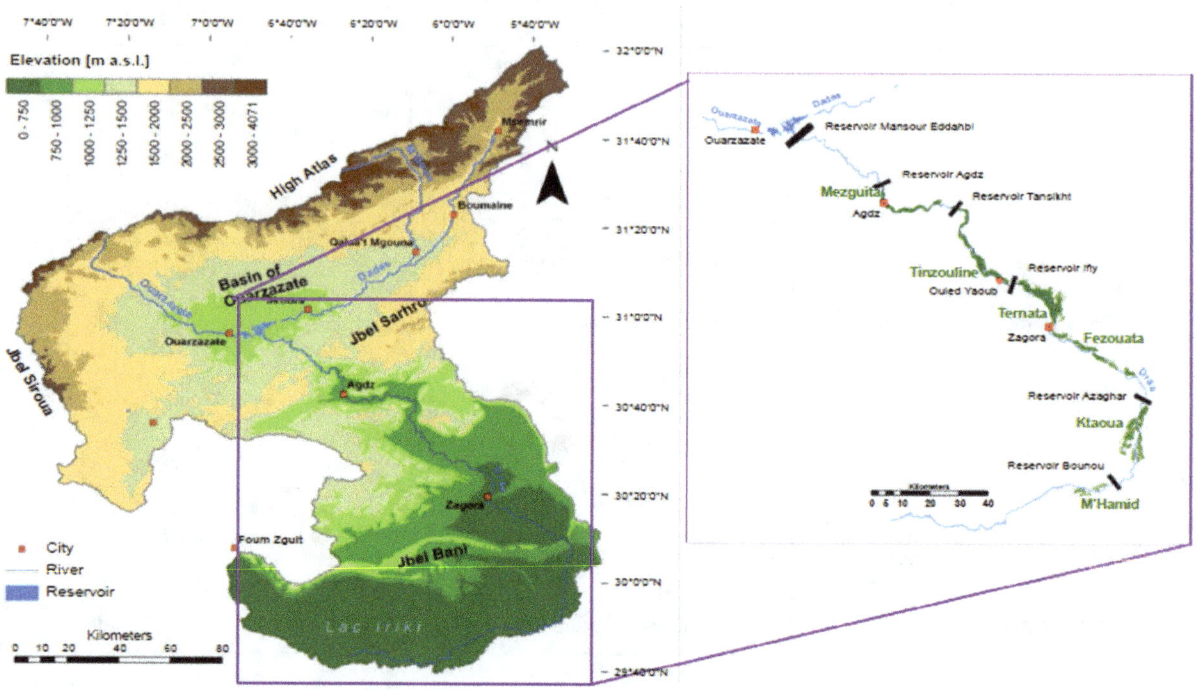

Another source for income in the oases is tourism, which comprises camel safaris into the Sahara and oases scenery (Heidecke, 2009). About 280,000 people live in the Middle Drâa catchment, distributed over the six oases and the city of Ouarzazate which has about 100,000 inhabitants (Platt, 2008). Inhabitants of the rural oases are mostly farmers living off self-sufficient husbandry, which does not yield much income. The little income earned due to cash crops, such as henna and dates, are sold at local markets in order to afford fertilizers and fuel for the groundwater pumps (Heidecke and Schmidt, 2008). The calculation of incomes for agriculture is hard to define or evaluate, due to the farmer's prevailing illiteracy and lack of bookkeeping (Heidecke, 2009). In comparison to the water consumption of agriculture, the households' water needs are much more limited (Heidecke and Schmidt, 2008). In the Drâa valley, only about 2% of the water withdrawal is used for human water consumption.

Assessment of Economic Effects of Climate Change in the Middle Drâa Oases Using the MIVAD Model

The results of this study provide an assessment of the impact of Climate Change, and hence of changing water availability, on water use for irrigation and other farm-based activities for income generation in six Drâa oases (Mezguita, Tinzouline, Ternata, Fezouata, Ktaoua, M'hamid).

Climate Change Scenarios Using the MIVAD Modeling Approach

To simulate the impact of Climate Change on water use, the MIVAD model (Integrated Model of the Dràa Valley) was modified to include the dynamically downscaled IPCC Climate Change scenarios A1B and B1 using the REMO model. Recursive-dynamic simulations with MIVAD were carried out for the years 2000-2020, which is the period of the national development strategy of Morocco. The results of the study are based on averages of simulations based on data from 1972 until 2000. Furthermore, an alternative water distribution was discussed as an intervention scenario, and water pricing as one option for groundwater resource preservation was examined.

Impact of Climate Change on Modeling Results

Both the record of historical inflows from 1972 (the year the Mansour Eddahbi dam was put into operation) until 2000, as well as the estimated inflow scenarios on the basis of the climate scenarios A1B and B1 in figure 8, suggest that inflows are likely to decrease in the future. The inflows into the dam are an exogenous variable in the MIVAD model and constitute a decisive factor for the model results, as they determine the major part of the water availability in the system. Lateral inflows and groundwater aquifers also contribute to water availability. However, most groundwater is extracted from the shallow aquifers below the riverbed.

Figure 9 summarizes the impact of Climate Change until 2020 on water use and farm income. It is obvious that farm income is likely to decline in the future as a result of lower water availability in the Mansour Eddahbi dam. The decreasing water availability is also reflected in the agricultural surface water use, which represents the water drawn from the Drâa river for agricultural irrigation. Consequently, groundwater use is likely to increase nearly twofold depending on the underlying Climate Change scenario. As IPCC A1B represents a stronger greenhouse-gas climate forcing and therefore a stronger projected rainfall decline, water availability is the lowest in this scenario, and model results are the most drastic under this scenario as well.

Figure 8. Observed and predicted inflows into the dam across all decades from 1972 to 2050 (Kuhn et al., 2010)

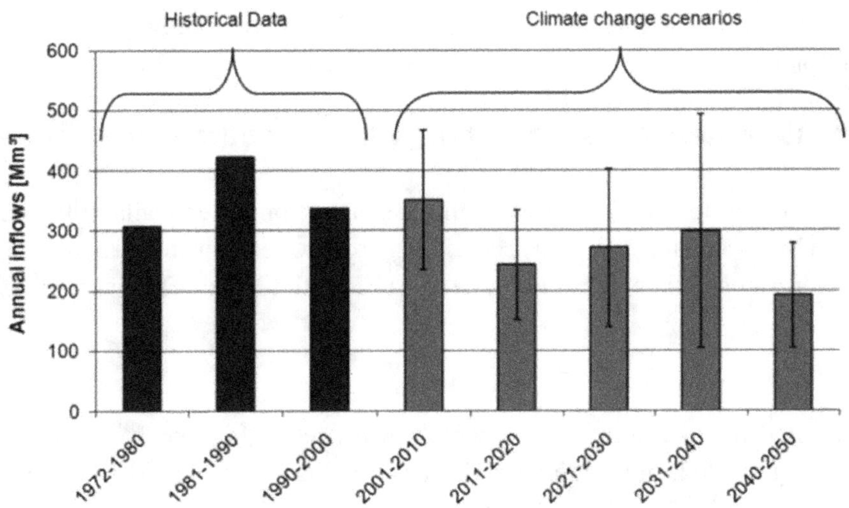

Figure 9. Development of agricultural income, surface water (SW) and groundwater (GW) use: comparison of base line (averages from 1972 to 2000) to B1 and A1B scenarios (calculated from 2000 to 2020) (Kuhn et al., 2010)

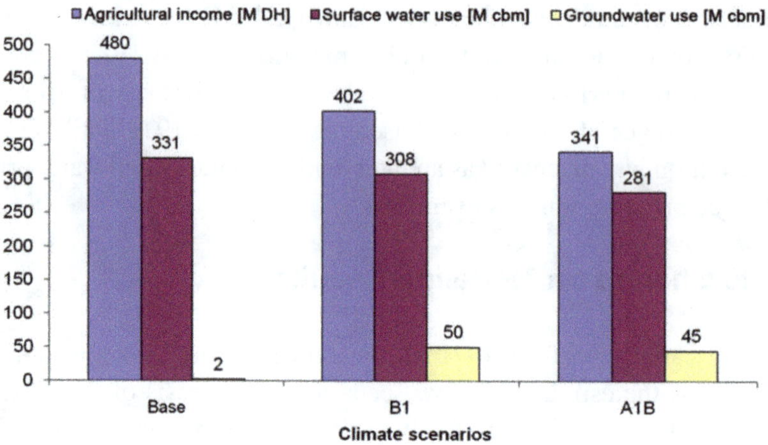

Figure 10 depicts the fill levels of the aquifers in meters. In the simulations with no charge, groundwater use is still more expensive than surface water due to the running costs of the motor pumps. Nevertheless, with an additional charge of 40 Moroccan cents, groundwater use is even less attractive than surface water and only used for irrigation if surface water is scarce and groundwater use is economically efficient. Consequently, farmers use less groundwater for irrigation than in the case with no charges, and groundwater levels are therefore more stable over the years. This higher stability of groundwater levels over the years has a direct effect on farm income in the six oases.

Figure 11 displays farm income over the 20 years horizon for the six oases, from north to south along the Drâa river. It can be seen that in the long run, a groundwater charge leads to more stable farm incomes.

The effect of a groundwater charge on groundwater use and income is summarized in Table 1. The implementation of a groundwater charge induces a lower level of groundwater use, presented as the average over 20 years. Surprisingly, agricultural income is also slightly higher than with no extra charge, although the farmers incur extra costs for using irrigation water. This cost is due to the more stable groundwater availability over the 20 years time span, which then leads to a more stable income. The greater stability of water availability is also presented in the lower shadow price for irrigation water, which can be interpreted as a factor representing the scarcity of irrigation water.

Planning and Integrated Management of Water Resources in Tafilalet Oases

This study provide an overview of the impact of climate change on water availability and water demand in Tafilalet oases in Morocco under varying scenarios in order to see how investments in better irrigation techniques may alter water availability in the water system of the Middle Tafilalet valley.

Case Study

Tafilalet plain is located in the extreme center east of the Moroccan Kingdom. It extends over an estimated area of 700 km² (Margat, 1962). It is ordered to the north by the Ziz River, to the south by the lower Ziz

Figure 10. Fill levels of the six aquifers (GW1 to GW6) of the six oases in the Middle Drâa valley (depth of water table in meter)
(Kuhn et al., 2010)

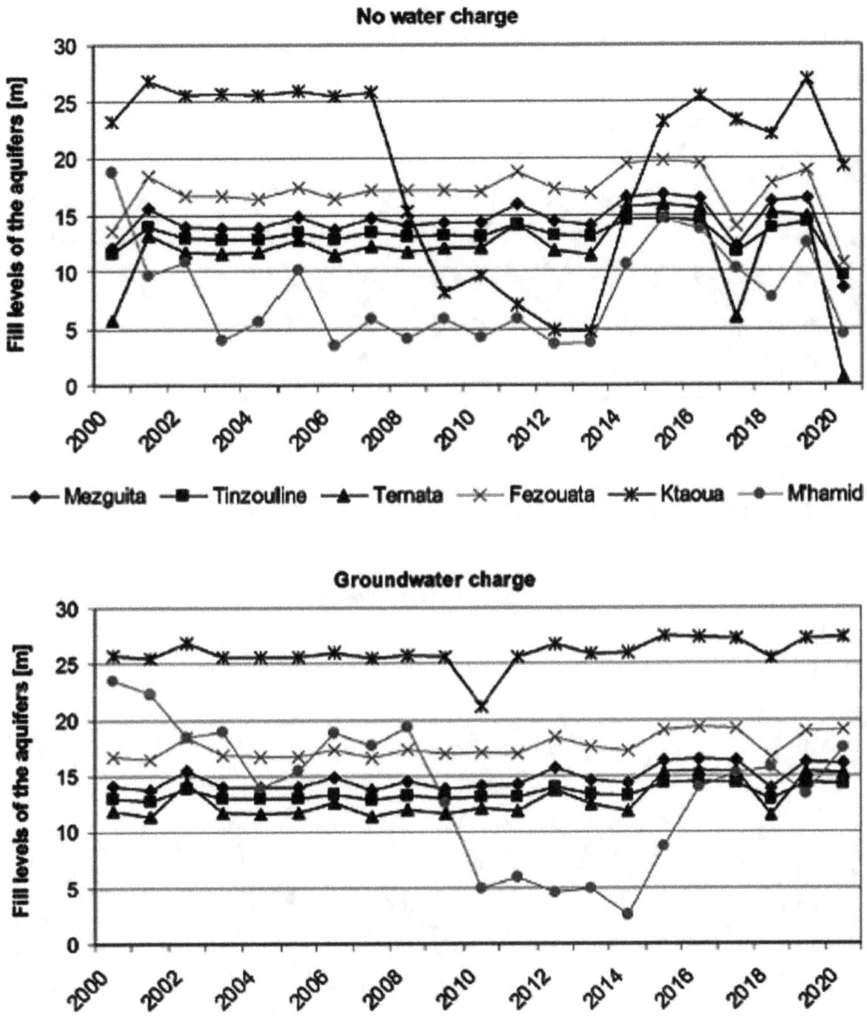

River, to the east by the river Amerbouh and to the west by the river Ghris (Figure 12). It is part of the UNESCO biosphere reserve and is a site of biological and ecological interest. The plain is crossed by the Ziz valley and the river Ghris to ensure irrigation of oases vegetation, and supply of water table. The cultivated area is of 21,300 ha of which 19,290 ha is irrigated. It is subject to a desert climate, strong continental influence, with very low rainfall not exceeding 85 mm, and having a great spatiotemporal irregularity. The annual rainfall regime shows the existence of two distinct rainy seasons with major floods, Autumn and Spring with a very limited number of rainy days per year (10 to 15 days). The annual average temperature is around 20 °C and is characterized by very high amplitudes going up to 50 °C (Kabiri, 2004). Average potential evaporation amounts to 2,500 mm/year. However, the existence of date palm trees creates a microclimate, which soothes the harsh climatic conditions and consequently reduces the potential evapotranspiration to 1,200 mm/yr. (Bouhlassa and Paré, 2006). The maximum

Figure 11. Development of agricultural income in the six oases (A1 to A6) of the Middle Drâa valley in Million Moroccan Dirham (MDH)
(Kuhn et al., 2010)

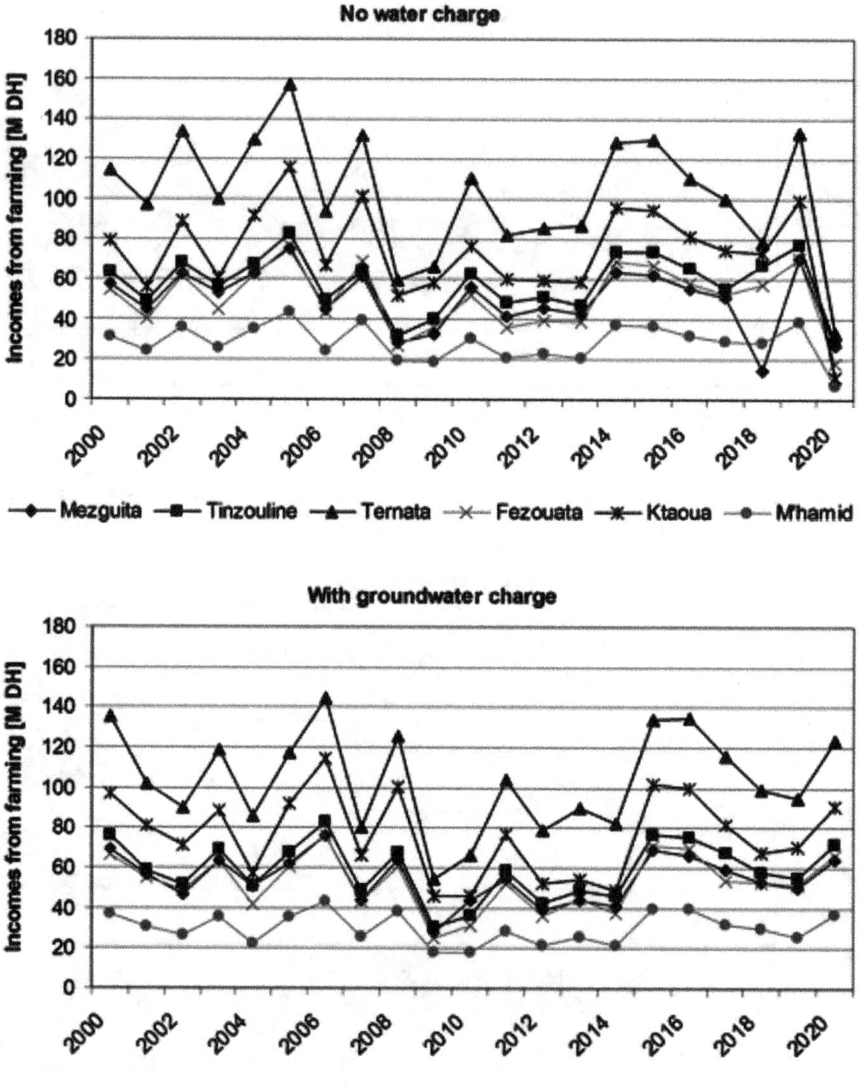

Table 1. Impact of a groundwater charge in contrast to no charges (average of 20 years assuming climate change scenario A1B)

	No Charge	Groundwater Charge
Agricultural river water use [Mm^3]	280.5	303.7
Agricultural groundwater use [Mm^3]	44.6	25.6
Agricultural income basin-wide [M DH]	341.4	351.3
Shadow agricultural water price [DH/ m^3]	1.79	1.76

(Kuhn et al., 2010)

evapotranspiration is reached in July (220 mm) and the minimum in January (12 mm) (Bouhlassa and Paré, 2007). The low rainfall and potential evapotranspiration ratio requires use of irrigation (pumping, spreading floodwaters, etc.)

Tafilalet plain soils belong to the poorly developed sub-desert and alluvial soil class. Cultivated soils are of silty alluvial clay due to irrigation with flood waters for many previous centuries (Guemimi, 1991). Salinization is widespread in the Tafilalet. The deposition of salts due to evaporation results in a higher surface salinity and a trend towards alkalinization (Bousfoul, 2008). Apart from irrigated areas, soils are poorly developed because of severe weather conditions.

The Tafilalet oasis consists of a set of neighboring palm groves. The applied agricultural system includes two to three layers: date palm trees, fruit trees and surface cropping which are generally cereals, legumes, fodder and vegetables. The number of layers depends on the availability of water. Similarly, intercropped agriculture depends on the water quality (Renevot et *al*., 2010). In this area, the most spread is flood irrigation system. Dripping irrigation is practiced very little (Keddal, 2006).

Figure 12. Tafilalet region
(Ben Salem et al., 2011)

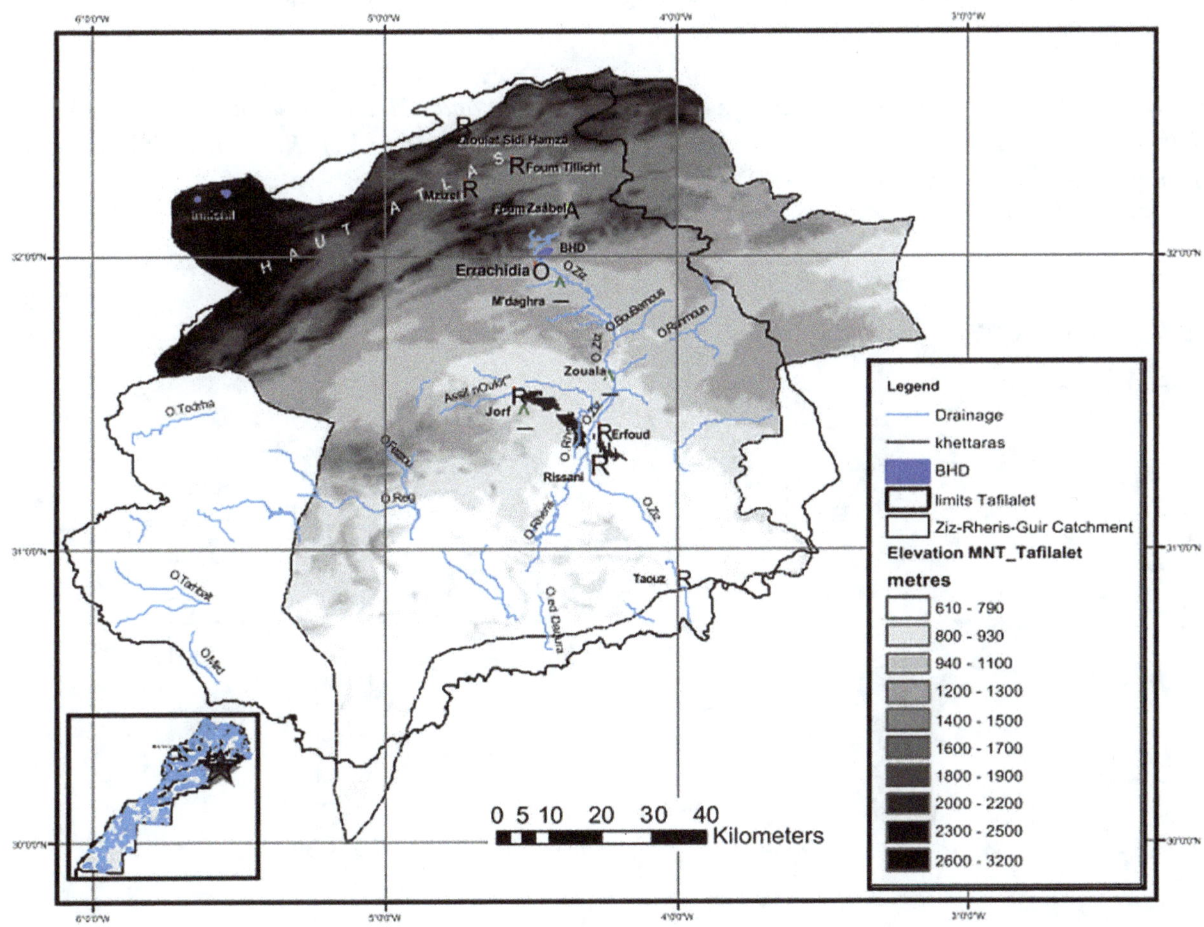

Evaluation of Changes on Water Requirements Under Climate Change in the Middle Taifilalet Oases Using the WEAP Model

The upper Ziz hydrology was represented by the model WEAP (Water Evaluation and Planning System) (Figure 13). The model has been adapted, calibrated and validated for the Ziz Basin in the framework of the MIOS (Integrated Model of the Southern Oases) and permits the analysis of different hydrological parameters in different climate and policy scenarios. Water sustainability assessment requires a scenario approach for taking a long wide view that considers futures with fundamentally different development and environmental assumptions and policies. The reference scenario (1970-2008) is the base scenario that uses the actual data, to help in understanding the best estimates about the studied period. Two scenarios of high level of growth population were used in this study: Scenario A: growth rates between 1993 and 2030 were as follows: domestic 1.12-1.2%, livestock 1.2% and irrigation 1.2% (IRATE, 2006); Scenario B: proposed scenario of high level of growth population (1993 to 2050) with a growth rate of 3% for domestic, livestock and irrigation. A Modern Irrigation Scenario (1970 - 2050) was also used and it is presented by the introduction of new technologies, modern pumps and drip system developed lately in the region with the support of farmers by the Green Morocco Plan. Regarding Climate change scenario, four scenarios of climate change (1970 - 2050) can be adapted for the Tafilalet region: +1 °C, -10% P; +1 °C, -20% P; +2 °C, -10% P; +2 °C, -20% P (Tabet-Aoul, 2008). Two rainfall reduction scenarios were used in the WEAP model for the Ziz basin to evaluate water demand trends.

Figure 13. a) WEAP conceptual model for water management and irrigated area in Tafilalet, b) Water distribution in main parts of the Ziz oasis. BHD: Hassan Addakhil dam; KH: Khettaras; Z: Zone, GW: Groundwater
(Ben Salem et al., 2011).

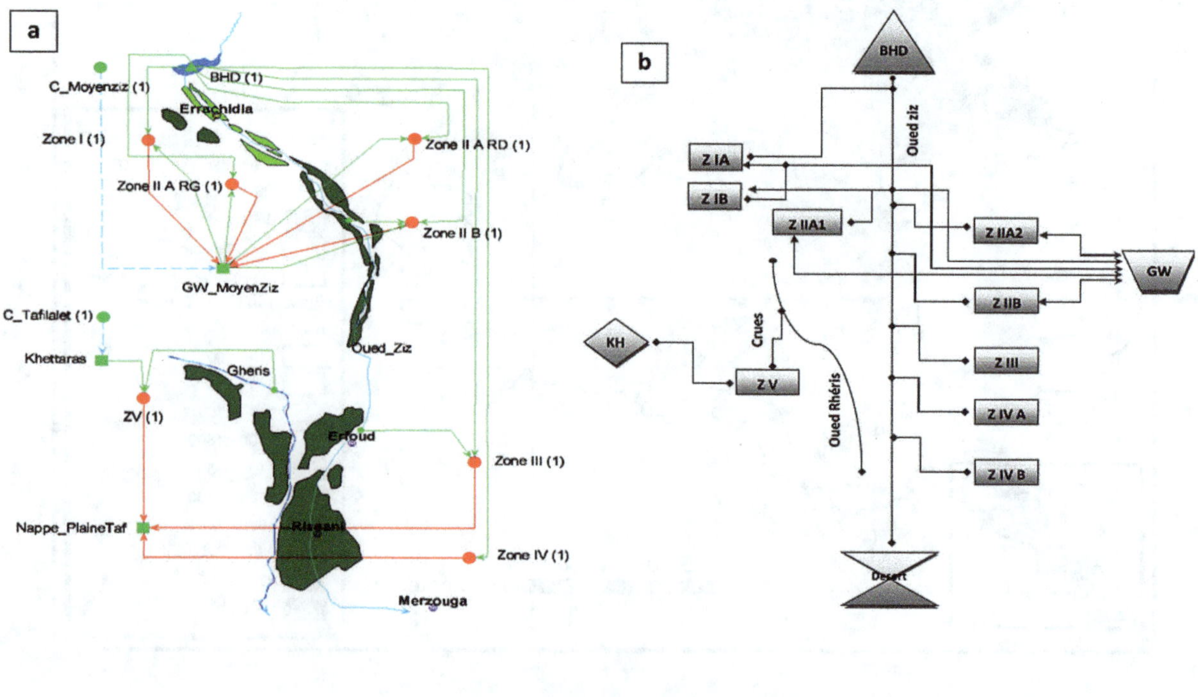

Impacts of Different Scenarios on Modeling Results

Based on Figure 14, all scenarios showed a higher annual water demand compared to Reference scenario. With the normal demographic growth, the demands will pass from 450 Mm^3 to 700 Mm^3 in 2050; however, a high rate of growth water demand can reach 1,000 Mm^3 in 2050. Those water quantities can not be satisfied in the region of Tafilalet because of its arid climatic conditions, and its limited water resources. We can also see a clear effect of climate change on annual water demand especially under climate change sc1, which showed a trend of 1,000 Mm^3 in 2050. The largest proportion of increasing water demand is required for irrigational purposes for agriculture subject to demographic pressure.

The result of figure 15 showed the effect of increased population growth on the water demand of irrigated area. The effect was much more visible with high level of growth population (3%). The use of modern irrigation technics could decrease the water demand compared to reference scenario. Ghallabi et *al.* (2011), who studied groundwater resources in the Moroccan Tensift-Basin, suggested that a change in irrigation efficiency by investments in better techniques is not privileged by farmers due to the financial risk. It is especially changes in traditional adaptation practices of cultivation area or in the choice of crops that show large differences in water demand in their scenarios (Ghallabi et *al.*, 2011).

Figure 14. Annual water demand according to several scenarios for 1970-2050, in millions of m^3 *(Ben Salem, 2014)*

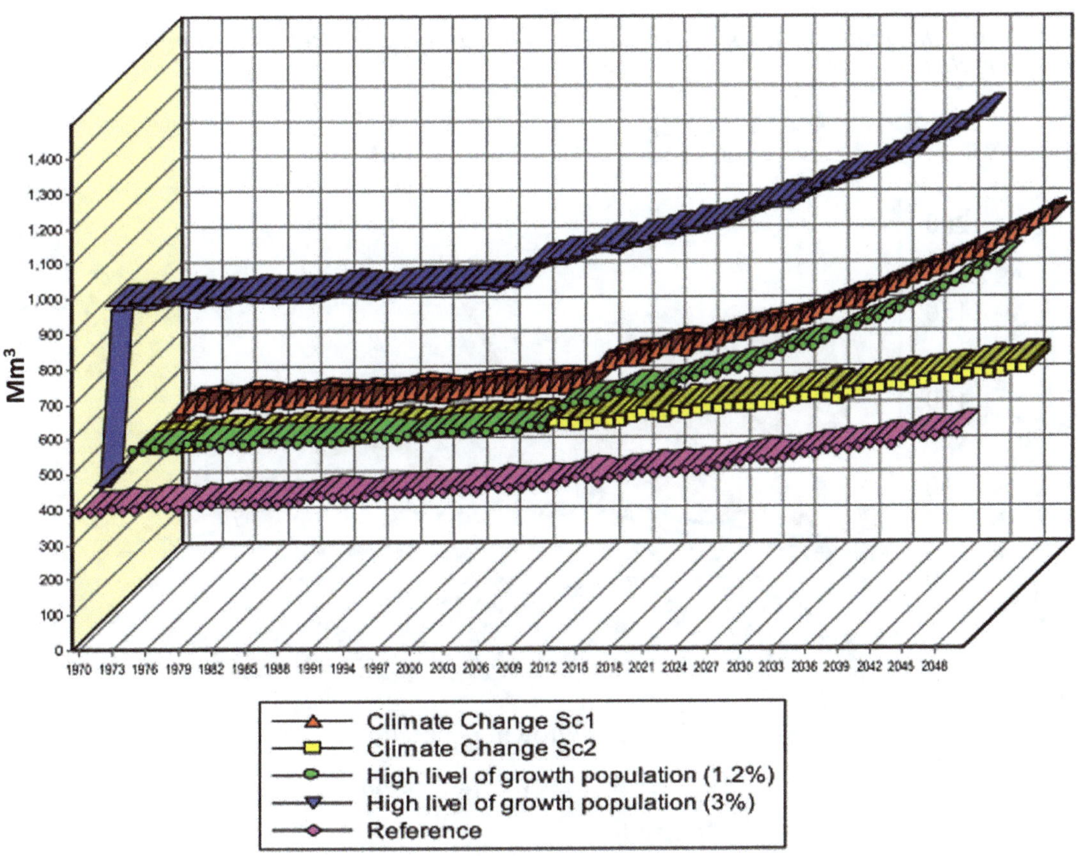

Based on the same figure, it was clear that the water demand for irrigation increases in summer months unlike this demand during the other months of the year, which means that it becomes more important in parallel with the increase in temperature.

Figures 16 presents unmet water demand depending on the high population growth and modern irrigation scenarios. The comparison of the three scenarios shows that the distributed water will increase from 400 million m^3 to 1,000 million m^3 up to 2050 for the high population growth rate scenario. This increase goes from stage to stage. From 2000 to 2008: the unset water demand is the same for all the three scenarios. From 2008 to 2020: the unset water demand for high growth rate scenario increases from 400 million m^3 to 600 million m^3 (50% increase), and the reference scenario remains on average of 450 million m^3. From 2025 to 2050, unset demand will reach maximum volume of about 1,000 million m^3.

There is a trend of increasing unset water demand because the dam's storage capacity is not enough to compensate for longer droughts and even the groundwater reserves are not capable to buffer parts of the lacking surface water in all oases. The farmer's choice of enlarging cultivated field sizes lead to a growing unmet water demand. In the year 2050, the maximum unmet demand is caused by an extreme low inflow into the dam due to climate change. The unmet water demand reflects the interactions between groundwater and dam storage, as well as the cultivated field size (Johannsen et *al.*, 2016).

The modern irrigation scenario (Figure 16), with the introduction of new technologies in agriculture will decrease unmet demand water in a very significant way. For the 2013 - 2030 horizon, the reduction will be 25% of water resources, which means 100 Mm^3 per year.

Figure 15. Water demand scenarios of irrigated area in Tafilalet (1970-2050), in millions of m^3 *(Ben Salem et al., 2011)*

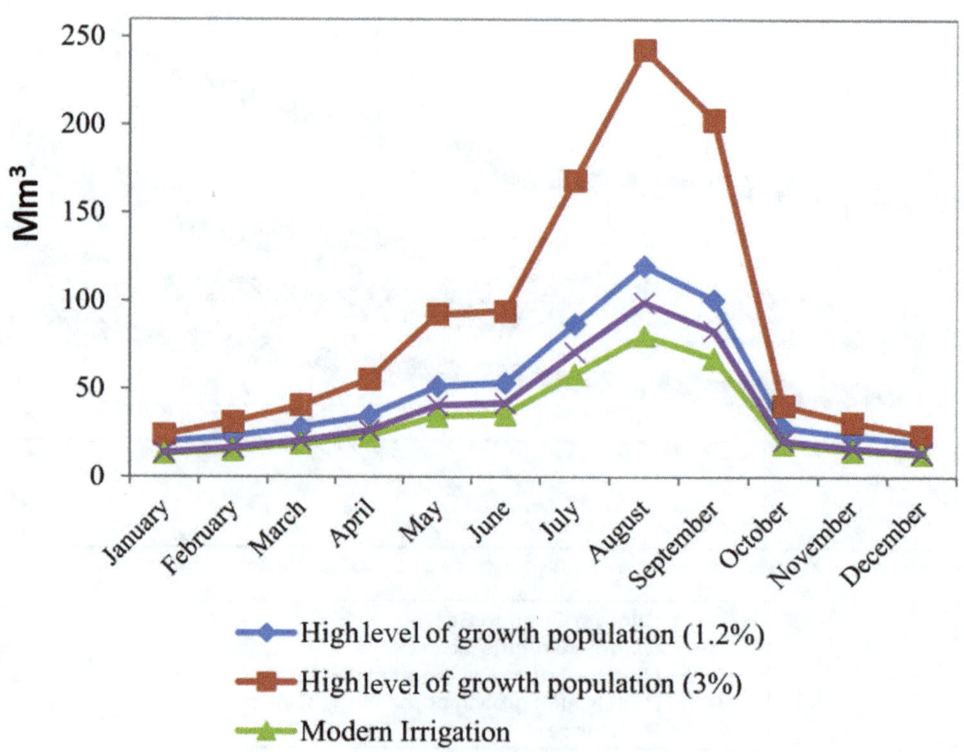

Figure 16. Scenarios of unmet water demand in the Tafilalet area in millions of m³
(Ben Salem, 2014)

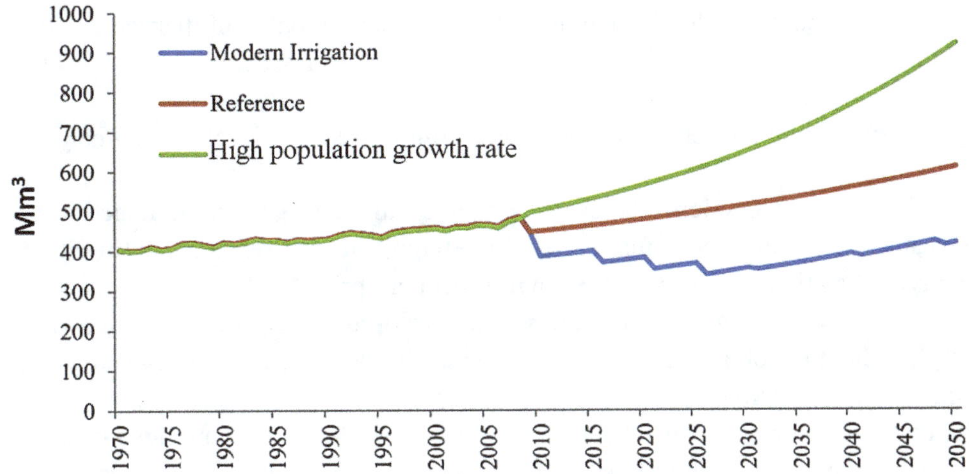

The saving of dam water by the new techniques is still not enough to satisfy the crops' water demand. Still, additional groundwater has to be used. The saving of water by new irrigation methods must be efficient enough to satisfy irrigational purposes in 'regular dry' years. The aim should be to use groundwater only in very dry years in order to give the aquifers time to recover and to provide a buffer in years of need. To reach this stable and sufficient state, the supply-demand imbalance must be overcome. Since no additional water can be provided, the water needed must be reduced (Hellegers et *al*., 2013), either by installing better irrigation techniques or by reducing field size.

MANAGEMENT OF CLIMATE CHANGE IMPACTS ON OASIS ECOSYSTEMS

With a pessimistic assumption of a climatic runaway in the future, accompanied by a greater worsening of drought in duration and intensity, the oases, already very vulnerable space, will therefore be the first to be threatened by climate change. Different strategies have been adopted to mitigate the adverse effect of climate change on the oasis ecosystem.

Oasis Rehabilitation Programs

In November 2000 and to preserve the heritage of the oases, the Ministry of Agriculture in collaboration with UNESCO through the partnership with the National Committee of Man and the Biosphere Programme (MAB), have launched the realization of the Southern Oases Biosphere Reserve at the scale of two major natural and historical regions of Drâa and Tafilalet (Khardi, 2003).

The integration of oases regions into the network of biosphere reserves allows:

- To reduce the loss of biodiversity.
- Improve people's livelihoods.

- Promote the social, economic and cultural conditions essential to the sustainability of sustainable development.
- And thus, contribute to the Millennium Development Goals, in particular to sustainable development.

To this purpose, biosphere reserves combine three complementary functions, namely:

1. The conservation of ecosystems, landscapes, species and their genetic heritage must be ensured, both in natural areas and those that are exploited agroforestry, fishing, hunting, tourism or any other activity. Practices that respect the environment are privileged.
2. Reserves also play a role in economic and social development respecting nature and local culture. This implies that the population takes an active part in the sustainable management of territories and is involved in decision-making.
3. Finally, particular importance is attached to research, studies and continuous observation of the environment, training and public education - especially young people - as they provide real support to consider in a more enlightened way on the future of the territory and its inhabitants.

Another initiative concerned the launch of these oases rehabilitation programs, such as the one initiated by the Mohammed VI Foundation for the protection of the environment for the benefit of the Marrakesh palm grove. This project has resulted in the planting of 561,000 palm trees and the maintenance of 81,000 adults in this palm grove (Meddich et *al.*, 2017).

Biotechnology on Agriculture

It is in the oasis that ingenious processes of biotechnology and agronomy have been invented and applied for centuries. The best known is the binary or ternary strata concept (Toutain, 1979; Resagro, 2016). It combines low crops (market gardening/forage), sometimes with a second stage of arboriculture, all protected by a biological date palm interface, resistant to the pressure of the surrounding dry and windy climate (role of windbreaker and temperature and humidity control device).

As part of the Green Morocco Plan for the revival of agriculture, the oases are already heavily involved in the use of biotechnology and agricultural technology in general, with the aim of improving the use of available water resources. Recent projects of the concrete rosebush and medicinal and aromatic plants in the Dades Valley, equipped with drip and tomato greenhouse projects in Dakhla, with indeterminate varieties very rewarding unit of water volume consumed (250-300 T/ha) are some examples. However, the largest project in Morocco remains the program to accelerate the rehabilitation of palm groves in the Oriental, Souss-Massa, Goulimine-Oued Noun and Draa-Tafilalet (Table 2). Thanks to the vitroplant production laboratories, 2.9 million plants are planned by 2020, with Mejhoul as the leading variety because of its quality and its high price on the market, and also the other varieties of slightly lesser quality but tolerant to Bayoud disease, such as Boufegouss, Bousekri, Nejda, Elfayda, Degulet Nour, Soukkari, NP3, NP4 (Guennouni, 2012; Sedra, 2012).

Table 2. Program of rehabilitation and development of oases in Morocco

Total Area	Regions Concerned	Global Investment	Variety	Total Number of Plants	Deadlines Production	Production Expected in 2020	Markets
471,000 Km^2	4 of which Draa-Tafilalet	700 million €	Mejhoul, Boufegouss, Bousekri,…	2.9 millions	2020	160,000 T	Morocco and Export

(Resagro, 2016)

Reflected Choice of Cultures

For different aforementioned reasons, cereal monoculture for grain must continue in the oases, in cohabitation with livestock. The hope of preserving the sustainability of the system in place against a possible even more urgent drought seems to be more focused on biotechnology and genetic improvement than on agronomy. The first alternative would consist in procuring, if they already exist in the current catalogs, varieties of early cereals particularly resistant to drought or failing to engage the research to create some. Very short-cycle Y-Rojo wheat, imported from Latin America in the 1990s for testing would probably be suitable for these regions, although with a red appearance, it is little appreciated by the consumer. On the agronomic level, there is also the idea of the use of silica gels or potassium polyacrylate for the improvement of the useful soil reserve, but the very expensive cost of these products is incompatible with the small margin cereals in arid regions.

In the absence of the first alternative above, a second one would be a total reconversion of these areas to the benefit of fodder production with introduction instead of wheat for grain, more drought-tolerant species such as barley, triticale, oats, sown densely to produce earlier exploitable biomass, either directly or in immature form as hay. This is a difficult solution that must be accepted by the producer and proven to be economically more profitable.

In the arid regions, in particular, there has always been and there would be even more episodes of food shortage of 4-5 years in a row to the image of 1980-1985, where a third alternative for the safeguarding of livestock is inevitable. Traditionally, it is the use of xerophilic plants such as cactus, saw palmetto, asphodel but also the argan or green oak, when we are close to the forest estate. Having been used to these crises for a very long time, the state has already somewhat advanced the events by planting as demonstration, large areas of rocky land of cactus, resource usable in industry certainly, but also as animal feed in the ultimate cases. However, the question remains how to make it a broader program among the populations who, for the moment, do not give the impression of being convinced that we are on the eve of a real climate change where we would have to prepare on more draconian conditions than in the past.

Managing Water Scarcity

In terms of sustainability, it is important to distinguish between the surface water from which the oasis valleys benefit and the aquifer water around these valleys. The first is a renewable resource; its exploitation raises a priori no particular objection regarding its sustainability except to adjust the cultivation programs to the decline in availability that climate change will have produced in the future, in each oasis. The same is not true of aquifers where biotechnology and agricultural technology in general (micro-irrigation, fertigation, protection, etc.) are likely to be a double-edged sword. While providing the producer with excellent commercial varieties, increasing significantly productivity per hectare and improving water efficiency, there is concern that they may also be accelerating factors of groundwater depletion in the case of a mining exploitation like in the Guerdane area, in the Souss valley, 30 years ago (Nakhli, 2011, Kedima, 2014). This water is of particular value, either because the aquifer does not hide much of it, either because it is fossil and non-renewable, or because the aquifer contains an appreciable quantity of water, but represents a precious capital of the water ecosystem that makes it deserve a special valuation effort under surveillance. After Zagora, in the Drâa oasis, where the state has already decreed the stopping of major water consumer cultures like the early watermelon, given the scarcity of water, it is towards the valley of Erfoud and surroundings that the eyes are now turned as a result of the thousands of ha of date palm programmed around these oases (Resagro, 2016).

CONCLUSION

Water is the basic resource for survival in oases areas. However, this scarce resource is subjected to global, local and regional drastic changes. In the two last decades, the balance between supply and demand for water has become more difficult. Effectively, in response to climate change affecting the oases, the extracted water volumes are constantly increasing. This resulted in lower water levels and higher concentrations of salt in farmland and thus a decline in agricultural production. The gained insights into the oases future water demand and supply show that the way farming performed today in these areas will not work out in future and, therefore, needs management solutions in order to preserve water and food resources. This situation requires a review of adopted policies by integrating a global vision that takes into account all the components of the oasis ecosystem and more specifically the water resource as the first factor of production. Morocco communicated a lot about climate change at the 22nd COP United Nations Climate Change Conference (COP22) held on its territory in November 2016. Sustainable Oasis is one of the main initiatives being supported by Morocco for the (COP22), a program designed to get to the very core of climate change issues, to address the key problems and provide support by helping oases to adopt climate resilient practices for the future. Today, the debate should focus on important issues such as the new environmental responsibility of the citizens, their future relationship to water resources, food resources, energy resources, ... and, in a general way to the restrictive new way of life that climate change is likely to impose on it.

REFERENCES

Adger, W. N. (1999). Social vulnerability to climate change and extremes in coastal Vietnam. *World Development*, *27*(2), 249–269. doi:10.1016/S0305-750X(98)00136-3

Agoumi, A. (2003). *Vulnerability of North African countries to climatic changes: adaptation and implementation strategies for climatic change. In Developing Perspectives on climate change: Issues and analysis from developing countries and countries with economies in transition.* IISD/Climate Change Knowledge Network.

Agoumi, A., & Debbarh, A. (2006). Ressources en eau et bassins versants du Maroc: 50 ans de développement (1955-2005). *50 Ans de Développement Humains. Perspectives*, 2025.

Ait Houssa, A., Drissi, S., Asehraou, A., Asfers, A., Oubaki, L., & Chraibi, H. (2017). Changements climatiques au Maroc: Quels systèmes de culture et quelles biotechnologies pour s'y adapter? *Revue Marocaine des Sciences Agronomiques et Vétérinaires*, *5*(3).

Ait Lamqadem, A., Saber, H., & Rahimi, A. (2017). Spatiotemporal Changes of Vegetation in the Middle Draa Valley Oasis: A Study Case of M'hamid El Ghizlane Oasis (Morocco). *European Scientific Journal, 13*(24).

Almeida-Garcıa, F., & Chahine, S. (2015). Morocco, tourism. In Encyclopedia of Tourism (pp. 815–816). Academic Press.

Aus der Beek, T., Voß, F., & Flörke, M. (2011). Modelling the impact of global change on the hydrological system of the Aral Sea basin. *Physics and Chemistry of the Earth Parts A/B/C*, *36*(13), 684–695. doi:10.1016/j.pce.2011.03.004

Babqiqi, A., Kassami, A., & Messouli, M. (2011). Development of high resolution climate change scenarios for Morocco for the 21st century. In *Geoinformatics for Climate Change Studies*. New Delhi: TERI Press.

Bell, I., & Visbeck, M. (n.d.). *North Atlantic Oscillation*. Retrieved from https://www.ldeo.columbia.edu/res/pi/NAO/

Ben Salem, A. (2014). *Vulnerability and adaptation to climate change in the oasis region Tafilalet-Morocco* (Doctoral dissertation). Department of Biology, Faculty of Semlalia, Cadi Ayyad University, Marrakesh.

Ben Salem, A., Messouli, M., & Yacoubi-Khebiza, M. (2011). *Developing an oasis-based water management tool: Ecohydrologic approach and weap software for a large arid catchment in Morocco*. Academic Press.

Born, K., Fink, A. H., & Paeth, H. (2008). Dry and wet periods in the northwestern Maghreb for present day and future climate conditions. *Meteorologische Zeitschrift*, *17*(5), 533–551. doi:10.1127/0941-2948/2008/0313

Botes, A., & Zaid, A. (2002). The economic importance of date production and international trade. In A. Zaid & E. J. Arias-Jimenez (Eds.), *Date palm cultivation. FAO plant production and protection paper, 156 rev. 1*. Rome: FAO.

Bouhlassa, S., & Paré, S. (2006). Evapotranspiration de référence dans la région aride de Tafilalet au sud-est du Maroc. *African Journal of Environmental Assessment and Management, 11*, 1-16.

Bouhlassa, S., & Paré, S. (2007). Effect of climatic parameters on reference evapo-transpiration in the tafilalet arid area. *Physical and Chemical News, 12*(38), 1-6.

Bousfoul, M. (2008). Gestion intégrée des ressources en eau. *Revue Homme terre et eau, 140*, 80-83.

Brancic, B. (1968). *Sols de la Palmeraie de Fezouata*. Office Régional de Mise en Valeur Agricole Ouarzazate.

Christoph, M., Fink, A. H., Paeth, H., Born, K., Kerschgens, M., & Piecha, K. (2010). *Climate scenarios. In Impacts of global change on the hydrological cycle in west and northwest Africa* (pp. 402–425). Heidelberg, Germany: Springer.

De Grenade, R. (2013). Date palm as a keystone species in Baja California peninsula, Mexico oases. *Journal of Arid Environments*, *94*, 59–67. doi:10.1016/j.jaridenv.2013.02.008

Dong, W., Cui, B., Liu, Z., & Zhang, K. (2014). Relative effects of human activities and climate change on the river runoff in an arid basin in northwest China. *Hydrological Processes*, *28*(18), 4854–4864. doi:10.1002/hyp.9982

Driouech, F., Déqué, M., & Mokssit, A. (2009). Numerical simulation of the probability distribution function of precipitation over Morocco. *Climate Dynamics*, *32*(7-8), 1055–1063. doi:10.100700382-008-0430-6

Driouech, F., Déqué, M., & Sánchez-Gómez, E. (2010). Weather regimes—Moroccan precipitation link in a regional climate change simulation. *Global and Planetary Change*, *72*(1-2), 1–10. doi:10.1016/j.gloplacha.2010.03.004

Endlicher, W. (2000). Mittelmeerländer. In Regionale Klimatologie: Die Alte Welt: Europa, Afrika, Asien. Teubner.

FAO. (2018). FAO water reports: Drought characteristics and management in North Africa and the Near East. Rome: FAO.

Fertah, M., Belfkira, A., Taourirte, M., & Brouillette, F. (2017). Extraction and characterization of sodium alginate from Moroccan Laminaria digitata brown seaweed. *Arabian Journal of Chemistry*, *10*, S3707–S3714. doi:10.1016/j.arabjc.2014.05.003

Fink, A., Christoph, M. H., Paeth, H., Born, K., Kerschgens, M., & Piecha, K. (2010). *Climate. Impacts of global change on the hydrological cycle in west and northwest Africa*. Heidelberg, Germany: Springer.

Gerstengarbe, F. W., & Werner, P. C. (2007). Der rezente Klimawandel. In Der Klimawandel – Einblicke, Rückblicke und Ausblicke (pp. 34–43). Humboldt University.

Ghallabi, L. B., Messouli, M., & Yacoubi, M. (2011). Integrated approaches to the assessment of the impacts of climate and socio-economic change on groundwater resources in the Tensift basin, Morocco. *Int. J. Water Resour. Arid Environ*, *1*, 219–225.

Gibelin, A. L., & Déqué, M. (2003). Anthropogenic climate change over the Mediterranean region simulated by a global variable resolution model. *Climate Dynamics*, *20*(4), 327–339. doi:10.100700382-002-0277-1

Giorgi, F., & Bi, X. (2005). Regional changes in surface climate interannual variability for the 21st century from ensembles of global model simulations. *Geophysical Research Letters*, *32*(13), L13701. doi:10.1029/2005GL023002

Giorgi, F., & Lionello, P. (2008). Climate change projections for the Mediterranean region. *Global and Planetary Change*, *63*(2-3), 90–104. doi:10.1016/j.gloplacha.2007.09.005

Glueck, M. F., & Stockton, C. W. (2001). Reconstruction of the North Atlantic oscillation, 1429–1983. *International Journal of Climatology: A Journal of the Royal Meteorological Society, 21*(12), 1453-1465.

Guemimi, A. (1991). Drainage des sols salés dans le Tafilalet. *Revue Homme terre et eau*, 47-60.

Guennouni, A. (2012). Filière phoénicicole: Des acquis à consolide. *Agriculture du Maghreb*, *62*, 57–59.

Haddouch, M. (1996). Situation actuelle et perspectives de développement du palmier dattier au Maroc. *Options Mediterr*, *28*, 63–79.

Heidecke, C. (2009). *Economic Analysis of Water Use and Management in the Middle Drâa Valley in Morocco* (Ph.D. Thesis). University of Bonn, Bonn, Germany.

Heidecke, C., & Schmidt, T. (2008). Agricultural Land Use. In Impetus Atlas Morocco: Research Results 2000–2007 (pp. 67–68). University of Bonn.

Hellegers, P., Immerzeel, W., & Droogers, P. (2013). Economic concepts to address future water supply–demand imbalances in Iran, Morocco and Saudi Arabia. *Journal of Hydrology (Amsterdam)*, *502*, 62–67. doi:10.1016/j.jhydrol.2013.08.024

Herrera, R. G., Puyol, D. G., MartÍn, E. H., Presa, L. G., & Rodríguez, P. R. (2001). Influence of the North Atlantic oscillation on the Canary Islands precipitation. *Journal of Climate*, *14*(19), 3889–3903. doi:10.1175/1520-0442(2001)014<3889:IOTNAO>2.0.CO;2

Hertig, E. (2004). *Niederschlags und Temperaturabschätzungen für den Mittelmeerraum unter anthropogen verstärktem Treibhauseffekt*. Wuerzburg: University of Wuerzburg.

Hertig, E., & Jacobeit, J. (2008). Downscaling future climate change: Temperature scenarios for the Mediterranean area. *Global and Planetary Change*, *63*(2-3), 127–131. doi:10.1016/j.gloplacha.2007.09.003

Hu, G., & Jia, L. (2015). Monitoring of evapotranspiration in a semi-arid inland river basin by combining microwave and optical remote sensing observations. *Remote Sensing*, *7*(3), 3056–3087. doi:10.3390/rs70303056

Hübener, H., Schmidt, M., Sogalla, M., & Kerschgens, M. (2005). Simulating evapotranspiration in a semi-arid environment. *Theoretical and Applied Climatology*, *80*(2-4), 153–167. doi:10.100700704-004-0097-9

Hurrell, J. W., & Van Loon, H. (1997). Decadal variations in climate associated with the North Atlantic Oscillation. *Climatic Change*, *36*(3/4), 301–326. doi:10.1023/A:1005314315270

IRATE. (2006). *Atlas*. Région Meknès-Tafilalet, Inspection Régionale de l'Aménagement du Communication Territoire et de l'Environnement, Edition.

Jacobeit, J. (1994). Empirical estimations of Mediterranean rainfall changes in the transitional seasons due to enhanced greenhouse warming. In R. Brazdil & M. Kolnar (Eds.), *Contemporary Climatology (pp. 266–71)*. Academic Press.

Jarsj, Ö. J., Asokan, S. M., Shibuo, Y., & Destouni, G. (2008). Water Scarcity In The Aral Sea Drainage Basin: Contributions Of Agricultural Irrigation Anda Changing Climate. In Environmental Problems of Central Asia and their Economic, Social and Security Impacts (pp. 99-108). Springer.

Johannsen, I. M., Hengst, J. C., Goll, A., Höllermann, B., & Diekkrüger, B. (2016). Future of water supply and demand in the middle Draa Valley, Morocco, under climate and land use change. *Water (Basel), 8*(8), 313. doi:10.3390/w8080313

Kabiri, L. (2004). *Contribution à la connaissance, la préservation et la valorisation des Oasis du Sud marocain: cas de Tafilalt*. Thèse d'habilitation universitaire, Facultés des Sciences et Techniques, Errachidia, Université My Ismaïl, Maroc.

Karmaoui, A., Ifaadassan, I., Messouli, M., & Khebiza, M. Y. (2015). *Characterization of Common Environmental Indicators of the Moroccan Oasean Biome, Pilot Study in the Reserve Biosphere of Oases in Southern Morocco*. Adv in Res. doi:10.9734/AIR/2015/19369

Karmaoui, A., Messouli, M., Khebiza, Y. M., & Ifaadassan, I. (2014). Environmental Vulnerability to climate change and anthropogenic impacts in dryland,(Pilot study: Middle Draa Valley, South Morocco). *Journal of Earth Science & Climatic Change*, (S11), 1.

Keddal, H. (2006). *National study on date palm irrigation and associated crops in the kingdom of morocco*. FAO.

Kedima. (2014). *Rapport sur la situation de la nappe à Guerdane*. Rapport Interne. Providence verte. Rabat.

Khardi, A. (2003). Aperçu des problèmes d'environnement pour l'agriculture oasienne. *Revue Homme terre et eau*, 127.

Klose, A. (2009). *Soil Characteristics and Soil Erosion by Water in a Semi-arid Catchment (Wadi Drâa, South Morocco) under the Pressure of Global Change* (Dissertation). University of Bonn, Bonn, Germany.

Klose, S. (2012). *Regional Hydrogeology and Groundwater Budget Modeling in the Arid Middle Drâa Catchment* (Dissertation). University of Bonn, Bonn, Germany.

Knippertz, P., Christoph, M., & Speth, P. (2003a). Long-term precipitation variability in Morocco and the link to the large-scale circulation in recent and future climates. *Meteorology and Atmospheric Physics, 83*(1-2), 67–88. doi:10.100700703-002-0561-y

Knippertz, P., Ulbrich, U., Marques, F., & Corte-Real, J. (2003b). Decadal changes in the link between El Niño and springtime North Atlantic Oscillation and European–North African rainfall. *International Journal of Climatology: A Journal of the Royal Meteorological Society, 23*(11), 1293-1311.

Kuhn, A., Heidecke, C., Roth, A., Goldbach, H., Burkhardt, J., Linstädter, A., ... Gaiser, T. (2010). *II-5.1 Importance of resource management for livelihood security under Climate Change in Southern Morocco*. Academic Press.

Lionello, P., Malanotte-Rizzoli, P., & Boscolo, R. (Eds.). (2006). *Mediterranean climate variability* (Vol. 4). Elsevier.

Ma, L., Wu, J., Liu, W., & Abuduwaili, J. (2014). Distinguishing between anthropogenic and climatic impacts on lake size: A modeling approach using data from Ebinur Lake in arid northwest China. *Journal of Limnology, 73*(2). doi:10.4081/jlimnol.2014.852

Margat, J. (1962). *Mémoire explicative de la carte hydrogéologique au 1/50000 de la plaine de Tafilalet*. Academic Press.

Meddi, M. M., Assani, A. A., & Meddi, H. (2010). Temporal variability of annual rainfall in the Macta and Tafna catchments, Northwestern Algeria. *Water Resources Management*, *24*(14), 3817–3833. doi:10.100711269-010-9635-7

Meddich, A., Ait El Mokhtar, M., Bourzik, W., Mitsui, T., Baslam, M., & Hafidi, M. (2018). Optimizing growth and tolerance of date palm (Phoenix dactylifera L.) to drought, salinity, and vascular fusarium-induced wilt (Fusarium oxysporum) by application of arbuscular mycorrhizal fungi (AMF). In *Root Biology* (pp. 239–258). Cham: Springer. doi:10.1007/978-3-319-75910-4_9

Meddich, A., Ait El Mokhtar, M., Wahbi, S., & Boumezzough, A. (2017). Évaluation des potentialités mycorhizogènes en lien avec les paramètres physico-chimiques des sols de palmeraies du Maroc (Marrakech et Tafilalet). *Cahiers Agricultures*, *26*(4), 45012. doi:10.1051/cagri/2017044

Ministry of Environment of Morocco, (2001). *Rapport de l'Etat de l'environnement du Maroc, 292*. Author.

Nakhli, F. (2011). Ressources en eau/Souss: menaces sur la nappe phréatique. *L'économiste,* 3547.

Paeth, H., Born, K., Girmes, R., Podzun, R., & Jacob, D. (2009). Regional climate change in tropical and northern Africa due to greenhouse forcing and land use changes. *Journal of Climate*, *22*(1), 114–132. doi:10.1175/2008JCLI2390.1

Palutikof, J. P., & Wigley, T. M. L. (1996). Developing climate change scenarios for the Mediterranean region. In Climatic Change and the Mediterranean (pp. 27–56). Academic Press.

Patricola, C. M., & Cook, K. H. (2010). Northern African climate at the end of the twenty-first century: An integrated application of regional and global climate models. *Climate Dynamics*, *35*(1), 193–212. doi:10.100700382-009-0623-7

Pauling, A., Luterbacher, J., Casty, C., & Wanner, H. (2006). Five hundred years of gridded high-resolution precipitation reconstructions over Europe and the connection to large-scale circulation. *Climate Dynamics*, *26*(4), 387–405. doi:10.100700382-005-0090-8

Platt, S. (2008). Current Development of the Population in the Provinces of Ouarzazate and Zagora. In Impetus Atlas Morocco: Research Results 2000–2007 (pp. 59–60). University of Bonn.

PNUED. (2013). Ecosystem Assessment of Tafilalet Oasis. *Synthesis*, 1-24.

Renevot, G., Bouaziz, A., & Raki, T. R. (2010). Pratiques d'irrigation du palmier dattier dans les systèmes oasiens du Tafilalet, Maroc. *Revue Homme terre et eau, 146*.

Resagro. (2016). Le mensuel des décideurs. *Le palmier dattier face aux changements climatiques, 71*, 27-29.

Rowell, D. P. (2005). A scenario of European climate change for the late twenty-first century: Seasonal means and interannual variability. *Climate Dynamics*, *25*(7-8), 837–849. doi:10.100700382-005-0068-6

Royaume du Maroc. (2010). *Seconde communication nationale à la Convention-Cadre des Nations Unies sur les Changements Climatiques*. Rabat: Ministère de l'Environnement.

Saaidi, M. (1992). Comportement au champ de 32 cultivars de palmier dattier vis-à-vis du bayoud: 25 années d'observations. *Agronomie*, *12*(5), 359–370. doi:10.1051/agro:19920502

Schilling, J., Freier, K. P., Hertig, E., & Scheffran, J. (2012). Climate change, vulnerability and adaptation in North Africa with focus on Morocco. *Agriculture, Ecosystems & Environment*, *156*, 12–26. doi:10.1016/j.agee.2012.04.021

Schulz, O. (2006) *Analyse schneehydrologischer Prozesse und Schneekartierung im Einzugsgebiet des Oued M'Goun, Zentraler Hoher Atlas (Marokko)* (Dissertation). Mathematisch-Naturwissenschaftliche Fakultät, Universität Bonn.

Sedra, M. H. (2012). Bayoud et charançon rouge: Les ennemis les plus dangereux du palmier dattier. *Agriculture du Maghreb*, *62*, 62–65.

Sraïri, M. T., M'ghar, F. A., Benidir, M., & Bengoumi, M. (2017). Analyse typologique de la diversité et des performances de l'élevage oasien. *Cahiers Agricultures*, *26*(1), 15005. doi:10.1051/cagri/2017002

Tabet-Aoul, M. (2008). Impacts du changement climatique sur les agricultures et les ressources hydriques au Maghreb. *Les Notes d'Alerte du CIHEAM*, (48).

Touchan, R., Anchukaitis, K. J., Meko, D. M., Sabir, M., Attalah, S., & Aloui, A. (2011). Spatiotemporal drought variability in northwestern Africa over the last nine centuries. *Climate Dynamics*, *37*(1-2), 237–252. doi:10.100700382-010-0804-4

Toutain, G. (1979). Eléments d'agronomie saharienne: de la recherche au développement. *INRA-GRET*, 276.

Wanner, H., Brönnimann, S., Casty, C., Gyalistras, D., Luterbacher, J., Schmutz, C., ... Xoplaki, E. (2001). North Atlantic Oscillation–concepts and studies. *Surveys in Geophysics*, *22*(4), 321–381. doi:10.1023/A:1014217317898

Ward, M. N., Lamb, P. J., Portis, D. H., El Hamly, M., & Sebbari, R. (1999). Climate variability in Northern Africa: Understanding droughts in the Sahel and the Maghreb. In *Beyond el Niño* (pp. 119–140). Berlin: Springer. doi:10.1007/978-3-642-58369-8_6

WB/MAPM, FAO, INRA, & DMN. (2008). Workshop on WB/Morocco/FAO Climate Change Study. Rabat: Author.

Weber, B. (2004). *Untersuchungen zum Bodenwasserhaushalt und Modellierung der Bodenwasser-flüsse entlang eines Höhenund Ariditätsgradienten (SE Marokko)* (Dissertation). Mathematisch-Naturwissenschaftliche Fakultät. Universität Bonn.

Xoplaki, E. (2002). *Climate variability over the Mediterranean* (PhD thesis). University of Bern, Switzerland.

Yang, G., Ye, B., Xie, X., & Zhou, L. (2010, July). Dynamics of oasis landscape in inland Shule River basin in arid northwest China. In *Geoscience and Remote Sensing Symposium (IGARSS), 2010 IEEE International* (pp. 922-925). IEEE. 10.1109/IGARSS.2010.5653756

KEY TERMS AND DEFINITIONS

Adaptation: The process of modifying an ecosystem so as to remain functional under new conditions, as well as the result of this process.

Bayoud: A fungal disease of date palm caused by *Fusarium oxysporum f. sp. albedinis (Foa).*

Climatic Scenario: Projection of future greenhouse gas (GHG) emissions used as an input for studying the potential impacts of climate change.

Climate Change: Refers to all variations in climatic parameters of the global climate of the earth or its various regional climates over time.

COP22: 22nd COP United Nations Climate Change Conference.

Equilibrium of Ecosystems: Is an equilibrium reached when living beings live in harmony with their environment, they use it and modify it by exchanging with each other and with the environment, energy, and matter.

GCMs: General circulation models.

IPCC: Intergovernmental Panel on Climate Change.

MAB: Man and the Biosphere Programme.

MIOS: Modèle Intégré des Oasis du Sud.

MIVAD: Modèle Integrée du Vallée du Drâa.

NAO: North Atlantic oscillation.

Oases: An ecosystem consisting of a spot of isolated vegetation in a desert found near a water source.

REMO: The dynamical regional model.

Sustainability: The ability to be maintained at a certain rate or level over a period of time.

UNFCCC: United Nations Framework Convention on Climate Change.

Vulnerable Area: An area with low socioeconomic and environmental status which make it threatened by degradation.

WEAP: Water evaluation and planning system.

Section 4

Climate Change, Ecosystem Services, Food Sercurity, and Health

Chapter 13
Ecosystem Services, Climate Change, and Food Security

Samreen Siddiqui
https://orcid.org/0000-0001-8238-384X
Texas A&M University – Corpus Christi, USA

ABSTRACT

Food security is a rising concern around the world, especially in developing countries within arid and semi-arid regions. Ecosystem provides different services to support living and human survival, which includes some major food sources around the world like agriculture, fisheries, and livestock. With advancing times, humans improved these services and produced enough food to support the rising population. However, with increasing greenhouse gases, a new problem came into existence, commonly known as climate change (CC), which accelerated issues like food security and safety. Under such issues, people don't have access to basic facilities and food supply to survive, and with future population growth estimates, it is becoming even more difficult. Some major food sources together with alternate sources are discussed in this chapter. Upcoming CC impacts are discussed in detail in relation to major food sources and supported by world maps to provide a better picture. Major actions, government initiatives, and some suggestions are also provided to overcome this global crisis.

INTRODUCTION

Ecosystem Services (ES) are the benefits people obtain from the ecosystem itself. These include food and water (provisioning services); floods, droughts, degradation (regulating services); soil formation, nutrient cycles (supporting services) as well as recreational, spiritual and religious benefits (cultural services) (Fig.1). These services are regulated by overall biosphere, lithosphere, atmosphere and anthrosphere cycles. At present, anthrosphere causing a huge influence on atmosphere and environmental processes by increasing pollution, like ocean acidification and global warming, commonly, known as climate change. These effects causing impacts on ecosystem services at various levels. The major point of concern with increasing global population is food security, supported as provision services under a broad range of ecosystem services.

DOI: 10.4018/978-1-5225-7387-6.ch013

ES are under major threat in Arid and semi-arid (ASA) regions due to limited water and agriculture resources. ASA regions are characterised by low average rainfall (300-600mm), average high temperatures, resulting in fundamental limits on animal and plant population together with agriculture production (CSIRO 2011; Ludwig & Asseng, 2006; Ribot, Magalhães, & Panagides, 2005; John, Pannell, & Kingwell, 2005; Vörösmarty, Green, Salisbury, & Lammers, 2000). It covers major areas in the United States, Argentina, Southern Africa, North Africa, the Sahel, West, and Central Asia and Australia (Fig.1). ASA regions represent 15.7% and 22.6% global land area, respectively and together support 18.5% of the global population (UN, 2010).

The natural variability of climate together with anthropogenic impacts induced changes influencing the climate. Temperature and precipitation demonstrated a strong change in last decades as shown in Fig. 2. Natural climate variability includes the changes introduced by the short term external forces (Volcanic activity, short-term changes in solar radiation) as well as long-term external forces (tectonic movement, solar radiation, changes in the Earth's orbit and asteroid bombardment) of nature. Whereas anthropogenic climate change can include the human-influenced changes causes the emission of greenhouse gases by burning fossil fuels, deforestation and land use changes ultimately lead to warming of earth's surface (IPCC, 2007; 2013). Both types of climate variation when interacting with each other causes climate change (CC), which ultimately affect the global system.

Figure 1. Pie chart explaining a) Percentage share of different types of Arid regions globally; b) Arid regions globally by continental share; c) Semi- Arid region globally by continental share
(Data Source: UNSO/UNDP, 1997)

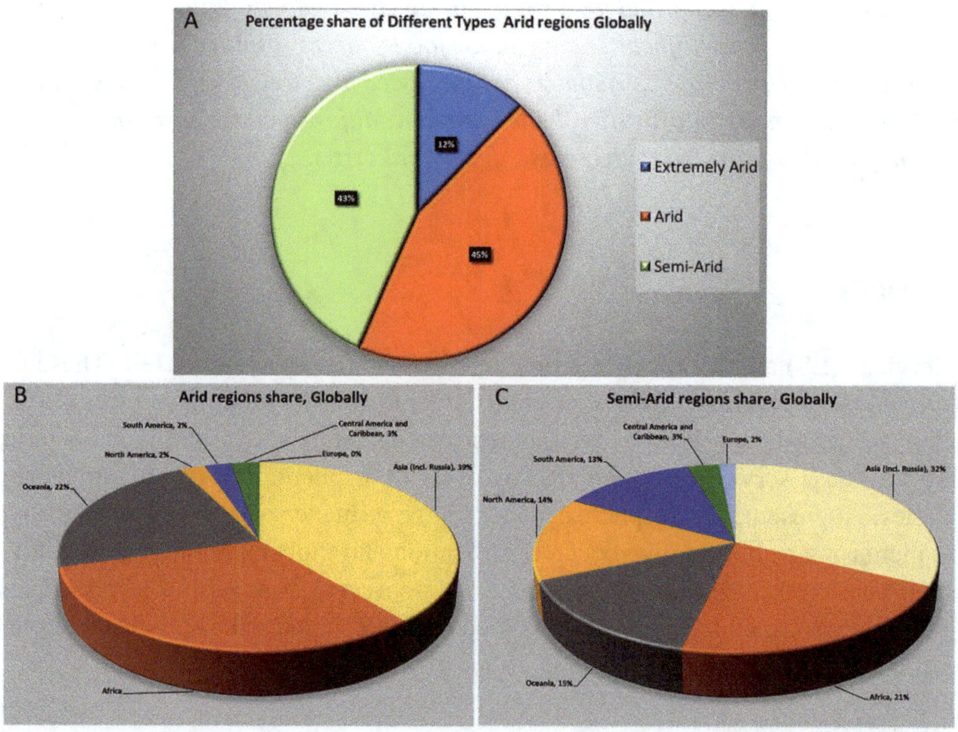

Source: UNSO/UNDP, 1997

Figure 2. a) world map representing annual mean temperature from 1960-1990; b) World map representing annual precipitation from 1960-1990
(Data Source: World Climate Organisation)

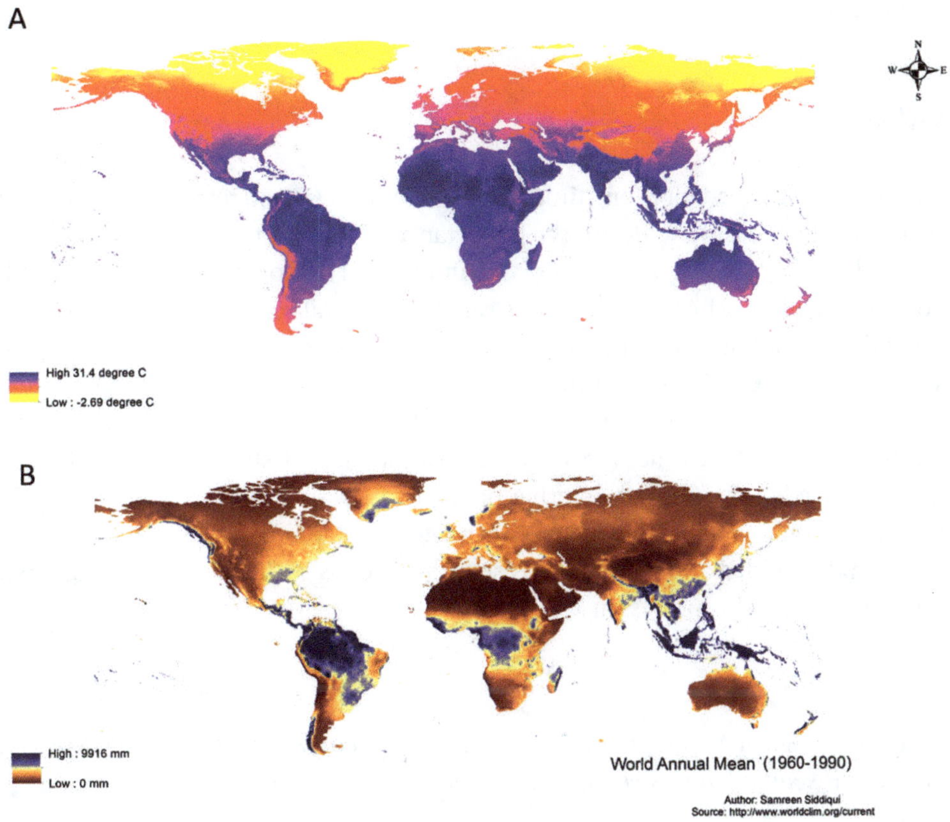

CC may cause several issues including food shortage leading towards food security issues around the world. World health organization (WHO, 2012) describes food security as "when all people at all times have access to sufficient, safe, nutritious food to maintain a healthy and active life". Major sectors to ensure food security are agriculture, livestock, and fisheries (aquaculture and capture fisheries) (Rosegrant & Cline, 2003). Based on this classification potential food sectors discussed further in the chapter are agriculture, capture fisheries, aquaculture, and livestock. All the above-discussed food sources come under the Provisions of ES and require proper care and attention towards sustainability. One of the prominent effects of CC is changing temperature and shifting seasons, which is causing issues with insect's life cycles and ultimately resulting in decreasing ES regulating and supporting. All the ES works together to support life on the earth. The environmental risks from the CC is the possible loss of ES. One such case study was performed on the South American coast to understand it in more detail, which has been discussed later in this chapter (Asmus, Nicolodi, Anello, & Gianuca, 2017). ES can provide a food security through supporting grazing cattle and wild animals. Which is also under threat due to the ongoing impacts of climate change as studied in the grassland and agro-pastoral transitional zone of China (Hao & Yu, 2018).

This chapter will include major food sources categorised under provisional services of ES, impacts of CC on each service and their security concerns. The overall impacts and suggested measures are described through a well-defined format and concise information which can lead towards the understanding biggest cause of concern.

BACKGROUND

The concept of ecosystem existed for a long time when institutions started realising about the relationship between human and nature. The first idea of the importance of this relationship was coined in the 1940s and in 1970 when the term 'ecosystem services' was introduced (Mooney et al., 1997). The ES concept entered in scientific agenda in 21 century in policy reforms. Several scientific projects used this concept in policy formulation including, Millennium Ecosystem Assessment (MEA) in 2005, The Economics of Ecosystems and Biodiversity in 2010 (TEEB), and the establishment of the Intergovernmental Panel on Biodiversity and Ecosystem Services (IPBES) in 2012 (Chaudhary, McGregor, Houston, & Chettri, 2015; Mace, 2014). During that time, the European Union started standardising ES by preparing a uniform definition named "the Common International Classification of Ecosystem Services (CICES)" (Haines-Young & Potschin, 2011). The EU has mandated a Mapping and Assessment of Ecosystem Services (MAES) and provided attention towards securing the ES with the role of biodiversity in ES as well as defined the limits of this provision.

Figure 3. a) Ecosystem Services categorised by Millennium Ecosystem Assessment (MEA) in 2005; b) Major Food sources discussed in this chapter under provisional services

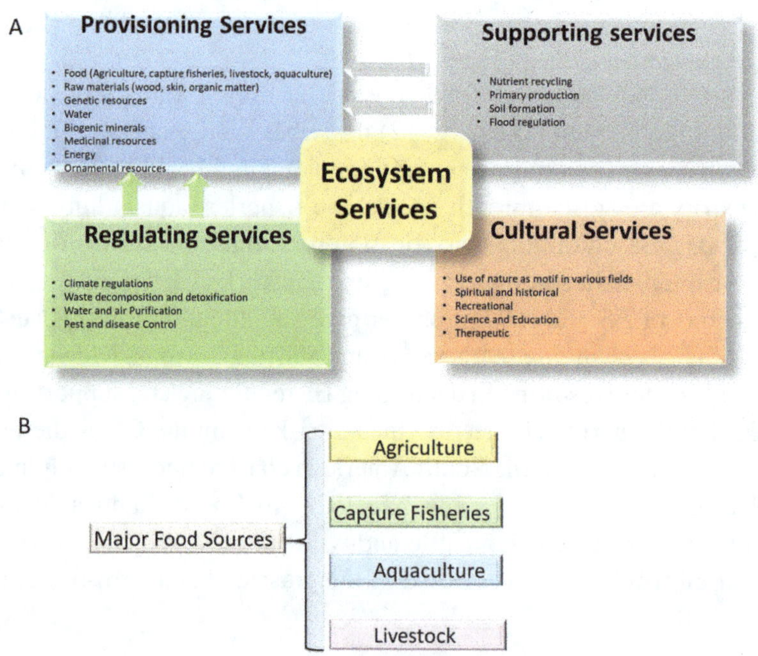

The ES concept has been implemented within the concept of planning and administration with some countries like, USA (Hansen et al., 2015), Australia, Sweden (Wilkinson, Saarne, Peterson, & Colding, 1998), Vietnam (Hanh et al., n.d.), China (Li, Ouyang, Tam, & Chen, 2008), Europe (Schleyer, Görg, Hauck, & Winkler, 2015) and Poland (Maczka et al., 2016) linked this concept to their legal system. However, the ES concept potential to improve climate mitigation actions are still needed to be explored. Some research initiatives such as OPERAs - Operational Potential of Ecosystem Research Applications (OPERAs, 2016) and OpenNESS - Operationalisation of Natural Capital and Ecosystem Services (European Centre for Nature Conservation, 2016) and the related joint knowledge hub OPPLA (OPPLA, 2015) are focusing on operationalizing the ES concept for application in praxis.

Recently, ES concepts are studied together with CC to understand and mitigate the rising environmental issues in the future. CC is a global phenomenon and includes a wide range of natural climate changes like variable tropical storm activity, accelerated sea-level rise, increasing sea surface and atmospheric temperature, ocean acidification, shifting timing and quantity of precipitation, forest fires, and littoral erosion/accretion (IPCC, 2005). IPCC (2014) defines climate change as "Any change in climate over time, whether due to natural variability or as a result of human activity". Globally, observed major impacts of climate change are:

- Since 1951, increased precipitation events within mid-latitude areas of the Northern hemisphere.
- Ocean uptake of CO_2 has increased since the industrial era resulting in decreased pH of Seawater by 0.1, a corresponding increase in acidity by 26%.
- Greenland and Arctic ice sheet have been losing a larger rate from 1992 to 2001, which increased drastically after 2002.
- The annual mean arctic sea ice extent decreased at a rate of 3.5-4.1% per decade, since 1979.
- Global mean sea level rose by 0.19 (0.17 to 0.21) m from 1901 to 2010.

According to FAO (1996), the food security is determined by three major factors: i) availability of sufficient food; ii) access to the food; iii) quality and utilisation of the food in terms of both nutritional and cultural perspectives. World hunger is on the rise, estimated undernourished people to 815 million in 2016 from 777 million in 2015 (FAO, 2017). Reports suggest that if the allocation of food is not improved now, it may cause severe food shortage with time (Rice & Garcia, 2011; OECD-FAO, 2014). The fifth assessment report of IPCC indicates that the food security is a growing concern due to the changes in global temperature and precipitation regimes which could impact different aspects of food security including access to food, its utilization and price stability (Porter et al., 2014). Food security has some direct impact together with indirect impacts from climate change. Direct impacts include an increase in temperature and change in precipitation on crops, livestock, and fishery systems. Indirect impacts include flooding, pollination, droughts and extreme events like storms on food system infrastructures (Lobell, Burke, Tebaldi, Mastrandrea, Falcon & Naylor, 2008; Porter et al., 2014).

Changing climate conditions may lead towards decreasing biodiversity and shifting community assemblages, causing threats to productivity and resilience to current food systems (Cabell & Oelofse, 2012; Oppenheimer et al., 2014; Khoury, 2014). Increasing evapotranspiration with decreasing rainfall together contributing to the desertification process. Furthermore, the World Meteorological Organization (WMO) suggested that CC together with the desertification can increase the levels of salinization and desertification of agriculture lands (UNCCD, 2007). Globally, semi-arid regions expanded by 7% during 1990-2004, compared to 1948-1962, out of which the expansion across the continents was greatest in Eastern Hemisphere (75% of total semi-arid area) followed by American continent (25% of total semi-arid area) (Huang et al., 2016).

With the rising impacts of CC, there is a need to identify alternate food sources such as fish farming or integrated fish farming. In case of fish as a food source, even a small amount of intake can supply adequate nutrition such as heat calories, protein, vitamins, unsaturated fatty acids, and carbohydrates (FAO, 2005, 2009; OECD-FAO, 2014; FAO, 2014). According to some projections, global seafood demand is expected to increase by 18% and 50% by 2020 and 2050 respectively (Rice & Garcia, 2011; OECD-FAO, 2014). However, there are rising concerns about the serious impacts of CC on fisheries and livestock production (Creighton, Hobday, Lockwood, & Pecl, 2016; Herrero et al., 2015). Food safety is also a point of concern under the CC scenario as elevated temperature can cause increased rates of microbial growth (Hammond et al., 2015), particularly in fresh fruits and vegetables (Liu, Hofstra, & Franz, 2013) as well as in fisheries supply chains (Marques, Nunes, Moore & Strom, 2010). Other than microbes, CC change may also cause increasing risks of vector-borne diseases, heat stress and natural disaster, which may affect people's nutrition and their ability to care children and dependents (Costello et al., 2009).

To conserve and efficiently produce food, there is a need for advanced mechanical equipment's, available presently. It includes mechanical tractors with additional ploughing facility, water sprinklers, and other agriculture uses. Another way to secure future food supply is by conserving water resources, which can be achieved by water recycling and conserving available water resources. Food security can also be achieved by recycling food and securing food wastage. Food security with rising CC is accelerating this problem at a faster rate, which needs major attention. CC is an inversible phenomenon, but it can be slowed down by cutting carbon footprints and reducing fossil fuel burning. Some government policies and effort in this area include the latest COP21 UNFCCC meeting in Paris 2015 points to the importance and urgency towards the CC and works required to mitigate this issue to ensure food safety and sustainability.

The major division to food source discussed in this chapter is; Agriculture, Aquaculture, Capture fisheries, and Livestock Farming. Each division is discussed with some form of direct or indirect climate change impacts in details. Livestock is often regarded as a subsector of agriculture by a various organisation like FAO. However, in this chapter, it is considered as a separate source of food security. Supported figures in this chapter use Arc GIS (Geographical information system) (version 10.1.4) platform and data from various government sources as reported in each figure. Some of the ways to secure food sources discussed at the end of this chapter.

MAJOR FOOD SOURCES

From the ancient times, agriculture remains the most available food source for the world. Beginning from the Nile river crescent, the cradle of civilization, agriculture practices were followed by most of the historic civilization. However, with increasing population and expansion of people to different parts of the world, where agriculture practices were not possible due to soil type or limited water supply, various other sources of food were identified like, fisheries and livestock production. People living in coastal and marine places found fisheries as one of the most efficient sources of livelihood and grew with these practices. Just the natural sources were not enough to sustain the world and during 1875 (industrial revolution), increasing technological advances brought fertilizers and other advanced ways to practice these major food sources in areas. Agriculture practices in ASA areas and even in some deserts started growing with the use of technology to extract groundwater and aquifers. With time, transportation connected these farms to the market. Most recently, aquaculture was introduced as an additional food source, where farmers impounded water from the main water source and started growing fishes in confined waters like ponds and ditches, rather than catching from wild. Similarly, livestock and cattle used for farming started making their own market and with rising meat and milk demand, they were cultured commercially. In this section, four major food sources are discussed in detail.

Agriculture

Agriculture (only crops) occupies 25% of Earth's terrestrial surface (excluding Antarctica), which makes it the largest land use commodity on the planet (FAOSTAT, 2018). In addition to food and raw material, agriculture provides a big share of the country's economy in the form of international trade and source of employment. According to FAO, it is a source of livelihood to approx. 70% of the people directly. China and India are the top agriculture producer and make the highest consumer at the same time due to high population size (IMF, 2015). It is important to note that 30-40% area in India classified as ASA region.

ASA lands are known to have scarce water resources, however, with increasing technology, water through canals and other engineered structure, used widely for agriculture purposes. Other than food crops, these regions provide an excellent variety of exotic fruits, carrying economic share. The reference growing period in ASA is of 30-120 days (IIASA/FAO, 2012). Semi-arid regions support 14.4% of the global population. Some ASA regions like West Asia and North Africa contains three of the eight centers of crop origin. Important crops like Barely, wheat, lentil, and Veth were also domesticated near East. International Centre for Agricultural Research in the Dry Areas (ICARDA) is working towards the distribution of wild relatives of wheat in these regions. Arid regions share almost 44% percent of global food crop share like North China plain produces one-fourth of the country's total grain yield (winter wheat and summer maize) (Changming, Jingjie, Kendy, 2001). Therefore, agriculture is one of the most popular food sources around the world.

Capture Fisheries

Capture fisheries (CF) includes fish and other food sources captured directly from natural sources, including marine and freshwater resources. The total fish, crustaceans, and mollusks production from CF comprise 57% globally, and the remaining 43% came from aquaculture in 2013 (FAO, 2016). Total CF production in 2013 was 92.6 million t, 1.4% higher than 2012 (FAO, 2015). The global total production remains stable in last 15 years ranging between 88.3 to 93.8 million tonnes every year (FAO, 2015). CF also is known to be the most diverse fisheries resource with FAO landing statistics refers, 2500 species, mostly including finishes. CF supports about 35 million people globally who are involved in fishing and fish processing, out of which 80% belongs to the small-scale fishery sectors (Béné, 2006).

Marine capture fisheries (MCF) is one of the vital contributors of food security by providing a direct source of proteins, essential fatty acids and micronutrients (Hall et al., 2013; McClanahan et al., 2015). It also provides a wide range of employment and thus helps people purchasing powers for food (Jonson and Welch, 2010). In terms of global share, MCF provides 80 million tons of protein and micronutrient-rich food for human consumption per year (Sumaila, Cheung, Lam, Pauly, Herrick, 2011). MCF provides a share of US$ 80–85 billion an estimated annual gross revenue to the global economies with an estimated

260 million people employment (FAO, 2014; Teh & Sumaila, 2013). MCF is a vital contributor in ASA regions with limited freshwater supply together with soil support to agriculture. As seen in Fig.1 most of the ASA like the East coast of the USA and South America, Australia, Saudi Arabia, and South Africa has a huge coastal belt to support MCF. For example, CMF is an important sector of the Australian economy, provided employment to 7,646 people in 2009/10 (ABARES, 2011). Similarly, Saudi Arabia produces 57, 385 tonnes of fish and support 5887 people directly and share USD 69.82 million GDP (FAO, 2003). Additionally, Eastern coast of USA is one of the most productive and diverse seafood regions of the world, producing 80,015 metric tonnes of MCF landing by California alone making it share of USD 2.1 million (NOAASTAT, 2018).

Inland capture Fisheries (ICF) included wild fish captured from freshwaters and does not include aquaculture in this chapter. FAO (2014), reported about a 6-fold increase in ICF since 1950, in which 156 contributed out of 230 towards wild fish catch from Inland waters. Africa and Asia dominate in ICF production with 71 low income- food deficit countries produce nearly 80% (7 million tons) of global ICF (FAO, 2012). Major food fish in this list are cyprinids (family Cyprinidae) and tilapias (*Sarotherodon, Oreochromis*, and *Tilapia* spp.) (FAO, 2014). ICF contributed to USD 5.5 billion in 2011, with an expected rival marine fish production when all the freshwater bodies (e.g. small streams, ponds, lakes, and rivers) globally combined and accounted together (Welcomme, 2011; FAO, 2012). However, due to limited water availability, ASA regions are not major contributors of ICF and use it only as a seasonal source of fish production. Some small ponds and wetlands collect water during rainy seasons are highly productive in fish production. These wetlands act as an important ecological zone for ICF by supporting freshwater fish production and provide refugia against predators (Denny 1985; Balirwa, 1998). In East African lakes these places are described as a typical habitat for *Tilapia* species (LoweMcConnell, 1975).

Apart from the food source, ICF also an important source of employment as reported by BNP (2009), more than 56 million people were directly involved in ICF from developing the world in 2009, out of which 54% were women (Welcomme et al., 2010). Another study performed in West and Central Africa involving 7 river basins found that ICF supported 227,000 full-time fishermen with USD 295.17 million revenue (Neiland & Béné, 2006). Similarly, Lower Mekong Basin, Indonesia (Largest inland fishery in the world), alone generates USD 17 billion per year (Nam et al., 2015) and provides food security, nutrition, and livelihoods for more than 60 million people in the basin (Orr, Pittock, Chapagain, & Dumaresq, 2012).

CMF contributes about 87% of the total global capture production compared to with ICF, (FAO, 2016). ICF fishes have lower levels of Omega-3 fatty acids, however, some fishes e.g. rainbow trout (*O. mykiss*), Siscowet lake trout *Salvelinus namaycush siscowet*), common carp (*Cyprinus carpio*), wild tilapia (*Oreochromis, Tilapia spp.*), highwaterman catfish (*Hypophthalmus edentatus*), and speckled pavon (*Cichla temensis*) can contain high amounts of eicosapentaenoic acid (EPA) and docosahexaenoic acid (DHA) (Gogus & Smith, 2010; Guler, Kiztanir, Aktumsek, Citil, & Ozparlak, 2008; Inhamuns & Franco, 2008; Steffens & Wirth, 2005; Wang, Miller, Perren, & Addis,1990; Young, 2009). EPA and DHA have been reported to associate with many health benefits (e.g. maintenance and growth of normal brain function) as well as in the reduction of several human diseases (e.g. Alzheimer's disease, cardiovascular disease, arthritis) (He et al., 2004; Horrocks & Yeo, 1999).

Aquaculture

Aquaculture is expected continue to be the world's fastest growing food sector (FAO, 2016). Animal protein from fish has been obtained by more than 1 billion people with that 41% fish consumed is coming from the aquaculture (Beveridge, Thilsted, Phillips, Metian, Troell, & Hall, 2013). About 33% of people (over 18 million) were engaged in aquaculture in 2014. By 2025, fish from aquaculture farms will be 52% out of which 60% belong to freshwater fish species (e.g., carp, catfish, tilapia, trout) and 95% will be produced in developing countries (FAO, 2016). One of the predicted modelled impact of climate change is intensification of hydrological cycle, which means dry regions become drier and wet regions become wetter raising risk of drought in ASA areas (Chou, Neelin, Chen, & Tu, 2009; Held & Soden, 2006; Seager, Naik, & Vecchi, 2010). In such cases, aquaculture can come up as an alternative to capture fisheries to support food security. Although ASA regions don't have enough freshwater supply from main sources, however, there are plenty of groundwater and aquifers available in some of the regions like Murray Basin, Australia and The San Pedro, Arizona (Herczeg, Dogramaci, & Leaney, 2001; Stromberg, Tiller, & Richter, 1996). Also, technology provided different ways to extract water from these sources, which could be used for aquaculture production.

The high mineral content with ambient temperature in ASA regions supports a high primary productivity, which makes it a suitable and favourable food-base for the fish (Pruginin, Fishelson, & Koren, 1988). Aquaculture in arid regions using artificial water reservoirs or advanced techniques in limited water available resources already going successfully in different parts of the world. For example, farming of two salt-tolerant fish species, the Japanese meagre or mulloway (*Argyrosomus japonicus*) and the rainbow trout (*Oncorynchus mykiss*) have been successfully carried out in Australia in the artificial water body. Similarly, farming of the Mozambique tilapia (*Oreochromis mossambicus*) in floating cages demonstrated a successful production technique in Namibia. Other commercially important organisms are gaining attraction in arid regions like small brine shrimp (*Artemia* sp.) and unicellular algae (*Dunaliella* sp.). Both these organisms are used as a feed for a variety of aquatic organisms. For example, *Dunaliella salina*, currently cultured in Australia, supplies over 60% of the world's natural ß-carotene and mainly produced in large saline evaporation ponds in South and Western Australia (Benemann, 2008).

Various ASA regions are using water-saving strategies including harvesting of run-off water, sharing water from reservoirs with green crops and the exploitation of saline water sources not fit for human consumption or agriculture to develop fish farms on a commercial scale (Mires, 2007). The suitable species for ASA regions include the tilapias (*Oreochromis spp.*), barramundi or the Asian seabass (*Lates calcarifer*), carps and mullets (*Mugil cephalus* and *Liza ramada*) and several catfish species (*Clarias gariepinus* and *Bagrus* spp.). The Indian white prawn (*Penaeus indicus*) is one of a good example of successful commercial production in marine aquaculture with Saudi Arabia producing the largest share of 11,300 tonnes in 2005 (Flegel, 2006). Similarly, Algeria is producing 500 tonnes/ year of red and Nile tilapias by harvesting groundwater resources (Crespi, 2009).

Livestock

Livestock (including Poultry) is a branch of Agriculture which includes cattle and other farmed animals. It is taken as a separate food source in this chapter due to its increasing production and demand. As projected, the Global demand for meat to increase 327 million tonnes in 2020 from 209 million tonnes in 1997 with the developing world as a major supplier (De Han, 2001). Meat and Milk consumption is expected to increase by 70-80% by 2050 (Herrero et al., 2015). FAO (2009b) estimates the livestock products provide roughly 1/5th of total caloric intake and half of the total protein consumed in developing countries. The GDP of livestock in 2014 was 8.7 billion USD with China contributing to the highest level of 1.2 million USD in 2014 (FAOSTAT, 2018). Poultry production has been increased to 68 million Metric tonnes in 2000 to 118,080 metric tonnes in 2017 globally and expected to increase up to 13 million metric tonnes by 2026 (FAOSTAT, 2018).

Globally, more than 60% of rural households keep livestock (FAO, 2009a). This livestock is an important source of income, food, fuel, building materials, draft power, and fertilizers for the general population (FAO, 2011). It is expected that livestock production will increase by 56% between 1997 and 2050 (Rosegrant, Meijer, & Cline, 2002). Arid areas are mostly used to support livestock production due to climatic conditions (UN, 2010). It can be seen from the Fig. 4 the most of ASA regions are supporting livestock from medium to high level, which can strengthen the feed security if managed properly. Livestock is often raised in ASA areas, due to their diversity and adaptation to certain climatic conditions. For example, camels in the desert are the best source of milk and meat for the indigenous people. Similarly, a goat can be raised in water-limited dry hilly areas, due to its habitat adaptation capability. According to FAO, livestock supports 40% of Somalia's economy with an increase in annual growth of 6% per year. Similarly, Australian livestock (Beef, cattle, sheep, pigs, and Poultry) industry together provides a share of USD 48.63 billion to the nation's economy each year and employ 200,000 people (ABS, 2017).

Figure 4. a) World map representing arid and semi-arid regions with livestock density in insight maps focusing on a) North America, b) middle eastern countries
(Data Source: FAOSTAT Data, 2012)

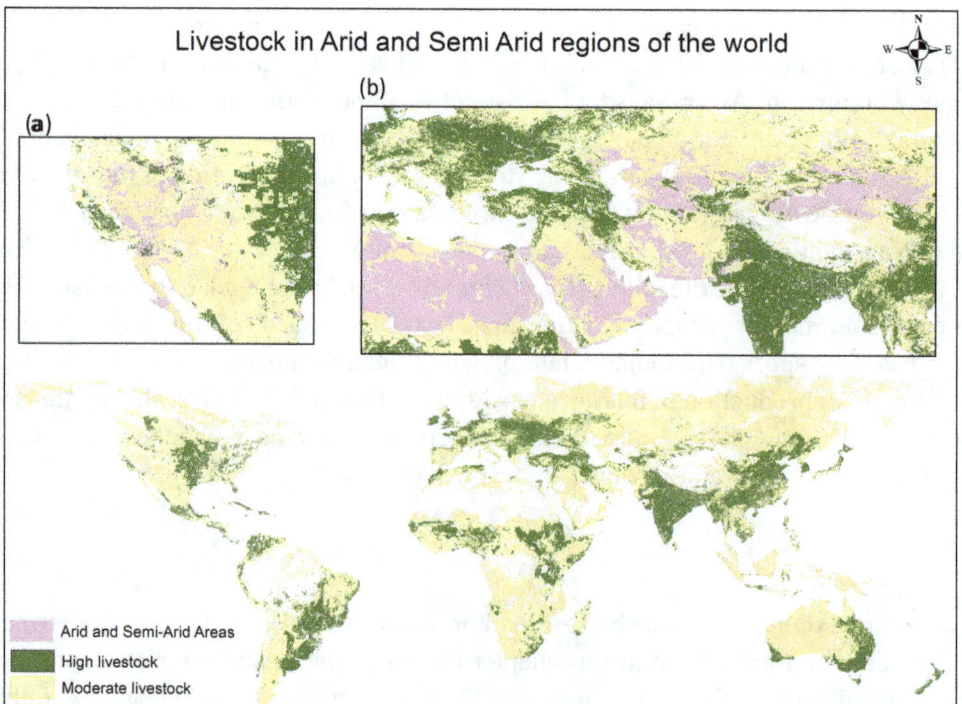

IMPACTS OF CLIMATE CHANGE ON EACH ECOSYSTEM SERVICES

The major impact of CC in ASA regions is projected in the form of land degradation estimated up to 14% due to high temperature and shifting precipitation patterns. The projected adverse impact on food and grass production is even up to 50% in some regions, especially in Africa (GEF, 2010). CC may impact ES services in various ways depending upon their type as explained in detail further. However, temperature and shifting precipitation are the major contributors to overall food security which can result in several other issues.

Impact of Climate Change on Agriculture

Being a greenhouse gas, increasing CO_2 levels in the atmosphere is one of the major reason for temperature rise and cause for anthropogenic CC. This is not favourable for other food sectors, but in terms of agriculture, CO_2 concentration can have a direct effect on plant growth and some studies suggest it may be favourable for the crop production and use term CO_2 fertilization for this (Kohlmaier, Siré, Janecek, Keeling, Piper, & Revelle, 1989). However, at the same time, changing CO_2 concentration (from 350-400 to 800-1200 ppb) with increasing temperature (2-5°C) coupled by shifting precipitation patterns, may have a profound impacts on pests and diseases, which ultimately affects yields (Bebber, Holmes, & Gurr, 2014; Bebber, Ramotowski, & Gurr, 2013; Gregory, Johnson, Newton, & Ingram, 2009). Additionally, decreasing crop biodiversity as a result of genetic and cultural loss is estimated as a result of CC (Ehrenfeld, 2005).

The CC may cause various detrimental impacts to agriculture like ripening in southern and central Europe will occur much earlier than at present. Some studies suggested that on the global scale, CC may be responsible for up to 1/3 of yield variability in key staple commodities (Ray, Gerber, MacDonald, & West, 2015). According to IPCC (2007), the suitable area for agriculture production, length of growing seasons and yield potential is expected to decrease, particularly along the margins of ASA areas. Semi-arid population relies mostly on rain-fed agriculture and a slight fluctuation in rainfall shifting may cause a huge impact towards agriculture sustainability (Liu & Xia 2004; Schwinning, Sala, Loik, & Ehleringer, 2004). Reduced water availability due can cause greater challenges to fulfill crop water needs, resulting in lower yields and higher food prices (Schmidhuber & Tubiello, 2007).

One of the major impact of CC on the agriculture industry is increasing pests and infectious diseases. Prevalence of pests and increase in the frequency of shock pest event can put the agricultural system at greater risks under CC condition (Bebber et al., 2013). One such infection is from mycotoxigenic moulds. The key mycotoxigenic fungi include *Alternaria, Aspergillus, Fusarium* and *Penicillium* species. There is a growing concern that three-way interactions of temperature, water stress, and elevated CO_2 may affect the growth patterns of mycotoxigenic fungi as studied in *F. graminearum* and *F. verticillioides* (Medina et al., 2015). Changing host-pathogen interactions with predicted fungal diseases can impact crop agronomy, yield, and quality (Ziska, Morris, & Goins, 2004; Dixon, 2012). CO_2 may influence the resilience of cereal crops and plant resistance to plant diseases (Płażek, Hura, Rapacz, & Żur, 2001).

Some studies focused on commercially important crops under CC conditions and its impacts to plant fungal infection and mycotoxin contamination. Such as Maize was studied under elevated CO_2 concentration, (approx. 800 ppm) with its susceptibility to *F. verticillioides* colonisation with no consideration to drought conditions (Vaughan et al., 2014). It was found that elevated CO_2 concentration, increased the maize susceptibility to *F. verticillioides*. Similarly, wheat was observed with some changes in leaf physiology and stomatal production on adaxial and abaxial surfaces under increased CO_2 conditions (780 ppm) (Váry, Mullins, Mcelwain, & Doohan, 2015) with previous studies suggested a significant effects on key gene (involved in fumonisin biosynthetic pathway (FUM 1)) by changing environmental stress (Marín, Magan, Vázquez, & González-Jaén, 2010).

Studies suggested that, due to high adaptability under various stress conditions, fungi have a high degree of plasticity and may evolve as a challenge to crop ecosystem under CC conditions (Vaughan et al., 2014; Váry, Mullins, Mcelwain, & Doohan, 2015). It was also suggested that with elevated CO_2 concentration the acclimatisation of *Septoria tritici blotch* (STB) disease and *Fusarium* Head Blight (FHB; *F. graminearum*) for 10-20 generations was also increased (Vary et al., 2015). With recent advances in technology and research, pest and disease management were controlled to a certain extent, yet it reduced the global harvest by 10-16% particularly in developing countries (Chakraborty & Newton, 2011; Grace, Bett, Lindahl, & Robinson, 2015).

Reduced food production with increased temperature is another growing concern and recently been reported by various studies. Asseng et al. (2013), reported the yield was reduced by 5% $^{o}C^{-1}$ for wheat when temperature raised above 32°C. Similarly, maize production was predicted to decrease by at least 10% in Latin America and Africa under different scenarios by 2055 (Jones & Thornton, 2003). There is a predicted reduction in wheat yield by 6% per degree of warming (Asseng et al., 2014). Increasing CO_2 concentrations and more variable and warmer climates may reduce the quality due to decreases in leaf and grain N, protein and macro- and micronutrient (Fe, Zn, Mn, Cu) concentrations (DaMatta, Grandis, Arenque, & Buckeridge, 2010). Lobell et al. (2008) suggested that, by 2030, Southern Africa could lose more than 30% of its main crop, maize due to CC. The recent literature reported that if the proper adaptive measure were not taken towards the agriculture production, then it may reduce three times faster in developing countries compared to developed countries by 2050 (IPCC, 2007; Calzadilla, Rehdanz, Betts, Falloon, Wiltshire, & Tol, 2013).

Impact of Climate Change on Capture Fisheries

The major climate challenges faced by MCF are Ocean acidification, ocean temperature rise and sea level rise (Jackson et al., 2001; Halpern et al., 2008; 2012; Brander, 2007; IPCC, 2007). There are various ways CC can affect target species, which may broadly have categorised under the direct and indirect mechanism. The direct mechanism includes changes in reproduction, distribution and population dynamics of fish stocks (Brander, 2007; Portner & Knust, 2007; Munday, Jones, Pratchett, & Williams, 2008; Cheung et al., 2010). Indirect mechanisms can include changes in the productivity and composition of the ecosystems (Hoegh-Guldberg et al., 2007; Blanchard et al., 2012). Changes in the ecosystem have a direct relation to fish survival, as it provides food and habitat to the fishes.

As a consequence of CC, alterations in global hydrology pattern is of major concern (Allan & Soden, 2008), with largescale changes in rainfall amounts (Donat, Lowry, Alexander, O'Gorman, & Maher 2016), river discharges (Gedney, Cox, Betts, Boucher, Huntingford, & Stott, 2006; Piao, Friedlingstein, Ciais, de Noblet-Ducoudré, Labat, & Zaehle, 2007), drought frequency (Dai, 2012) and intensity (Trenberth et al., 2013) as well as changes in dry season length (Murray-Tortarolo et al., 2016). These alterations can lead to reduced water availability predictions and shifting seasonality. Even though ASA regions are water-limited regions, however, ICF depends mostly on seasonal precipitation. In such cases, ICF may affect badly especially in ASA regions. Similarly, changing salinity also reported by various studies as an impact of changing temperature with melting ice resulting in global ocean freshening (Antonov,

Levitus and Boyer, 2002; Boyer et al., 2005; Bindoff et al., 2007). Changing salinity with temperature can together result in an overall reduction of surface density, expected an increase in vertical stratification and changes in surface mixing (Sarmiento et al., 2004). These changes together cause a decreased overall marine organism survival and growth in the longer run.

Fish catch in marine ecosystem is projected to decrease by 5-10% by 2050 (Barange et al., 2014). Changing fish and plankton distribution is also suggested with changing temperature, winds, ice thickness, pH and nutrient supply (Brander, 2010). Ding et al. (2017) identified 109 countries worldwide with high CF contribution in 2013 and calculated the overall vulnerability. Major countries identified for food security risks due to climate change is all most of the Oceanian countries with high exposure to Australia with low sensitivity and high adaptive capacity. On the other hand, countries like Fiji, Samoa, Solomon Islands, and Vanuatu heavily relied on marine fisheries for protein and livelihoods are under high exposure to the impacts of climate change with low adaptive capacity due to weak socio-economic conditions.

World Fish Centre suggested that temperature rise due to CC may results in wild fish stock reduction by affecting water quality, worsening dry season, mortality, increasing predators and new pathogens as well as by changing food availability (WFS, 2009). It may also effect on fish distribution pattern, as reported, more species gradually shifting towards higher latitudes (Cheung et al., 2009). For example, shifting of Grey mullet (*Mugil cephalus*) fishing grounds to northwards as a result of increasing sea surface temperature (Lan et al., 2014). Similarly, 45 fish species in Australia are documented to shift Southwards since the 1800s (Last et al., 2010) and 16 invertebrate species demonstrated range shift in Tasmania (Pitt, Poloczanska, & Hobday, 2010). These distributions may cause ecology of the system to collapse in the shorter term and lower fish production due to high exposure to new predators.

Impact of Climate Change on Aquaculture

Aquaculture mostly practiced in open areas with water impounded from either surface waters, rain waters or groundwater sources. Aquaculture practiced in coastal areas by using cages or impounding areas by using large nets to culture organisms. Temperature and water availability are the biggest requirement for growing aquaculture production. Water availability demonstrated a sharp decline and projected to decrease in the future with CC impacts. In such situations, aquaculture in ASA regions can be seen as the biggest concern, especially in rainfed aquaculture places. Water availability may cause decreased production, which is a big requirement for food security in a growing world. Fig. 5 shows a global baseline for water stress in the world, and ASA regions are mostly coming under extremely high (>80%) category, which demonstrates the future water availability concerns in this region.

Increasing temperature in lakes and rivers since the 1960s (0.2 to 2 °C) in surface waters of Europe, North America, and Asia raised the concerns for aquaculture industry (Rosenzweig et al., 2007). Longer ice-free seasons and higher temperatures raise the concerns in various physicochemical changes in aquaculture system like thermal stratification, decreasing DO concentration and varying ecosystem. Increasing temperature may also cause disease, pathogen and virus outbreak in ASA regions, with seasonal precipitation. Despite temperature and precipitation variation, saltwater inundation is one of the biggest threat to freshwater aquaculture, together with increasing frequency of extreme events such as coastal flooding, cyclones and storm surges in the low-lying deltaic regions (IPCC, 2014).

Figure 5. a) World map representing Baseline water stress in the world
(Data Source: Aqueduct Global maps 2.1, 2013)

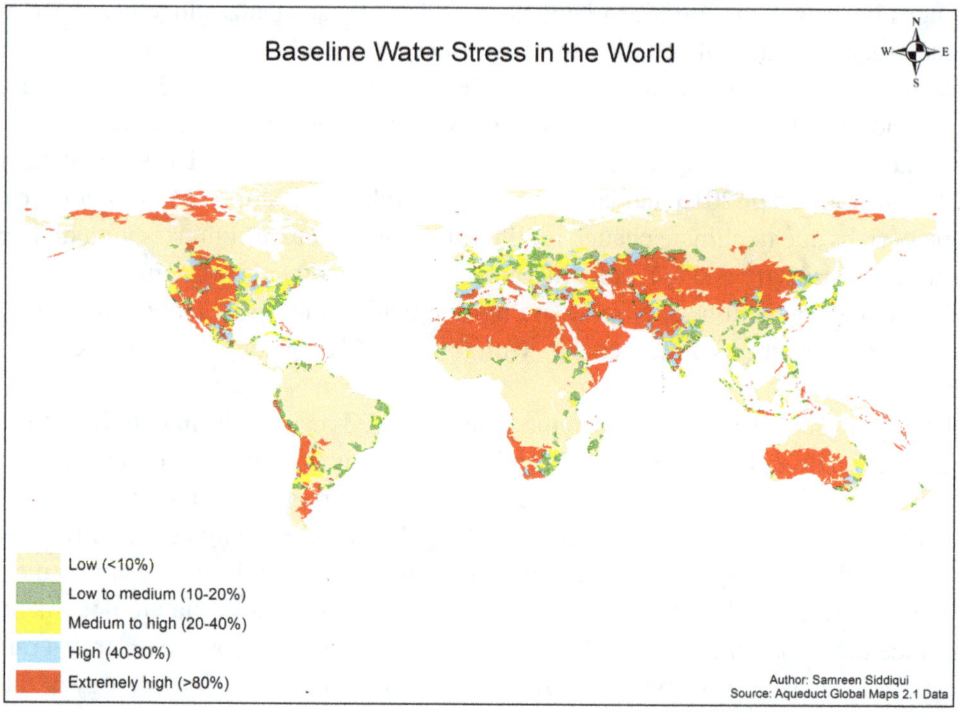

Coastal aquaculture is more vulnerable to the CC impacts compared to inland aquaculture (Ahmed & Diana, 2015). Extreme events like cyclones can have major impacts on coastal aquaculture. Due to increasing CO_2 concentrations, ocean acidity is increasing the threat to coastal aquaculture, especially for shellfish farming. Most of the ASA regions rely on coastal farming like Shrimp culture. Changing acidic levels may have adverse impacts on their shell calcium concentration, which can be vulnerable for their existence. Similarly, in Ocean acidification can cause severe physiological changes and effect on finfish growth and reproduction.

CC has a direct impact on some aquaculture organisms as reported in cultivated bivalves, because of temperature increase they demonstrated adverse effects on their morphometric characteristics, disease outbreak, growth rates and condition index (Hofmann, Ford, Powell, & Klinck, 2001; Mackenzie et al., 2014). Similarly, increasing temperature can affect the metabolism, calcium fixation ability as well as energy allocation to somatic and gonadic structures of cultivated mollusks (Kooijman, 2010; Hiebenthal, Philipp, & Eisenhauer, 2012, 2013). This effect may reduce the thickness of shell structure making it fragile in longer-term (Olson, Kozdon, Valley, & Gilbert, 2012; Briones, Rivadeneira, Fernândez, & Guiñez, 2014). At the same time, there are some positive effects of CC like Coastal upwelling which may affect regions like Eastern USA (California), North East Africa and Arabian sea. Warming between the ocean and land masses can intensify the alongshore wind stress on the ocean surface, lead to coastal upwelling acceleration (Bakun, 1990; Snyder, Sloan, Diffenbaugh, & Bell, 2003; Auad, Miller, & Di Lorenzo, 2006).

Other than biological and physiological issues, CC is a growing challenge for the aquaculture market, it is expected to affect the volume and price of production, operational costs together with revenues and profits of enterprises and households that depends upon them (Brander, 2007; Cheung et al., 2010; Sumaila, Cheung, Lam, Pauly, & Herrick, 2011). Increasing CC resulting in higher demand with low production that may cause higher food price, as a result, decreases the food security in growing population world. Due to extreme events like storms, flood, drought, and other calamities, transportation can get affected, which increases the waste rate and decreases the availability in the market. These problems together affect future aquaculture growth and production.

Impact of Climate Change on Livestock

Much like agriculture and fisheries, livestock is on risks of CC impacts due to unstable weather conditions, damage from extreme events (e.g. heat stress, droughts, and flooding) and shifting temperatures (Jones & Thornton, 2009). Increased frequency of such extreme events may affect water and food availability for livestock production as well as limited access to for the people who rely on the market exchange of animal products for grains (Battisti & Naylor, 2009). Impacts of CC on a livestock system can be assessed by a reduced feed quantity and quality, changes in pest and disease prevalence, and physiological stress (Campbell et al., 2016). Reduced feed intake, growth, meat, egg and milk yield, and quality may decrease as temperature goes beyond 30°C (Thornton & Gerber, 2010). High temperature and humid environment are known to be detrimental to livestock in various ways. Increasing heat stress can cause reduced fertility of female and male cattle, resulting in reduced reproductive performance (St-Pierre, Cobanov, & Schnitkey, 2003). Heat stress can also reduce the expression of estrous behaviour (Hansen et al., 2001), alters follicular development (Wolfenson et al., 1995) and the growth and function of the dominant follicle (Wilson et al., 1998), compromises oocyte competence (Wolfenson, Roth, & Meidan, 2000), and inhibits embryonic development (Drost et al., 1999). The incidence of udder infection and mastitis were also reported during hot summer months (Giesecke, 1985). In some extreme cases, cow mortality also reported due to heat stress (Hahn, 1985). In the case of broilers, reduced thyroid size, with undesirable meat characteristics were reported and male broiler breeders are shown to affect more by heat stress, compared to females (Dale & Fuller, 1980; McDaniel, Bramwell, Wilson, & Howarth Birkett, 1995; Sandercock, Hunter, Nute, Mitchell, & Hocking, 2018). Heat stress can cause increased mortality especially while transporting and marketing (Mitchell & Kettlewell, 1998; De Basilio Vilariño, Yahav, & Picard, 2001). In the case of hen, decreased egg weight and shell thickness are reported due to increased heat stress (Muiruri & Harrison, 1991; Wolfenson, Bachrach, Maman, Graber, & Rozenboim, 2001).

Pathogen and parasite which spend some of their life outside animal host can develop more with higher temperature (Harvell et al., 2002). With increasing population, changes to wind patterns can cause widespread of certain pathogens and vectors. Additionally, CC can have severed impacts on the vectors of diseases common in ASA regions (midges, flies, ticks, mosquitoes, and tsetse). Changing precipitation together with temperatures and increasing extreme events (e.g. El Niño-Southern Oscillation (ENSO)) can affect both the distribution and disease abundance. With increasing disease outbreak, the spread of these diseases may in also increase with higher temperature. Changing competitors, predators, and parasites of vectors patterns can change the ecological patterns due to CC and results in a new mixture species that may expose hosts to novel pathogens and vectors resulting in new diseases emergence (WHO, 1996). For example, CC impacts on the distribution of the brown-ear tick, *Rhipicephalus appendiculatus*, was studied by Rogers (1996), it is a primary vector of East Coast Fever, a disease that affects both grazing (LG) and mixed (MR) systems in eastern and southern Africa. It was reported that by 2050, its range may expand in western and central parts of southern Africa. Similarly, white et al. (2003) simulated an increased vulnerability of cattle tick (*Boophilus microplus*) in the Australian beef industry as a risk of CC.

FOOD SECURITY

There is a need of 60% increase agriculture by 2050 globally to meet the increasing demand and most of it can be achieved by increasing productivity (3). To increase the productivity in sustainable way Lipper et al. (2014) introduced climate-smart agriculture (CSA) practices to be applied on local to global scale over the short and long time period. The main objectives for CSA include (1) sustainably increasing agricultural productivity to support equitable increases in incomes, food security, and development; (2) adapting and building resilience to climate change from the farm to national levels; and (3) developing opportunities to reduce GHG emissions from agriculture compared with past trends. To achieve CSA objectives there is a need to build evidence and assessment tools; strengthen national and local institution; developing coordinated and evidence-based policies and increasing financing and its effectiveness. Similarly, Gebbers & Adamchuk (2010), described how precision agriculture can be used to secure food in the longer run. By identifying soil variation, crop conditions and right treatment using global navigation satellite systems (GNSSs), geographical information system (GIS) and microcomputers, can improve production.

Fisheries is one of the best alternatives to food security and facing major impacts of CC (Ocean acidification, global warming) in addition to human exploitation. Therefore, by identifying major impact areas, future research can be directed to develop the capacity for mitigation and adaptation initiatives. After identifying most vulnerable regions around the world, they can further be subclassified into three major categories according to their sources e.g. i) regions with dominant sources for fish consumption; regions with dominant aquaculture and inland fisheries consumption and iii) regions with dominant marine fisheries consumption. Ding et al. (2017) identified six countries in the first subcategory, three countries in second and eighteen countries in a third subcategory. There is a need to identify major sensitivity and required adaptation in these regions to secure future food resources. Aquaculture and culture fisheries are alternatives to capture fisheries and which needs better ways to provide sustainable fisheries production. McClanahan, Allison, & Cinner (2015) pointed out various food and social security faced by fisheries sector which needs attention like: i) limited government subsidies; ii) uncontrolled and illegal fishing (mainly in developing countries); iii) weak and inappropriate access and property rights; iii) development and trade policies ignoring poor people; iv) human settlement and infrastructural developments in coastal areas that can reduce the area for fisheries production; and (vii) a growing, urbanizing and increasingly affluent population making increasing demands for seafood, especially among the emerging middle classes of developing and transitional economies.

Livestock can also face similar CC issue including flood, drought and heat stress. According to AgMIP Workshop in Colorado (2015), to mitigate such issues there is a need to develop food policies which can target import/export control together with subsidies. There is a need to improve adaptation practices by further research on genetic improvements and resource management. Additionally, food security can be achieved by improving diet standards, increasing land use and identifying major disease issues. Food security in the modern world can also be achieved by combining fisheries with agriculture and livestock, commonly termed integrated fish farming (Paul, Mandal, & Datta, 2018). Some examples of this type of farming are Tilapia- paddy fish farming; duck-fish farming and Fish-duck- paddy farming commonly practiced in South Asian countries (Majhi, 2018; Nayak et al., 2018). The major benefit of this type of farming is reduced fertilizer cost, smaller area, and planned farming. This sustainable way can provide maximum food production with limited water and feed usage.

CONCLUSION AND FUTURE ACTIONS

In this changing climate conditions and challenges to food security policy makes have enough information on CC and variability to understand that it is a serious problem which requires immediate attention (Heal & Millner, 2014). Global total water withdrawal is increasing and creating stress in ASA regions, due to which access to improved drinking water is also under stress (Fig. 6). There is a strong research-implementation gap appears, especially in terms of action (Knight, Cowling, Rouget, Balmford, Lombard, & Campbell, 2008). With growing threats of CC, the sustainable use of water resource, environmental conservation, technological appropriate usage, economic viability and social acceptability of development issue is a priority for agriculture in ASA areas (Periera, 2003).

Planned water recycling is a big addition to solve such food security problems. Wastewater treatment plants (WWTP) in big cities are used widely for such purposes to solve water crisis issues for the future. The water recycling is carried out in such a way that can nurture the water cycle and cause a minimal environmental impact. Another way to recycle water is by recharging dried aquifer system artificially using partially treated water. Such water recycling is possible in selective lithological zones (vadose zones) that can have some groundwater resources. Theses unsaturated lithology zones are present in nature with such structure that can act as a natural filter, by removing essentially all suspended solids, biodegradable materials, bacteria, viruses and other microorganisms (Fadlelmawla, Abdel-Halen, & Vissers, 1999). This process is also called Soil-Aquifer Treatment (SAT) and presently in use around the world. Whittier Narrows, Los Angeles, USA, is one of the largest area using SAT technology to recharge area of approximately 279 ha forms a small basin is divided into 15 sub-basins. Rainwater harvest is another way to conserve water effectively and efficiently. Rainwater harvest involves various steps including receiving, collection, storage, and use of rainwater in the various household, agriculture and industrial purposes. This rainwater can also use to recharge an aquifer as described above. There are various types of rainwater harvesting structures built to collect it efficiently like, construction of small ridge and furrow structures in agriculture fields (in situ), diversion of surface runoff water into a collection basin (Ex situ). Rainwater can also be collected from rooftops, courtyards, compacted or treated surfaces for domestic uses.

Water conservation is essential for the growth and support of food resources, however, there is a growing need to adopt advanced practices to support crop production without harming the environment to strengthen food security. Such as use of new mechanical and advanced tractors systems with water sprinkle facilities in farms to reduce water wastage and maximum use of water by increasing surface area. Crop diversification is an important practice which can help in managing risks from against various

Figure 6. a) World map representing a) Total water withdrawal (million cubic metre) for the year 2013 b) Global Access to improved drinking water sources index from low to extremely high for the year 2013 (Data Source: Aqueduct Global maps 2.1, 2013)

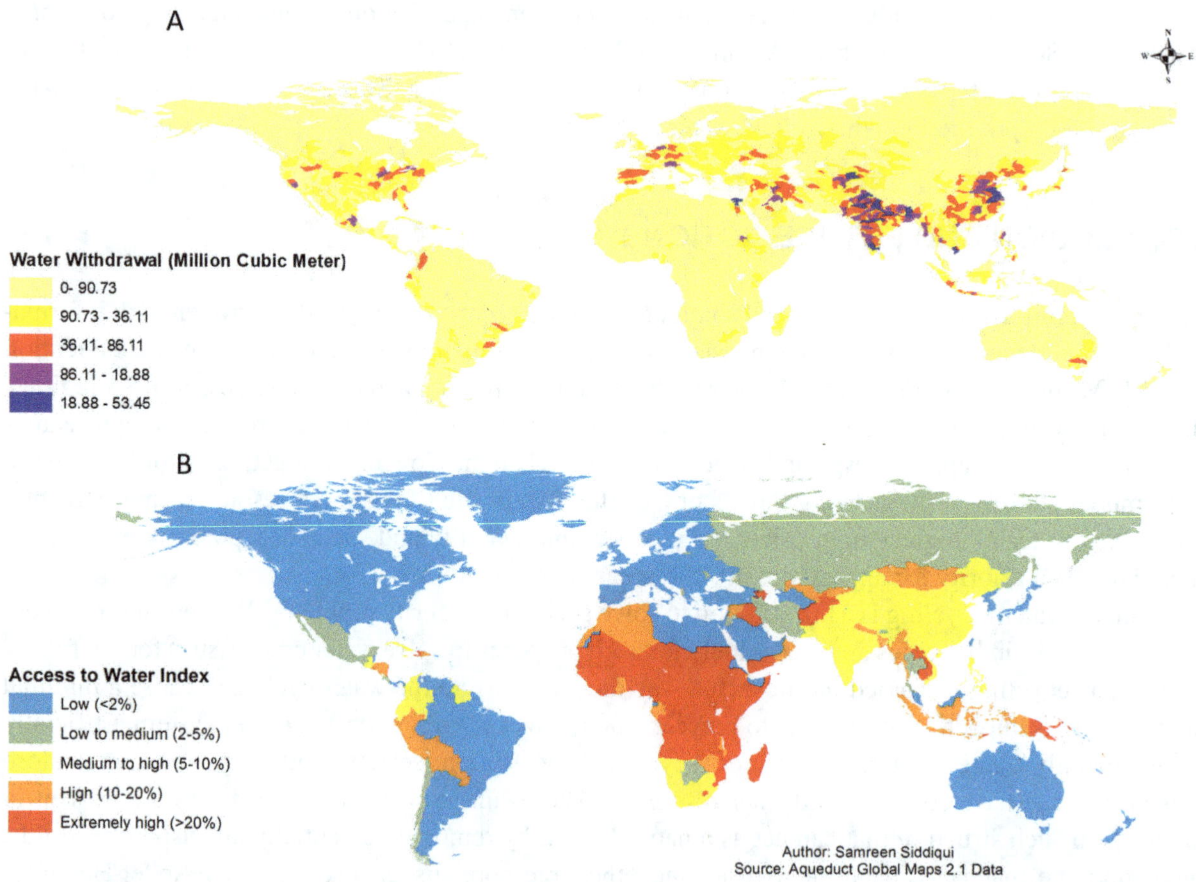

environmental issues. As studied (2003-2004) in semi-arid savannah zone, Africa, where the authors identified 50 varieties of yam *Dioscorea sp*. and 33 varieties of rice, *Oryza glaberrima* practiced in this area. This crop diversification underscores the importance of crop variety and security against unpredictable rainfall, pest attack, fluctuating market and another such variable environmental and socioeconomic conditions (Gyasi et al., 2004).

Food resources are one of the best ES which needs to be secure for the survival of increasing population. World population has been increasing the faster rate and food availability is not enough to support population by 2030. There is a need to use technology responsibly and precisely to secure food resources. Increasing research and using food resources sustainably can result in a longer-term food security. Policies are required to facilitate to transition grazing systems areas to potential mixed farming systems in the semiarid and sub-humid tropics through integrating crops and livestock (Steinfeld, De Haan, & Blackburn, 1997). Additionally, integrated farming, reducing food water and minimising water usage are some of the best practices need to be followed.

REFERENCES

ABARES. (2011). *Australian fisheries statistics 2010*. Canberra: Australian Bureau of Agricultural and Resource Economics and Sciences.

ABS. (2017). *The Australian Beauro of Statistics*. Retrieved April 04, 2018, from http://www.abs.gov.au/Agriculture

AgMIP Workshop. (2015). Retrieved October 06,2018 from https://eos.org/meeting-reports/assessing-impacts-of-climate-change-on-food-security-worldwide

Ahmed, N., & Diana, J. S. (2015). Threatening "white gold": Impacts of climate change on shrimp farming in coastal Bangladesh. *Ocean and Coastal Management*, *114*, 42–52. doi:10.1016/j.ocecoaman.2015.06.008

Alexandratos, N., & Bruinsma, J. (2012). *World Agriculture Towards 2030/2050: The 2012 Revision ESA Working Paper No. 12–03* (FAO, 2012). FAO.

Allan, R. P., & Soden, B. J. (2008). Atmospheric Warming and the Amplification of Precipitation Extremes. *Science*, *321*(5895), 1481 LP-1484. Retrieved from http://science.sciencemag.org/content/321/5895/1481.abstract

Antonov, J. I., Levitus, S., & Boyer, T. P. (2002). Steric sea level variations during 1957-1994: Importance of salinity. *Journal of Geophysical Research: Oceans*, *107*(C12), SRF 14-1-SRF 14-8. doi:10.1029/2001JC000964

Asseng, S., Ewert, F., Martre, P., Rötter, R. P., Lobell, D. B., Cammarano, D., ... Zhu, Y. (2014). Rising temperatures reduce global wheat production. *Nature Climate Change*, *5*(2), 143–147. doi:10.1038/nclimate2470

Asseng, S., Ewert, F., Rosenzweig, C., Jones, J. W., Hatfield, J. L., Ruane, A. C., ... Wolf, J. (2013). Uncertainty in simulating wheat yields under climate change. *Nature Climate Change*, *3*(9), 827–832. doi:10.1038/nclimate1916

Auad, G., Miller, A., & Di Lorenzo, E. (2006). Long-term forecast of oceanic conditions off California and their biological implications. *Journal of Geophysical Research. Oceans*, *111*(9), 1–14. doi:10.1029/2005JC003219 PMID:20411040

Bakun, A. (1996). *Patterns in the ocean. Ocean processes and marine population dynamics*. California Sea Grant College System.

Balirwa, J. S. (1998). *Lake Victoria wetlands and the ecology of the Nile Tilapia Oreochromis niloticus Linne´. A.A.* Rotterdam, The Netherlands: Balkema Publishers.

Barange, M., Merino, G., Blanchard, J. L., Scholtens, J., Harle, J., Allison, E. H., ... Jennings, S. (2014). Impacts of climate change on marine ecosystem production in societies dependent on fisheries. *Nature Climate Change*, *4*(3), 211–216. doi:10.1038/nclimate2119

Battisti, D. S., & Naylor, R. L. (2009). Historical Warnings of Future Food Insecurity with Unprecedented Seasonal Heat. *Science, 323*(5911), 240 LP-244. Retrieved from http://science.sciencemag.org/content/323/5911/240.abstract

Bebber, D. P., Holmes, T., & Gurr, S. J. (2014). The global spread of crop pests and pathogens. *Global Ecology and Biogeography*, *23*(12), 1398–1407. doi:10.1111/geb.12214

Bebber, D. P., Ramotowski, M. A. T., & Gurr, S. J. (2013). Crop pests and pathogens move polewards in a warming world. *Nature Climate Change*, *3*(11), 985–988. doi:10.1038/nclimate1990

Béné, C. (2006). *Small-scale fisheries: assessing their contribution to rural livelihoods in developing countries.* FAO Fisheries Circular No. 1008 FIPL/C1008 (En).

Benemann, J. R. (2008). *Opportunities and challenges in algae biofuels production.* Position paper by Dr. John R. Benemann in line with Algae World 2008. Retrieved from www.futureenergyevents.com/algae/whitepaper/algae_positionpaper.pdf

Beveridge, M. C. M., Thilsted, S. H., Phillips, M. J., Metian, M., Troell, M., & Hall, S. J. (2013). Meeting the food and nutrition needs of the poor: The role of fish and the opportunities and challenges emerging from the rise of aquaculture. *Journal of Fish Biology*, *83*(4), 1067–1084. doi:10.1111/jfb.12187 PMID:24090563

Bindoff, N. L., Willebrand, J., Artale, V., Cazenave, A., Gregory, J., Gulev, S., ... Unnikrishnan, A. (2007). Observations: Oceanic climate change and sea level. In S. Solomon, D. Qin, M. Manning, Z. Chen, M. Marquis, K. B. Averyt, ... H. L. Miller (Eds.), *Climate Change 2007: The physical science basis. Contribution of Working Group I to the Fourth Assessment Report of the Intergovernmental Panel on Climate Change*. Cambridge, UK: Cambridge University Press.

Blanchard, J. L., Jennings, S., Holmes, R., Harle, J., Merino, G., Allen, J. I., ... Barange, M. (2012). Potential consequences of climate change for primary production and fish production in large marine ecosystems. *Philosophical Transactions of the Royal Society of London. Series B, Biological Sciences*, *367*(1605), 2979–2989. doi:10.1098/rstb.2012.0231 PMID:23007086

BNP. (2009). Big Number Program: Intermediate Report. Penang, Malaysia: BNP.

Boyer, T. P., Levitus, S., Antonov, J. I., Locarnini, R. A., & Garcia, H. E. (2005). Linear trends in salinity for the World Ocean, 1955-1998. *Geophysical Research Letters*, *32*(1), 1–4. doi:10.1029/2004GL021791

Brander, K. (2010). Impacts of climate change on fisheries. *Journal of Marine Systems*, *79*(3), 389–402. doi:10.1016/j.jmarsys.2008.12.015

Brander, K. M. (2007). Global fish production and climate change. *Proceedings of the National Academy of Sciences, 104*(50), 19709-19714. Retrieved from http://www.pnas.org/content/104/50/19709.abstract

Briones, C., Rivadeneira, M. M., Fernández, M., & Guiñez, R. (2014). Geographical Variation of Shell Thickness in the Mussel *Perumytilus purpuratus* Along the Southeast Pacific Coast. *The Biological Bulletin*, *227*(3), 221–231. doi:10.1086/BBLv227n3p221 PMID:25572210

Cabell, J. F., & Oelofse, M. (2012). An Indicator Framework for Assessing Agroecosystem Resilience. *Ecology and Society*, *17*(1), art18. doi:10.5751/ES-04666-170118

Calzadilla, A., Rehdanz, K., Betts, R., Falloon, P., Wiltshire, A., & Tol, R. S. J. (2013). Climate change impacts on global agriculture. *Climatic Change, 120*(1), 357–374. doi:10.100710584-013-0822-4

Campbell, B. M., Vermeulen, S. J., Aggarwal, P. K., Corner-Dolloff, C., Girvetz, E., Loboguerrero, A. M., … Wollenberg, E. (2016). Reducing risks to food security from climate change. *Global Food Security, 11*, 34–43.

Chakraborty, S., & Newton, A. C. (2011). Climate change, plant diseases and food security: An overview. *Plant Pathology, 60*(1), 2–14. doi:10.1111/j.1365-3059.2010.02411.x

Changming, L., Jingjie, Y., & Kendy, E. (2001). Groundwater exploitation and its impact on the environment in the North China Plain. *Water International, 26*(2), 265–272. doi:10.1080/02508060108686913

Chaudhary, S., McGregor, A., Houston, D., & Chettri, N. (2015). The evolution of ecosystem services: A time series and discourse-centered analysis. *Environmental Science & Policy, 54*, 25–34. doi:10.1016/j.envsci.2015.04.025

Cheung, W. W. L., Lam, V. W. Y., Sarmiento, J. L., Kearney, K., Watson, R., & Pauly, D. (2009). Projecting global marine biodiversity impacts under climate change scenarios. *Fish and Fisheries, 10*(3), 235–251. doi:10.1111/j.1467-2979.2008.00315.x

Cheung, W. W. L., Lam, V. W. Y., Sarmiento, J. L., Kearney, K., Watson, R., & Pauly, D. (2009). Projecting global marine biodiversity impacts under climate change scenarios. *Fish and Fisheries, 10*(3), 235–251. doi:10.1111/j.1467-2979.2008.00315.x

Cheung, W. W. L., Lam, V. W. Y., Sarmiento, J. L., Kearney, K., Watson, R., Zeller, D., & Pauly, D. (2010). Large-scale redistribution of maximum fisheries catch potential in the global ocean under climate change. *Global Change Biology, 16*(1), 24–35. doi:10.1111/j.1365-2486.2009.01995.x

Chou, C., Neelin, J. D., Chen, C.-A., & Tu, J.-Y. (2009). Evaluating the rich-get-richer mechanism in tropical precipitation change under global warming. *Journal of Climate, 22*(8), 1982–2005. doi:10.1175/2008JCLI2471.1

Costello, A., Abbas, M., Allen, A., Ball, S., Bell, S., Bellamy, R., . . . Patterson, C. 2009. Managing the health effects of climate change. *Lancet, 373*, 1693–1733. Retrieved from https://www.ucl.ac.uk/global-health/project-pages/lancet1/ucl- lancet-climate-change.pdf

Creighton, C., Hobday, A. J., Lockwood, M., & Pecl, G. T. (2016). Adapting Management of Marine Environments to a Changing Climate: A Checklist to Guide Reform and Assess Progress. *Ecosystems (New York, N.Y.), 19*(2), 187–219. doi:10.100710021-015-9925-2

Crespi, V. (2009). Support to the development of desert aquaculture and management of the brackish water lakes in Algeria. *FAO Aquaculture Newsletters, 43*, 20–22.

CSIRO. (2011). *Sustainability in Australia's Arid Lands*. Retrieved from http://www.csiro.au/Outcomes/Water/Rural-and-regional-water/arid-land-sustainability.aspx

Dai, A. (2012). Increasing drought under global warming in observations and models. *Nature Climate Change, 3*(1), 52–58. doi:10.1038/nclimate1633

Dale, N. M., & Fuller, H. L. (1980). Effect of Diet Composition on Feed Intake and Growth of Chicks Under Heat Stress.II. Constant vs. Cycling Temperatures. *Poultry Science, 59*(7), 1434–1441. doi:10.3382/ps.0591434 PMID:7393854

DaMatta, F. M., Grandis, A., Arenque, B. C., & Buckeridge, M. S. (2010). Impacts of climate changes on crop physiology and food quality. *Food Research International, 43*(7), 1814–1823. doi:10.1016/j.foodres.2009.11.001

De Basilio, V., Vilariño, M., Yahav, S., & Picard, M. (2001). Early Age Thermal Conditioning and a Dual Feeding Program for Male Broilers Challenged by Heat Stress. *Poultry Science, 80*(1), 29–36. doi:10.1093/ps/80.1.29 PMID:11214333

De Han, C. (2001). *Livestock development: Implications for rural poverty, the environment, and global food security*. The World Bank. doi:10.1596/0-8213-4988-0

Denny, P. (Ed.). (1985). *The ecology and management of African wetland vegetation. Geobotany 6.* Dordrecht, The Netherlands: Dr. W. Junk Publishers. doi:10.1007/978-94-009-5504-2

Ding, Q., Chen, X., Hilborn, R., & Chen, Y. (2017). Vulnerability to impacts of climate change on marine fisheries and food security. *Marine Policy, 83*, 55–61. doi:10.1016/j.marpol.2017.05.011

Dixon, G. R. (2012). Climate change – impact on crop growth and food production, and plant pathogens. *Canadian Journal of Plant Pathology, 34*(3), 362–379. doi:10.1080/07060661.2012.701233

Donat, M. G., Lowry, A. L., Alexander, L. V., O'Gorman, P. A., & Maher, N. (2016). More extreme precipitation in the world's dry and wet regions. *Nature Climate Change, 6*(5), 508–513. doi:10.1038/nclimate2941

Drost, M., Ambrose, J. D., Thatcher, M. J., Cantrell, C. K., Wolfsdorf, K. E., Hasler, J. F., & Thatcher, W. W. (1999). Conception rates after artificial insemination or embryo transfer in lactating dairy cows during summer in Florida. *Theriogenology, 52*(7), 1161–1167. doi:10.1016/S0093-691X(99)00208-3 PMID:10735094

Ehrenfeld, D. (2005). The environmental limits to globalization. *Conservation Biology, 19*(2), 318–326. doi:10.1111/j.1523-1739.2005.00324.x

European Centre for Nature Conservation. (2016). Retrieved April 21, 201 from OpenNESS project: http://www.openness-project.eu/

Fadlelmawla, A., Abdel-Halen, M., & Vissers, M. (1999). Preliminary plans for artificial recharge with exclaimed wastewater in Egypt. *9th Biennial Symposium on Artificial recharge of groundwater*, Tempe, AZ.

Falcon, W. P., Naylor, R. L., & Asia, S. (2008). Needs for Food Security in 2030 Region. *Science, 319*, 607–610. doi:10.1126cience.1152339 PMID:18239122

FAO. (2003). Retrieved April 7, 2018 from http://www.fao.org/fi/oldsite/FCP/en/SAU/profile.htm

FAO. (2005). *The State of World Fisheries and Aquaculture 2005*. Rome: Food and Agriculture Organisation.

FAO. (2009a). *Climate Change Implications for Fisheries and Aquaculture 2009*. Food and Agriculture Organisation.

FAO. (2009b). *Livestock, food security and poverty reduction*. Rome: The State of Food and Agriculture. FAO.

FAO. (2009c). *Livestock in the balance*. Rome: State of Food and Agriculture. FAO.

FAO. (2009c). *The State of Food Insecurity in the World: Economic crises—Impacts and Lessons Learned 8–12*. Food and Agriculture Organization of the United Nations (FAOSTAT). Retrieved March 05, 2015 from http://faostat.fao.org/site/567/default.aspx#ancor

FAO. (2011). *World Livestock 2011-Livestock in food security*. Rome: FAO.

FAO. (2012). *The State of World Fisheries and Aquaculture*. Rome, Italy: FAO.

FAO. (2014). *The State of the World Fisheries and Aquaculture 2014*. Food and Agriculture Organization.

FAO. (2014). FAO Yearbook: Fishery and Aquaculture Statistics. FAO.

FAO. (2014). *FIGIS*. FishStat (Database). Retrieved from http://data. fao.org/ref/babf3346-ff2d-4e6c-9a40-ef6a50fcd422.html?version=1.0

FAO. (2015). *FAO Global Capture Production database updated to 2013 – Summary information*. FAO.

FAO. (2016). *The State of World Fisheries and Aquaculture – 2016 (SOFIA)*. Rome, Italy: FAO.

Flegel, T. W. (2006). Detection of major penaeid shrimp viruses in Asia, a historical perspective with emphasis on Thailand. *Aquaculture (Amsterdam, Netherlands)*, *258*(1), 1–33. doi:10.1016/j.aquaculture.2006.05.013

Gebbers, R., & Adamchuk, V. I. (2010). Precision Agriculture and Food Security. *Science, 327*(5967), 828-831. Retrieved from http://science.sciencemag.org/content/327/5967/828.abstract

Gedney, N., Cox, P. M., Betts, R. A., Boucher, O., Huntingford, C., & Stott, P. A. (2006). Detection of a direct carbon dioxide effect in continental river runoff records. *Nature*, *439*(7078), 835–838. doi:10.1038/nature04504 PMID:16482155

Giesecke, W. H. (1985). The effect of stress on udder health of dairy cows. *Journal of Veterinary Research (Pulawy)*, *52*, 175–193. PMID:3911132

Gogus, U., & Smith, C. (2010). N-3 omega fatty acids: A review of current knowledge. *International Journal of Food Science & Technology*, *45*(3), 417–436. doi:10.1111/j.1365-2621.2009.02151.x

Grace, D., Bett, B., Lindahl, J., & Robinson, T. (2015). *Climate and livestock disease: assessing the vulnerability of agricultural systems to livestock pests under climate change scenarios*. CCAFS Working Paper no.116. CGIAR Research Program on Climate Change, Agriculture and Food Security (CCAFS), Copenhagen. Retrieved from https://cgspace.cgiar.org/rest/bitstreams/55094/retrieve

Gregory, P. J., Johnson, S. N., Newton, A. C., & Ingram, J. S. I. (2009). Integrating pests and pathogens into the climate change/food security debate. *Journal of Experimental Botany*, *60*(10), 2827–2838. doi:10.1093/jxb/erp080 PMID:19380424

Guler, G. O., Kiztanir, B., Aktumsek, A., Citil, O. B., & Ozparlak, H. (2008). Determination of the seasonal changes on total fatty acid composition and ω3/ω6 ratios of carp (Cyprinus carpio L.) muscle lipids in Beysehir Lake (Turkey). *Food Chemistry*, *108*(2), 689–694. doi:10.1016/j.foodchem.2007.10.080 PMID:26059149

Gyasi, E. A., Kranjac-Berisavljevic, G., Blay, E. T., & Oduro, W. (Eds.). (2004). *Managing agrodiversity the traditional way: Lessons from West Africa in sustainable use of biodiversity and related natural resources*. Tokyo: UN Univ. Press.

Hahn, G. L. (1985). Management and housing of farm animals in hot environment. In M. K. Yousef (Ed.), *Stress Physiology of Livestock. Ungulates* (Vol. 2, pp. 151–176). Boca Raton, FL: CRC Press.

Haines-Young, R., & Potschin, M. (2011). Common International Classification of Ecosystem Services (CICES): 2011 Update. Report to the European Environmental Agency, Nottingham, UK.

Hall, S. J., Hilborn, R., Andrew, N. L., & Allison, E. H. (2013). Innovations in capture fisheries are an imperative for nutrition security in the developing world. *Proceedings of the National Academy of Sciences of the United States of America*, *110*(21), 8393–8398. doi:10.1073/pnas.1208067110 PMID:23671089

Halpern, B. S., Longo, C., Hardy, D., McLeod, K. L., Samhouri, J. F., Katona, S. K., ... Zeller, D. (2012). An index to assess the health and benefits of the global ocean. *Nature*, *488*(7413), 615–620. doi:10.1038/nature11397 PMID:22895186

Halpern, B. S., Walbridge, S., Selkoe, K. A., Kappel, C. V., Micheli, F. D., Agrosa, C., … Watson, R. (2008). A Global Map of Human Impact on Marine Ecosystems. *Science, 319*(5865), 948-952. Retrieved from http://science.sciencemag.org/content/319/5865/948.abstract

Hammond, S. T., Brown, J. H., Burger, J. R., Flanagan, T. P., Fristoe, T. S., Mercado-Silva, N., ... Okie, J. G. (2015). Food Spoilage, Storage, and Transport: Implications for a Sustainable Future. *Bioscience*, *65*(8), 758–768. doi:10.1093/biosci/biv081

Hanh, V. T., Moore, P., & Emerton, L. (n.d.). Review of laws and policies related to payment for ecosystem services in Viet Nam. *Viet Nam Ctry. Program.*

Hansen, P. J., Drost, M., Rivera, R. M., Paula-Lopes, F. F., Al-Katanani, Y. M., Kringinger, C. E., & Chase, C. C. (2001). Adverse impact of heat stress on embryo production. *Theriogenology*, *55*(00), 91–103. doi:10.1016/S0093-691X(00)00448-9 PMID:11198091

Hansen, R., Frantzeskaki, N., McPhearson, T., Rall, E., Kabisch, N., Kaczorowska, A., … Pauleit, S. (2015). The uptake of the ecosystem services concept in planning discourses of European and American cities. *Ecosystem Services, 12*, 228–246.

Harvell, C. D., Kim, K., Burkholder, J. M., Colwell, R. R., Epstein, P. R., & Grimes, D. J. (1505). … Vasta, G. R. (1999). Emerging Marine Diseases--Climate Links and Anthropogenic Factors. [abstract]. *Science*, *285*(5433), LP-1510. Retrieved from http://science.sciencemag.org/content/285/5433/1505

He, K., Song, Y., Daviglus, M. L., Liu, K., Van Horn, L., Dyer, A. R., & Greenland, P. (2004). Accumulated evidence on fish consumption and coronary heart disease mortality: A meta-analysis of cohort studies. *Circulation*, *109*(22), 2705–2711. doi:10.1161/01.CIR.0000132503.19410.6B PMID:15184295

Heal, G., & Millner, A. (2014). ReflectionsUncertainty and Decision Making in Climate Change Economics. *Review of Environmental Economics and Policy*, *8*(1), 120–137. doi:10.1093/reep/ret023

Held, I. M., & Soden, B. J. (2006). Robust Responses of the Hydrological Cycle to Global Warming. *Journal of Climate*, *19*(21), 5686–5699. doi:10.1175/JCLI3990.1

Herczeg, A. L., Dogramaci, S. S., & Leaney, F. W. J. (2001). Origin of dissolved salts in a large, semi-arid groundwater system: Murray Basin, Australia. *Marine & Freshwater Research*, *52*(1), 41–52. doi:10.1071/MF00040

Herrero, M., Wirsenius, S., Henderson, B., Rigolot, C., Thornton, P., Havlík, P., ... Gerber, P. J. (2015). Livestock and the Environment: What Have We Learned in the Past Decade? *Annual Review of Environment and Resources*, *40*(1), 177–202. doi:10.1146/annurev-environ-031113-093503

Hiebenthal, C., Philipp, E., Eisenhauer, A., & Wahl, M. (2012). Interactive effects of temperature and salinity on shell formation and general condition in Baltic Sea *Mytilus edulis* and Arctica islandica. *Aquatic Biology*, *14*(3), 289–298. doi:10.3354/ab00405

Hiebenthal, C., Philipp, E. E. R., Eisenhauer, A., & Wahl, M. (2013). Effects of seawater pCO_2 and temperature on shell growth, shell stability, condition and cellular stress of Western Baltic Sea *Mytilus edulis* (L.) and *Arctica islandica* (L.). *Marine Biology*, *160*(8), 2073–2087. doi:10.100700227-012-2080-9

Hoegh-Guldberg, O., Mumby, P. J., Hooten, A. J., Steneck, R. S., Greenfield, P., & Gomez, E. (1737). … Hatziolos, M. E. (2007). Coral Reefs Under Rapid Climate Change and Ocean Acidification. [abstract]. *Science*, *318*(5857), LP-1742. Retrieved from http://science.sciencemag.org/content/318/5857/1737

Hofmann, E., Ford, S., Powell, E., & Klinck, J. (2001). Modeling studies of the effect of climate variability on MSX disease in eastern oyster (Crassostrea virginica) populations. *Hydrobiologia*, *460*(1), 195–212. doi:10.1023/A:1013159329598

Horrocks, L. A., & Yeo, Y. K. (1999). Health benefits of docosahexaenoic acid (DHA). *Pharmacological Research*, *40*(3), 211–225.

Huang, J., Ji, M., Xie, Y., Wang, S., He, Y., & Ran, J. (2016). Global semi-arid climate change over last 60 years. *Climate Dynamics*, *46*(3–4), 1131–1150. doi:10.100700382-015-2636-8

IIASA/FAO. (2012). *Global Agro-ecological Zones (GAEZ v3.0)*. IIASA and FAO.

Inhamuns, A. J., & Franco, M. R. B. (2008). EPA and DHA quantification in two species of freshwater fish from Central Amazonia. *Food Chemistry*, *107*(2), 587–591. doi:10.1016/j.foodchem.2007.07.032

International Assessment of Agricultural Knowledge (IAASTD). (2009). *Agriculture at a Crossroads, Global Report Chs 1, 4*. Island Press. Retrieved from http:// www.agassessment.org/reports/IAASTD/EN/Agriculture at a Crossroads_Global Report (English).pdf

IPCC. (2007). The Fourth Assessment Report: climate change (AR4). The AR4 Synthesis Report. In *The Working Group I Report: The Physical Science Basis (ISBN 9780521 88009–1 Hardback; 978 0521 70596-7 Paperback), the Working Group II Report Impacts: Adaptation and Vulnerability, the Working Group III Report: Mitigation of Climate Change*. Cambridge University Press.

IPCC. (2013). Climate change 2013: The physical science. In *Contribution of Working Group l to the Fifth Assessment Report of the Intergovernmental Panel on Climate Change*. Cambridge, UK: Cambridge University Press.

Jackson, J. B. C., Kirby, M. X., Berger, W. H., Bjorndal, K. A., Botsford, L. W., Bourque, B. J., ... Warner, R. R. (2001). Historical Overfishing and the Recent Colllapse of Coastal Ecosystems. *Science*, *293*(5530), 629–638. doi:10.1126cience.1059199 PMID:11474098

John, M., Pannell, D., & Kingwell, R. (2005). Climate change and the economics of farm management in the face of land degradation: Dryland salinity in western Australia. *Canadian Journal of Agricultural Economics-Revue Canadienne DAgroeconomie, 53*(4), 443–459. doi:10.1111/j.1744-7976.2005.00029.x

Johnson, J. E., & Welch, D. J. (2010). Marine fisheries management in a changing climate: A review of vulnerability and future options. *Reviews in Fisheries Science*, *18*(1), 106–124. doi:10.1080/10641260903434557

Jones, P. G., & Thornton, P. K. (2003). The potential impacts of climate change on maize production in Africa and Latin America in 2055. *Global Environmental Change*, *13*(1), 51–59. doi:10.1016/S0959-3780(02)00090-0

Jones, P. G., & Thornton, P. K. (2009). Croppers to livestock keepers: Livelihood transitions to 2050 in Africa due to climate change. *Environmental Science & Policy*, *12*(4), 427–437. doi:10.1016/j.envsci.2008.08.006

Khoury, C. K., Bjorkman, A. D., Dempewolf, H., Ramirez-Villegas, J., Guarino, L., Jarvis, A., … Struik, P. C. (2014). Increasing homogeneity in global food supplies and the implications for food security. *Proceedings of the National Academy of Sciences, 111*(11), 4001-4006. Retrieved from http://www.pnas.org/content/111/11/4001.abstract

Knight, A. T., Cowling, R. M., Rouget, M., Balmford, A., Lombard, A. T., & Campbell, B. M. (2008). Knowing but not doing: Selecting priority conservation areas and the research-implementation gap. *Conservation Biology*, *22*(3), 610–617. doi:10.1111/j.1523-1739.2008.00914.x PMID:18477033

Kohlmaier, G. H., Siré, E. -O, Janecek, A., Keeling, C. D., Piper, S. C., & Revelle, R. (1989). Modelling the seasonal contribution of a CO2 fertilization effect of the terrestrial vegetation to the amplitude increase in atmospheric CO2 at Mauna Loa Observatory. *Tellus B, 41 B*(5), 487–510. doi:10.1111/j.1600-0889.1989.tb00137.x

Kooijman, S. A. L. M. (2010). *Dynamic Energy Budget Theory for Metabolic Organisation*. Cambridge, UK: Cambridge University Press.

Lan, K.-W., Lee, M.-A., Zhang, C. I., Wang, P.-Y., Wu, L.-J., & Lee, K.-T. (2014). Effects of climate variability and climate change on the fishing conditions for grey mullet (Mugil cephalus L.) in the Taiwan Strait. *Climatic Change*, *126*(1–2), 189–202. doi:10.100710584-014-1208-y

Last, P. R., White, W. T., Gledhill, D. C., Hobday, A. J., Brown, R., Edgar, G. J., & Pecl, G. (2010). Long-term shifts in abundance and distribution of a temperate fish fauna: A response to climate change and fishing practices. *Global Ecology and Biogeography*, *20*(1), 58–72. doi:10.1111/j.1466-8238.2010.00575.x

Lipper, L., Thornton, P., Campbell, B. M., Baedeker, T., Braimoh, A., Bwalya, M., ... Torquebiau, E. F. (2014). Climate-smart agriculture for food security. *Nature Climate Change*, *4*(12), 1068–1072. doi:10.1038/nclimate2437

Liu, C., Hofstra, N., & Franz, E. (2013). Impacts of climate change on the microbial safety of pre-harvest leafy green vegetables as indicated by Escherichia coli O157 and Salmonella spp. *International Journal of Food Microbiology*, *163*(2), 119–128. doi:10.1016/j.ijfoodmicro.2013.02.026 PMID:23558195

Liu, C., & Xia, J. (2004). Water problems and hydrological research in the Yellow River and the Huai and Hai River basins of China. *Hydrological Processes*, *18*(12), 2197–2210. doi:10.1002/hyp.5524

Liu, J., Li, S., Ouyang, Z., Tam, C., & Chen, X. (2008). Ecological and socioeconomic effects of China's policies for ecosystem services. *Proceedings of the National Academy of Sciences of the United States of America*, *105*(28), 9477–9482. doi:10.1073/pnas.0706436105 PMID:18621700

Lobell, D. B., Burke, M. B., Tebaldi, C., Mastrandrea, M. D., Falcon, W. P., & Naylor, R. L. (2008). Prioritizing Climate Change Adaptation Needs for Food Security in 2030. *Science*, *319*(5863), 607-610. Retrieved from http://science.sciencemag.org/content/319/5863/607.abstract

Lowe-McConnell, R. H. (1979). Ecological aspects of seasonality in fishes of tropical waters. *Symp Zool Soc Lond London*, *44*, 219–241.

Ludwig, F., & Asseng, S. (2006). Climate change impacts on wheat production in a Mediterranean environment in Western Australia. *Agricultural Systems*, *90*(1), 159–179. doi:10.1016/j.agsy.2005.12.002

Mace, B. G. M. (2014). Changes in the perception and goals of nature conservation require a solid scientific basis. *Science*, *345*(6204).

Mace, G. M. (2014). Whose conservation? *Science*, *345*(6204), 1558-1560. Retrieved from http://science.sciencemag.org/content/345/6204/1558.abstract

Mackenzie, C. L., Ormondroyd, G. A., Curling, S. F., Ball, R. J., Whiteley, N. M., & Malham, S. K. (2014). Ocean Warming, More than Acidification, Reduces Shell Strength in a Commercial Shellfish Species during Food Limitation. *PLoS One*, *9*(1), e86764. doi:10.1371/journal.pone.0086764 PMID:24489785

Maczka, K., Matczak, P., Pietrzyk-Kaszyńska, A., Rechciński, M., Olszańska, A., Cent, J., & Grodzińska-Jurczak, M. (2016). Application of the ecosystem services concept in environmental policy—A systematic empirical analysis of national level policy documents in Poland. *Ecological Economics*, *128*, 169–176. doi:10.1016/j.ecolecon.2016.04.023

Majhi, A. (2018). Review article role of integrated duck cum fish farming and its economic efficiency : a study in Purulia district, West Bengal. Academic Press.

Marín, P., Magan, N., Vázquez, C., & González-Jaén, M. T. (2010). Differential effect of environmental conditions on the growth and regulation of the fumonisin biosynthetic gene FUM1 in the maize pathogens and fumonisin producers Fusarium verticillioides and Fusarium proliferatum. *FEMS Microbiology Ecology*, *73*(2), 303–311. doi:10.1111/j.1574-6941.2010.00894.x PMID:20491926

Marques, A., Nunes, M. L., Moore, S. K., & Strom, M. S. (2010). Climate change and seafood safety: Human health implications. *Food Research International*, *43*(7), 1766–1779. doi:10.1016/j.foodres.2010.02.010

Mcclanahan, T., Allison, E. H., & Cinner, J. E. (2015). Managing fisheries for human and food security. *Fish and Fisheries*, *16*(1), 78–103. doi:10.1111/faf.12045

McDaniel, C. D., Bramwell, R. K., Wilson, J. L., & Howarth Birkett, J. R. (1995). Fertility of Male and Female Broiler Breeders Following Exposure to Elevated Ambient Temperatures1. *Poultry Science*, *74*(6), 1029–1038. doi:10.3382/ps.0741029 PMID:7644414

Medina, A., Rodriguez, A., & Magan, N. (2015). Changes in environmental factors driven by climate change: effects on the ecophysiology of mycotoxigenic fungi. In *Climate Change and Mycotoxins* (pp. 71–90). De Gruyter. doi:10.1515/9783110333619-006

Mires, D. (2000). Development of inland aquaculture in arid climates: Water utilization strategies applied in Israel. *Fisheries Management and Ecology*, *7*(1–2), 189–195. doi:10.1046/j.1365-2400.2000.00186.x

Mitchell, M. A., & Kettlewell, P. J. (1998). Physiological stress and welfare of broiler chickens in transit: Solutions not problems! *Poultry Science*, *77*(12), 1803–1814. doi:10.1093/ps/77.12.1803 PMID:9872583

Mooney, H. A., Ehrlich, P. R., & Daily, G. (1997). *Ecosystem Services: A Fragmentary History*. Washington, DC: Island Press.

Muiruri, H. K., & Harrison, P. C. (1991). Effect of Roost Temperature on Performance of Chickens in Hot Ambient Environments. *Poultry Science*, *70*(11), 2253–2258. doi:10.3382/ps.0702253 PMID:1754541

Munday, P. L., Jones, G. P., Pratchett, M. S., & Williams, A. J. (2008). Climate change and the future for coral reef fishes. *Fish and Fisheries*, *9*(3), 261–285. doi:10.1111/j.1467-2979.2008.00281.x

Murray-Tortarolo, G., Friedlingstein, P., Sitch, S., Seneviratne, S. I., Fletcher, I., Mueller, B., ... Zeng, N. (2016). The dry season intensity as a key driver of NPP trends. *Geophysical Research Letters*, *43*(6), 2632–2639. doi:10.1002/2016GL068240

Nam, S., Phommakone, S., Vuthy, L., Samphawamana, T., Son, N. H., Khumsri, M., ... Starr, P. (2015). Lower Mekong fisheries estimated to be worth around $17 billion a year. *Catch Cult.*, *21*, 4–7.

Nayak, P. K., Nayak, A. K., Panda, B. B., Lal, B., Gautam, P., Poonam, A., … Jambhulkar, N. N. (2018). Ecological mechanism and diversity in rice based integrated farming system. *Ecological Indicators, 91*, 359–375.

Neiland, A. E., & Béné, C. (2006). Tropical River Fisheries Valuation: A Global Synthesis and Critical Review. Colombo, Sri Lanka: Academic Press.

NOAASTAT. (2019). Retrieved April 06, 2018, from https://www.st.nmfs.noaa.gov/commercial-fisheries/commercial-landings/annual-landings/index

OECD-FAO (OECD/Food and Agriculture Organization of the United Nations). (2014). *OECD- FAO Agricultural Outlook 2014*. OECD Publishing. doi:10.1787/agr_outlook-2014-

Olson, I. C., Kozdon, R., Valley, J. W., & Gilbert, P. U. P. A. (2012). Mollusk Shell Nacre Ultrastructure Correlates with Environmental Temperature and Pressure. *Journal of the American Chemical Society*, *134*(17), 7351–7358. doi:10.1021/ja210808s PMID:22313180

OPERAs. (2016). *OPERAs- Ecosystem Science for Policy & Practice*. Retrieved April 21, 2018, from http://www.operas-project.eu/

Oppenheimer, M., Campos, M., Warren, R., Birkmann, J., Luber, G., O'Neill, B., Takahashi, K. (2014). Emergent risks and key vulnerabilities. In *Climate Change 2014: Impacts, Adaptation, and Vulnerability. Part A: Global and Sectoral Aspects. Contribution of Working Group II to the Fifth Assessment Report of the Intergovernmental Panel on Climate Change*. Cambridge University Press. Retrieved from http://www.ipcc.ch/report/ar5/wg2/

OPPLA. (2015). Retrieved May 27, 2018, from http://oppla.eu/

Orr, S., Pittock, J., Chapagain, A., & Dumaresq, D. (2012). Dams on the Mekong River: Lost fish protein and the implications for land and water resources. *Global Environmental Change*, *22*(4), 925–932. doi:10.1016/j.gloenvcha.2012.06.002

Paul, I., Mandal, L., & Datta, S. (2018). *Feasibility study of the small-scale poultry-cum-fish farming in homestead ponds for empowerment of rural women: A case study in the new alluvial zone of west*. Academic Press.

Pereira, L. S. (2003). Irrigation Demand Management to Cope with Drought and Water Scarcity BT - Tools for Drought Mitigation in Mediterranean Regions. Dordrecht: Springer Netherlands; doi:10.1007/978-94-010-0129-8_2

Piao, S., Friedlingstein, P., Ciais, P., de Noblet-Ducoudré, N., Labat, D., & Zaehle, S. (2007). Changes in climate and land use have a larger direct impact than rising CO2 on global river runoff trends. *Proceedings of the National Academy of Sciences, 104*(39), 15242 LP-15247. Retrieved from http://www.pnas.org/content/104/39/15242.abstract

Pitt, N. R., Poloczanska, E. S., & Hobday, A. J. (2010). Climate-driven range changes in Tasmanian intertidal fauna. *Marine & Freshwater Research*, *61*(9), 963–970. doi:10.1071/MF09225

Płażek, A., Hura, K., Rapacz, M., & Żur, I. (2001). The influence of ozone fumigation on metabolic efficiency and plant resistance to fungal pathogens. *Journal of Applied Botany and Food Quality*, *75*, •••.

Porter, J. R., Xie, L., Challinor, A. J., Cochrane, K., Howden, S. M., & Iqbal, M. M. (2014). Food security and food production systems. In Climate Change 2014: Impacts, Adaptation, and Vulnerability. Part A. Global and Sectoral Aspects. Contribution of Working Group II to The Fifth Assessment Report of the Intergovernmental Panel on Climate Change. Academic Press.

Pörtner, H. O., & Knust, R. (2007). Climate Change Affects Marine Fishes Through the Oxygen Limitation of Thermal Tolerance. *Science, 315*(5808), 95 LP-97. Retrieved from http://science.sciencemag.org/content/315/5808/95.abstract

Ray, D. K., Gerber, J. S., MacDonald, G. K., & West, P. C. (2015). Climate variation explains a third of global crop yield variability. *Nature Communications*, *6*(1), 5989. doi:10.1038/ncomms6989 PMID:25609225

Ribot, J. C., Magalhães, A. R., & Panagides, S. (2005). *Climate Variability, Climate Change and Social Vulnerability in the Semi-arid Tropics*. Cambridge University Press.

Rice, J. C., & Garcia, S. M. (2011). Fisheries, food security, climate change, and biodiversity: Characteristics of the sector and perspectives on emerging issues. *ICES Journal of Marine Science*, *68*(6), 1343–1353. doi:10.1093/icesjms/fsr041

Rogers, D. J. (1996). Changes in disease vector distributions. In M. Hulme (Ed.), *Climate Change and Southern Africa: An Exploration of Some Potential Impacts and Implications in the SADC Region. Climatic Research Unit*. Norwich: University of East Anglia.

Rosegrant, M. W., & Cline, S. A. (2003). Global Food Security: Challenges and Policies. *Science, 302*(5652), 1917-1919. Retrieved from http://science.sciencemag.org/content/302/5652/1917.abstract

Rosenzweig, C., Casassa, G., Karoly, D. J., Imeson, A., Liu, C., Menzel, A., . . . Hanson, C. E. (2007). Assessment of observed changes and responses in natural and managed systems. In Climate Change 2007: Impacts, Adaptation and Vulnerability. Contribution of Working Group II to the Fourth Assessment Report of the Intergovernmental Panel on Climate Change. Cambridge University Press.

Sandercock, D. A., Hunter, R. R., Nute, G. R., Mitchell, M. A., & Hocking, P. M. (2018, April). Acute Heat Stress-Induced Alterations in Blood Acid-Base Status and\r Skeletal Muscle Membrane Integrity in Broiler Chickens at Two Ages:\r Implications for Meat Quality. *Environmental Toxicology and Chemistry,* 418–425.

Sarmiento, J. L., Slater, R., Barber, R., Bopp, L., Doney, S. C., Hirst, A. C., ... Stouffer, R. (2004). Response of ocean ecosystems to climate warming. *Global Biogeochemical Cycles*, *18*(3), n/a. doi:10.1029/2003GB002134

Schleyer, C., Görg, C., Hauck, J., & Winkler, K. J. (2015). Opportunities and challenges for mainstreaming the ecosystem services concept in the multi-level policy-making within the EU. *Ecosystem Services*, *16*, 174–181. doi:10.1016/j.ecoser.2015.10.014

Schmidhuber, J., & Tubiello, F. N. (2007). Global food security under climate change. *Proceedings of the National Academy of Sciences, 104*(50), 19703-19708. Retrieved from http://www.pnas.org/content/104/50/19703.abstract

Schwinning, S., Sala, O. E., Loik, M. E., & Ehleringer, J. R. (2004). Thresholds, memory, and seasonality: Understanding pulse dynamics in arid/semi-arid ecosystems. *Oecologia*, *141*(2), 191–193. doi:10.100700442-004-1683-3 PMID:15300489

Seager, R., Naik, N., & Vecchi, G. A. (2010). Thermodynamic and Dynamic Mechanisms for Large-Scale Changes in the Hydrological Cycle in Response to Global Warming. *Journal of Climate*, *23*(17), 4651–4668. doi:10.1175/2010JCLI3655.1

Snyder, M. A., Sloan, L. C., Diffenbaugh, N. S., & Bell, J. L. (2003). Future climate change and upwelling in the California Current. *Geophysical Research Letters*, *30*(15), 1–4. doi:10.1029/2003GL017647

St-Pierre, N. R., Cobanov, B., & Schnitkey, G. (2003). Economic Losses from Heat Stress by US Livestock Industries[1]. *Journal of Dairy Science*, *86*, E52–E77. doi:10.3168/jds.S0022-0302(03)74040-5

Steffens, W., & Wirth, M. (2005). Freshwater fish - an important source of n-3 polyunsaturated fatty acids: a review. *Archives of Polish Fisheries/Archiwum Rybactwa Polskiego, 13*(1), 5–16. Retrieved from http://search.proquest.com/docview/17136587?accountid=14643%5Cnhttp://mlbsfx.sibi.usp.br:3410/sfxlcl41?url_ver=Z39.88-2004&rft_val_fmt=info:ofi/fmt:kev:mtx:journal&genre=article&sid=ProQ:ProQ:asfaaquaculture&atitle=Freshwater+fish+-+an+important+source+of

Steinfeld, H., De Haan, C., & Blackburn, H. (1997). *Livestock–Environment Interactions: Issues and options*. Brussels: European Commission Directorate-General for Development.

Stromberg, J. C., Tiller, R., & Richter, B. (1996). Effects of Groundwater Decline on Riparian Vegetation of Semiarid Regions: The San Pedro, Arizona. *Ecological Applications*, *6*(1), 113–131. doi:10.2307/2269558

Sumaila, U. R., Cheung, W. W. L., Lam, V. W. Y., Pauly, D., & Herrick, S. (2011). Climate change impacts on the biophysics and economics of world fisheries. *Nature Climate Change*, *1*(9), 449–456. doi:10.1038/nclimate1301

Teh, L. C. L., & Sumaila, U. R. (2013). Contribution of marine fisheries to worldwide employment. *Fish and Fisheries*, *14*(1), 77–88. doi:10.1111/j.1467-2979.2011.00450.x

Thornton, P. K., & Gerber, P. J. (2010). Climate change and the growth of the livestock sector in developing countries. *Mitigation and Adaptation Strategies for Global Change*, *15*(2), 169–184. doi:10.100711027-009-9210-9

Trenberth, K. E., Dai, A., van der Schrier, G., Jones, P. D., Barichivich, J., Briffa, K. R., & Sheffield, J. (2013). Global warming and changes in drought. *Nature Climate Change*, *4*(1), 17–22. doi:10.1038/nclimate2067

United Nations Convention to Combat Desertification (UNCCD). (2007). *Desertification, exacerbated by climate change, represents one of the greatest environmental challenges of our times*. Retrieved from www.unccd.int/documents/Desertificationandclimatechange.pdf

Università, R. V, Cuore, S., Agriculture, F., Istituto, L. A., Nazionale, C., Nazionale, C., … Elena, V. R. (2012). *Scientific report submitted to EFSA*. Prepared by Battilani P ./ Università Cattolica del Sacro Cuore, Faculty of Agriculture.

Váry, Z., Mullins, E., Mcelwain, J. C., & Doohan, F. M. (2015). The severity of wheat diseases increases when plants and pathogens are acclimatized to elevated carbon dioxide. *Global Change Biology*, *21*(7), 2661–2669. doi:10.1111/gcb.12899 PMID:25899718

Vaughan, M. M., Huffaker, A., Schmelz, E. A., Dafoe, N. J., Christensen, S., Sims, J., ... Teal, P. E. A. (2014). Effects of elevated [CO_2] on maize defence against mycotoxigenic *Fusarium verticillioides*. *Plant, Cell & Environment*, *37*(12), 2691–2706. doi:10.1111/pce.12337 PMID:24689748

Vörösmarty, C. J., Green, P., Salisbury, J., & Lammers, R. B. (2000). Global Water Resources: Vulnerability from Climate Change and Population Growth. *Science, 289*(5477), 284-288. Retrieved from http://science.sciencemag.org/content/289/5477/284.abstract

Walther, G.-R., Post, E., Convey, P., Menzel, A., Parmesan, C., Beebee, T. J. C., ... Bairlein, F. (2002). Ecological responses to recent climate change. *Nature*, *416*(6879), 389–395. doi:10.1038/416389a PMID:11919621

Wang, Y. J., Miller, L., Perren, M., & Addis, P. B. (1990). Omega-3 Fatty Acids in Lake Superior Fish. *Journal of Food Science*, *55*(1), 71–73. doi:10.1111/j.1365-2621.1990.tb06018.x

Welcomme, R. L. (2011). An overview of global catch statistics for inland fish. *ICES Journal of Marine Science*, *68*(8), 1751–1756. doi:10.1093/icesjms/fsr035

Welcomme, R. L., Cowx, I. G., Coates, D., Bene, C., Funge-Smith, S., Halls, A., & Lorenzen, K. (2010). Inland capture fisheries. *Philosophical Transactions of the Royal Society of London. Series B, Biological Sciences*, *365*(1554), 2881–2896. doi:10.1098/rstb.2010.0168 PMID:20713391

White, N., Sutherst, R. W., Hall, N., & Whish-Wilson, P. (2003). The vulnerability of the Australian beef industry to impacts of the cattle tick (Boophilus microplus) under climate change. *Climatic Change*, *61*(1–2), 157–190. doi:10.1023/A:1026354712890

WHO. (1996). *Climate Change and Human Health*. Geneva: World Health Organization.

Wilkinson, C., Saarne, T., Peterson, G. D., & Colding, J. (1998). Strategic Spatial Planning and the Ecosystem Services Concept: an Historical Exploration. *Ecology and Society*, *18*(1), art37. doi:10.5751/ES-05368-180137

Wilson, S. J., Kirby, C. J., Koenigsfeld, A. T., Keisler, D. H., & Lucy, M. C. (1998). Effects of Controlled Heat Stress on Ovarian Function of Dairy Cattle. 2. Heifers. *Journal of Dairy Science*, *81*(8), 2132–2138. doi:10.3168/jds.S0022-0302(98)75789-3 PMID:9749377

Wolfenson, D., Bachrach, D., Maman, M., Graber, Y., & Rozenboim, I. (2001). Evaporative cooling of ventral regions of the skin in heat-stressed laying hens. *Poultry Science*, *80*(7), 958–964. doi:10.1093/ps/80.7.958 PMID:11469662

Wolfenson, D., Roth, Z., & Meidan, R. (2000). Impaired reproduction in heat-stressed cattle: basic and applied aspects. *Animal Reproduction Science, 60–61*, 535–547.

Wolfenson, D., Thatcher, W. W., Badinga, L., Savio, J. D., Meidan, R., Lew, B. J., ... Berman, A. (1995). Effect of heat stress on follicular development during the estrous cycle in lactating dairy cattle. *Biology of Reproduction*, *52*(5), 1106–1113. doi:10.1095/biolreprod52.5.1106 PMID:7626710

World Fish Center (WFS). (2009). *Climate Change: Research to Meet the Challenges Facing Fisheries and Aquaculture*. Issues Brief 1915 World Fish Center.

World Health Organization (WHO). (2014). *Food Security: Trade, foreign policy, diplomacy, Heal*. Retrieved from http://www.who.int/trade/glossary/story028/en/

Young, K. (2009). Omega-6 (n-6) and omega-3 (n-3) fatty acids in tilapia and human health: a review. *International Journal of Food Sciences and Nutrition*, *60*(sup5), 203–211. doi:10.1080/09637480903140503

Ziska, L. H., Morris, C. F., & Goins, E. W. (2004). Quantitative and qualitative evaluation of selected wheat varieties released since 1903 to increasing atmospheric carbon dioxide: Can yield sensitivity to carbon dioxide be a factor in wheat performance? *Global Change Biology*, *10*(10), 1810–1819. doi:10.1111/j.1365-2486.2004.00840.x

KEY TERMS AND DEFINITIONS

Aquaculture: Farming of aquatic organisms and plants known as aquaculture.

Capture Fisheries: Aquatic organisms captured directly from ocean or rivers.

Climate Change: Changing climate due to increased greenhouse gases known as climate change.

Ecosystem: An ecosystem is a community made up of living organisms and nonliving components such as air, water, and mineral soil.

Ecosystem Services: When different component of ecosystem works together to support each other known as ecosystem services.

Food Security: All people at all times have access to sufficient, safe, nutritious food to maintain a healthy and active life.

Grassland: Areas dominated by grasses in an open area of a country.

Ocean Acidification: Addition of carbon dioxide to the ocean, which causes slight acidic environment known as ocean acidification.

Chapter 14
Ecosystem Services–Climate–Health Associations:
Water–Climate–Leishmaniasis Nexus in an Endemic Focus of Zoonotic Cutaneous Leishmaniasis

Ahmed Karmaoui
https://orcid.org/0000-0003-3881-4029
Southern Center for Culture and Sciences, Morocco

Siham Zerouali
Southern Center for Culture and Sciences, Morocco

Ashfaq Ahmad Shah
National University of Sciences and Technology (NUST), Pakistan

Mohammed Yacoubi-Khebiza
Cadi Ayyad University, Morocco

Fadoua El Qorchi
Cadi Ayyad University, Morocco

ABSTRACT

Water is the main ecosystem service that supports the oasis system. Middle Draa Valley is an oasis zone located in the south of Morocco. The water availability in this area is the key element of vegetation cover change. This change added to other factors can cause some parasitic diseases. The zoonotic cutaneous leishmaniasis is one of these diseases. In this chapter, an analysis of the interaction between some key risk factors and the disease transmission was carried out. The outputs of this work rivaled that there is a very strong correlation between this disease and ecosystem services such as water stored and the dam outflow (directed to the oasis for the irrigation), and the groundwater availability. Regarding the correlation between this vector-borne disease and the cropping area, a strong correlation was recorded. However, for the relationship between leishmaniasis and the precipitation and the dam inflow, average correlations were found. Basically, in MDV, the water availability is the first element that affects an ensemble of processes that cause the disease infection.

DOI: 10.4018/978-1-5225-7387-6.ch014

INTRODUCTION

An oasis is an anthropogenic agro-system whose origin is a permanent source of water in most cases an underground groundwater source (Chelleri et al., 2014; Minucci & Karmaoui, 2017; Karmaoui et al., 2015a, 2015b, & 2016a). The water stored in the aquifer is fed mainly by a seasonal and sometimes permanent stream and supports several associated ecosystem services (Food, fiber, timber, drinking water, soil fertility…). Climate change and human intervention through over-exploitation and/or diversion of the watercourse for introduced crop irrigation causes radical spatial-temporal changes in flow (Karmaoui et al., 2015c). Consequently, this can upset the life cycles of animal and plant life. Additionally, this change causes the introduction and sometimes invasion by new plant and animal species including parasitic species. Environmental changes can affect host immune responses and parasite activity (Brunner & Eizaguirre, 2016). Some parasitic species can affect humans. The water parasitic diseases are associated with water for propagation or as a habitat for animal hosts (Lv et al., 2013). In the pre-Saharan zone of Morocco, precisely, the Middle Draa Valley (MDV), parasitic species associated with water are malaria, bilharzias, and leishmaniasis... In fact, these diseases are expanding due to several factors including climate change, intensive farming, dams, irrigation, rapid unplanned urbanization and others (WHO/DCO/WHD/2014.1).

The most cases of zoonotic cutaneous leishmaniasis (ZCL) are localized in the Moroccan oases (Karmaoui, 2018). This is valid also in oases ecosystem in other countries like Algeria, Tunisia, and Libya (WHO, 2010), and in the oases of southern Turkmenia and Uzbekistan (Saf'Janova, 1985).

According to data from the Moroccan Ministry of health (2015), leishmaniasis is the first water-borne disease that affects the oasis population in Morocco. Among the most affected sites in Morocco by this disease is MDV. This valley experienced in 2009, 1882 people affected (Moroccan Ministry of health, 2015), but it is one of the least studied sites in North Africa.

There are many risk factors. In this chapter, only the elements directly associated with water like dam water, cropping area, animal production, precipitation, and temperature were explored. After Ashford (2000), the leishmaniasis is related to the state of the environment. In addition to the environmental changes, the disturbance of the natural balance favoring the presence of the vector and/or the reservoir causes the transmission of leishmaniasis (OMS, 1990). According to Hunter (2003), the increased rainfall supports vegetation growth animal hosts; and then the probability of cutaneous Leishmaniasis cases. Rispail et al., (2002) reported the role of the bio-climate variable on Leishmaniasis risk. Climate change is a factor that impacts the leishmaniasis distribution through the temperature that influences the parasite development in the vector (the sand fly) (Ready, 2010). Furthermore, the environment change and the climate, there are others like the social and economic determinants (Karmaoui & Zerouali, 2018; El-Badry et al., 2009).

Regarding the association of animal owning and the leishmaniasis transmission, in the rural agglomerations, the oasis population keeps animals (livestock) such as goats, sheep and cattle, and sometimes camels. These species support the sand fly, the vector of this disease. The generated organic matter (cow dung) provides a suitable habitat for sand flies, both for spawning and breeding sites (OMS, 2002). In the Millennium Ecosystem Assessment, Campbell-Lendrum et al., (2005) specify that the larvae of these vectors have a terrestrial preference in organic matter moisture.

The main objective of this paper is to analyze the interaction between environmental (water, climate, and vegetation) change and ZCL transmission.

Because of climate, land cover change, and water diversion are risk-factors, the MDV could then be affected by the leishmaniasis transmission. In this chapter, the relationship between these risk factors and ZCL was explored in this area. We attempt to confirm or refute this affirmation basing on data and statistical analysis.

METHODS

Study Area

In order to contribute to the understanding of the functioning of the ecosystems of the region in relation to the vector-borne disease, the environment, physically, biologically and socio-economically was characterized in this section. Situated in the south-east of Morocco, (between the 5°-7°W and 29°-31°N) MDV extends for more than 200 km with an area of about 15,000 km^2 (Figure 1). Overall, it corresponds administratively to the province of Zagora. The oasis is a system artificialized by the man by the introduction of irrigation, which allowed developing a dense vegetation cover. The water requirements of crops and the choice of irrigation techniques depend on production systems.

MDV is characterized by an arid climate characterized by irregular rainfall, high temperatures, and high evapotranspiration. In this climatic environment, water resources are very scarce and water saving is a major constraint.

The increase in current and projected temperature and decreased precipitation (Karmaoui et al., 2019; 2016a) are enough to disrupt the ecosystems of the region and thus expose the welfare of the oasis population to considerable risks. This climate change (CC) has a catastrophic effect on the environment, habitats, economies, and populations. Floods and droughts will become more frequent and more intense as the earth's atmosphere continues to warm up (Karmaoui et al., 2016b).

According to Figure 1, most part of the population (urban cities and the villages) is concentrated along the oases and the Draa Wadi (Temporary River). The water is the essential factor in the concept of the oasis. Traditional oases have been established from easily "mobilizable" water sources (springs, wells, and surface water). The hydrographic network is organized around the Wadi (more than 200 km). The Mansour Eddahbi Dam in upstream of MDV allows the storage of the waters of the two main tributaries of Draa, Dadès, and Ouarzazate Wadis.

Methodology

The methodology is based on a retrospective descriptive study of ZCL epidemiology in MDV. ArcGIS10.1 software was used to carry out the geographical distribution of leishmaniasis patients.

After having mapped the distribution of the population, according to the distribution of palms and water surface source, a trend of six risk factors of ZCL was studied. In order to correlate these variables with the annual ZCL transmission, the available data of the period 1997-2003 were collected and compiled (Figure 2). For vegetation, and animal data, a series of years were chosen which coincides with the selected period (1997-2003). For reasons of comparison, the gathered data were standardized for the period 1997-2003. The correlations were done using Pearson correlation coefficient at a=0.05, following the equation 1 (Eq. 1), using the Arab processor in social statistics software (APSS 1.0).

Figure 1. Localization of the Middle Draa Valley (MDV), showing the geographic distribution of the villages, urban cities, the Mansour Eddahbi Dam, the Draa Wadi (temporary river), and the middle Draa Oases
Source: (Authors)

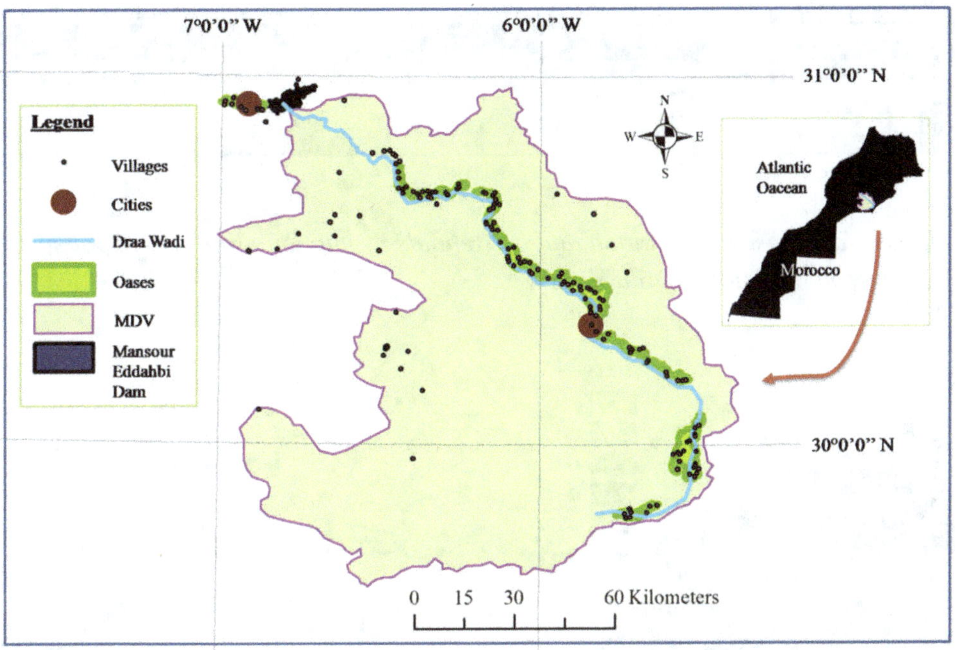

$$r = \frac{1}{n-1}\sum \frac{(x_i - \overline{X})(y_i - \overline{Y})}{s_x s_y} \tag{1}$$

$\overline{X}$ and $\overline{Y}$ are the mains of the X and Y (correlated variables), and S_x and S_y are standard deviations.

RESULTS AND DISCUSSION

The findings show that the most affected people are concentrated in the northwest of the MDV. This repartition is explained by the proximity to the main water resource of the region (Mansour Eddahbi Dam and Draa Wadi). In this zone both population and households are concentrated (Figure 3 a & b).

The Trend of Water-Climate-Zoonotic Cutaneous Leishmaniasis Cases in MDV

The main variables related to ZCL transmission used in this chapter are summarized in Figure 5. It depicts the water, climate, agriculture, and ZCL cases in MDV in the period 1996 to 2003. According to Figure 5, generally, a regressive trend was found in these variables.

Concerning the water, MDV is directly linked with the Mansour Eddahbi Dam that was constructed in 1972 with a capacity of about 536 million m^3 and allows regularizing some 250 Mm^3 per year. The

Figure 2. Data availability of the Middle Draa Valley on 1, Water stored in the Dam; 2, Dam in/outflow; 3, Climatic data; 4, Animal production; 5, Cropping area; 6, Groundwater data, and 7, Leishmaniasis cases
Source: (Authors)

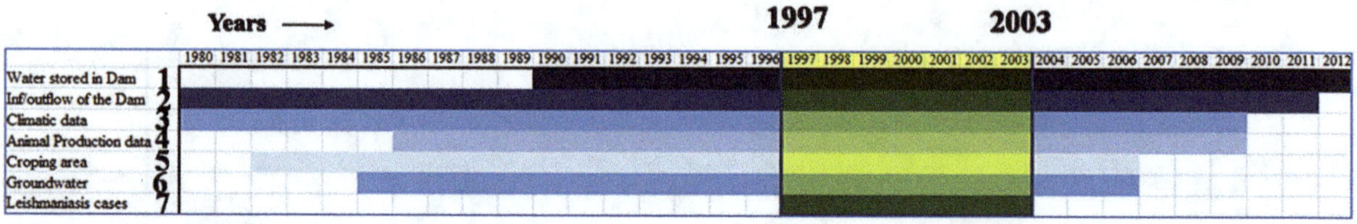

Figure 3. a, Draa Wadi, oases, population and households by municipalities (pink polygons); b, distribution of ZCL cases in the Middle Draa Valley
Source: (Authors)

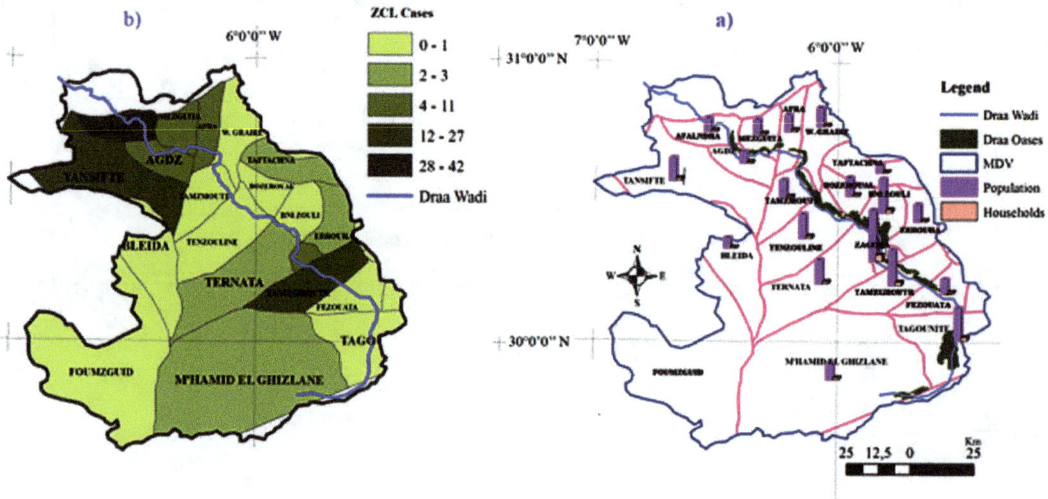

water of this dam can feed groundwater, regulate the flow of Oued Draa and allow the irrigation of six palm groves (oases) aligned from upstream to downstream of the study area as shown in Figure 1. The impact of decreasing precipitation in the upstream part has affected the dam balance (inflow, outflow, and storage) which has repercussions on groundwater level (Figure 4).

The same trend was recorded regarding the evolution of the cropping area and animal production. The water availability is an essential element that supports the vegetation and animal productivity (Figure 4). The negative trend of precipitation can explain the decrease of biomass in this period (represented by vegetation and milk productivity). Talking about ZCL cases, Figure 4 depicts also a negative trend from 1996 to 2003 in this region.

It should be noted that this ZCL cases trend was recorded equally in two other oases provinces as showed in the Figure 5.

The following section shows the correlations between the above-mentioned used indicators and ZCL cases using statistical analysis.

Figure 4. The trend of water-agriculture-climate-Zoonotic Cutaneous Leishmaniasis cases in the Middle Draa Valley. The vegetal productivity represented by cropland area and the animal production represented by the milk production
Source: (Authors)

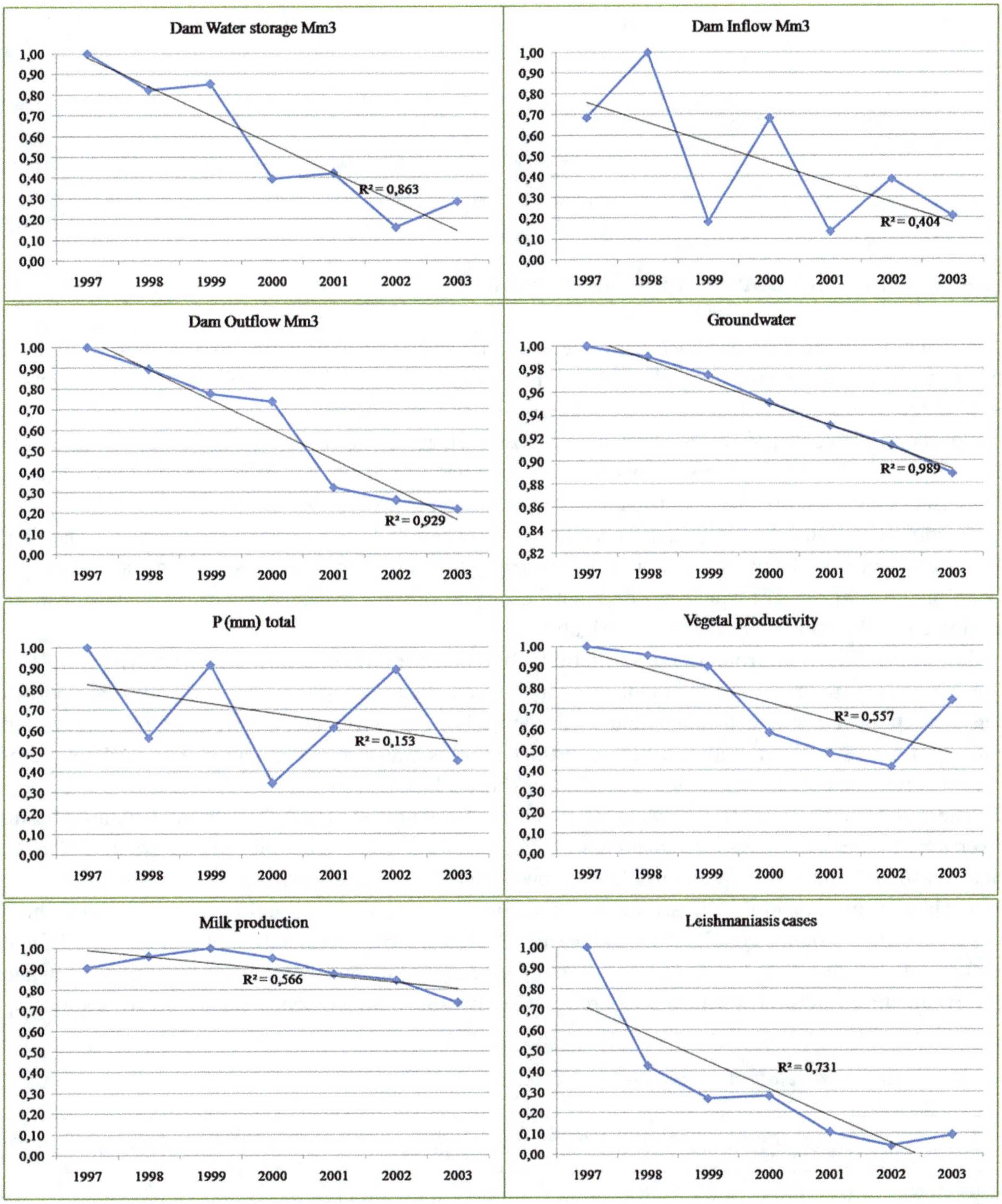

Figure 5. Zoonotic cutaneous leishmaniasis cases trend between 1997 and 2003
Source: (Anon, 2003)

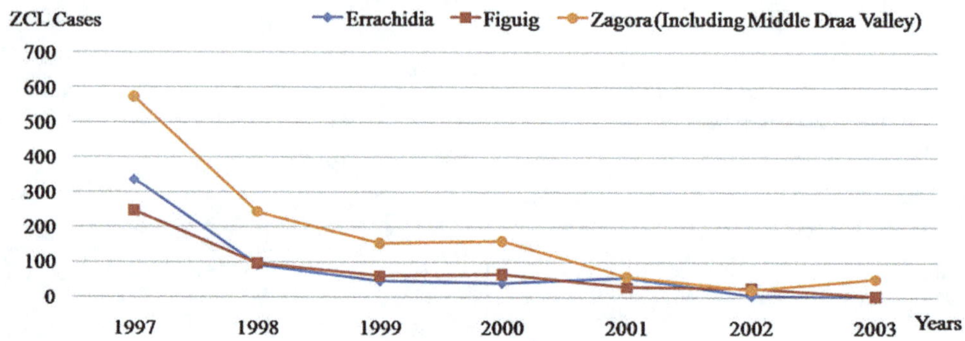

The Relationship Between the Used Factors

Water stored and the dam outflow/inflow, vegetation cover (cropland), animal production (estimated using the milk production), main climatic parameters (precipitation and temperature) are the aspects highlighted in this chapter that can be associated with ZCL expansion.

For the available data, a series of years was chosen where the available data coincide. The available data of the period 1997-2003 was standardized and correlated with the ZCL cases. The correlations were done using Pearson correlation coefficient at a=0.05. The results are shown in Table 1.

According to Table 1, a very strong correlations between ZCL and water stored (r=0,819) and dam outflow (directed to the oasis for the irrigation) (r=0,824) were recorded. This finding was found also by Talmoudi et al., (2017) which reported that ZCL cycle of transmission depends on climate changes originated by the building of dams and irrigation projects.

Regarding ZCL and groundwater availability, a very strong relationship was recorded (r=0,819). However, really, the relationship is indirect, which means that the groundwater availability supports a dense cropping area and then a favorable condition for both reservoir and vectors. This is in agreement with the findings of Mili et al., (2012), reporting that this water source enhances the food source of *Psammomys*, one of the main biological reservoir (host) of ZCL.

There is a strong correlation between ZCL and the cropping area (r=0,736). In the hydraulic basin, diversion of water into canals for irrigation, well digging and dam construction. On the agricultural level, an increase in the quantity of water available allows irrigation of larger agricultural areas throughout the year. The transmission of the disease was recorded in years with high vegetation cover. The same thing was explored by Bounoua et al., 2013, in the pre-Saharan region, and also by Yates et al., (2002), and Faulde et al., (2008). These researchers have found also a positive correlation between precipitation and the leishmaniasis incidence, which agrees relatively with our finding that depict an average relationship

Table 1. Pearson correlation coefficient at a=0.05

	Water Stored Mm3	Inflow Mm3	Outflow Mm3	P (mm) Total	T (°C)	Milk Prod	Vegetal Prod	Groundwater
Leishmaniasis	0,819	0,568	0,824	0,428	-0,386	0,371	0,736	0,819

Source: (Authors)

Figure 6. a, invidious of sand fly (Zoonotic cutaneous leishmaniasis vector); b, Reservoir of the ZCL disease; c, Aspect of a Ksar in Bni-Zouli center (the study area)
Source: (Authors)

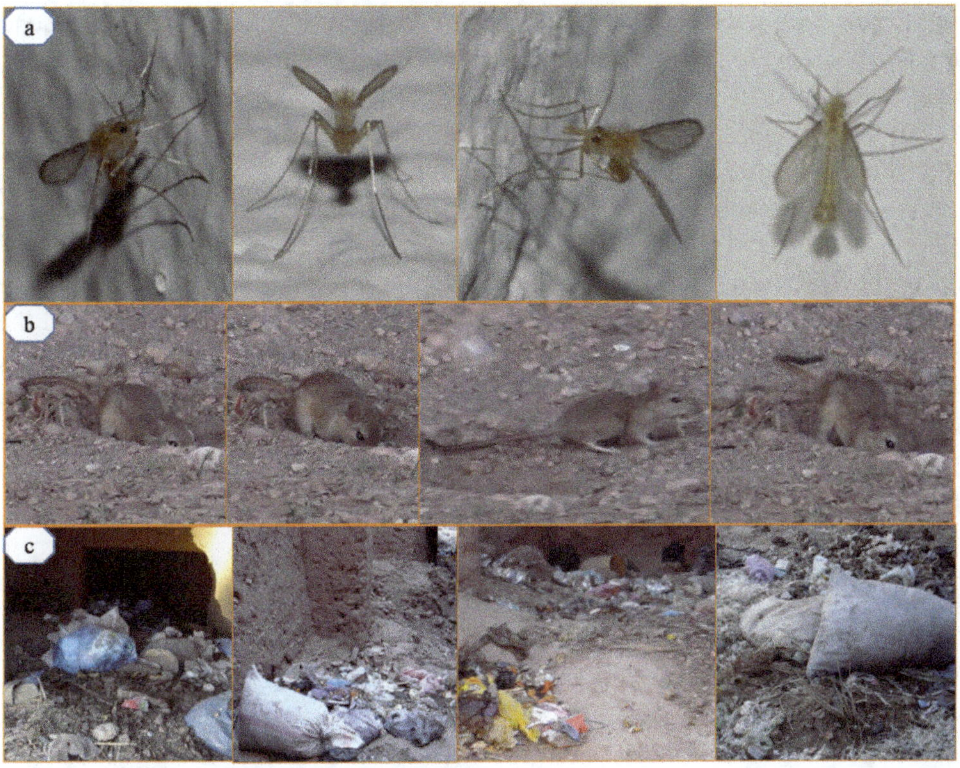

(r=0,428) between these last variables. The same correlation (average correlation, r=0,568) was recorded between ZCL disease and the dam inflow because the incidence depends firstly on the water stored and the flow or releases for irrigation.

Finally, the findings show a low correlation of this disease with the animal production represented by milk production variable (r=0,371), and a negative low relationship between ZCL and temperature (r=-0,386).

The geographic distribution of the population (villages) near the oases and the water surface source increase the risk of the vector-borne disease like leishmaniasis. Karmaoui et al., (2014), recorded a lack of sanitation and household waste management, especially in rural areas. The problem of waste (Figure 6c) creates favorable condition for the vector (Figure 6a) of ZCL (especially, the *Phlebotomus papatasi*) and the host spread (*Meriones shawi,* and *Psammomys obesus*) (Figure 6b).

In this chapter, we note that the weakness of this methodology is the short-term of the used data and space of the study area. Additionally, the large number of cases recorded in this valley is not only favored by to physical factors (climate and ecologic), but also by the socio-economic causes. The social and economic vulnerability and the health infrastructure can be other aspects associated with the spread of ZCL. The uses of aromatic and medicinal plants remain fairly widespread in this area to cure diseases. In all villages, generally, the oldest people, practice traditional medicine in a poor, landlocked environment where health coverage is low.

CONCLUSION

In Morocco, the oases are agrosystems located in the pre-Saharan region. They are the main sources of subsistence for the local population. The structure of the oases is based on the dependence between human, water, and palm trees. The agricultural activity and the water availability in this valley are seriously affecting the biodiversity, climate, and the health. The chapter shows ZCL expansion in a new potential Zoonotic leishmaniasis focus, the MDV. The main objective is to analyze the interaction between environmental change and leishmaniasis transmission. The chapter investigates several risk factors (water availability, vegetation, climatic…) impacting the spatiotemporal distribution of the cutaneous leishmaniasis.

The findings show very strong correlations between ZCL and water stored and the dam outflow, and the groundwater availability. For the correlation between this vector-borne disease and the cropping area, a strong correlation was recorded. Basically, in MDV, the water availability is the first ecosystem services that affect an ensemble of processes that may induce ZCL transmission.

ACKNOWLEDGMENT

We would like to thank Ibn El-Mahdi Eljirari high school for the technical assistance and the anonymous reviewers for the valuable comments.

REFERENCES

Abonnenc, E. (1972). Les phlébotomes de la région éthiopienne (Diptera, Psychodidae). *Cahiers de l'ORSTOM, série Entomologie médicale et Parasitologie, 55*, 1-239.

Anon. (2003). *Etat d'avancement des programmes de lutte contre les maladies parasitaires. Direction de l'épidémiologie et de lutte contre les maladies. Rapport Annuel de la Ministère de la Santé*. Rabat, Maroc: Ministère de la Santé.

Ashford, R. W. (1996). Leishmaniasis reservoirs and their significance in control. *Clinics in Dermatology, 14*(5), 523–532. doi:10.1016/0738-081X(96)00041-7 PMID:8889331

Bounoua, L., Kahime, K., Houti, L., Blakey, T., Ebi, K. L., Zhang, P., ... Messouli, M. (2013). Linking climate to incidence of zoonotic cutaneous leishmaniasis (L. major) in pre-Saharan North Africa. *International Journal of Environmental Research and Public Health, 10*(8), 3172–3191. doi:10.3390/ijerph10083172 PMID:23912199

Brunner, F. S., & Eizaguirre, C. (2016). Can environmental change affect host/parasite-mediated speciation? *Zoology (Jena, Germany), 119*(4), 384–394. doi:10.1016/j.zool.2016.04.001 PMID:27210289

Campbell-Lendrum, D., Molyneux, D., Amerasinghe, F., Davies, C., Fletcher, E., Schofield, C., ... Sinkins, S. (2005). Ecosystems and vector-borne disease control. *Ecosystems and human well-being: policy responses, 3*, 353-372.

Chelleri, L., Minucci, G., Ruiz, A., & Karmaoui, A. (2014). Responses to Drought and Desertification in the Moroccan Drâa Valley Region: Resilience at the Expense of Sustainability? *International Journal of Climate Change: Impacts & Responses*, *5*(2).

El-Badry, A. A., Al-Juhani, A. M., Ibrahim, E. K. D., & Al-Zubiany, S. F. (2009). Sand Flies Distribution and Bionomics, in Al-Madinah Al-Munawarah. *Research Journal of Parasitology*, *4*(1), 1–11. doi:10.3923/jp.2009.1.11 PMID:18576887

Faulde, M., Schrader, J., Heyl, G., Amirih, M., & Hoerauf, A. (2008). Zoonotic cutaneous leishmaniasis outbreak in Mazar-e Sharif, northern Afghanistan: An epidemiological evaluation. *International Journal of Medical Microbiology*, *298*(5-6), 543–550. doi:10.1016/j.ijmm.2007.07.015 PMID:17888728

Karmaoui, A. (2018). The cutaneous leishmaniasis vulnerability index (CLVI). *Acta Ecologica Sinica*, *38*(4), 288–295. doi:10.1016/j.chnaes.2018.01.001

Karmaoui, A., Balica, S. F., & Messouli, M. (2016b). Analysis of applicability of flood vulnerability index in Pre-Saharan region, Morocco. *Natural Hazards and Earth System Sciences*, *2016*. doi:10.5194/nhess-2016-96

Karmaoui, A., Ifaadassan, I., Babqiqi, A., Messouli, M., & Khebiza, M. Y. (2016a). Analysis of the Water Supply-demand Relationship in the Middle Draa Valley, Morocco, under Climate Change and Socio-economic Scenarios. *Journal of Scientific Research & Reports*, *9*(4), 1–10. doi:10.9734/JSRR/2016/21536

Karmaoui, A., Ifaadassan, I., Messouli, M., & Khebiza, M. Y. (2015a). Sustainability of the Moroccan oasean system (Case study: Middle Draa Valley). *Global J Technol Optim*, *6*(01), 170. doi:10.4172/2229-8711.1000170

Karmaoui, A., Ifaadassan, I., Messouli, M., & Khebiza, M. Y. (2015b). Monetarization of Ecosystem Services of Oasean Biome (Case Study: Provisioning Services of Middle Draa Valley Oases, Morocco). doi:10.9734/BJAST/2015/19689

Karmaoui, A., Messouli, M., Khebiza, Y. M., & Ifaadassan, I. (2014). Environmental Vulnerability to climate change and anthropogenic impacts in dryland (Pilot study: Middle Draa Valley, South Morocco). *J Earth Sci Clim Change, S11*, 2. doi:10.4172/2157-7617.S11-002

Karmaoui, A., Messouli, M., & Yacoubi, M. (2015c). Vulnerability of Ecosystem Services to Climate Change and Anthropogenic Impacts in South East of Morocco (Case study: Drying up of Iriki lake). *The International Journal of Climate Change: Impacts and Responses.*, *7*(3), 1835–7156.

Karmaoui, A., Minucci, G., Messouli, M., Khebiza, M. Y., Ifaadassan, I., & Babqiqi, A. (2019). Climate Change Impacts on Water Supply System of the Middle Draa Valley in South Morocco. In *Climate Change, Food Security and Natural Resource Management* (pp. 163–178). Cham: Springer. doi:10.1007/978-3-319-97091-2_8

Karmaoui, A., & Zerouali, S. (2018). Modeling the vulnerability to Zoonotic Cutaneous Leishmaniasis at the local scale. *Asian Journal of Applied Sciences*. Retrieved from https://scialert.net/abstract/?doi=ajaps.0000.91815.91815

Lv, S., Tian, L. G., Liu, Q., Qian, M. B., Fu, Q., Steinmann, P., ... Zhou, X. N. (2013). Water-related parasitic diseases in China. *International Journal of Environmental Research and Public Health, 10*(5), 1977–2016. doi:10.3390/ijerph10051977 PMID:23685826

Mili, A. F., Saghrouni, F., BenSaid, Z., Saadi-BenAoun, Y., Guizani, I., & BenSaid, M. (2012). Retrospective Analysis of Leishmaniasis in Central Tunisia: An Update on Emerging Epidemiological Trends. In Current Topics in Tropical Medicine. InTech.

Minucci, G., & Karmaoui, A. (2017). Exploring the Water-Food-Energy and Climate Nexus: Insights from the Moroccan Draa Valley. In *Peri-Urban Areas and Food-Energy-Water Nexus* (pp. 89–97). Cham: Springer. doi:10.1007/978-3-319-41022-7_11

Moroccan Ministry of Health. (2015). *Santé en Chiffre*. Rabat: Moroccan Ministry of Health.

Ready, P. D. (2010). Leishmaniasis emergence in Europe. *Eurosurveillance, 15*(10), 1–11. PMID:20403308

Saf'janova, V. M. (1985). Leishmaniasis in USSR. In *Leishmaniasis* (pp. 427–435). Amsterdam: Elsevier.

Talmoudi, K., Bellali, H., Ben-Alaya, N., Saez, M., Malouche, D., & Chahed, M. K. (2017). Modeling zoonotic cutaneous leishmaniasis incidence in central Tunisia from 2009-2015: Forecasting models using climate variables as predictors. *PLoS Neglected Tropical Diseases, 11*(8), e0005844. doi:10.1371/journal.pntd.0005844 PMID:28841642

WHO Expert Committee on the Control of the Leishmaniases, Meeting, & World Health Organization. (2010). Control of the Leishmaniases: Report of a Meeting of the WHO Expert Committee on the Control of Leishmaniases, Geneva, 22-26 March 2010 (Vol. 949). World Health Organization.

Yates, T. L., Mills, J. N., Parmenter, C. A., Ksiazek, T. G., Parmenter, R. R., Vande Castle, J. R., ... Peters, C. J. (2002). The ecology and evolutionary history of an emergent disease: Hantavirus pulmonary syndrome. *Bioscience, 52*(11), 989–998. doi:10.1641/0006-3568(2002)052[0989:TEAEHO]2.0.CO;2

KEY TERMS AND DEFINITIONS

Climate: Is the set of atmospheric parameters such as temperature, precipitation, wind speed, relative humidity.

Disease Transmission: Is the passage of a disease from an infected species to another non-infected.

Leishmaniasis: Considered a neglected tropical disease, caused by a protozoan pathogen. Zoonotic Cutaneous Leishmaniasis is among the types of this disease.

Oasis System: Man-made agro-ecosystem located generally in desert zones.

Parasitic Diseases: Infections inducing diseases caused by parasites.

Vector-Borne Disease: Refers to a disease transmitted by a vector, in the case of ZCL is an insect called *Phlebotomus papatasi*.

Zoonotic Cutaneous Leishmaniasis: A specific type of cutaneous leishmaniasis diseases caused by *Leishmania Major* transmitted by the vector *P. papatasi* from the *Meriones shawi* (main reservoir).

Chapter 15
Relationships Between Climate Parameters and the Density of *Phlebotomus papatasi*, the Main Vector of Zoonotic Cutaneous Leishmaniasis

Ahmed Karmaoui
https://orcid.org/0000-0003-3881-4029
Southern Center for Culture and Sciences, Morocco

Siham Zerouali
Southern Center for Culture and Sciences, Morocco

ABSTRACT

This chapter presents an analysis of the seasonal activity of the Phlebotomus papatasi (leishmaniasis vector) through a correlation between its monthly activity and the monthly meteorological parameters. Three sites from three leishmaniasis endemic countries were used Morocco, Iran, and Greece. The high density of P. papatasi was observed in May in Greece, in July in Morocco and September in Iran. Before May and after September, an important decrease of P. papatasi activity was observed with a decline of maximum and minimum temperatures and an increase of precipitations. Secondly, strong associations between P. papatasi density and minimum and maximum temperatures were explored, third, no significant association have been detected between P. papatasi abundance and precipitations in the three sites. There is an average correlation between relative humidity and P. papatasi density for Isfahan and Creater Athens. Basically, P. papatasi is sensitive to low temperatures and high relative humidity, which makes these two variables better indicators for the possible emergence of cutaneous leishmaniasis.

DOI: 10.4018/978-1-5225-7387-6.ch015

INTRODUCTION

Previous studies have investigated partially or mainly on the distribution on *Phlebotomus papatasi* because of its gravity as a vector of leishmaniasis (Cross et al., 1996; Yaghoobi-Ershadi et al., 2010;Soltani et al., 2015). During last five decades, this species was considered as the vector responsible for *L. major* in Mediterranean region mainly with arid climatic conditions (Bally-choumara et al., 1971; Rioux et al., 1984) and in Iran (Yaghoobi-Ershadi et al., 1995; Oshaghi et al., 2010; and Rassi et al., 2011),

According to the available literature, *P. papatasi* is the most abundant species in several countries, in Morocco (Zouirech et al., 2013), in Saudi Arabia (Mustafa et al. 1994; Doha & Samy, 2010), Libya (Dokhan et al., 2016), Iran (Abedi-Astaneh et al., 2015), Egypt (Ali et al., 2016), Tunisia (Chelbi et al., 2007), and in Turkey (Belen et al., 2011).

This species like others insects are climatic sensitive. Variation in climatic variables such as temperature, precipitation and humidity can impact on species distribution and abundance of leishmaniasis vectors (Maltezou, 2008). Topographically, the *P. papatasi* is localized in a large interval of altitude, between 814–1749m (Abedi-Astaneh et al., 2015) and 8–1756m (Yaghoobi-Ershadi, 2012). However, Doha & Samy (2010) recorded a distribution of the same species between 0-200 m and 800-1200 m. It is reported that currently, the leishmaniasis has a large geographical repartition compared to before (Dawit et al., 2013), and favored especially by human and environmental factors (Karmaoui, 2018; Karmaoui & Zerouali, 2018). Abedi-Astaneh et al., (2015) confirmed the high impact of human activities on leishmaniasis repartition.

Concerning the climatic factors, the impact of the mean temperature and relative humidity have been explored by Cross et al., (1996), Tarallo et al., (2010), and Reza & Mansour (2006), the maximum and minimum temperature were studied by Abedi-Astaneh et al., (2015) &Toprak et al., (2007), and the impact of the precipitation by Abdel-Dayem et al., (2012) &Abedi-Astaneh et al., (2015). Always in relation to the environmental factors, *P. papatasi* is frequent in the arid climate(Zhioua et al 2007) and attracted by the light (Volf et al., 2002). Additionally, the activity of *P. papatasi* in summer period was recorded by several authors, Alten et al., (2016) identified its activity in countries from south of Europe (Spain, France, Italy, Greece, and Cyprus), Boussaa et al., (2005), Veysi et al., (2012), and Sari et al., (2015) explored the abundance of *P. papatasi* respectively, in Kars province (Northeastern of Turkey),in Esfahan County(Iran), and in Marrakech city (Morocco). This seasonal activity can be used to explore the leishmaniasis risk (time and space).In this paper, the followings questions were investigated:

1. Is the insect behaves the same way in different areas and in different climatic parameters?
2. What is the common development period of the three sites?
3. Which variables have the greatest correlation with *P. papatasi* activity, and which will be used as relevant indicators of Leishmaniasis disease?

In order to address these issues, a statistic correlations between monthly *P. papatasi* density and monthly precipitation, temperature (maximum and minimum), relative humidity and wind speed were done between three sites from the three selected countries among the affected by leishmaniasis. The three selected sites are presents relatively in different climatic and geographical conditions. A comparison of climatic parameters was done in order to give a global vision of the vector behavior.

MATERIALS AND METHODS

Study Area

For the localization of the study area, three sites from three selected countries of Africa (Morocco), Asia (Iran), and the Europe (Greece) were used (Table 1 & Figure 1). The selection of the studied sites was based on the representativeness of three regions (Climatically and geographically different), where the *P. papatasi* is active (North Africa, South Europe, and Asia).

In order to visualize the study sites, a map was done showing the localization of these sites in the interval between latitude 31° and 36° (Figure 1).

Data and Methodology

Monthly climatic parameters (precipitation, maximum & minimum temperatures, relative humidity, and wind speed) of the three selected sites were acquired from (Tutiempo Network, S.L.) and used. Supplementary information on geographic coordinate, altitude, and annual rainfall were considered.

Data on *P. papatasi* abundance were extracted and normalized from the considered papers (see table 1). The normalized values vary in an interval ranging from 0 to 1 (0 indicates the absence of density and 1 indicates high density).

Climatic data were associated with the normalized values of *P. papatasi* density using determination correlation coefficient (r^2) (Equation 1) and Pearson correlation (r) at a=0.05 (Equation 2), which determines the association between variables defined by:

Table 1. Study sites from three selected countries, Morocco from Africa, Iran from Asia, and Greece from Europe, with the spatial temporal and climatic characteristics

	Country	Morocco	Iran	Greece
Geographic Information	Zone	Marrakech (Urban)	Badrood rural, Isfahan	Crete/Heraklion /Fodele
	Latitude	31°36N	33°44' N	35°22'52"N
	Longitude	8°02W	52°2' E	24°57'29"E
	Altitude	471m	1056m	40-70m
	Period	2002-2003	2001	2011-2012
	Reference	Boussaa et al., (2005)	Yaghoobi-Ershadi et al., (2005)	Veysi et al., (2012)
Climatic Information	Maximum temperature	27,32	25,19	21,61
	Minimum temperature	14,28	7,72	13,23
	Precipitation	19,23	6,26	21,23
	Relative humidity	51,63	33,98	58,38
	Wind speed	9,86	10,03	12,79

Source: (Tutiempo Network, S.L. for climatic data)

Figure 1. Selected sites, Marrakech city in Morocco (North Africa), Esfahan in Iran (Asia), and the Creator Athens Greece (South of Europe)
Source: (Authors)

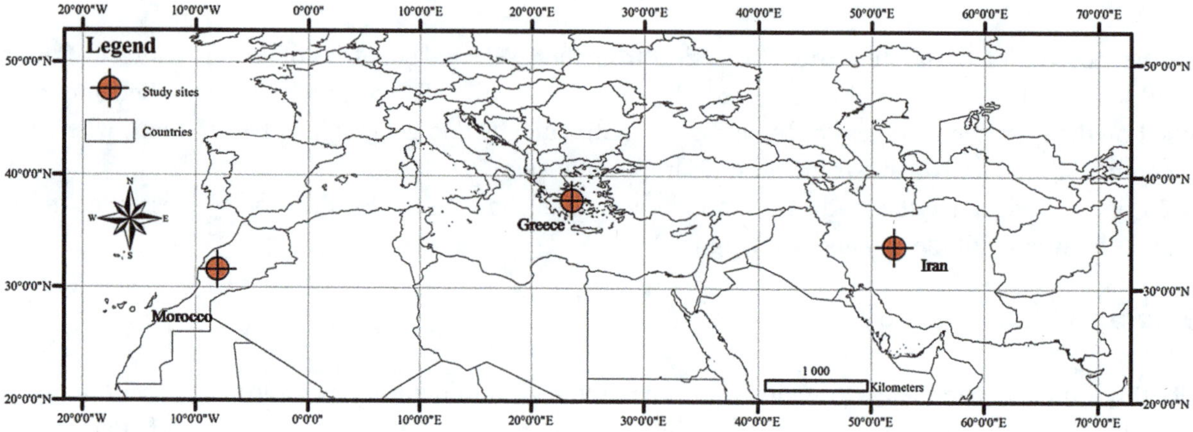

$$r^2 = 1 - \frac{\sum_{i=1}^{n}(yi - \hat{y}i)^2}{\sum_{i=1}^{n}(yi - \overline{y})^2} \in [0,1] \quad (1)$$

where *n* is the number of measurements, yi the measurement value, and $\hat{y}i$ the corresponding predicted value and $\overline{y}$ the average of the measurements.

$$r = \frac{1}{n-1}\sum\frac{(x_i - \overline{X})(y_i - \overline{Y})}{s_x s_y} \quad (2)$$

X and Y are the correlated variables, $\overline{X}$ and $\overline{Y}$, the averages, and S_x and S_y, the standard deviations.

RESULTS AND DISCUSSION

In order to correlate the climatic parameters with the activity of *P. papatasi* (same period of climatic data), three sites from three regions (North Africa, South Europe, and central Asia) were studied. The normalized values of *P. papatasi* activity in the three sites show a common peak in the period ranging from May to September (The hot season). The monthly abundance varies from a site to another. For Esfahan (Iran), the very high activity was recorded in September (1.0) when the maximum temperature was 32.8°C and the minimum temperature was 13,7°C, and the monthly precipitation of about 4.9mm, with a relative humidity of 21,6% and a speed wind of 10km/h (Figure 2). Concerning the Marrakech site (Morocco), the density is bi-modal and the peaks with very high activity were recorded in both May (0,92) and July (1.0) when the maximum temperature in May was 29.50°C and in July was 36.95°C and the minimum temperature is 16.0°C in May and 20.9°C in July and the monthly precipitation was about 13.08mm in May and 2mm in July. In the same time, the relative humidity was about 44.8% in May and was 38.70% in July. Additionally, the wind speed was about 11.55km/h in May and 10.90 km/h in July. Regarding Creater Athens site (Greece), the density is bi-modal and the peaks indicate very high activity in both May (1.0) and August (0,8) when the maximum temperature in May was 22.90°C and in August was 33.9°C and the minimum temperature was 13.9°C in May and 23.1°C in August. However, the monthly recorded precipitation was about 42.16mm in May and zero precipitation in July, with a relative humidity of about 54.7% in May and 40.9 in August. The wind speed was 14.4km/h in May and 15.20 km/h in July.

The big peaks of *P. papatasi* abundance were the principal, observed in three months, May, July, and September. The high density of *P. papatasi* was recorded in May in Greece, in July in Morocco and September in Iran. Before May and after September, an important decrease of *P. papatasi* activity was observed and conjugated with a decline of maximum and minimum temperatures and an increase of precipitations. After these results, basically, the *P. papatasi* is very active relatively in months with high minimum and maximum temperatures. This is consistent with Dokhan et al., (2016) where low temperatures decline the *P. papatasi* activity. Benkova and Volf (2007) explored that at low temperatures, this vector metabolic processes are also low. The *P. papatasi* is active relatively in months with moderate humidity and wind speed, and relative low precipitations. This vector prefers areas with high temperatures and moderate relative humidity was reported by (Lane et al. 1988, Benkova & Volf, 2007).

Talking about the relation between *P. papatasi* and the Cutaneous Leishmaniasis, as mentioned in the introduction section and reported by Ibrahim & Abdoon (2005) and El-Badry et al., (2008), *P. papatasi* raises the risk of Cutaneous Leishmaniasis transmission. The high density coincided with the high Cutaneous Leishmaniasis incidence (Dokhan et al., (2016). The identification of the period peak can be considered as the main control measure of this cutaneous disease.

Regarding the associations between climatic parameters and the *P. papatasi*, a difference between the three sites is shown in figure 3. This latter depicts a correlation between *P. papatasi* abundance and in one hand the maximum and minimum temperature, and in another hand with the precipitation.

The association between temperature (minimum and maximum) and abundance of *P. papatasi* is showed in figure 3. After this later, these two climatic parameters were positively correlated with the *P. papatasi* abundance. In fact, there are average to strong correlation between the maximum and the minimum temperatures and *P. papatasi* density regarding the three selected sites. Figure 3 shows a negative association and a weak correlation concerning the associations between *P. papatasi* density and the precipitations for all sites.

The wind speed (Figure 4) varies positively compared to the *P .papatasi* density but no significant associations have been detected in the three sites.

Usually, the findings depict that minimum and maximum temperatures were positively and strongly correlated with the *P. papatasi* activity. This activity is reverse and negatively correlated with the relative humidity in for Esfahan and Creater Athens, and weakly correlated in Marrakech site (Morocco). Consequently, minimum temperature, maximum temperature, and relative humidity are the most relevant climatic variables of *P. papatasi* activity in these sites and must be used as indicators of Cutaneous Leishmaniasis disease. The vector behavior was confirmed by previous studies. Dokhan et al., (2016) advanced that the sand flies densities including mainly the *P. papatasi* are directly correlated to the temperature and relative humidity. This was consistent with Abedi-Astaneh et al., (2015), which reported that the climatic factors with the highest gain were the monthly max temp–min temp.

Figure 2. Association between P. Papatasi density and the climatic parameters in the three selected sites; the multiple panels are listed as: (1) Monthly P. papatasi activity; (2 and 3) Monthly Maximum and Minimum temperatures in °C; (4) Monthly Precipitation in mm; (5) Monthly relative humidity in %; (6) Monthly wind speed in Km/h
Source: (Authors)

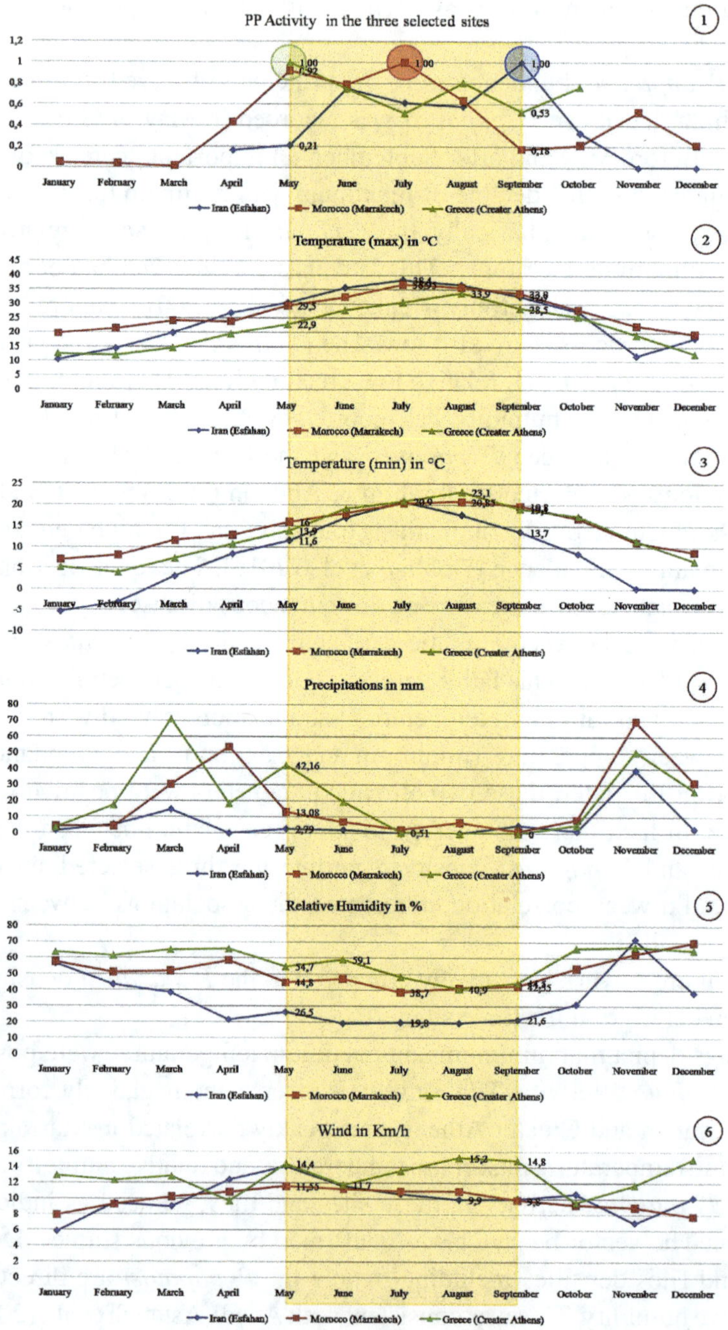

Figure 3. Association between P. papatasi density and maximum, minimum temperature and precipitations in the three selected sites
Source: (Authors)

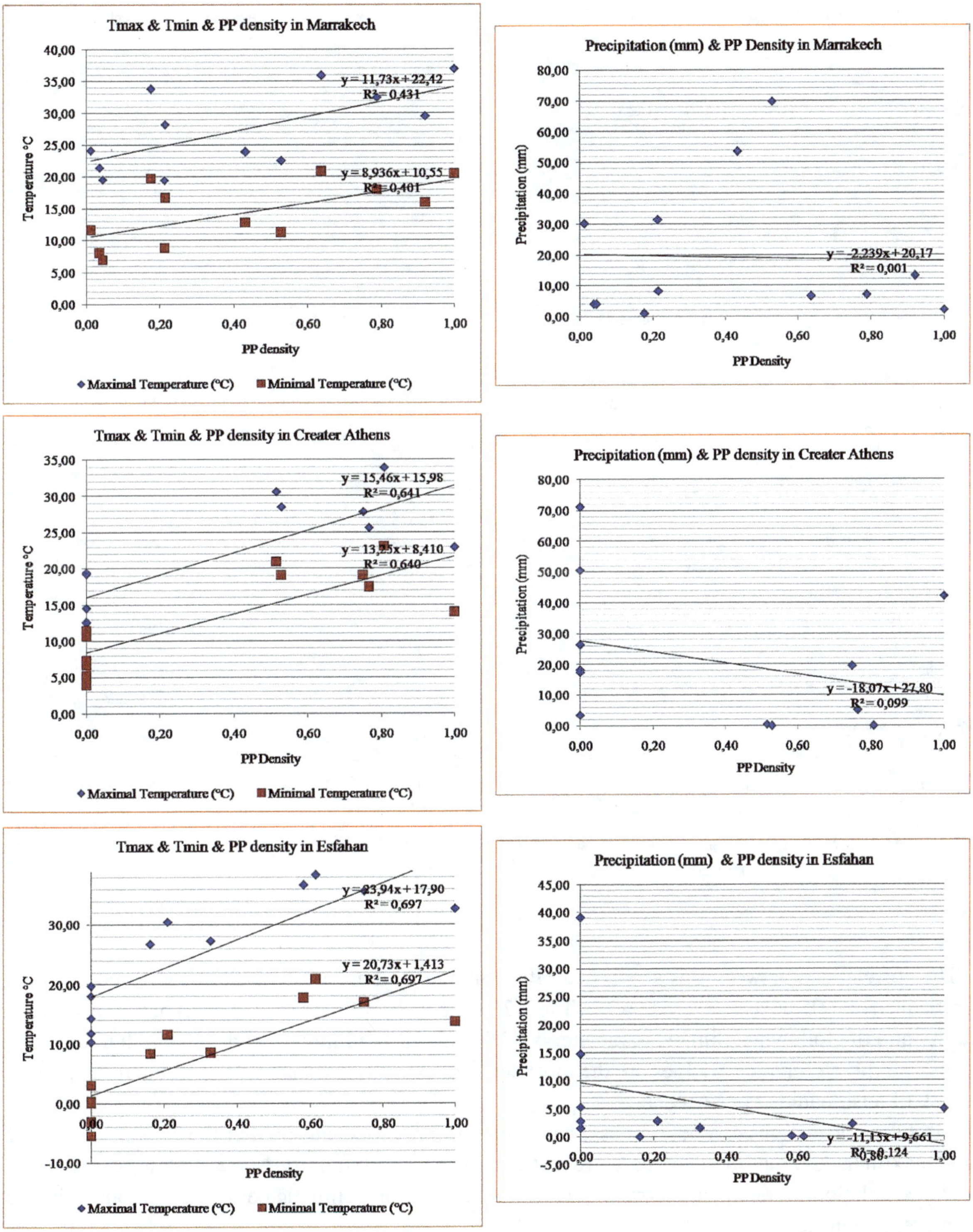

Figure 4. Association between P. papatasi density and relative humidity, and wind speed in the three selected sites
Source: (Authors)

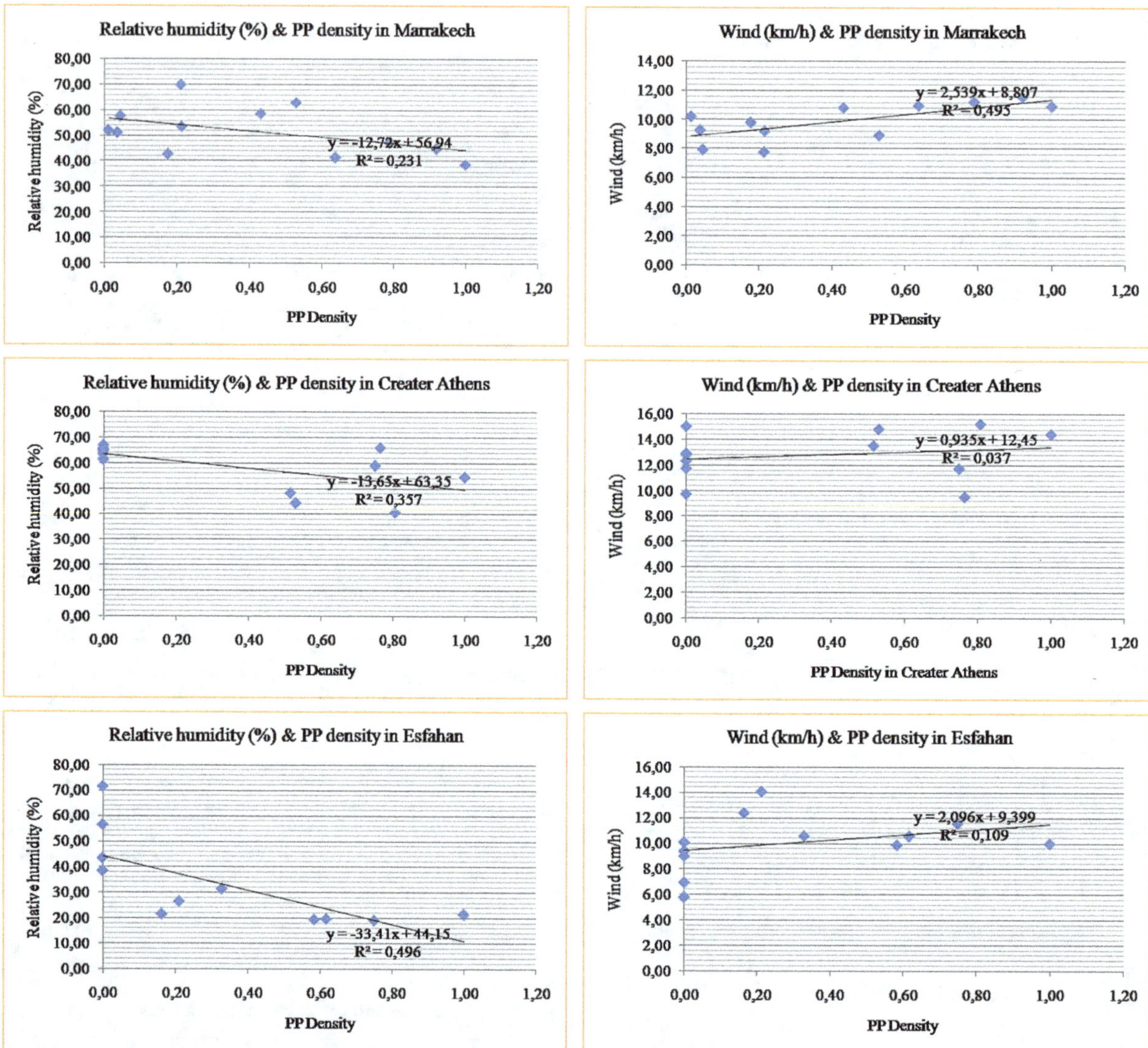

The correlations between this species and the climatic parameters using the Pearson coefficient at significance of 0.05 were done as shown in Table 2. It shows significant associations between the vector activity and the following climatic parameters: T Maximum, T minimum, and Relative humidity. This means that these climatic variables are the most suitable to use as indicators of insect activity and then the risk of leishmaniasis occurrence.

All these correlations can be used as a basis to conceive models to predict the *P. papatasi* activity using maximum and minimum temperature and also the relative humidity. More work is needed regarding the correlation between long term *P. papatasi* abundance and climatic parameters in other sites and in different countries.

Table 2. Pearson correlation between climatic parameters and P. papatasi activity at a=0.05

	TM	Tm	Pre	H	V
Iran	0.835	0.835	-3.52	-0.703	0.33
Correlation	Very strong	Very strong	- Low	- Strong	- Low
Morocco	0.658	0.621	-0.038	-0.484	0.704
Correlation	Very strong	Strong	- Very low	- Average	Very strong
Greece	0.802	0.81	-0.318	-0.598	0.193
Correlation	Very strong	Very strong	- Low	- Average	- Low

Source: (Authors)

Limitations

Two main limitations are associated to the current study:

- The small period of the study,
- Important spatial differences between the selected sites,

CONCLUSION

Monthly activity *P. papatasi* and the monthly meteorological parameters were correlated and compared. Three sites from three leishmaniasis endemic countries from three regions, Morocco from North Africa, Greece from south of Europe, and Iran from Asia were used. The high activity of this species was recorded in May in Greece, in July in Morocco, and in September in Iran. Before May and after September, an important decrease of *P. papatasi* activity was observed with a decline of maximum and minimum temperatures and an increase of precipitations. Secondly, strong associations between *P. papatasi* density and minimum and maximum temperatures were explored, third, no significant association have been detected between this vector abundance and precipitations in the three sites. Basically, the *P. papatasi* is sensitive to low temperatures and high relative humidity, which makes these two variables better indicators for the possible emergence of cutaneous leishmaniasis.

REFERENCES

Abdel-Dayem, M. S., Annajar, B. B., Hanafi, H. A., & Obenauer, P. J. (2012). The potential distribution of Phlebotomus papatasi (Diptera: Psychodidae) in Libya based on ecological niche model. *Journal of Medical Entomology*, *49*(3), 739–745. doi:10.1603/ME11225 PMID:22679884

Abedi-Astaneh, F., Akhavan, A. A., Shirzadi, M. R., Rassi, Y., Yaghoobi-Ershadi, M. R., Hanafi-Bojd, A. A., ... Raufi, H. (2015). Species diversity of sand flies and ecological niche model of Phlebotomus papatasi in central Iran. *Acta Tropica*, *149*, 246–253. doi:10.1016/j.actatropica.2015.05.030 PMID:26071647

Ali, R. M., Loutfy, N. F., Awad, O. M., & Suliman, N. K. (2016). Bionomics of phlebotomine sandfly species in west Alexandria,Egypt. *Journal of Entomology and Zoology Studies*, *4*, 349–353.

Alten, B., Maia, C., Afonso, M. O., Campino, L., Jiménez, M., González, E., ... Toty, C. (2016). Seasonal dynamics of phlebotomine sand fly species proven vectors of Mediterranean leishmaniasis caused by Leishmaniainfantum. *PLoS Neglected Tropical Diseases*, *10*(2), e0004458. doi:10.1371/journal.pntd.0004458 PMID:26900688

Bailly-Choumara, H., Abonnenc, E., & Pastre, J. (1971). Contribution a l'etude des Phlebotomes du Maroc (Diptera, Psychodidae). donnees fauntistiques et ecologiques. *Entomol Med Cah*. Retrieved from http://agris.fao.org/agris-search/search.do?recordID=US201302313429

Belen, A., & Alten, B. (2011). Seasonal dynamics and altitudinal distributions of sand fly (Diptera: Psychodidae) populations in a cutaneous leishmaniasis endemic area of the Cukurova region of Turkey. *Journal of Vector Ecology*, *36*(s1). doi:10.1111/j.1948-7134.2011.00116.x PMID:21366785

Benkova, I., & Volf, P. (2007). Effect of temperature on metabolism of Phlebotomus papatasi (Diptera: Psychodidae). *Journal of Medical Entomology*, *44*(1), 150–154. doi:10.1093/jmedent/41.5.150 PMID:17294933

Boussaa, S., Guernaoui, S., Pesson, B., & Boumezzough, A. (2005). Seasonal fluctuations of phlebotomine sand fly populations (Diptera: Psychodidae) in the urban area of Marrakech, Morocco. *Acta Tropica*, *95*(2), 86–91. doi:10.1016/j.actatropica.2005.05.002 PMID:15985259

Chelbi, I., Derbali, M., Al-Ahmadi, Z., Zaafouri, B., El Fahem, A., & Zhioua, E. (2007). Phenology of Phlebotomus papatasi (Diptera: Psychodidae) relative to the seasonal prevalence of zoonotic cutaneous leishmaniasis in central Tunisia. *Journal of Medical Entomology*, *44*(2), 385–388. doi:10.1093/jmedent/44.2.385 PMID:17427714

Cross, E. R., Newcomb, W. W., & Tucker, C. J. (1996). Use of weather data and remote sensing to predict thegeographic and seasonal distribution of *Phlebotomus papatasi* in southwest Asia. *The American Journal of Tropical Medicine and Hygiene*, *54*(5), 530–536. doi:10.4269/ajtmh.1996.54.530 PMID:8644911

Dawit, G., Girma, Z., & Simenew, K. (2013). A review on biology, epidemiology andpublic health significance of leishmaniasis. *Journal ofBacteriology & Parasitology*, *4*, 166. doi:10.4172/2155-9597.1000166

Doha, S. A., & Samy, A. M. (2010). Bionomics of phlebotomine sand flies (Diptera: Psychodidae) in the province of Al-Baha, Saudi Arabia. *Memorias do Instituto Oswaldo Cruz*, *105*(7), 850–856. doi:10.1590/S0074-02762010000700002 PMID:21120352

Dokhan, M. R., Kenawy, M. A., Doha, S. A., El-Hosary, S. S., Shaibi, T., & Annajar, B. B. (2016). Entomological studies of phlebotomine sand flies (Diptera: Psychodidae) in relation to cutaneous leishmaniasis transmission in Al Rabta, North West of Libya. *Acta Tropica*, *154*, 95–101. doi:10.1016/j.actatropica.2015.11.004 PMID:26589378

El-Badry, A., Al-Juhani, A., Ibrahim, E. K., & Al-Zubiany, S. (2008). Distribution of sand flies in El-Nekheil province, in Al-Madinah Al-Munawwarah region, western of Saudi Arabia. *Parasitology Research*, *103*(1), 151–156. doi:10.100700436-008-0942-3 PMID:18350318

Ibrahim, A. A., & Abdoon, A. M. (2005). Distribution and population dynamics of Phlebotomus sand flies (Diptera: Psychodidae) in an endemic area of cutaneous leishmaniasis in Asir Region, Southwestern Saudi Arabia. *Journal of Entomology*, *2*(Suppl 1), 102–108. doi:10.3923/je.2005.102.108

Karmaoui, A. (2018). The cutaneous leishmaniasis vulnerability index (CLVI). *Acta Ecologica Sinica*, *38*(4), 288–295. doi:10.1016/j.chnaes.2018.01.001

Karmaoui, A., & Zerouali, S. (2018). (in press). Modeling the vulnerability to Zoonotic Cutaneous Leishmaniasis at the local scale. *Asian Journal of Applied Sciences*.

Lane, R. P., Abdel-Hafez, S., & Kamhawi, S. (1988). The distribution of phlebotomine sandflies in the principal ecological zones of Jordan. *Medical and Veterinary Entomology*, *2*(3), 237–246. doi:10.1111/j.1365-2915.1988.tb00190.x PMID:2980180

Maltezou, H. C. (2008). Visceral leishmaniasis: Advances in treatment. *Recent Patents on Anti-infective Drug Discovery*, *3*(3), 192–198. doi:10.2174/157489108786242341 PMID:18991801

Mustafa, M. B., Hussein, S. M., Ibrahim, E. A., Al-Seghayer, S. M., Al Amri, S. A., & Gradoni, L. (1994). Phlebotomus papatasi (Scopoli), vector of zoonotic cutaneous leishmaniasis in Riyadh province, Saudi Arabia. *Transactions of the Royal Society of Tropical Medicine and Hygiene*, *88*(1), 40. doi:10.1016/0035-9203(94)90489-8 PMID:8153996

Oshaghi, M. A., Rasolian, M., Shirzadi, M. R., Mohtarami, F., & Doosti, S. (2010). First report on isolation of Leishmania tropica from sandflies of a classical urban Cutaneous leishmaniasis focus in southern Iran. *Experimental Parasitology*, *126*(4), 445–450. doi:10.1016/j.exppara.2010.05.020 PMID:20570590

Reza, F. M., & Mansour, N. (2006). Entomological studies of Phlebotomus papatasi and P. sergenti (Diptera: Psychodidae) as vectors of cutaneous leishmaniasis in Shiraz, Iran. *The Southeast Asian Journal of Tropical Medicine and Public Health*, *37*, 115–117. PMID:17547064

Rioux, J. A., Rispail, P., Lanotte, G., & Lepart, J. (1984). Relations Phlébotomes-bioclimats en écologie des leishmanioses Corollaires épidémiologiques. L'exemple du Maroc. *Bulletin de la Société Botanique de France. Actualités Botaniques*, *131*(2-4), 549–557.

Sari, B., Limoncu, M. E., Balcioglu, I. C., Aldemir, A., Tasci, G. T., Kiliç, Y., ... Olgen, M. K. (2015). Seroepidemiological and entomological survey in a new focus of zoonotic visceral leishmaniasis in Kars province, Northeastern Turkey. *Veterinary Parasitology*, *209*(3), 179–187. doi:10.1016/j.vetpar.2015.02.021 PMID:25769470

Soltani, Z., Fakoorziba, M. R., Moemenbellah-Fard, M. D., Kalantari, M., Akbarpour, M., Faramarzi, H., & Azizi, K. (2015). Phlebotomus papatasi (Diptera: Psychodidae) as the Vector of Leishmania Major in Kharameh District, Southern Iran. *Journal of Health Sciences and Surveillance System, 3*(4), 160-164.

Tarallo, V. D., Dantas-Torres, F., Lia, R. P., & Otranto, D. (2010). Phlebotomine sand fly population dynamics in a leishmaniasis endemic peri-urban area in southern Italy. *Acta Tropica*, *116*(3), 227–234. doi:10.1016/j.actatropica.2010.08.013 PMID:20816927

Toprak, S., & Ozer, N. (2007). Distribution of sand fly (Diptera: Psychodidae) species and efficiency of capturing methods in Sanliurfa Province, *Turkey. Journal of Medical Entomology*, *44*(1), 23–28. doi:10.1093/jmedent/41.5.23 PMID:17294917

Veysi, A., Vatandoost, H., Yaghoobi-Ershadi, M. R., Arandian, M. H., Jafari, R., & Hosseini, M. (2012). Comparative study on the effectiveness of Coumavec® and zinc phosphide in controlling zoonotic cutaneous leishmaniasis in a hyperendemic focus in central Iran. *Journal of Arthropod-Borne Diseases*, *6*(1), 18. PMID:23293775

Volf, P., Ozbel, Y., Akkafa, F., Svobodova, M., Votýpka, J., & Chang, K. P. (2002). Sand flies (Diptera: Phlebotominae) in Sanliurfa, Turkey: relationship of Phlebotomus sergenti with the epidemic of anthroponotic cutaneous leishmaniasis. *Journal of Medical Entomology*, *39*(1), 12–15. doi:10.1603/0022-2585-39.1.12 PMID:11931243

Yaghoobi-Ershadi, M. R., Akhavan, A. A., Zahraei-Ramazani, A. R., Jalali-Zand, A. R., & Piazak, N. (2005). Bionomics of Phlebotomuspapatasi (Diptera: Psychodidae) in an endemic focus of zoonotic cutaneous leishmaniasis in central Iran. *Journal of Vector Ecology*, *30*(1), 115. PMID:16007964

Yaghoobi-Ershadi, M. R., Hakimiparizi, M., Zahraei-Ramazani, A. R., Abdoli, H., Akhavan, A. A., Aghasi, M., ... Ranjbar, A. A. (2010). Sand fly surveillance within an emerging epidemic focus of cutaneous leishmaniasis in southeastern Iran. *Journal of Arthropod-Borne Diseases*, *4*(1), 17. PMID:22808384

Yaghoobi-Ershadi, M. R., Hakimiparizi, M., Zahraei-Ramazani, A. R., Abdoli, H., Akhavan, A. A., Aghasi, M., ... Ranjbar, A. A. (2010). Sand fly surveillance within an emerging epidemic focus of cutaneous leishmaniasis in southeastern Iran. *Journal of Arthropod-Borne Diseases*, *4*(1), 17. PMID:22808384

Yaghoobi-Ershadi, M. R., Javadian, E., & Tahvildare-Bidruni, G. H. (1995). Leishmania major MON-26 isolated from naturally infected Phlebotomus papatasi (Diptera: Psychodidae) in Isfahan Province, Iran. *Acta Tropica*, *59*(4), 279–282. doi:10.1016/0001-706X(95)92834-3 PMID:8533663

Yavar, R., Abedin, S., Reza, A. M., Ali, O. M., Sina, R., Mehdi, M., ... Babak, F. (2011). Phlebotomus papatasi and Meriones libycus as the vector and reservoir host of cutaneous leishmaniasis in Qomrood District, Qom Province, central Iran. *Asian Pacific Journal of Tropical Medicine*, *4*(2), 97–100. doi:10.1016/S1995-7645(11)60045-X PMID:21771429

Zhioua, E., Kaabi, B., & Chelbi, I. (2007). Entomological investigations following the spread of visceral leishmaniasis in Tunisia. *Journal of Vector Ecology*, *32*(2), 371–374. doi:10.3376/1081-1710(2007)32[371:EIFTSO]2.0.CO;2 PMID:18260530

Zouirech, M., Belghyti, D., El Kohli, M., & Faraj, C. (2013). Entomological investigation of an emerging leishmaniasis focus in Azilal province, Morocco. *Pakistan Entomologist*, *35*, 11–15.

Section 5

Biodiversity and Ecosystem Services Adaptation to Climate Change

Chapter 16
Effectiveness of Local Community Policy Responses to Climate Change Impact on Ecosystem Services for Biodiversity Conservation in the Semi-Arid Zones

Wilson Truman Okaka
Kyambogo University, Uganda

ABSTRACT

The Ugandan government has decentralized climate change policy strategies at the local government district development planning levels to build local community adaptive capacity for accelerated action for disaster resilience. This chapter focuses on the local community response by Isingiro local government district authority. The purpose is to provide a context of the significance of eco-services and natural biodiversity resources in the semi-arid district, located in the South-Western Uganda, elaborate on the district climate change action plan for community-based adaptation strategies, and the biodiversity conservation district development plan. For decades, most households and local communities have perpetually suffered from the severe climatic stress of galloping heatwaves, shocking floods, ferocious frequent whirlwinds and wind storms, catastrophic droughts, perennial food insecurity, malnutrition, migrations, and famines. The district has adopted strategies to enhance biodiversity conservation for eco-services for food security and sustainable community livelihoods.

DOI: 10.4018/978-1-5225-7387-6.ch016

INTRODUCTION

Climate change and variability impact have degraded the fragile ecosystem services and eroded the supply of the natural biodiversity resources in the semi- arid zones of Uganda known as the cattle corridor which occupies almost 40 percent of the total land area in the country. As a result, the Ugandan government launched development planning levels to build local community adaptive capacity for accelerated action for disaster resilience. The purpose is to provide a context of significance of ecoservices and natural biodiversity resources in the semi-arid district, located in the South –Western Uganda, elaborate on the district climate change action plan for community based adaptation strategies, and the biodiversity conservation interventions guided by a five – year district development plan (2020). For decades, most households and local communities have perpetually suffered from the severe climatic stress of galloping heat waves, shocking floods, ferocious frequent whirlwinds and wind storms, catastrophic droughts, food insecurity, and famines. The semi-arid district has adopted accelerated responses to climate change and variability impact; enhance biodiversity conservation and biodiversity based ecoservices for local adaptation strategy for sustainable community livelihoods in Uganda.

ECOSERVICES IN CLIMATE CHANGE REGIME

Furthermore, some of the common social and cultural ecoservices experienced by the local communities in these dry savannah lands are: cultural diversity, arts and crafts, furniture, game hunting, fishing, curios, tourism, settlements, conservation, entertainments, dresses, dances, music, songs, stories or fables or legends, farming, communication, information, ideas, skills, knowledge, roads, power lines, and pipelines, trading centers or markets with a focus on agro-products. Environmental ecoservices: food, medicine, drinks, colors, shades, wind breaks, flood controls, soil conservation, water conservation, climate services, shelter or settlement, beauty, ecotourism, flowers, wood fuel, minerals, raw materials, and political ecoservices: rules and regulations, policies, names, logos, slogans, mottos, emblems, flags, technology, or innovations.

Mainstreaming ecoservices in the implementation of the national and decentralized climate change policy strategy in Uganda (GU, 2015), which is gradually beginning to get off the ground with early adoption of the climate being taken by the most climate change and variability stressed districts like Isingiro district in the South- Western Uganda region. Isingiro district has taken a practical role to mainstream ecoservices in the advent of climate change and variability uncertainty by developing a community based plan to enhance local ecoservices (GU, 2017). The ecosystems services are characterized by different components of the environment or the natural resources which are seen as the local communities' natural, cultural, or fundamental human rights, social, economic, ecological, and local policy participation and their welfare (AU, 2O15). The ecoservices supply in the societal, economic and political structures and systems represent communal livelihoods (INASP, 2016). In abid to enhance the local ecoservices in the district has decided to promote gender equality for women empowerment (GoU, 2012).

Despite the importance of the ecosystems services in Uganda, the arid zones face both the risks and opportunities derived from climate change for socio-cultural, economic, and environmental resources. The climate change and global warming has increased the risks of disasters and opportunities conditions they face. The areas experience the bimodal climatic and weather conditions. The community services in this decentralized local government districts such as Isingiro district have developed plans to enhance ecoservices among the local communities in collaboration with the National Planning Authority of Uganda (NPA). Climate services include a variety of socio-cultural, economic, and environmental functions, values, and benefits such as:

- Promotion of ecotourism (local and international tourists),
- Leisure and entertainment,
- Agriculture and farming services from planting to post-harvesting activities,
- More sources of biomass and solar energy for cooking, lighting, space heating, boiling,
- Seasoning or drying hides/skins, pottery, timber, woodfuels, briquettes, crops, clothes, meat,
- Water and wetlands services (ground, surface, and atmospheric water sources),
- Socio-cultural or traditional ecoservices,
- Markets for food (cereals or crops), fish, livestock (cattle, goats, sheep, donkeys, and poultry)
- Hunting for sports, trade, and food ecoservices
- Human settlements, road infrastructures, water storage systems for production and sanitation
- Local roads or foot paths and water transport services
- Private or communities pasture land for grazing livestocks;

Gender inequality works against the promotion of sustainable ecosystems in most arid zones in Uganda because the bulk of the women's roles are not harnessed for community development. In addition, gender equality, which is defined as equal enjoyment by women and men of socially valued goods, rights, opportunities and resources can be achieved, should be the primary ethical issue in decision on any subject in our day-to-day lifestyle (UNDP, 2005). It is critical to raise awareness of all the practical and strategic gender specific needs are created, raised, developed, addressed, and sustained at individual, household (HH), institutional, and community levels in all sectors of socio-economic, political, and environmental development (IDPU, 2014). Public awareness of ecoservices is improved by climate information services. The current perception and uncertainty is that, the poor in Isingiro district are highly vulnerable to extreme shocks such as the ravages of droughts to crops, livestock, vegetation, and homes (IDLG, 2017). The vulnerability of the district to the stressful effects of climate change depends on the magnitude of its impact, gender sensitivity, and adaptive capacity of the vulnerable communities to combat climate hazards and risks.

Due to severe climate variability, food security remains elusive in Isingiro district which records semi arid bi-modal climate pattern (IDLG, 2017). Agricultural drought has undermined vital nutrition for socio economic HH and community welfare development of the district but acute deficiency of dietary intake is largely caused by HH food insecurity, all of which are directly related to lack of or poor accessibility, availability, utilization, and stability of food including humanitarian emergency food relief (IDLG, 2017). Now, Isingiro district has come up with the concept of gender equality integration in ecoservices, the district community development plan was approved (GU, 2017). Policy evidence of awareness is that, the district has defined gender as the socio - cultural constructs of roles, responsibilities, attributes, opportunities, privileges, status, access to and control over resources and benefits between women and men, boys, and girls (IDLG, 2016, 2017). It has identified key gender specific discrimination, exclusion, deprivation, subordination, oppression, domination, vulnerability, and unfavorable inclusion (IDLG, 2017). Lake Mburo National Park is one of Uganda's eco-tourism hotspots of global significance.

Some of the most common ecoservices are provided by animals (both domestic and wildlife animals). The wildlife animals attract and promote tourism in places like Lake Mburo National Park (adjacent to Isingiro local communities in the South Western Uganda), Kidepo National Park (in Karamoja sub-region in the North Western Uganda), and Karuma National Park in Northern and Western Uganda). The animals provide cultural services, social services, and economic services. Some cultural services include animal totems, bride price for bride price or bride wealth in marriage, exchange of value in payments, celebrations or cultural rites, food and nutrition services, hunted for trophies or meat, skins and hides are used for household services and for making shields. Animals provide political values and services because they act as national symbols. For example, the crested crane bird is Uganda's national bird which is depicted in the national flag and emblem. Likewise, the kob is the national animal which is featured in the national emblem of Uganda. Animals are the livelihoods of the semi-nomadic tribes in the aridity.

ACCESS TO ECOSERVICES IN CLIMATE UNCERTAINTY

In addition, the land based eco-benefits are derived from soil in agriculture or farming, clays for pottery (clay pots), building settlements or homes, cattle feeds (salty mud clays), stones or pebbles, rocks and rock surfaces for drying or seasoning things including agricultural produce, fresh meat, vegetables, entertainment purposes, rituals or cultural rites, and use of hills for grazing livestock (transhumance) including games and spots, cultural rites, fuelwood harvesting, and making gender based acquaintances for social relations, and rites of passage into manhood for boys or womanhood for girls. The undulating hills are also famous for their natural aesthetics and they provide resources for both local and international tourism attractions. Land based ecoservices are a combination of multiple socio-cultural, economic, and environmental services: farming or agricultural services such as the livestock, crops, and fisheries resources. Due to the ferocity of climate change related heavy rains (El Nino) two new lakes emerged in this zone alone bring to three the total number of lakes in the area. The two El Nino lakes were formed in May 2009 in another district called Kiruhura. They now provide water ecoservices to at 3000 livestock belonging to the local pastoralists. The main natural ecoservices here are provided throughout the year by a local small shallow lake called Lake Mburo.

Most households and local communities in Isingiro district are already participating in the local government district development plan implementation to conserve biodiversity, forests, shrubs, pastureland, natural resources, and protect environmental degradation through gender equality awareness-based community climate adaptation programmes (IDLG, 2016-2020). In addition, it will ensure that all traditional and modern ecoservices are well addressed to avert the adverse effects of climate variability risks on the local welfare status (IDLG, 2016, 2017, 2018).

One of the critical needs for effective climate information communication services is the need to bridge the critical information and knowledge gaps that will help to create, raise, promote and sustain public awareness of the functions, values, and benefits of climate change and variability services for HH welfare, incomes, nutrition, and food security.

To this end, gender equality infusion into the ecoservices awareness campaigns assumes the equal enjoyment by women and men of socially valued goods, rights, opportunities, and resources are achieved, as the primary ethical issue in decision on any subject in our day-to-day lifestyle (UNDP, GoU, 2012). Public awareness of both the practical gender needs (PGNs) and strategic gender needs (SGNs) need to be unpacked for all to appreciate (UN, 2018). Awareness should cut across all socio-economic demographics. (Okaka, 2013) to promote effective implementation of the national climate change policy which is based on the following top priority areas (GoU, 2015): adaptation; mitigation; research and observation (including monitoring, detection, attribution, and prediction). This is in line with the East African climate change policy (EAC, 2011) policy that stresses climate change adaptation and mitigation of emissions by members of the sub-regional member states. The locals now say that gender is the social and cultural construct of roles, responsibilities, attributes, opportunities, privileges, status, access to and control over resources and benefits to women and men, boys, and girls (IDLG, 2016, 2017).

Likewise, ecoservices and climate information communication services data are required for the analysis or measuring local climate adaptation performance to: provide feed- back on system behavior and policy performance; improve chances of successful adaptations and implementation; and increase accountability (NEMA, 2003, 2014). The district has adopted a public communication strategy that employs the community media channels, including: those largely oriented to print, those that can be delivered through radio and television, internet-based reporting, and use of alternative communication tools such as posters, discussions, theatre and songs (Okaka, 2017). For example, data are usually collected to measure climate charge service performance as the critical aspects of the local ecoservices which are to be desegregated to show gender inputs, benefits, effects, opportunities, needs, or concerns. Gender analysis perspectives examine the differences in gender roles, activities, needs, opportunities, rights, and their effects on men, women, girls, and boys in a given context. In the past few decades, the local communities in Isingiro district have experienced the bi-modal climatic patterns. Climate change phenomenon is still a subject of community perceptions, myths, anxiety, and uncertainty. For example, the overnight formation of the two local lakes in Kiruhura district in May 2009, which displaced about 170 families, was immediately attributed by some local communities to the presence or actions of the traditional semi or demi-gods called Bachwezi by the local tribes known as the Bahima. But climate variability has increased the stressful intensities, frequencies, or ferocities of extreme weather conditions and climatic events like severe deluges flood, hailstorms, windstorms, torrential rains, and all the types of droughts like: agricultural, metrological, hydrological, and socio-economic droughts experienced intermittently by most of the vulnerable households and communities around the arid zone around the Lake Mburo National Park ecosystem.

DECENTRALISED CLIMATE AND ECOSERVICES

The district has developed a comprehensive local community focused plan to promote access to ecoservices, natural resources, and food security in tandem with decentralized policy process. The issues of local vulnerability to climate change and climate variability have depended on the socio-economic systems in adapting to the impacts of climate change on ecoservices supply. So, some of the most common ecoservices are provided by animals (both domestic and wildlife animals). The wildlife animals attract and promote tourism in places like Lake Mburo National Park (adjacent to Isingiro local communities in the South Western Uganda), Kidepo National Park (in Karamoja sub-region in the North Western Uganda), and Karuma National Park in Northern and Western Uganda). Ecoservices are the lifeblood of all households in arid zones.

The animals provide cultural services, social services, and economic services. Some cultural services include animal totems, bride price for bride wealth in marriage, exchange of value in payments, celebrations or cultural rites, food and nutrition services, hunted for trophies or meat, skins and hides are used for household services and for making shields. Animals provide political values and services because they act as national symbols. For example, the crested crane bird is Uganda's national bird which is depicted in the national flag and emblem. Likewise, the kob is the national animal which is featured in the national emblem of Uganda. Animals are critical to livelihoods of the semi-nomadic tribes there. These include health, access to water, homes, transport, education, agriculture, ecosystem, markets, land, and food security. Most women traditionally depend heavily on climate sensitive natural resources and agriculture which are vulnerable to climate change and variability. The capacity to adapt to the effects of climate change depends mostly on socio-economic factors like: wealth, technology, education, institutions, information, skills, infrastructure, and financial resources or higher incomes.

COMMUNITY-BASED CLIMATE AND ECOSERVICES

Besides, rainfall itself is a ritualistic or symbolic cultural service to the local communities as it at times provides very important cultural events or rites for the families and communities. On the other hand, droughts too call for special the special roles of elders or rain-makers to perform to attrack rainfall through communal payer services. In communities where both livestock and crop production form the bed-rock of the socio-economic life style, water supply is at the centre of all their dreams. Wetlands have been very useful to the local individuals, households, and communities especially for learning life skills or sporting activities like swimming, diving, boating or canoeing, fishing, leisure, and ecotourism. The local authorities are responding to the disastrous effects of climate variability in the district and water projects have been launched by the government and non-governmental organizations. These include: drilling ground water to use water pumps, protecting spring wells drilling shallow wells, and digging valley dams for water livestock like cattle, goats, pigs, camels, and donkeys.

Vulnerability to climate change and climate variability will depend on the socio-economic systems in adapting to the impacts of climate change as climate change is projected to devalue the vulnerable community with respect to socio-economic ecoservices for the livelihoods. These include health, access to water, homes, transport, education, agriculture, ecosystem, markets, land, and food security. Most women traditionally depend heavily on climate sensitive natural resources and agriculture which are vulnerable to climate change and variability. The capacity to adapt to the effects of climate change depends mostly on socio-economic factors like: wealth, technology, education, institutions, information, skills, infrastructure, and financial resources or higher incomes. Ecoservices gender equality (SIDA, 2015) is achieved when women and men, girls and boys, have equal rights, life prospects, opportunities, skills, experience, competence, influence, resources, and power for productive or gainful human employment, food, full human and democratic rights (WHO, 2012).

Engendering ecoservices is a wise option for communities. The Gender Analysis Matrix (GAM) is an analytical tool that uses participatory methodology to facilitate analysis of gender issues by target communities (Rana, 1993). GAM provides a realistic identification of gender specific issues based bottom-up gender analysis (GA) capacity from the grassroots level. Gender specific issues include (UN, 2018) gender mainstreaming (GMS), practical gender needs (PGNs), mainstreaming gender equality (MGE), and strategic gender needs (SGNs). Isingiro locals are mow trying to internalise these concepts. Uganda relies heavily on climate-vulnerable sectors like agriculture, water, wetlands, forests, and biodiversity for socio-economic development. It is vital to manage climate change risks as part of the current district development plan. The integration of climate change risks and opportunities in development activities will enhance resilience to climate change. Mainstreaming climate change into district plans, programmes, and development projects will reduce vulnerability to climate impacts and variability, and increasing the adaptive capacity of communities and activities facing climate impacts on the ecoservices.

Likewise, ecoservices sustainability will ensure progress in the implementation of the SDGs in Uganda. It will also promote socio-economic community development Experience shows the vital contribution that better environmental management can make to improving health, resilience to environmental risks, economic development, and livelihood opportunities, especially for the poor, and fight poverty and to preserve the ecosystems that form the foundation of poor people's livelihoods, pro-poor economic growth, climate and ecoservices, and environmental sustainability are included in the district LGDP. Poverty has heightened anxiety, desperation, and exclusion in segments of the local communities. The district administration units will have developed an integrated community-based plan to mainstream climate change adaptation into its decentralized policy-making, budgeting, programs, policy implementation and monitoring processes at all levels by 2020. The local community-based climate and ecoservices for 2030 SDGs, have has been well identified and the current district development plan identified up to 14 climate change issues with specific sectorial interventions from 2016-2020 and ecoservices are considered as top on the priorities.

For example, administrative services support towards climate change adaptation interventions, will focus on the appointment key environment and climate change focal persons (ECCFPs) and climate change adaptation funds will be raised to facilitate all the socio-economic development sectors, mobilize budget (financial) resources for mitigation of climate change and climate variability effects on the vulnerable local communities in the district. In addition, Isingiro has made bye -laws and ordinances on climate change adaptation strategies aimed at climate change mitigation activities like devegetation restoration, greening degraded bare-hills, and environment protection. The following climate change and variability issues were identified and planned for interventions for community-based climate variability adaptation (IDPU, 2014):

- Production sector interventions have been planned to cover key socio-economic and environmental services.
- Soil fertility restoration and enhancement management
- Water resources sub-sector
- Educational services
- Health services
- Energy supply and demand management strategies
- Roads, buildings, and transport sub-sector infrastructure
- Planning, monitoring, and evaluation process
- Auditing across all sectors based on checklist
- Community services

Likewise, the current planned interventions for community based services will focus on public awareness campaigns on climate change (climate variability) effects on the district; raising public awareness among the local community members on climate change (variability) socio-economic implications on different sectors, mobilizing, and advocating for community-based climate change adaptation (CBCA) strategies, mobilizing, and advocating climate change and climate variability adaptation interventions in local community mobilization activities with a focus on mainstreaming gender equality for local community based climate adaptation strategies for women empowerment in the semi-arid cattle corridor district. A gender issue involves three main elements: gender discrimination, gender gaps and gender oppression. Gender issues affect access to and control over resources, benefits and opportunities within every sector. Gender issues affect the success and sustainability of development plans and programmes within all sectors.

Today, it has also emerged that some of the needs or lessons learned from the effective climate change interventions call for a variety of coordinated interventions (Okaka, 2012): In this regard, the AfDB has launched a fund-drive to build local capacities in climate information gathering, dissemination for adaptation, and mitigation practices to curb the challenges of climate change (Kofi, 2014). In addition, climate information services will better integrate water, land use, energy, food security, and prospects for the youth and women are key inputs for African agricultural value chain for prosperity (Denton, 2014). Access to climate change policy information is one of the key communication challenges facing most communities in all the five east Africa community (EAC) member states. Awareness levels on the opportunities or benefits of EAC integration among the citizens are still too low (Eyotaru, 2013).

More importantly, the vulnerability of human health and socio-economic systems depends on economic and institutional infrastructure (IPCC & UNFCCC, 2014). Technological innovations have increased adaptation options for managed systems. Adaptation options for fresh water resources, efficient management of existing supplies, and infrastructure; institutional arrangements to promote conservation; improved monitoring and forecasting systems for floods or droughts; rehabilitation of watersheds, and construction of new reservoir capacity. Isingiro Adaptation options for agriculture include changes in varieties of crops, improved water-management, irrigation systems, and changes in planting schedules and tillage practices (IPCC & UNFCCC, 2014). The community vulnerability and adaptation assessments indicate that the poor, mostly women are most affected by climate change and variability especially women, girls, or children who face more gender inequality attitudes (GU, 2013).

DEPLOYMENT OF WATER AND WETLANDS ECOSERVICES

It is very important to state that at all times, water resources (wetlands) provide several socio-cultural, economic, and environmental ecoservices in all the Ugandan arid zones especially within the cattle corridor districts which are predominantly rural communities. The following ecoservices are provided by the utilities of natural water resources (surface water, atmospheric precipitation, and ground water): human and animal drinking, foods and drinks preparations, household and community sanitation systems, personal and family or household hygiene such as bathing, cleaning utensils and households or homes. Water is the main source of fishing or fisheries which contributes to 50 percent of the animal proteins in the households. In addition, the wetlands provide entertainments for socio-cultural rites, routines, or events to the local communities where they are found. They provide climate services as they regulate the micro-climates of places where they are located. They are often the natural avenues for local social gatherings or meetings among various peer groups like: boys, girls, women, or men for passing time or negotiating social relationships, market places, and so they are quite therapeutic indeed. Many economic benefits are derived from water services and wetlands are also used for demarcating political maps for administrative boundaries or units from village to district levels.

Furthermore, engendered community radio can play a specific and crucial role in encouraging (Basnet, 2012):public participation; facilitating community-level debates; strengthening cultural and linguistic diversity; providing access to information; the most effective tool for assisting people in poor communities in agriculture, education, health services, building infrastructure, community forestry, environmental conservation, natural disaster preparedness, empowering the marginalized, and establishing good governance; helping to localize information and messages; giving people tools for cultural expression, news, and information, dialogue, and development.

ECOSERVICES FOR HOUSEHOLD AND COMMUNITY LIVELIHOODS

Likewise, the plants provide a variety of critical ecoservices to the local households and communities in the semi-arid cattle corridor of Uganda with respect to social, economic, and environmental benefits. Both domesticated and wildlife plants are characterized as shrubs, grasses or dry savannah grasslands, trees, forests, shrubs, grasses, and crops. The vegetation are very valuable as they enhance food and nutrition security for the sustainable livelihoods of the local population. The domestic and wildlife plants have continued to provide the semi-arid local dwellers with food and nutrition obtained from vegetables, tubers, grains, fruits, leaves, and flowers. Vegetation or plant biodiversity also provide natural aesthetics for beautification of homes, institutions, for social ceremonies or church or religious services. Trees or plants (local forests) provide shade for both people and animals, fuels for cooking, lighting, heating space or boiling water, local herbal medicines for both human and animal health, conserve water in the soil, control flooding or erosion of the soil, and act as a source of building materials. Trees modify the micro-climatic conditions, control or bounce incoming noise and reflect sounds or echoes, and timber or wood. In addition, plants like cotton and coffee crops provide political services to the local communities and the country as whole because they are used as popular national symbols in the Ugandan emblem. They are used because of their socio-economic significance. .

CONCLUSION

Although ecoservices are the lifeblood of all the local communities in the cattle corridor semi-arid districts of Uganda with respect to all manners of socio-cultural, economic, and environmental benefits; almost all these services have been dealt a death blow by the adverse effects of climate change and variability in arid zones hitherto. For example, one of the typical arid zones located in the South Western stretch of the cattle corridor is, Isingiro district which has literally weathered the storms of climate change disasters that have wrecked heavy havoc and destructions to the vital local communal ecoservices. The districts have developed plans to restore and enhance the supply of the socio-ecological ecosystems for sustainable ecoservices for non-conventional or traditional uses and conventional or commercial applications. Today, the most common market products in local markets come from ecoservices outputs.

It is imperative that communities are empowered so that both men and women participate equally in planning, testing, and rolling out adaptation and mitigation activities in rural and urban areas. There are vital gender inequality and equity information, policy, research, knowledge, and practice gaps across the district due to: poor or wrong: attitudes, behavior, planning, perceptions, policies, management, budgeting, leadership, education, communication, practice, staffing, methods, values, and strategies experienced by most countries where lip service is the norm. Most climate change adaptation and mitigation policy weaknesses, failures, illiteracy, and vulnerabilities to the adverse impacts of climate change could be attributed to huge information and knowledge gaps, lack of research, and adaptation activities by the key stakeholders involved.

Besides, the majority of the local communities in the cattle corridor are basically the agrarian people who rely heavily on agricultural livelihoods (livestock farming /animal rearing, crop production, and fisheries). In most cases, they practice mixed farming with many local communities practicing normadism as well as peasant or subsistence agriculture for food and cash needs. The authors examine some of the most common ecoservices in this famous corridor which occupies about 40% of the Ugandan arable land surface but which continues to promote the values, functions, and benefits of ecoservices for sustainable livelihoods and food security.

Ecoservice advocacy communication should be coordinated to achieve a faster rate of adoption and diffusion of innovations, capacity building, funding, technology transfer, collaborative research, and global networking. It should examine: gender blind bureaucracy in decision making; lack of competence in project planning and management; database gaps and weaknesses; institutional capacity weaknesses; national debt crisis; procurement procedures; technology transfer and knowledge sharing; conflicting policies and laws; inherent technical, political and administrative problems; project delays or budget overruns. Ecoservices will thrive on a prudent mitigation, adaptation, and innovation steps can reduce the risks of climate change to ecosystems, food security, water resources, human health, natural, and socio-economic systems. Likewise, household and community access to climate information services is critical for improved ecoservices for most families or communities in the Ugandan semi-arid zones.

REFERENCES

Day, B. A., & Monroe, C. M. (1999). *Environmental Education and Communication for Sustainable World (Handbook for International Practitioners)*. Washington, DC: GreenCom.

Denton, F. (2013). *Africa on the rise: can the opportunities from climate change spring the continent to transformative development?* An inaugural speech at the 3rd annual climate change and development, Addis Ababa, Ethiopia.

Denton, F. (2014). Does climate – smart agriculture hold the key to Africa banishing hunger by 2025? ACPC/UNECA.

European Union. (2000). *Caring for Our Future: Action for Europe's Environment* (3rd.ed.). Luxembourg: EU.

Eyotaru, O. (2013, July 23). East African Integration Hindered by Information Gaps – Minister. Daily Monitor Newspaper, p. 1.

Gollifer, S., & Gorman, S. (2016). *An introductory Workshop Report on Gender Mainstreaming in Higher Education at University of Dodoma*. INASP.

Government of Uganda. (1999). *A Simplified Version of the United Nations Convention on the Elimination of all forms of Discrimination against Women (CEDAW)*. Gov. of Uganda.

Government of Uganda. (2013). *Climate Change Policy 2015. Ministry of Water and Environment.* Kampala, Uganda: Government of Uganda.

Government of Uganda. (2017). *National Population and Houwsing Census (2014): Area Specific Profiles for Isingiro Diistrict*. Kampala, Uganda: Government of Uganda.

IPCC. (2014). *IPCC Synthesis of Scientific-Technical Information Relevant to Interpreting Article 2 of the UN Framework Convention on Climate Change*. IPCC.

Isingiro District Local Government. (2017). *Five Year District Local Government Development*. Author.

Kiruhura District Local Government. (2016). *Five Year District Local Government Development*. Author.

Kofi, A. D. (2014). CCDA- Live. New initiative will help finance efforts to harness information to fight climate change. Daily news from 4th CCDA-8-10 October 2014. *Friday*, *10*, 1.

Lopes, C. (2013). *Can Climate Change Spring Africa's Transformative Development? A Statement by the UN Undersecretary-General and ECA Executive Secretary on 21 October 2013,on the 3rd Annual Conference on Climate Change and Development in Africa from 21-23 October 2013 at United Nations conference centre*. Addis Ababa, Ethiopia: ECA, ACPC.

Ministry of Water, Lands and Environment. (2010). Wetland Sector Strategic Plan 2001-2010, Kampala. Government of Uganda.

National Environment Management Agency. (2012). *State of the Environment Report for Uganda 2002. Ministry of Water, Lands and Environment*. Kampala: Government of Uganda.

National Environment Management Authority. (2000/2001). *State of Environment Report 2000/2001. Government of Uganda, Uganda*. Kampala: NEMA.

National Environment Management Authority. (2003). *Guidelines for District State of Environment Report (DSOER). Preparation in Uganda* [Final Report]. Kampala: NEMA.

Okaka, W. (2002). An Environmental Education Programme: Uganda Polytechnic Kyambogo.Applied environmental education and communication. Taylor & Francis.

Okaka, W. (2007). *The Role of Media Communications in Developing Tourism Policy and Cross-Cultural Communication for Peace, Security for Sustainable Tourism Industry in Africa.* A Paper at the 4th International Institute of Peace through Tourism (IIPT) African Conference on Peace through Tourism at Educators' Forum, Uganda, Kampala.

Okaka, W. (2010). Developing Regional Communications Campaigns Strategy For Environment and Natural Resources Management Policy Awareness For The East African Community. *Research Journal of Environmental and Earth Sciences*, *2*(2), 106–111.

Okaka, W. (2013). *The dawn of community radio in Bangladesh*. Lambert Academic publishers.

Plan, I. I. 2010/2011-2015/2016. Uganda, Kampala: Government of Uganda.

Plan I for 2010/2011-2015. (2016). Kampala, Uganda: Isingiro District Local Government.

Plan II for 2015/2016-2019/2020. Uganda, Kampala: Government of Uganda Isingiro District Local Government. (2017). Five Year District Local Government Development.

Plan II for 2015/2016-2019/2020. Uganda, Kampala: Government of Uganda Isingiro District Local Government. (2017). Five Year District Local Government Development.

Plan II for 2015/2016-2019/2020. Uganda, Kampala: Government of Uganda Isingiro District Planning Unit. (2014). Five Year District Local Government Development.

Rani Parker, A. (1993). *Another Point of View: A Manual on Gender Analysis Training for Grassroots Workers*. New York: Ink. Publishers.

SIDA. (2015). *Gender Amalysis – Principles and Elements. Tool Box 1 (March 2015)*. SIDA.

The United Nations Statistics Division (UNSD). (2018). *Global Gender Statistics Programme*. New York: UN. Available at: http://unstats.un.org/unsd/genderstatmanual/Glossary.ashx

UNDP. (2005). *Uganda human development report 2005. Linking environment to development: A deliberate choice*. Kampala, Uganda: UNDP.

UNECA/ACPC. (2014). *Fourth conference on climate change and development in Africa summary and recommendations*. Marrakech, Morocco: UNECA/ACPC.

Wafula, D. (2012). Where do policy-makers get information? Science Communication Training Workshop for Bio- Innovate Projects, Kigali, Rwanda.

WHO. (2012). *Mainstreaming gender equality in heallth adaptation to climate change programmes. User's guide*. Geneva, Switzerland: WHO.

Chapter 17
Adaptive Strategies of Small Family Farms in the Face of Climate Change:
The Case of the Tleta Watershed in Northern Morocco

Mohamed El Amrani
National School of Agriculture of Meknes, Morocco

ABSTRACT

Climate change is now an accepted reality. It is felt globally through increased temperature and precipitation regime disruption and increased frequency of extreme events such as floods and droughts. In the absence of effective mitigation and adaptation actions, these changes could have significant negative impact on the sustainability of agriculture and the resilience of populations especially in areas with fragile ecology. However, these changes remain an issue that is difficult to grasp and still not well integrated into management strategies at the farm, sector, and territory levels. The objectives of this research are to describe the production systems, and to study the resilience strategies, perception, and adaptive practices of farms in the Tleta watershed in Northwest Morocco in the face of climate change and landscape dynamics. It describes farming systems and activities, attempts to analyze how farmers perceive global changes in their landscape, and adopts innovative strategies and practices to adapt to them. It also shows that the actions of institutional actors in the area that can contribute to the resilience of the populations are numerous but remain fragmentary and lack integration.

DOI: 10.4018/978-1-5225-7387-6.ch017

INTRODUCTION

In recent decades, climate change is a major challenge facing humanity. Although difficult to define, the consequences of this phenomenon are multiple, irreversible and could exceed the response capacity of ecosystems and humans who are likely to be highly affected.

Actions that can be taken with regard to climate change are organized around two main components: reduction of emissions (mitigation) and adaptation (IFPRI, CIRAD). Adaptation to climate change covers very different forms of actions depending on the zones, geographical scales, and sectors. Its implementation challenges what is called "adaptive capacity", which is not the same from one region to another or from one country to another (Driouech, 2010).

Morocco's fragile ecosystems are affected by climate change and are influenced by several factors, including anthropogenic factors. This is the case of the northern zone of Morocco, which is experiencing an erosion rate of 20 to 30 m^3/ ha/ year (Hammouda, 2010).

A study was carried out by FAO, the World Bank and national institutions (CIRAD, IFIPRI) on the impact of climate change on 50 agricultural productions, in the main agro-ecological zones and according to several scenarios of climate change. This study revealed that the drier and warmer climate expected in Morocco would have negative effects on the main rain-fed; whereas, by 2050, the yields of common wheat would decrease by 10% in the wet year, compared with the current situation, but by 33% during the dry years; that several important irrigated crops would also be affected and that the negative impact would grow (FAO, 2008).

The Tleta watershed is a landscape mosaic with rain-fed agro-systems. This basin is located in the northern area of Morocco, in the heart of the Tangier Basin, between the two major urban centers Tangier and Tetouan. This basin has experienced in recent years low agricultural yields and damage to vegetation cover (Kabafing Kourouma, 2017). In addition, erosion losses are enormous and alarming and the Ibn Battouta Dam, which supplies the city of Tangier with drinking water, is continually sinking. Thus, the siltation of this dam over the period 1977-2003 was evaluated 2340m3 / km^2 / year. As a result, its storage capacity is reduced by 70% after only 32 years of activity (El Kemmoune, 2009).

It is from these observations that the problem of this research has emerged, which attempts to describe and explain the adaptive practices of farms in rain-fed agro-systems in the Tleta zone in the face of climate change.

CONTEXT OF THE STUDY AND METHODOLOGY

This chapter was guided by the following objectives:

1. To describe farming systems in the Tleta basin;
2. To analyze what would be the dynamics of the landscape in the future;
3. To study the practices adopted by the farms to adapt to climate change;
4. To analyze the roles of institutions in helping people to be resilient; and
5. To study farmers' perceptions of the dynamics and evolution of their agrarian landscape.

The Tleta watershed is a hydrologic unit located geographically within the western Rif in northern Morocco (Figure 2). It is located in the provinces of Tangier and Tetouan. This watershed occupies an

Figure 1. Example of erosion in the Tleta watershed
Source: (The Author, 2018)

area of 19,961 ha. It is characterized by the presence of the Ibn Batouta Dam in upstream area with an initial storage capacity of 43.6 million m³ in 1977. This capacity decreased to 31 million m³ in 2010.

The study is based on field surveys using questionnaire on a random sample of 101 (51 in 2014 and 50 in 2016) small family farms sample taken at random but taking into account the diversity of situations (production systems, area, upstream or downstream location) in the seven different rural communes of the area. In addition, interviews were conducted with a dozen institutional actors concerned with the sustainable development of the Tleta watershed.

The main axes of the questionnaires focused on the identification of the farm, characterization of the production system, identification of the adaptive practices, perception of climate and landscape changes, and the projects and future prospects of the farms.

The crops grown mainly in the region are: cereals, legumes, market gardening, arboriculture, and fodder (Table 3) and the livestock includes mainly cattle, sheep, and goats (Table 4).

RESULTS

Perception of Global Changes By Farmers Over the Last 10 to 15 Years

In the last 10 to 15 years, the watershed has undergone several changes. In fact, 98% of respondents agreed on the findings of the decrease in rainfall and changes in flora, fauna, and landscape. Decreasing rainfall is considered the major problem since the agriculture practiced is rain-fed, and consequently affects other factors (fauna, flora, and landscape), the decrease of forest areas and the disappearance of certain species (Figure 3).

Farmers also claim to observe a change in the landscape including some areas planted by fruit arboriculture project as part of the projects under the "Moroccan Green Plan", infrastructure (rural roads,

Figure 2. Location of the Tleta watershed
Source: (Report of the Advisory Committee on Advanced Regionalization of the Kingdom of Morocco, 2013)

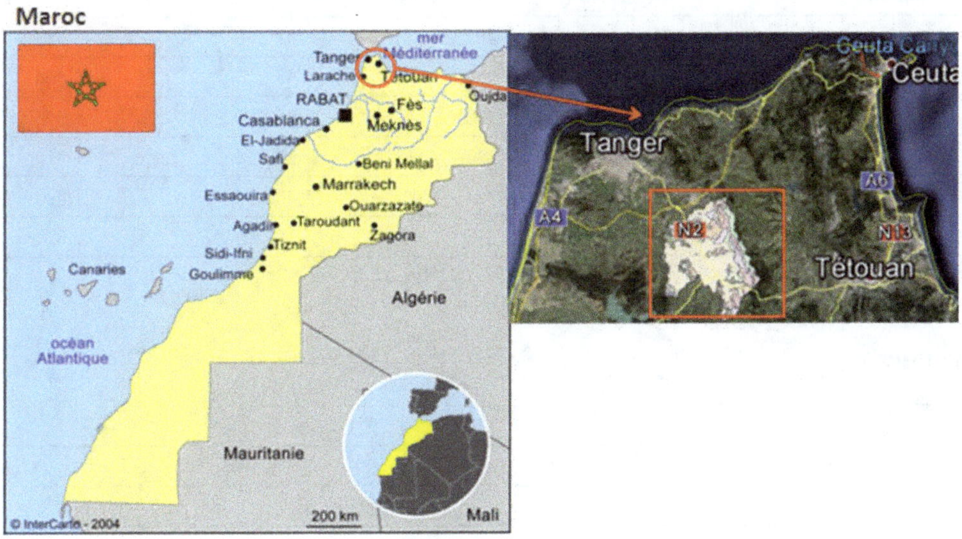

Table 1. Areas of the territorial communes of the Tleta watershed

Province or Prefecture	Caïdats	Area in Ha
Fahs Anjra	Caïdat Anjra	108
	Caïdat Jouamaa	6817
Tanger Assilah	Caïdat Dar Chaoui	302
Tétouan	Caïdat Aïn Lahsen	5180
	Caïdat Ben Karich	13
	Caïdat Jbel Lahbib	4672
	Caïdat Ain Lahcen	599
Total	-	17961

Source: (Authors)

Table 2. Land distribution

Communes	Agricultural Area (ha)		
	Irrigated	Pluvial	Total
Anjra	183	3.018	3201
Jouamaâ	9	2793	2802
Souk Kdim	340	2277	2617
Ain Lahcen	18	4555	4573
Bni Harchen	110	5473	5583
Bghaghza	122	2535	2657

Source: (DRA Tanger-Tétouan, 2016)

Table 3. Areas devoted to different crops grown

Communes	Cereals (ha)	Pulses	Gardening	Fodder
Anjra	11151	455	97	778
Jouamaâ	1146	205	33	406
Souk Kdim	1390	430	350	375
Ain Lahcen	2950	720	200	310
Bni Harchen	3100	780	400	400
Bghaghza	1035	246	169	82

Source: (DRA Tanger-Tétouan, 2016)

Table 4. Livestock

Communes	Livestock Numbers		
	Cattle	Sheep	Goats
Anjra	3690	7695	5700
Jouamaâ	3674	7600	2772
Souk Kdim	2899	3584	2514
Ain Lahcen	2725	3671	3122
Bni Harchen	3654	5354	6068
Bghaghza	2242	9698	8205

Source: (DRA Tanger-Tétouan, 2016)

Figure 3. Perception of global changes by Tleta farmers
Source: (Authors)

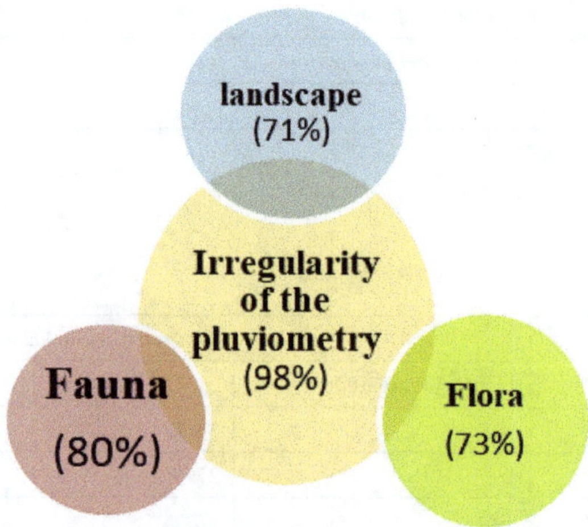

electrification), the construction of the new city "Charafat" as Tangier's urban sprawl that shapes the landscape of Tleta Basin next to the installation of wind turbines and the construction of Renault's industrial plant.

Perception of Climate Change

The climate at the regional level conditions the systems practiced by farmers. Its characterization depends on a farmer to another.

According to respondents (66%), the climate of the region is normal, 32% considers it too dry, 12% sees it too cold, 10% thinks it is too wet, and only 2% thinks it is hot.

In addition, 84% of respondents consider the climatic year as a normal year on the temperature side and 92% on the rainfall side, 14% considers it too cold.

These results are in line with previous work. In fact, in terms of climate, the Tleta basin recorded a decrease in precipitation compared to the average cumulation, calculated over the 1961-1990 period. This decrease is 8% per decade over the winter period and 4% per decade over the fall period. At the annual scale, negative trends in mean precipitation are recorded, representing decennial declines of 5%. In addition, the warming of the climate during the period 1961-2008 is of the order of + 0.1 ° C per decade (Driouech, 2010).

Perception of the Changing Situation of Agriculture and Landscape in the Tleta Watershed

According to farmers, the changes in the Tleta basin during the past decade are linked to olive plantations, reforestation, progress in urbanization, and the progress of cleared land (Figure 4).

The perception of the evolution of agriculture at the zone level differs from a farmer to another. 70% of respondents consider the practice of agriculture has become increasingly difficult (Figure 6)

Figure 4. Perception of Tleta landscape dynamics by farmers
Source: (Authors)

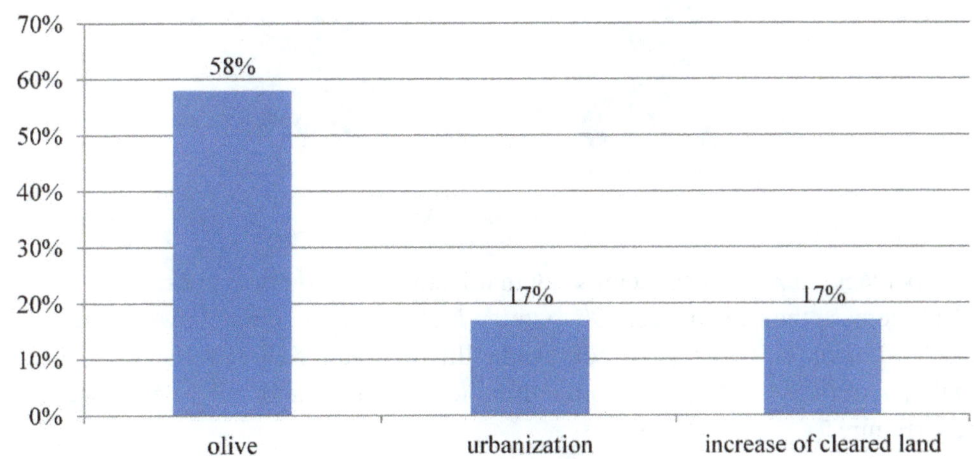

Figure 5. Landscapes of Tleta
Source: (Authors)

Figure 6. Perception of the agricultural situation at the regional level
Source: (Authors)

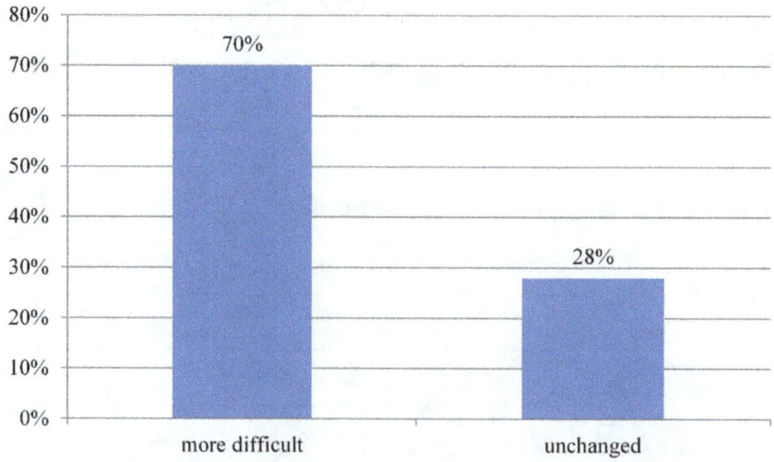

Ninety-one percent (91%) of respondents say that institutions do not support them in order to help their farms and increase their resilience. 89% cite the high cost and complexity of farming techniques and low household incomes. 63% reports non-availability of water, 51% have land issues such as fragmentation and lack of titles, 46% reports competition for access to land and 29% for security and theft due to poverty (Figure 7).

Figure 7. Reasons Explaining Perception of the Difficulty of Farming Practice by Farms
Source: (Authors)

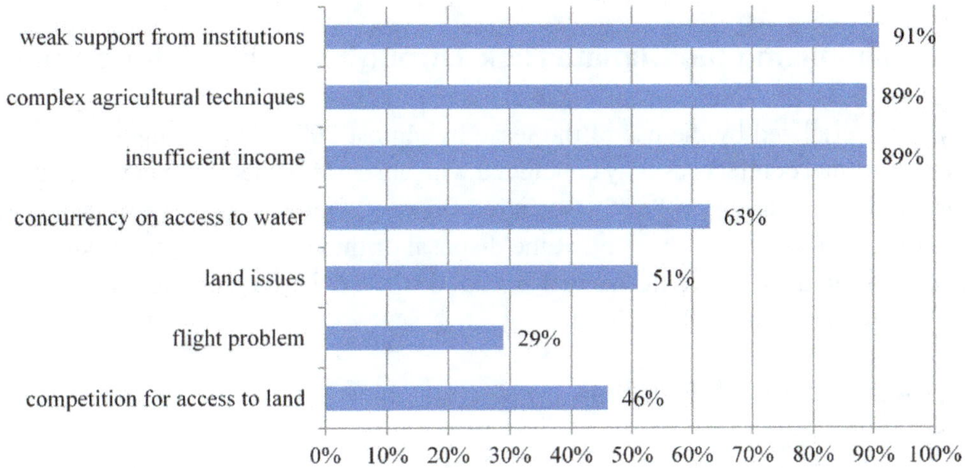

Future Perception of the Agriculture in the Region According to the Respondents

Ninety-four percent (94%) of farmers believe that the future of agriculture in the region is dependent on state support and its administrations.

They state that they need financial support and ongoing advisory support, the creation of improved varieties of cereals and other valuable crops at affordable prices, the implementation of rural development programs and projects where the benefits of which are accessible to the poor rather than to certain "fundraisers" that drip the various sources of funding (INDH, ADS, and PMV). 92% consider that the future of agriculture is dependent on the creation and support of professional agricultural organizations, which are currently very little present. 86% think private companies are almost absent, 80% believe that access to the national and international market is necessary to be able to market and value their production.

Adaptive Strategies of Farms to Cope With Climate Change

In these very constraining biophysical and climatic conditions, farmers of the Tleta watershed have had to develop several adaptation strategies over time. Most are "defensive" to allow them to increase resilience.

Agricultural Land Valorization Strategy

The land status of the area is dominated by the Melk (private property) (95%). The collective represents about 5%. Most farmers do not own a land title.

The land structure of the farms is characterized by the dominance of the micro-property which is grafted on the enclosed of the landscape and soil erosion since 42% of the farms surveyed have an area below 5 ha, 36% have between 5 and 10 ha, 8% have between 10 and 15 ha. The largest area in the area is 22ha while the smallest area is 0.4 ha Micro-ownership is a real handicap to the mechanization of agricultural work and the use of agricultural inputs to increase crop yields. But it is also a factor encouraging certain explanations to intensify cropping such as the small market gardening or corn.

Strategy for Minimizing the Climate Risk Through the Use of the Lease

This strategy is materialized by the use of the lease by almost 20% of the farms surveyed. These are either informal or formal contracts usually concluded with absentee owners who have inherited plots but who live outside the zone, or the mode of association Bennesf, that is to say, half of the harvest. In this type of association, the owner puts his land at the disposal of the farmer; the latter takes care of all the inputs necessary for production and the harvest is shared half late in the growing season. crop residues, such as cereal straws, stubble are rarely used.

Life Strategy in Extended Families

In the study area, the expanded nature of families living under the same roof reflects social cohesion. It is a singular practice in this area. Indeed, extended families consist of two households (20%), three households (15%) and four households (5%).

The workforce used in the farming tasks mostly made of the family members (70% full-time). Employees are usually engaged in farm work for older farmers who do not have sons present in the area. They work full time with 11%, part time with 6% and seasonal workers with 13%. The wages in the agricultural sector are considered very low paid.

Diversification Strategy of the Sources of Incomes

Forty three percent (43%) of farmers resort to the processing of dairy products (cheese, beaten milk ..) according to local know-how, 13% take advantage of the forests containing aromatic and medicinal plants for the breeding of bees and the production of honey, 2% uses the rental of their tractor.

20% of the respondents' family members use off-farm activities that allow them to participate in household and farm expenses, i.e. building construction, handicrafts (baskets or *Doum* hats) or in the small business. They say that their strategy of survival in this constraining context consists in "putting a foot in the city and a foot in the countryside". This expression of the farmers is eloquent and testifies to the feeling of insecurity among the people about the future of their farms.

Strategy for Diversification of Agricultural Production Systems

All farms combine the practice of cultivation and animal husbandry, the latter is considered as mobile cash which is used whenever is needed to meet the households need. The rotation is characterized by the diversification of vegetable productions. Most of the land is planted with cereals, fodder, market gardening and food legumes, fig and olive trees (Table 5).

Livestock is the main source of financing for farms. It is practiced by 60% of the farms and the size of the herd varies from 5 to 100 heads. Goats are present in 40% of the farms and the size of the herd varies from 2 to 50 heads. Local cattle are present in 30% of the farms and the number of heads ranges from 2 to 20. The size structure of the different animal species is presented in Figures 8, 9, and 10.

Table 5. Crop rotation by the farms surveyed (average 2014/2015/2016)

Crops	Area (%)	Farmers (%)
Cerals	65%	90%
Fodder	13%	42%
pulses	15%	71%
Gardening	5%	43%
fig tree	12%	36%
Olive	11%	52%
fallow	8%	37%

Figure 8. Size of the sheep flock
Source: (Authors)

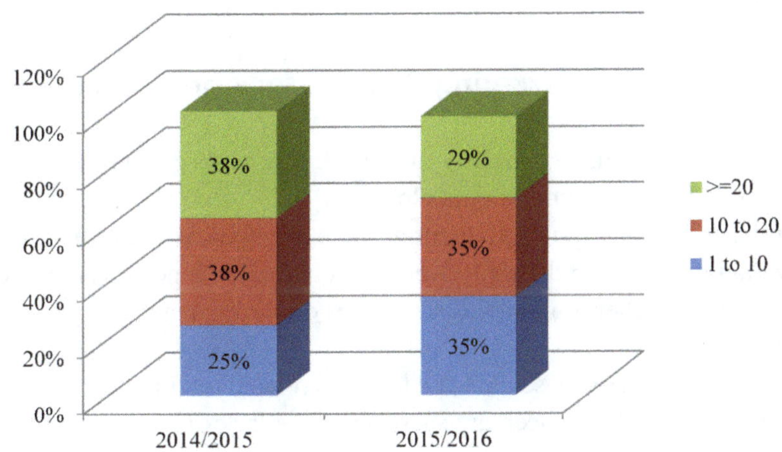

Figure 9. Size of the local cattle herd
Source: (Authors)

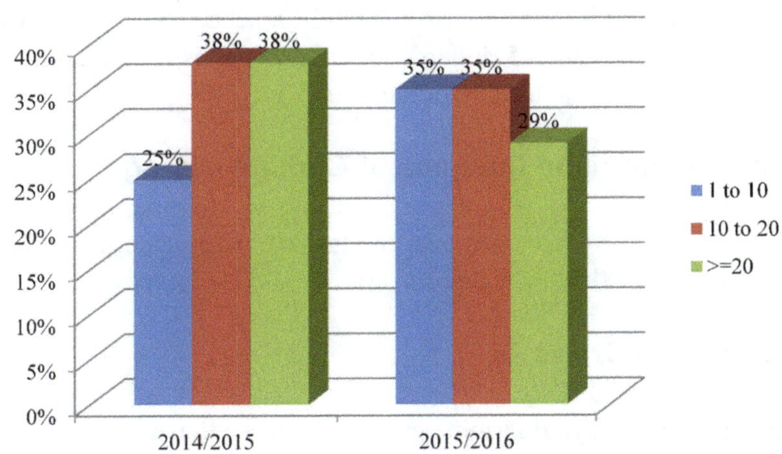

Figure 10. Goat herd size
Source: (Authors)

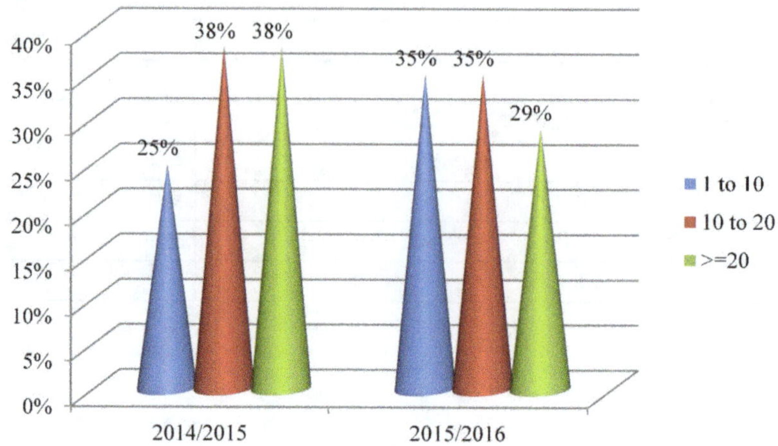

Diversification Strategy for Marketing Circuits for Agricultural Products

The production of cereals and pulses is mostly on-farm consumed. For gardening and arboriculture, farmers opt either for sale on the ground or on a plot.

For the commercialization of animal production, there are two main circuits: intermediary, where the farmer brings his animals back to the souk (weekly local market) and sells them to butcher or horse traders, and runs where the butcher or intermediary comes on the farm and buys the animal to slaughter it, usually the same day.

The marketing generally undergoes several fluctuations related to climatic conditions or the feast of Eid Al Adha (Muslim festival where sheep are sacrifice). The lack of organization of marketing channels leads to the proliferation of many speculators who take advantage of the lack of information on the prices practiced by the breeder.

For the marketing of honey, the farmers surveyed expose their products to the souk and it is the law of supply and demand that defines the price. Honey is sold to friends (regular and loyal customers). These sell them to intermediaries in Tangier, Tetouan and Larache. The price of a liter of honey can vary between 150 and 400 DH (20 and 45 dollars). Production varies according to the season and the flower. The price of honey depends on feeding of the bees.

Adaptive Practices of Farms in the Face of Climate Change

To cope with the changes that have occurred over the last 10 to 15 years, many practices have been developed by farms over time to adapt to changes in climate and landscape.

Intensification by Stolen Crops

About 10% of farms use some backyard farming crops to increase the area harvested. These are crops grown between two main crops for 3-4 months of the same crop year, such as market gardening and maize in summer after harvesting cereals. This practice contributes to the intensification of rotations.

The Multi-Functionality of Crops

This is the case of feed barley which is depressed (cut by mowing) at the beginning of the crop cycle to serve as a source of forage and then at the tillering stage it is allowed to continue its development to give a grain yield which will be used to both to animal and human nutrition. This practice is present in 8% of the farms surveyed.

Introduction of New Animal Breeds

Two percent (2%) have introduced alpine goats to value derived products (milk, cheese), but in the absence of support and technical support, the production is very low and currently self-consumed

Adoption of Endogenous and Exogenous Innovations

Several innovations have been adopted over time by farms. Some are the result of local initiative by the farmers themselves. It is:

- The use of artisanal tools for tillage and sowing by 10% of respondents)., Well digging and motor pump purchase (5%),
- The manufacture of benches and olive plantation (by 8%)
- Windbreak installation: 20%
- And tillage leaving small bowls and / or stony surface (55%)

Other innovations are the result support of public institutions. This is the case of the practice of direct seeding by 4% of farms in the form of tests conducted by INRA Tangier and the CCA during a crop year) and the manufacture of gabion by the support of the DREF (5% farms surveyed).

Institutional Actors: Punctual Actions and Lack of Integration to Help People Adapt to Climate Change

In order to contribute to the adaptation of farmers to climate change, several actors intervene notably in the Loukkos watershed agency, the Provincial Water and Forests Directorates and the fight against the desertification of Tangiers and Tetouan, the Regional Directorate of Agriculture of Tangier-Tetouan, the Provincial Directorates of Agriculture of Tangiers and Tetouan, the Social Action Division of the province of Tetouan, Ben Karrich and Dar Ben Saddok Agricultural Advisory Centers et the territorial municipalities.

Over the last five years, these actors have carried out various projects and income-generating activities, and the fight against erosion.

Income-generating activities are aimed at enabling farmers to diversify income sources such as fruit trees, bee keeping, and goat farming.

They are accompanied by measures to strengthen the capacities of the professional organizations created in terms of administrative and financial management. The technical supervision, provided by the Agricultural Advisory Center, is constituted in its entirety by sessions of theoretical and practical training on technical conduct and study trips within the framework of the Moroccan Green Plan. In addition to these three activities, olive plantations are mainly aimed at increasing farmers' incomes but also at protecting the biophysical environment of the Tleta watershed against erosion. The plantations are also accompanied by water and soil conservation works, through the construction of catchments allowing the capture and better use of rainwater. Planting fruit trees such as olive trees is one of the measures to mitigate the effects of climate change and protect the environment through the preservation of local biodiversity.

According to the institutional actors interviewed, it is planned for the period 2017/2019 the construction of a local for the marketing of traditional cheese (Ain Lahsen), for the crushing of the olive (Jouamma and Anjra) and for the collection of Milk (Dar Chaoui) with awareness-raising and training on the collective management of units.

Reforestation operations carried out by the Directorate of Waters and Forests (DREFLD) in turn contribute to mitigating the effects of climate change through soil fixation.

According to the respondents, the constraints and limits that hamper the interventions of institutional actors are mainly related to weak governance (problems of coordination and consultation when the types of activities to be financed by the populations and actions of accompaniment and support council). The analysis of the type of actions carried out by the four main actors namely: the Ministry of the Interior (INDH), the Ministry of the Family and Solidarity (ADS) and the Ministry of Agriculture (MCA projects and pillar II of the PMV) and the High Commission for Water and Forests shows that they concern the financing of agricultural projects (apiculture, arboriculture) without any real coordination. These projects mainly benefit some leaders of the zones organized in NGOs according to the agricultural advisors of the CCA of Ben Sadol and the heads of the technical services of the rural communes surveyed.

This should challenge decision-makers to take an integrated approach to Tleta territory-wide actions in order to have lasting impacts on the resilience of populations.

CONCLUSION

The Tleta watershed knows both the dynamics of the agrarian landscape and the effects of climate change. These changes are perceptible by the populations and the impact is materialized mainly by the continuous erosion of soil.

In order to consolidate farms' resilience farms have developed over time several innovative practices and defensive adaptive strategiesDespite the diversification of production systems and activities, the future of agriculture and its sustainability in the basin, like other areas with fragile ecology of the country, is perceived as uncertain by the people. It is dependent on the implementation of an integrated strategy; the coherence and integration of public policies currently carried out whose actions remain crumbled not benefiting the populations surveyed.

Many assets are to be valorized by the development of certain local products such as honey and goat's milk and its by-products, the support of advice, the promotion of collective action and the creation of professional agricultural organizations, currently very rare and not very functional.

REFERENCES

Abbad, Z. (2016). *Etude des pratiques adaptatives des exploitations agricoles dans les mosaïques paysagères à agro-système pluvieux face aux changements climatiques: Cas du bassin versant Tleta (Région Tetouan, Tanger), PFE, option: Sciences et Techniques des productions végétales*. ENA-Meknès.

Belaich. (2014). *Cartographie diachronique et modélisation du ravinement au niveau du bassin versant Tleta* (Rif occidental). Mémoire de 3ème cycle, ENFI.

El Kamoune, I. (2009). *Application du modèle Stream pour la simulation de l'érosion dans le bassin versant Tleta* (Rif occidental, Maroc). Science du sol, IAV Hassan II.

El Katrani, H. (2014). *Impact des utilisations des terres sur les propriétés hydrologiques du sol au niveau de bassin versant Tleta Rif occidental. Mémoire de 3ème cycle*. ENFI.

El Yadari, H. (2012). *Evolution spatio-temporelle des conditions de surface dans le bassin versant Tleta (Rif Occidental). Management des Ressources en sols et en Eaux*. IAV Hassan II.

Hammouda, N. (2010). Evolution de l'occupation des sols et de l'érosion dans le bassin versant Tleta. Management des Ressources en sols et en Eaux, IAV Hassan II.

Houry, H. (2014). *Etude des stratégies et pratiques adaptatives des exploitations agricoles dans les mosaïques paysagères à agrosystème pluvio-méditéranéen: Cas du bassin versant Tleta », PFE, option: Ingénierie du développement*. ENA-Meknès.

Kourouma. (2017). *Evaluation économique du service de stabilisation des sols et de contrôle de l'érosion d'un agrosystème pluvial méditerranéen* (Maroc). Mémoire de recherche Supagro de Montpellier. Retrieved from http://www.fao.org/nr/climpag/pub/FAO_WorldBank_Study_CC_Morocco_2008.pdf

Leye, M. L. (2015). *Etude de la perception de l'évolution paysagère des écosystèmes fragiles et du rôle des acteurs institutionnels dans l'adaptation des agriculteurs aux changements climatiques: Cas du Bassin versant de Tleta (Tanger-Tetouan). Ingénierie du développement*. ENA-Meknès.

KEY TERMS AND DEFINITIONS

Adaptative Practices: The common agricultural adaptation strategies used by farmers.

Climate Change: The catch-all term for the shift in worldwide weather phenomena associated with an increase in global average temperatures.

Climate Change Adaptation: A response to global warming, that seeks to reduce the vulnerability of social and biological systems to relatively sudden change and thus offset the effects of global warming.

Climate Change Mitigation: Consists of actions to limit the magnitude or rate of long-term global warming and its related effects. Climate change mitigation generally involves reductions in human (anthropogenic) emissions of greenhouse gases (GHGs).

Climate Resilience: The capacity for a socio-ecological system to absorb stresses and maintain function in the face of external stresses imposed upon it by climate change and adapt, reorganize, and evolve into more desirable configurations that improve the sustainability.

Landscape Dynamics: Refers to every change that occurs in the physical, biological, and cognitive assets of a landscape.

Perception: The subjective judgement that people make about the characteristics of climate change.

Watershed: The area that drains into a single river is the watershed for that river.

Compilation of References

Aafi, A. (2007). Rôle des Cactus dans la Restauration de l'Arganeraie et de ses Composantes Floristique et Faunistique dans la Province de Tiznit. Annales de la recherche forestière au Maroc, 38, 69-76.

ABARES. (2011). *Australian fisheries statistics 2010*. Canberra: Australian Bureau of Agricultural and Resource Economics and Sciences.

Abbad, Z. (2016). *Etude des pratiques adaptatives des exploitations agricoles dans les mosaïques paysagères à agrosystème pluvieux face aux changements climatiques: Cas du bassin versant Tleta (Région Tetouan, Tanger), PFE, option: Sciences et Techniques des productions végétales*. ENA-Meknès.

Abdel-Dayem, M. S., Annajar, B. B., Hanafi, H. A., & Obenauer, P. J. (2012). The potential distribution of Phlebotomus papatasi (Diptera: Psychodidae) in Libya based on ecological niche model. *Journal of Medical Entomology*, *49*(3), 739–745. doi:10.1603/ME11225 PMID:22679884

Abedi-Astaneh, F., Akhavan, A. A., Shirzadi, M. R., Rassi, Y., Yaghoobi-Ershadi, M. R., Hanafi-Bojd, A. A., ... Raufi, H. (2015). Species diversity of sand flies and ecological niche model of Phlebotomus papatasi in central Iran. *Acta Tropica*, *149*, 246–253. doi:10.1016/j.actatropica.2015.05.030 PMID:26071647

Abonnenc, E. (1972). Les phlébotomes de la région éthiopienne (Diptera, Psychodidae). *Cahiers de l'ORSTOM, série Entomologie médicale et Parasitologie, 55*, 1-239.

Aboutayeb, H. (2014). The biosphere reserve of the argan tree: eco-tourism a new territory South of Morocco. *PASOS: Revista de Turismo y Patrimonio Cultural, 12*(4), 915-922. Retrieved from https://www.cabdirect.org/cabdirect/abstract/20143365231

ABS. (2017). *The Australian Beauro of Statistics*. Retrieved April 04, 2018, from http://www.abs.gov.au/Agriculture

Adejuwon, J. O. (2006). *Food security, climate variability and climate change in Sub- Saharan West Africa*. Assessments of impacts and adaptations to climate change. Working paper series. Project No. AF 23. International START Secretariat.

Adger, W. N. (1999). Social vulnerability to climate change and extremes in coastal Vietnam. *World Development*, *27*(2), 249–269. doi:10.1016/S0305-750X(98)00136-3

Adger, W. N., Arnell, N. W., & Tompkins, E. L. (2005). Successful adaptation to climate change across scales. *Global Environmental Change*, *15*(2), 77–86. doi:10.1016/j.gloenvcha.2004.12.005

AgMIP Workshop. (2015). Retrieved October 06,2018 from https://eos.org/meeting-reports/assessing-impacts-of-climate-change-on-food-security-worldwide

Agoumi, A. (2003). *Vulnérabilité des pays du Maghreb face aux changements climatiques; Besoin réel et urgent d'une stratégie d'adaptation et de moyens pour sa mise en œuvre*. Available on http://www.cckn.net/pdf/north_africa_fr.pdf

Agoumi, A. (2003). *Vulnerability of North African countries to climatic changes: adaptation and implementation strategies for climatic change. In Developing Perspectives on climate change: Issues and analysis from developing countries and countries with economies in transition*. IISD/Climate Change Knowledge Network.

Agoumi, A., & Debbarh, A. (2006). Ressources en eau et bassins versants du Maroc: 50 ans de développement (1955-2005). *50 Ans de Développement Humains. Perspectives*, 2025.

Ahmed, N., & Diana, J. S. (2015). Threatening "white gold": Impacts of climate change on shrimp farming in coastal Bangladesh. *Ocean and Coastal Management*, *114*, 42–52. doi:10.1016/j.ocecoaman.2015.06.008

Ainsworth, E. E., Yendrek, C. R., Sitch, S., Collins, W. J., & Emberson, L. D. (2012). The effects of tropospheric ozone on net primary productivity and implications for climate change. *Annual Review of Plant Biology*, *63*(1), 637–661. doi:10.1146/annurev-arplant-042110-103829 PMID:22404461

Ait Boughrous, A. (2007). *Biodiversité, écologie et qualité des eaux souterraines dans deux régions arides du Maroc: le Tafilalet et la région de Marrakech* (PhD Thesis). Cadi Ayyad University.

Ait Boughrous, A. (2007). *Biodiversité, écologie et qualité des eaux souterraines de deux regionsarides du Maroc: Le Tafilalet et la région de Marrakech* (PhD Thesis). Cadi Ayyad University, Marrakech, Morocco.

Ait Houssa, A., Drissi, S., Asehraou, A., Asfers, A., Oubaki, L., & Chraibi, H. (2017). Changements climatiques au Maroc: Quels systèmes de culture et quelles biotechnologies pour s'y adapter? *Revue Marocaine des Sciences Agronomiques et Vétérinaires*, *5*(3).

Ait Lamqadem, A., Saber, H., & Rahimi, A. (2017). Spatiotemporal Changes of Vegetation in the Middle Draa Valley Oasis: A Study Case of M'hamid El Ghizlane Oasis (Morocco). *European Scientific Journal, 13*(24).

Akçakaya, H. R., Butchart, S. H., Mace, G. M., Stuart, S. N., & Hilton-Taylor, C. (2006). Use and misuse of the IUCN Red List Criteria in projecting climate change impacts on biodiversity. *Global Change Biology*, *12*(11), 2037–2043. doi:10.1111/j.1365-2486.2006.01253.x

Akey, A., & Maden, S. (2006). A new approach to fire blight control: *Mycorihiza*. Selcuk University, Faculty of Agriculture, Department of Plant Protection, Campus, Konya. Turkey. *HortScience*, *41*, 1309–1312.

Akpomi, M. E., & Vipene, J. (2016). Promoting Knowledge of Climate Change (CC) amongst Nigerians: Implications for Education Managers. *Journal of Education and Practice*, *7*(32), 132–138.

Akter, T., Quevauviller, P., Eisenreich, S. J., & Vaes, G. (2018). Impacts of climate and land use changes on flood risk management for the Schijn River, Belgium. *Environmental Science & Policy*, *89*, 163–175. doi:10.1016/j.envsci.2018.07.002

Akuru, U. B., Okoro, O. I., & Chikuni, E. (2013, September). Impact of renewable energy deployment on climate change in Nigeria. In AFRICON, 2013 (pp. 1-8). IEEE. doi:10.1109/AFRCON.2013.6757649

Alali, A. B. (2013). La plaine du Tafilalet (Sud-est, Maroc) face aux problèmes d'environnement. *Rev. Mar. Sci. Agron. Vét.*, *2*, 68–72.

Albers, M., & Deppisch, S. (2013). Resilience in the light of climate change: Useful approach or empty phrase for spatial planning? *European Planning Studies*, *21*(10), 1598–1610. doi:10.1080/09654313.2012.722961

Alexander, B., Park, R. J., Jacob, D. J., Li, Q. B., Yantosca, R. M., Savarino, J., ... Thiemens, M. H. (2005). Sulfate formation in sea-salt aerosols: Constraints from oxygen isotopes. *Journal of Geophysical Research: Atmospheres, 110*(D10).

Alexandratos, N., & Bruinsma, J. (2012). *World Agriculture Towards 2030/2050: The 2012 Revision ESA Working Paper No. 12–03* (FAO, 2012). FAO.

Ali, R. M., Loutfy, N. F., Awad, O. M., & Suliman, N. K. (2016). Bionomics of phlebotomine sandfly species in west Alexandria,Egypt. *Journal of Entomology and Zoology Studies*, *4*, 349–353.

Alkemade, R., van Oorschot, M., Miles, L., Nellemann, C., Bakkenes, M., & Ten Brink, B. (2009). GLOBIO3: A framework to investigate options for reducing global terrestrial biodiversity loss. *Ecosystems (New York, N.Y.)*, *12*(3), 374–390. doi:10.100710021-009-9229-5

Allan, R. P., & Soden, B. J. (2008). Atmospheric Warming and the Amplification of Precipitation Extremes. *Science*, *321*(5895), 1481 LP-1484. Retrieved from http://science.sciencemag.org/content/321/5895/1481.abstract

Allan, J. D., Abell, R., Hogan, Z. E. B., Revenga, C., Taylor, B. W., Welcomme, R. L., & Winemiller, K. (2005). Overfishing of inland waters. *Bioscience*, *55*(12), 1041–1051. doi:10.1641/0006-3568(2005)055[1041:OOIW]2.0.CO;2

Allen, L. H. Jr, Barker, J. T., & Boote, K. J. (1996). The CO2 Fertilization Effect: Higher Carbohydrate Production and Retention as Biomass and Seed Yield. In *Global Climate Change and Agricultural Production. Direct and Indirect Effects of Changing Hydrological, Pedological and Plant Physiological Processes*. Rome, Italy: Food and Agricultural Organization of the United Nations.

Allison, E. H., Adger, W. N., & Badjeck, M. C. (2005). *Effects of climate change on the sustainability of capture and enhancement fisheries important to the poor: Analysis of the vulnerability and adaptability of fisherfolk living in poverty*. Fisheries Management Science Programme, DfID, UK. No. R4778J. Retrieved from http://www.fmsp.org.uk

Allison, E. H., Perry, A. L., Badjeck, M., Adger, W. N., Brown, K., Conway, D., ... Dulvy, N. K. (2009). Vulnerability of national economies to the impacts of climate change on fisheries. *Fish and Fisheries*, *10*(2), 173–196. doi:10.1111/j.1467-2979.2008.00310.x

Almeida-Garcıa, F., & Chahine, S. (2015). Morocco, tourism. In Encyclopedia of Tourism (pp. 815–816). Academic Press.

Alonso, A., & Raman, S. (2016). *Tropical Conservation Perspectives on Local and Global Priorities*. Oxford University Press.

Alten, B., Maia, C., Afonso, M. O., Campino, L., Jiménez, M., González, E., ... Toty, C. (2016). Seasonal dynamics of phlebotomine sand fly species proven vectors of Mediterranean leishmaniasis caused by Leishmaniainfantum. *PLoS Neglected Tropical Diseases*, *10*(2), e0004458. doi:10.1371/journal.pntd.0004458 PMID:26900688

Aly, A. H., Debbab, A., & Proksch, P. (2011). Fungal endophytes: Unique plant inhabitants with great promises. *Applied Microbiology and Biotechnology*, *90*(6), 1829–1845. doi:10.100700253-011-3270-y PMID:21523479

Amelung, B., Moreno, A., & Scott, D. (2008). The place of tourism in the IPCC fourth assessment report: A review. *Tourism Review International*, *12*(1), 5–12. doi:10.3727/154427208785899984

Andersen, Z., Hvidberg, M., Jensen, S., Ketzel, M., Loft, S., Sørensen, M., ... Raaschou-Nielsen, O. (2011). Chronic obstructive pulmonary disease and long-term exposure to traffic related air pollution: A cohort study. *American Journal of Respiratory and Critical Care Medicine*, *183*(4), 455–461. doi:10.1164/rccm.201006-0937OC PMID:20870755

Anderson, E. P., Marengo, J., Villalba, R., Halloy, S., Young, B., Cordero, D., & Martinez, R. (2017). Consequences of climate change for ecosystems and ecosystem services in the tropical. *Andes (Salta)*.

Anderson, T. A., Kruger, E. L., & Coats, J. R. (1994). Enhanced degradation of a mixture of three herbicides in the rhizosphere of a herbicide-tolerant plant. *Chemosphere*, *28*(8), 1551–1557. doi:10.1016/0045-6535(94)90248-8

Andersson-Skold, Y., Thorsson, S., Rayner, D., Lindberg, F., Janhall, S., Jonsson, A., ... Granberg, M. (2015). An integrated method for assessing climate related risks and adaptation alternatives in urban areas. *Climate Risk Management*, *7*, 31–50. doi:10.1016/j.crm.2015.01.003

Andrews, C. (2008). Green house gas emissions along the rural-urban gradient. *Journal of Environmental Planning and Management*, *51*(6), 847–870. doi:10.1080/09640560802423780

Anenberg, S. C., Talgo, K., Arunachalam, S., Dolwick, P., Jang, C., & West, J. J. (2011). Impacts of global, regional and sectoral black carbon emission reductions on surface air quality and human mortality. *Atmospheric Chemistry and Physics*, *11*(14), 7253–7267. doi:10.5194/acp-11-7253-2011

Anenberg, S., Horowitz, L., Tong, D., & West, J. (2010). An estimate of the global burden of anthropogenic ozone and fine particulate matter on premature human mortality using atmospheric modeling. *Environmental Health Perspectives*, *118*(9), 1189–1195. doi:10.1289/ehp.0901220 PMID:20382579

An, G. H., Cha, S. L., & Kim, J. K. (2018). Modification of the long-term deformation models for steel reinforced concrete columns. *Construction & Building Materials*, *189*, 245–252. doi:10.1016/j.conbuildmat.2018.08.095

Angelo, J., Andrade, L., Madeira, L. M., & Mendes, A. (2013). An overview of photocatalysis phenomena applied to NOx abatement. *Journal of Environmental Management*, *129*, 522–539. doi:10.1016/j.jenvman.2013.08.006 PMID:24018117

Annesi-Maesano, I., Baiz, N., Banerjee, S., Rudnai, P., & Rive, S. (2013). Indoor air quality and sources in schools and related health effects. *J. Toxicol. Environ. Health B*, *16*(8), 491–550. doi:10.1080/10937404.2013.853609 PMID:24298914

Anon. (2003). *Etat d'avancement des programmes de lutte contre les maladies parasitaires. Direction de l'épidémiologie et de lutte contre les maladies. Rapport Annuel de la Ministère de la Santé*. Rabat, Maroc: Ministère de la Santé.

Anonymous. (2000). *Fish collection database of the Gulf Coast Research Laboratory (GCRL). The Gulf Coast Research Laboratory*. Ocean Springs, MI: GCRL.

Antonov, J. I., Levitus, S., & Boyer, T. P. (2002). Steric sea level variations during 1957-1994: Importance of salinity. *Journal of Geophysical Research: Oceans*, *107*(C12), SRF 14-1-SRF 14-8. doi:10.1029/2001JC000964

Araújo, M. B., Alagador, D., Cabeza, M., Nogués-Bravo, D., & Thuiller, W. (2011). Climate change threatens European conservation areas. *Ecology Letters*, *14*(5), 484–492. doi:10.1111/j.1461-0248.2011.01610.x PMID:21447141

Archer, D., Eby, M., Brovkin, V., Ridgwell, A., Cao, L., Mikolajewicz, U., ... Tokos, K. (2009). Atmospheric lifetime of fossil fuel carbon dioxide. *Annual Review of Earth and Planetary Sciences*, *37*(1), 117–134. doi:10.1146/annurev.earth.031208.100206. doi:10.1146/annurev.earth.031208.100206

Archer, V. E. (1990). Air Pollution and Fatal Lung Disease in Three Utah Counties. *Archives of Environmental Health*, *45*(6), 325–334. doi:10.1080/00039896.1990.10118751 PMID:2270951

Arden Pope, C. III, & Dockery, D. W. (2006). Health Effects of Fine Particulate Air Pollution: Lines that Connect. *Journal of the Air & Waste Management Association*, *56*(6), 709–742. doi:10.1080/10473289.2006.10464485 PMID:16805397

Arnold, A. (2008). Hidden within our botanical richness, a treasure trove of fungal endophytes. Plant Press.

Arnold, A. E. (2007). Understanding the diversity of foliar endophytic fungi: Progress, challenges, and frontiers. *Fungal Biology Reviews*, *21*(2-3), 51–66. doi:10.1016/j.fbr.2007.05.003

Arnold, A. E., & Lutzoni, F. (2007). Diversity and host range of foliar fungal endophytes: Are tropical leaves biodiversity hotspots. *Ecology*, *88*(3), 541–549. doi:10.1890/05-1459 PMID:17503580

Arrus, R., & Rousset, N. (2007*). L'agriculture du Maghreb au défi du changement climatique: Quelles stratégies d'adaptation face à la raréfaction des ressources hydriques?* Available on https://halshs.archives-ouvertes.fr/halshs-00134115

Ashford, R. W. (1996). Leishmaniasis reservoirs and their significance in control. *Clinics in Dermatology*, *14*(5), 523–532. doi:10.1016/0738-081X(96)00041-7 PMID:8889331

Asseng, S., Ewert, F., Martre, P., Rötter, R. P., Lobell, D. B., Cammarano, D., ... Zhu, Y. (2014). Rising temperatures reduce global wheat production. *Nature Climate Change*, *5*(2), 143–147. doi:10.1038/nclimate2470

Asseng, S., Ewert, F., Rosenzweig, C., Jones, J. W., Hatfield, J. L., Ruane, A. C., ... Wolf, J. (2013). Uncertainty in simulating wheat yields under climate change. *Nature Climate Change*, *3*(9), 827–832. doi:10.1038/nclimate1916

Auad, G., Miller, A., & Di Lorenzo, E. (2006). Long-term forecast of oceanic conditions off California and their biological implications. *Journal of Geophysical Research. Oceans*, *111*(9), 1–14. doi:10.1029/2005JC003219 PMID:20411040

Aunan, K., Pátzay, G., Aaheim, H. A., & Seip, H. M. (1998). Health and environmental benefits from air pollution reductions in Hungary. *The Science of the Total Environment*, *212*(2-3), 245–268. doi:10.1016/S0048-9697(98)00002-3 PMID:9573631

Aus der Beek, T., Voß, F., & Flörke, M. (2011). Modelling the impact of global change on the hydrological system of the Aral Sea basin. *Physics and Chemistry of the Earth Parts A/B/C*, *36*(13), 684–695. doi:10.1016/j.pce.2011.03.004

Avery, M. A. (2010). Increasings pringtime ozone mixing ratios in the free troposphere over western North. *American Naturalist*, *463*, 344–348.

Azovedo, J. L., Maccheroni, W., Pereira, J. O., & De Araujo, W. L. (2000). Endophytic microorganisms: A review on insect control and recent advances on tropical plants. *Electronic Journal of Biotechnology*, *3*, 40–65.

Babqiqi, A. (2008). *Scenarios de changements climatiques à haute résolution au Maroc*. Casablanca, Maroc: Rapport de la Direction de la Météorologie Nationale.

Babqiqi, A., Kassami, A., & Messouli, M. (2011). Development of high resolution climate change scenarios for Morocco for the 21st century. In *Geoinformatics for Climate Change Studies*. New Delhi: TERI Press.

Badjeck, M. C., Allison, E. H., Halls, A. H., & Dulvy, N. K. (2009). Impacts of climate variability and change on fishery-based livelihoods. *Marine Policy*. doi:10.1016/j.marpol.2009.08.007

Badman, D. G., & Jaffe, E. R. (1996). Blood and air pollution: State of knowledge and research needs. *Otolaryngology - Head and Neck Surgery*, *114*(2), 205–208. doi:10.1016/S0194-5998(96)70166-3 PMID:8637733

Baillie, J., Hilton-Taylor, C., & Stuart, S. N. (Eds.). (2004). *IUCN red list of threatened species: a global species assessment*. Iucn.

Bailly-Choumara, H., Abonnenc, E., & Pastre, J. (1971). Contribution a l'etude des Phlebotomes du Maroc (Diptera, Psychodidae). donnees fauntistiques et ecologiques. *Entomol Med Cah*. Retrieved from http://agris.fao.org/agris-search/search.do?recordID=US201302313429

Baker, K. S., Benson, B. J., Henshaw, D. L., Blodgett, D., Porter, J. H., & Stafford, S. G. (2000). Evolution of a multisite network information system: The LTER information management paradigm. *A.I.B.S. Bulletin*, *50*(11), 963–978. doi:10.1641/0006-3568(2000)050[0963:EOAMNI]2.0.CO;2

Bakkenes, M., Alkemade, J. R. M., Ihle, F., Leemans, R., & Latour, J. B. (2002). Assessing effects of forecasted climate change on the diversity and distribution of European higher plants for 2050. *Global Change Biology*, *8*(4), 390–407. doi:10.1046/j.1354-1013.2001.00467.x

Bakun, A. (1996). *Patterns in the ocean. Ocean processes and marine population dynamics*. California Sea Grant College System.

Balirwa, J. S. (1998). *Lake Victoria wetlands and the ecology of the Nile Tilapia Oreochromis niloticus Linne´. A.A.* Rotterdam, The Netherlands: Balkema Publishers.

Balmes, J. R., Fine, J. M., & Sheppard, D. (1987). Symptomatic bronchoconstriction after short-term inhalation of sulfur dioxide. *The American Review of Respiratory Disease*, *136*(5), 1117–1121. doi:10.1164/ajrccm/136.5.1117 PMID:3674573

Barange, M., Merino, G., Blanchard, J. L., Scholtens, J., Harle, J., Allison, E. H., ... Jennings, S. (2014). Impacts of climate change on marine ecosystem production in societies dependent on fisheries. *Nature Climate Change*, *4*(3), 211–216. doi:10.1038/nclimate2119

Barbour, D. A. (1986). *Why are there so few butterflies in Liverpool? An answer*. London: Royal Entomological Society.

Barnosky, A. D., Matzke, N., Tomiya, S., Wogan, G. O., Swartz, B., Quental, T. B., ... Mersey, B. (2011). Has the Earth's sixth mass extinction already arrived? *Nature*, *471*(7336), 51–57. doi:10.1038/nature09678 PMID:21368823

Barrie, L. A., & Hoff, R. M. (1984). The oxidation rate and residence time of sulphur dioxide in the Arctic atmosphere. *Atmospheric Environment (1967)*, *18*(12), 2711-2722.

Bartomeus, I., Ascher, J. S., Wagner, D., Danforth, B. N., Colla, S., Kornbluth, S., & Winfree, R. (2011). Climate-associated phenological advances in bee pollinators and bee-pollinated plants. *Proceedings of the National Academy of Sciences of the United States of America*, *108*(51), 20645–20649. doi:10.1073/pnas.1115559108 PMID:22143794

Barua, S. K., Uusivuori, J., & Kuuluvainen, J. (2012). Impacts of carbon-based policy instruments and taxes on tropical deforestation. *Ecological Economics*, *73*, 211–219. doi:10.1016/j.ecolecon.2011.10.029. doi:10.1016/j.ecolecon.2011.10.029

Battisti, D. S., & Naylor, R. L. (2009). Historical Warnings of Future Food Insecurity with Unprecedented Seasonal Heat. *Science*, *323*(5911), 240 LP-244. Retrieved from http://science.sciencemag.org/content/323/5911/240.abstract

Baughman, A. V., Gadgil, A. J., & Nazaroff, W. W. (1994). Mixing of a point source pollutant by natural convection flow within a room. *Indoor Air*, *4*(2), 114–122. doi:10.1111/j.1600-0668.1994.t01-2-00006.x

Bebber, D. P., Holmes, T., & Gurr, S. J. (2014). The global spread of crop pests and pathogens. *Global Ecology and Biogeography*, *23*(12), 1398–1407. doi:10.1111/geb.12214

Bebber, D. P., Ramotowski, M. A. T., & Gurr, S. J. (2013). Crop pests and pathogens move polewards in a warming world. *Nature Climate Change*, *3*(11), 985–988. doi:10.1038/nclimate1990

Becken, S., & Hay, J. (2007). *Tourism and climate change – risks and opportunities*. Cleveland: Channel View Publications.

Beckett, K. P., Freer-Smith, P., & Taylor, G. (1998). Urban woodlands: Their role in reducing the effects of particulate pollution. *Environmental Pollution*, *99*(3), 347–360. doi:10.1016/S0269-7491(98)00016-5 PMID:15093299

Beelen, R., Raaschou-Nielsen, O., Stafoggia, M., Andersen, Z. J., Weinmayr, G., Hoffmann, B., ... Hoek, G. (2014). Effects of long-term exposure to air pollution on natural-cause mortality: An analysis of 22 European cohorts within the multicentre ESCAPE project. *Lancet (London, England)*, *383*(9919), 785–795. doi:10.1016/S0140-6736(13)62158-3 PMID:24332274

Beevers, S. D., Westmoreland, E., de Jong, M. C., Williams, M. L., & Carslaw, D. C. (2012). Trends in NOx and NO2 emissions from road traffic in Great Britain. *Atmospheric Environment*, *54*, 107–116. doi:10.1016/j.atmosenv.2012.02.028

Beger, M., Linke, S., Watts, M., Game, E., Treml, E., Ball, I., & Possingham, H. P. (2010). Incorporating asymmetric connectivity into spatial decision making for conservation. *Conservation Letters*, *3*(5), 359–368. doi:10.1111/j.1755-263X.2010.00123.x

Belaich. (2014). *Cartographie diachronique et modélisation du ravinement au niveau du bassin versant Tleta* (Rif occidental). Mémoire de 3ème cycle, ENFI.

Belarbi, A., Boayad, A., Diaou, M., Kaassis, N., & Tidjani Miliki, M. (2004). Agrobiodiversité et durabilité des systèmes de production oasiens dans la palmeraie d'Aoufous. ICRA, INRA.

Belen, A., & Alten, B. (2011). Seasonal dynamics and altitudinal distributions of sand fly (Diptera: Psychodidae) populations in a cutaneous leishmaniasis endemic area of the Cukurova region of Turkey. *Journal of Vector Ecology*, *36*(s1). doi:10.1111/j.1948-7134.2011.00116.x PMID:21366785

Bell, I., & Visbeck, M. (n.d.). *North Atlantic Oscillation*. Retrieved from https://www.ldeo.columbia.edu/res/pi/NAO/

Bellard, C., Bertelsmeier, C., Leadley, P., Thuiller, W., & Courchamp, F. (2012). Impacts of climate change on the future of biodiversity. *Ecology Letters*, *15*(4), 365–377. doi:10.1111/j.1461-0248.2011.01736.x PMID:22257223

Bell, G., & Gonzalez, A. (2009). Evolutionary rescue can prevent extinction following environmental change. *Ecology Letters*, *12*(9), 942–948. doi:10.1111/j.1461-0248.2009.01350.x PMID:19659574

Bellinger, D. C. (2005). Teratogen update: Lead and pregnancy. *Birth Defects Research. Part A, Clinical and Molecular Teratology*, *73*(6), 409–420. doi:10.1002/bdra.20127 PMID:15880700

Bell, M. L., Ebisu, K., & Belanger, K. (2007). Ambient air pollution and low birth weight in Connecticut and Massachusetts. *Environmental Health Perspectives*, *115*(7), 1118–1124. doi:10.1289/ehp.9759 PMID:17637932

Ben Salem, A. (2014). *Vulnérabilité et adaptation aux changements climatiques dans les oasis de la région de Tafilalet-Maroc* (Thèse de doctorat). Département de biologie, Faculté de Semlalia. Université Cadi Ayyad, Marrakesh, Morocco.

Ben Salem, A. (2014). *Vulnérabilité et adaptation aux changements climatiques dans les oasis du Tafilalet- Maroc* (Doctoral thesis). University Cadi Ayyad.

Ben Salem, A. (2014). *Vulnerability and adaptation to climate change in the oasis region Tafilalet-Morocco* (Doctoral dissertation). Department of Biology, Faculty of Semlalia, Cadi Ayyad University, Marrakesh.

Ben Salem, A. B., Messouli, M., & Yacoubi-Khebiza, M. (2011). *Developing an oasis-based water management tool: Ecohydrologic approach and weap software for a large arid catchment in Morocco*. Academic Press.

Ben Salem, A., Messouli, M., & Yacoubi-Khebiza, M. (2011). *Developing an oasis-based water management tool: Ecohydrologic approach and weap software for a large arid catchment in Morocco*. Academic Press.

Ben Salem, A., Messouli, M., &Yacoubi-Khebiza, M. (2011). *Developing an oasis-based water management tool: Ecohydrologic approach and weap software for a large arid catchment in Morocco*. Academic Press.

Benabid, A., & Fennane, M. (1999). Main forest formations in the great Book of the Forest Moroccan. Mardaga.

Béné, C. (2006). *Small-scale fisheries: assessing their contribution to rural livelihoods in developing countries*. FAO Fisheries Circular No. 1008 FIPL/C1008 (En).

Benemann, J. R. (2008). *Opportunities and challenges in algae biofuels production*. Position paper by Dr. John R. Benemann in line with Algae World 2008. Retrieved from www.futureenergyevents.com/algae/whitepaper/algae_positionpaper.pdf

Benkova, I., & Volf, P. (2007). Effect of temperature on metabolism of Phlebotomus papatasi (Diptera: Psychodidae). *Journal of Medical Entomology*, *44*(1), 150–154. doi:10.1093/jmedent/41.5.150 PMID:17294933

Benmohammadi, A., Benmohammadi, L., Ballais, J. L., & Riser, J. (2000). Analysis of anthropic and natural interrelations: their impact on the recrudescence of the phenomena of sand deposition and desertification in south east Morocco (Drâa valley and Ziz valley). *Sécheresse*, *11*(4), 297-308. Available on https://www.cabdirect.org/cabdirect/abstract/20013044152

Bennani, A., Buret, J. & Senhaji, F. (2001). *Communication Nationale Initiale a la Convention Cadre des Nations Unies sur les changements climatiques*. Ministère de l'Aménagement du Territoire, de l'Urbanisme de l'Habitat et de l'Environnement.

Bentayeb M, Simoni M, Norback D, Baldacci S, Maio S, Viegi G.(2013). Indoor air pollution and respiratory health in the elderly. *J Environ Sci Health A Tox Hazard Subst Environ Eng, 48*, 1783-9.

Berg, G., Eberl, L., & Hartmann, A. (2005). The rhizosphere as a reservoir for opportunistic human pathogenic bacteria. *Environmental Microbiology*, *7*(11), 1673–1685. doi:10.1111/j.1462-2920.2005.00891.x PMID:16232283

Beveridge, M. C. M., Thilsted, S. H., Phillips, M. J., Metian, M., Troell, M., & Hall, S. J. (2013). Meeting the food and nutrition needs of the poor: The role of fish and the opportunities and challenges emerging from the rise of aquaculture. *Journal of Fish Biology*, *83*(4), 1067–1084. doi:10.1111/jfb.12187 PMID:24090563

Bharali, S., & Khan, M. L. (2011). Climate change and its impact on biodiversity; some management options for mitigation in Arunachal Pradesh. *Current Science*, 855–860.

Bhattacharjee, R. B., Singh, A., & Mukhopadhyay, S. N. (2008). Use of nitrogen- fixing bacteria as bio fertilize for non-legumes: Prospects and challenges. *Applied Microbial Biotechnology*, *80*(2), 199–209. doi:10.100700253-008-1567-2 PMID:18600321

Bigazzi, A. Y., & Figliozzi, M. A. (2015). Roadway determinants of bicyclist exposure to volatile organic compounds and carbon monoxide. *Transportation Research Part D, Transport and Environment*, *41*, 13–23. doi:10.1016/j.trd.2015.09.008

Bindoff, N. L., Willebrand, J., Artale, V., Cazenave, A., Gregory, J., Gulev, S., ... Unnikrishnan, A. (2007). Observations: Oceanic climate change and sea level. In S. Solomon, D. Qin, M. Manning, Z. Chen, M. Marquis, K. B. Averyt, ... H. L. Miller (Eds.), *Climate Change 2007: The physical science basis. Contribution of Working Group I to the Fourth Assessment Report of the Intergovernmental Panel on Climate Change*. Cambridge, UK: Cambridge University Press.

Birch, M. B. L., Gramig, B. M., Moomaw, W. R., Doering, O. C. III, & Reeling, C. J. (2011). Why metrics matter: Evaluating policy choices for reactive nitrogen in the chesapeake bay watershed. *Environmental Science & Technology*, *45*(1), 168–174. doi:10.1021/es101472z PMID:20853823

Blanchard, J. L., Jennings, S., Holmes, R., Harle, J., Merino, G., Allen, J. I., ... Barange, M. (2012). Potential consequences of climate change for primary production and fish production in large marine ecosystems. *Philosophical Transactions of the Royal Society of London. Series B, Biological Sciences*, *367*(1605), 2979–2989. doi:10.1098/rstb.2012.0231 PMID:23007086

BNP. (2009). Big Number Program: Intermediate Report. Penang, Malaysia: BNP.

Bobak, M. (2000). Outdoor air pollution, low birth weight, and prematurity. *Environmental Health Perspectives*, *108*(2), 173–176. doi:10.1289/ehp.00108173 PMID:10656859

Boffetta, P., Merler, E., & Vainio, H. (1993). Carcinogenicity of mercury and mercury compounds. *Scandinavian Journal of Work, Environment & Health*, *19*(1), 1–7. doi:10.5271jweh.1510 PMID:8465166

Böhner, J., Gross, J., & Riksen, M. (2004). Impact of land use and climate change on wind erosion: prediction of wind erosion activity for various land use and climate scenarios using the WEELS wind erosion model. In D. Goossens & M. Riksen (Eds.), *Wind Erosion and Dust Dynamics: Observations, Simulations, Modelling*. Wageningen, The Netherlands: ESW Publications.

Bonan, G. B. (2008). Forests and climate change: Forcings, feedbacks, and the climate benefits of forests. *Science*, *320*(5882), 1444–1449. doi:10.1126cience.1155121 PMID:18556546

Boonen, E., Akylas, V., Barmpas, F., Boréave, A., Bottalico, L., Cazaunau, M., ... Beeldens, A. (2015). Construction of a photocatalytic De-polluting field site in the Leopold II tunnel in Brussels. *Journal of Environmental Management, 155*, 136–144. doi:10.1016/j.jenvman.2015.03.001 PMID:25863437

Born, K., Fink, A. H., & Paeth, H. (2008). Dry and wet periods in the northwestern Maghreb for present day and future climate conditions. *Meteorologische Zeitschrift, 17*(5), 533–551. doi:10.1127/0941-2948/2008/0313

Bot, A., & Benites, J. (2005). *The importance of soil organic matter Key to drought-resistant soil and sustained food and production. In FAO Soils Bulletin 80* (p. 78). Rome: Food and Agriculture Organization of the United Nations.

Botes, A., & Zaid, A. (2002). The economic importance of date production and international trade. In A. Zaid & E. J. Arias-Jimenez (Eds.), *Date palm cultivation. FAO plant production and protection paper, 156 rev. 1*. Rome: FAO.

Botkin, D. B., Saxe, H., Araujo, M. B., Betts, R., Bradshaw, R. H., Cedhagen, T., & Ferrier, S. (2007). Forecasting the effects of global warming on biodiversity. *A.I.B.S. Bulletin, 57*(3), 227–236.

Bouaamlat, I., Larbi, A., & Faouzi, M. (2016). Hydrogeological investigation of an oasis-system aquifer in arid southeastern Morocco by development of a groundwater flow model. *Hydrogeology Journal, 24*(6), 1479–1496. doi:10.100710040-016-1409-8

Boucher, Y., Douady, C. J., Papke, R. T., Walsh, A., Boudreau, M. E. R., Nesbo, C. L., ... Doolittle, W. F. (2003). Lateral gene transfer and the origins of prokaryotic groups. *Annual Review of Genetics, 37*(1), 283–328. doi:10.1146/annurev.genet.37.050503.084247 PMID:14616063

Bouhlassa, S., & Paré, S. (2006). Evapotranspiration de référence dans la région aride de Tafilalet au sud-est du Maroc. *African Journal of Environmental Assessment and Management, 11*, 1-16.

Bouhlassa, S., & Paré, S. (2007). Effect of climatic parameters on reference evapo-transpiration in the tafilalet arid area. *Physical and Chemical News, 12*(38), 1-6.

Bounoua, L., Kahime, K., Houti, L., Blakey, T., Ebi, K. L., Zhang, P., ... Messouli, M. (2013). Linking climate to incidence of zoonotic cutaneous leishmaniasis (L. major) in pre-Saharan North Africa. *International Journal of Environmental Research and Public Health, 10*(8), 3172–3191. doi:10.3390/ijerph10083172 PMID:23912199

Bousfoul, M. (2008). Gestion intégrée des ressources en eau. *Revue Homme terre et eau, 140*, 80-83.

Boussaa, S., Guernaoui, S., Pesson, B., & Boumezzough, A. (2005). Seasonal fluctuations of phlebotomine sand fly populations (Diptera: Psychodidae) in the urban area of Marrakech, Morocco. *Acta Tropica, 95*(2), 86–91. doi:10.1016/j.actatropica.2005.05.002 PMID:15985259

Bowker, G. C. (2000). Biodiversity datadiversity. *Social Studies of Science, 30*(5), 643–683. doi:10.1177/030631200030005001. doi:10.1177/030631200030005001

Boyer, T. P., Levitus, S., Antonov, J. I., Locarnini, R. A., & Garcia, H. E. (2005). Linear trends in salinity for the World Ocean, 1955-1998. *Geophysical Research Letters, 32*(1), 1–4. doi:10.1029/2004GL021791

Brack, C. L. (2002). Pollution mitigation and carbon sequestration by an urban forest. *Environ. Pollut, 116*(Suppl. 1), 195–200. doi:10.1016/S0269-7491(01)00251-2 PMID:11833907

Brancic, B. (1968). *Sols de la Palmeraie de Fezouata*. Office Régional de Mise en Valeur Agricole Ouarzazate.

Brander, K. M. (2007). Global fish production and climate change. *Proceedings of the National Academy of Sciences, 104*(50), 19709-19714. Retrieved from http://www.pnas.org/content/104/50/19709.abstract

Brander, K. (2010). Impacts of climate change on fisheries. *Journal of Marine Systems*, *79*(3), 389–402. doi:10.1016/j.jmarsys.2008.12.015

Brasseur, G. P., Muller, J.-F., Tie, X., & Horowitz, L. (2001). Tropospheric ozone and climate: past, present and future. In T. Matsuno & H. Kida (Eds.), *Present and Future of Modeling Global Environmental Change: Toward Integrated Modeling* (pp. 63–75). Tokyo: TERRAPUB.

Brauer, M., Amann, M., Burnett, R. T., Cohen, A., Dentener, F., Ezzati, M., ... Thurston, G. D. (2012). Exposure assessment for estimation of the global burden of disease attributable to outdoor air pollution. *Environmental Science & Technology*, *46*(2), 652–660. doi:10.1021/es2025752 PMID:22148428

Brazil, G. M., Kenefick, L., Callanan, M., Haro, A., De Lorenzo, V., Dowling, D. N., & O'gara, F. (1995). Construction of a rhizosphere pseudomonad with potential to degrade polychlorinated biphenyls and detection of bph gene expression in the rhizosphere. *Applied and Environmental Microbiology*, *61*(5), 1946–1952. PMID:7646029

Brekke, L. D., Miller, N. L., Bashford, K. E., Quinn, N. W. T., & Dracup, J. A. (2004). Climate change impacts uncertainty for water resources in the San Joaquin River Basin, California. *Journal of the American Water Resources Association*, *40*(1), 149–164. doi:10.1111/j.1752-1688.2004.tb01016.x

Bretagnolle, V., Berthet, E., Gross, N., Gauffre, B., Plumejeaud, C., Houte, S., ... Gaba, S. (2018). Towards sustainable and multifunctional agriculture in farmland landscapes: Lessons from the integrative approach of a French LTSER platform. *The Science of the Total Environment*, *627*, 822–834. doi:10.1016/j.scitotenv.2018.01.142 PMID:29426207

Briones, C., Rivadeneira, M. M., Fernández, M., & Guiñez, R. (2014). Geographical Variation of Shell Thickness in the Mussel *Perumytilus purpuratus* Along the Southeast Pacific Coast. *The Biological Bulletin*, *227*(3), 221–231. doi:10.1086/BBLv227n3p221 PMID:25572210

BROSM. (2008). *Management Framework of the Reserve Biosphere Oasis of southern Morocco*. BRSMO.

Brunekreef, B., & Holgate, S. T. (2002). Air pollution and health. *Lancet (London, England)*, *360*(9341), 1233–1242. doi:10.1016/S0140-6736(02)11274-8 PMID:12401268

Brunner, F. S., & Eizaguirre, C. (2016). Can environmental change affect host/parasite-mediated speciation? *Zoology (Jena, Germany)*, *119*(4), 384–394. doi:10.1016/j.zool.2016.04.001 PMID:27210289

Burke, M., Hsiang, S. M., & Miguel, E. (2015). Global non-linear effect of temperature on economic production. *Nature*, *527*(7577), 235–239. doi:10.1038/nature15725 PubMed. doi:10.1038/nature15725 PMID:26503051

Butchart, S. H., Walpole, M., Collen, B., Van Strien, A., Scharlemann, J. P., Almond, R. E., & Carpenter, K. E. (2010). Global biodiversity: Indicators of recent declines. *Science*. PMID:20430971

Butler, R. (2008). 55% Of The Amazon May Be Lost By 2030. *Conservation News*. Retrieved from https://news.mongabay.com/2008/01/55-of-the-amazon-may-be-lost-by-2030/

Bytnerowicz, A., Arbaugh, M., Schilling, S., Fraczek, W., & Alexander, D. (2008). Ozone distribution and phytotoxic potential in mixed conifer forests of the San Bernardino Mountains, southern California. *Environ. Pollut*, *155*(3), 398–408. doi:10.1016/j.envpol.2008.01.046 PMID:18359541

Cabell, J. F., & Oelofse, M. (2012). An Indicator Framework for Assessing Agroecosystem Resilience. *Ecology and Society*, *17*(1), art18. doi:10.5751/ES-04666-170118

Callahan, J. T. (1984). Long-term ecological research. *Bioscience*, *34*(6), 363–367. doi:10.2307/1309727

Calzadilla, A., Rehdanz, K., Betts, R., Falloon, P., Wiltshire, A., & Tol, R. S. J. (2013). Climate change impacts on global agriculture. *Climatic Change*, *120*(1), 357–374. doi:10.100710584-013-0822-4

Campbell, B. M., Vermeulen, S. J., Aggarwal, P. K., Corner-Dolloff, C., Girvetz, E., Loboguerrero, A. M., … Wollenberg, E. (2016). Reducing risks to food security from climate change. *Global Food Security*, *11*, 34–43.

Campbell, N. A. (1996). Biology (4th ed.). The Benjamin/Cummings Publishing Company.

Campbell-Lendrum, D., Molyneux, D., Amerasinghe, F., Davies, C., Fletcher, E., Schofield, C., ... Sinkins, S. (2005). Ecosystems and vector-borne disease control. *Ecosystems and human well-being: policy responses, 3*, 353-372.

Cappy, S., Reichert, B., & Thein, J. (2002). Hydrogeological framework of the northern Drâa-catchment:Results of local and regional scale investigations. *Afrikagruppe deutscher Geowissenschaftler,Jahrestreffen 2002 – Forschungs- und Entwicklungszusammenarbeit in Afrika. Proceedings.*

Card, S. D., Johnson, L. J., & Teasdale, S. (2016). Deciphering microbial behaviour—the link between endophyte biology and efficacious biological control agents. *FEMS Microbiology Ecology,* 92.

Cardelino, C. A., & Chameides, W. L. (1990). Natural hydrocarbons, urbanization, and urban ozone. *Journal of Geophysical Research*, *95*(D9), 13,971–13,979. doi:10.1029/JD095iD09p13971

Cardozo, J., & Comp, L. (2003). Formulating an International Legal Protection for Genetic Resources, Traditional Knowledge and Folklore: Challenges for the Intellectual Property System. *Weerawit Weeraworawit*, *11*, 769.

Carle, J., & Holmgren, P. (2008). Wood from planted forests. *Forest Products Journal*, *58*(12), 6.

Carpenter, K. E., Abrar, M., Aeby, G., Aronson, R. B., Banks, S., Bruckner, A., ... Wood, E. (2008). One-third of reef-building corals face elevated extinction risk from climate change and local impacts. *Science*, *321*(5888), 560–563. doi:10.1126cience.1159196 PMID:18653892

Carslaw, K. S., Boucher, O., Spracklen, D. V., Mann, G. W., Rae, J. G. L., Woodward, S., & Kulmala, M. (2010). A review of natural aerosol interactions and feedbacks within the Earth system. *Atmospheric Chemistry and Physics*, *10*(4), 1701–1737. doi:10.5194/acp-10-1701-2010

Catcott, E. J. (1961). Effects of air pollution on animals. In *World Health Organization Monograph series, no.41* (pp. 221–231). Geneva: WHO.

Catska, V. (1994). Interrelationship between *vesicular-arbuscular mycorrhiza* and rhizospheremicroflora in apple replant disease. *Biologia Plantarum*, *36*(1), 99–104. doi:10.1007/BF02921276

Cavanagh, J. E. (2006). *Potential of Vegetation to Mitigate Road-generated Air Pollution*. Retrieved April 21st, 2016 from https://www.niwa.co.nz/sites/niwa. co.nz/files/import/attachments/co1x0405 p05.pdf

Cecchi, L. D., Amato, G., & Annesi-Maesano, I. (2013). Climate, urban air pollution, and respiratory allergy. In R. A. Pielke (Ed.), *Climate vulnerability: understanding and addressing threats to essential resources* (pp. 105–113). Waltham, MA: Academic Press. doi:10.1016/B978-0-12-384703-4.00116-7

Cesaroni, G., Forastiere, F., Stafoggia, M., Andersen, Z. J., Badaloni, C., Beelen, R., & … . (2014). Long term exposure to ambient air pollution and incidence of acute coronary events: Prospective cohort study and meta-analysis in 11 European cohorts from the ESCAPE project. *BMJ (Clinical Research Ed.)*, *348*, f7412. doi:10.1136/bmj.f7412 PMID:24452269

CFCIM. (2015). *Chambre Française de Commerce et d'Industrie du Maroc (CFCIM).* L'énergie éolienne au Maroc.

Chakraborty, S., & Newton, A. C. (2011). Climate change, plant diseases and food security: An overview. *Plant Pathology*, *60*(1), 2–14. doi:10.1111/j.1365-3059.2010.02411.x

Chambers, R., & Conway, G. R. (1992). *Sustainable rural livelihoods: practical conceptsfor the 21st century.* Discussion paper 296, Institute of Development Studies, Brighton, UK.

Chander, S. (2012). Impact of Climate Change on Insects. In H. Pathak & P. K. Aggarwal (Eds.), *Climate Change Impact, Adaptation and Mitigation in Agriculture: Methodology for Assessment and Application* (p. 131). New Delhi, India: Indian Agricultural Research Institute.

Chandra, S. V., & Garg, N. (1991). A cross-sectional study of pulmonary function among workers exposed to multimetals in the glass bangle industry. *American Journal of Industrial Medicine, 20*(3), 391–399. doi:10.1002/ajim.4700200311 PMID:1928114

Chanel, O., Henschel, S., Goodman, P. G., Analitis, A., Atkinson, R. W., Le Tertre, A., & Medina, S. (2014). Economic valuation of the mortality benefits of a regulation on SO2 in 20 European cities. *European Journal of Public Health, 24*(4), 631–637. doi:10.1093/eurpub/cku018 PMID:24567289

Change, D. P. O. C. (2003). *Vulnerability of North African Countries to Climatic Changes.* Available online: www.iisd.org/cckn/pdf/north_africa.pdf

Changming, L., Jingjie, Y., & Kendy, E. (2001). Groundwater exploitation and its impact on the environment in the North China Plain. *Water International, 26*(2), 265–272. doi:10.1080/02508060108686913

Chapin, F. S. III, Walker, B. H., Hobbs, R. J., Hooper, D. U., Lawton, J. H., Sala, O. E., & Tilman, D. (1997). Biotic control over the functioning of ecosystems. *Science, 277*(5325), 500–504. doi:10.1126cience.277.5325.500

Charmantier, A., McCleery, R. H., Cole, L. R., Perrins, C., Kruuk, L. E., & Sheldon, B. C. (2008). Adaptive phenotypic plasticity in response to climate change in a wild bird population. *Science, 320*(5877), 800–803. doi:10.1126cience.1157174 PMID:18467590

Chaturvedi, R. K., Gopalakrishnan, R., Jayaraman, M., Bala, G., Joshi, N. V., Sukumar, R., & Ravindranath, N. H. (2011). Impact of climate change on Indian forests: A dynamic vegetation modeling approach. *Mitigation and Adaptation Strategies for Global Change, 16*(2), 119–142. doi:10.1007/s11027-010-9257-7. doi:10.100711027-010-9257-7

Chaudhary, S., McGregor, A., Houston, D., & Chettri, N. (2015). The evolution of ecosystem services: A time series and discourse-centered analysis. *Environmental Science & Policy, 54*, 25–34. doi:10.1016/j.envsci.2015.04.025

Chauhan, A. J., Krishna, M. T., Frew, A. J., & Holgate, S. T. (1998). Exposure to nitrogen dioxide (NO2) and respiratory disease risk. *Reviews on Environmental Health, 13*, 73. PMID:9718623

Chelbi, I., Derbali, M., Al-Ahmadi, Z., Zaafouri, B., El Fahem, A., & Zhioua, E. (2007). Phenology of Phlebotomus papatasi (Diptera: Psychodidae) relative to the seasonal prevalence of zoonotic cutaneous leishmaniasis in central Tunisia. *Journal of Medical Entomology, 44*(2), 385–388. doi:10.1093/jmedent/44.2.385 PMID:17427714

Chelleri, L., Minucci, G., Ruiz, A., & Karmaoui, A. (2014). Responses to Drought and Desertification in the Moroccan Drâa Valley Region: Resilience at the Expense of Sustainability? *International Journal of Climate Change: Impacts & Responses, 5*(2).

Chenchen, L., Jiantao, S., Yuewei, L., Hui, L., Minsheng, W., Chunhong, W., & Tingming, S. (2017). Different exposure levels of fine particulate matter and preterm birth: A meta-analysis based on cohort studies. *Environmental Science and Pollution Research International, 24*(22), 17976–17984. doi:10.100711356-017-9363-0 PMID:28616740

Cheng, L., Trenberth, K. E., Fasullo, Boyer, T., Abraham, J., & Zhu, J. (2017). *Improved estimates of ocean heat content from 1960 to 2015.* Retrieved from doi:10.1126ciadv.1601545

Chen, I. C., Hill, J. K., Ohlemüller, R., Roy, D. B., & Thomas, C. D. (2011). Rapid range shifts of species associated with high levels of climate warming. *Science*, *333*(6045), 1024–1026. doi:10.1126cience.1206432 PMID:21852500

Chen, L., Liu, C., Zou, R., Yang, M., & Zhang, Z. (2016). Experimental examination of effectiveness of vegetation as bio-filter of particulate matters in the urban environment. *Environ. Pollut*, *208*, 198–208. doi:10.1016/j.envpol.2015.09.006 PMID:26385643

Chen, W., & Jim, C. Y. (2008). Assessment and valuation of the ecosystem services provided by urban forests. In M. Carreiro, Y. C. Song, & J. Wu (Eds.), *Ecology, Planning, and Management of Urban Forests* (pp. 53–83). New York: Springer. doi:10.1007/978-0-387-71425-7_5

Chestnut, L., & Mills, D. M. (2005). Afresh look at the benefits and costs of the US acid rain program. *Journal of Environmental Management*, *77*(3), 252–266. doi:10.1016/j.jenvman.2005.05.014 PMID:16171931

Cheung, W. W. L., Lam, V. W. Y., Sarmiento, J. L., Kearney, K., Watson, R., & Pauly, D. (2009). Projecting global marine biodiversity impacts under climate change scenarios. *Fish and Fisheries*, *10*(3), 235–251. doi:10.1111/j.1467-2979.2008.00315.x

Cheung, W. W. L., Lam, V. W. Y., Sarmiento, J. L., Kearney, K., Watson, R., Zeller, D., & Pauly, D. (2010). Large-scale redistribution of maximum fisheries catch potential in the global ocean under climate change. *Global Change Biology*, *16*(1), 24–35. doi:10.1111/j.1365-2486.2009.01995.x

Chevin, L. M., Lande, R., & Mace, G. M. (2010). Adaptation, plasticity, and extinction in a changing environment: Towards a predictive theory. *PLoS Biology*, *8*(4), e1000357. doi:10.1371/journal.pbio.1000357 PMID:20463950

Chou, C., Neelin, J. D., Chen, C.-A., & Tu, J.-Y. (2009). Evaluating the rich-get-richer mechanism in tropical precipitation change under global warming. *Journal of Climate*, *22*(8), 1982–2005. doi:10.1175/2008JCLI2471.1

Christian, R. R., French, C., Gosz, J., & Waide, R. (1999). Perspectives on international long-term ecological research. In *Perspectives in Ecology: A Glance from the VII International Congress of Ecology*. Backhuys.

Christoph, M., Fink, A. H., Paeth, H., Born, K., Kerschgens, M., & Piecha, K. (2010). *Climate scenarios. In Impacts of global change on the hydrological cycle in west and northwest Africa* (pp. 402–425). Heidelberg, Germany: Springer.

Clancy, L., Goodman, P., Sinclair, H., & Dockery, D. W. (2002). Effect of air-pollution control on death rates in Dublin, Ireland: An intervention study. *Lancet*, *360*(9341), 1210–1214. doi:10.1016/S0140-6736(02)11281-5 PMID:12401247

Claussen, M., Brovkin, V., Ganopolski, A., Kubatzki, C., & Petoukhov, V. (2003). Climate change in Northern Africa: The past is not the future. *Climatic Change*, *57*(1/2), 99–118. doi:10.1023/A:1022115604225

Clifford, A., Lang, L., Chen, R., Anstey, K. J., & Seaton, A. (2016). Exposure to air pollution and cognitive functioning across the life course–A systematic literature review. *Environmental Research*, *147*, 383–398. doi:10.1016/j.envres.2016.01.018 PMID:26945620

Cocking, E. C. (2003). Endophytic colonization of plant roots by nitrogen- fixing bacteria. *Plant and Soil*, *252*(1), 169–175. doi:10.1023/A:1024106605806

Cohen, A. J., Anderson, H. R., Ostro, B., Pandey, K. D., Krzyzanowski, M., Kunzli, N., ... Smith, K. (2005). The global burden of disease due to outdoor air pollution. *Journal of Toxicology and Environmental Health, Part A: Current Issues*, *68*(13-14), 1301–1307. doi:10.1080/15287390590936166 PMID:16024504

Cohen, M. F., Williams, J., & Yamasaki, H. (2002). Biodegradation of diesel fuel by an *Azolla*-derived bacterial consortium. *Journal of Environmental Science and Health. Part A*, *37*(9), 1593–1606.

Collado, A. D. (2000). *Spatio-temporal dynamics of dune patterns in semiarid Argentina: A neural network analysis. Edaphomatics Bull. no. 20.* Buenos Aires, Argentina: AICET-INTA.

Collado, A. D., Chuvieco, E., & Camarasa, A. (2000). Satellite remote sensing analysis to monitor desertifi cation processes in the crop-rangeland boundary of Argentina. *Journal of Arid Environments*, *52*(1), 121–133. doi:10.1016/S0140-1963(01)90980-2

Compant, S., Clément, C., & Sessitsch, A. (2010). Plant growth-promoting bacteria in the rhizo and endosphere of plants: Their role, colonization, mechanisms involved and prospects for utilization. *Soil Biology & Biochemistry*, *42*(5), 669–678. doi:10.1016/j.soilbio.2009.11.024

Cooke, S., Allison, E., Beard, T. D., Arlinghaus, R., Arthington, A., Bartley, D., ... Welcomme, R. (2016). On the sustainability of inland fisheries: Finding a future for the forgotten. *Ambio*, *45*(7), 753–764. doi:10.100713280-016-0787-4 PMID:27312662

Cooper, O. R., Parrish, D. D., Stohl, A., Trainer, M., Nédélec, P., Thouret, V., ... Leblanc, T. (2010). Increasing springtime ozone mixing ratios in the free troposphere over western North America. *Nature*, *463*(7279), 344–348. doi:10.1038/nature08708 PMID:20090751

Costanza, R., de Groot, R., Sutton, P., Van der Ploeg, S., Anderson, S. J., Kubiszewski, I., ... Turner, R. K. (2014). Changes in the global value of ecosystem services. *Global Environmental Change*, *26*, 152–158. doi:10.1016/j.gloenvcha.2014.04.002

Costello, A., Abbas, M., Allen, A., Ball, S., Bell, S., Bellamy, R., . . . Patterson, C. 2009. Managing the health effects of climate change. *Lancet, 373*, 1693–1733. Retrieved from https://www.ucl.ac.uk/global-health/project-pages/lancet1/ucl- lancet-climate-change.pdf

Coulthard, S. (2009). Adaptation and conflict within fisheries: insights for living with climate change. In W. N. Adger, I. Lorenzoni, & K. O'Brien (Eds.), *Adapting to climate change: thresholds, values, governance* (pp. 255–268). Cambridge, UK: Cambridge University Press. doi:10.1017/CBO9780511596667.017

Cowling, R. M., Rundel, P. W., Lamont, B. B., Arroyo, M. K., & Arianoutsou, M. (1996). Plant diversity in Mediterranean-climate regions. *Trends in Ecology & Evolution*, *11*(9), 362–366. doi:10.1016/0169-5347(96)10044-6 PMID:21237880

CPCB. (2012). *Epidemiological study on effect of air pollution on human health (adults) in Delhi.* Central pollution control board, environmental health management series: EHMS/01/2012.

Creighton, C., Hobday, A. J., Lockwood, M., & Pecl, G. T. (2016). Adapting Management of Marine Environments to a Changing Climate: A Checklist to Guide Reform and Assess Progress. *Ecosystems (New York, N.Y.)*, *19*(2), 187–219. doi:10.100710021-015-9925-2

Crespi, V. (2009). Support to the development of desert aquaculture and management of the brackish water lakes in Algeria. *FAO Aquaculture Newsletters*, *43*, 20–22.

Crimmins, S. M., Dobrowski, S. Z., Greenberg, J. A., Abatzoglou, J. T., & Mynsberge, A. R. (2011). Changes in climatic water balance drive downhill shifts in plant species' optimum elevations. *Science*, *331*(6015), 324–327. doi:10.1126cience.1199040 PMID:21252344

Cross, E. R., Newcomb, W. W., & Tucker, C. J. (1996). Use of weather data and remote sensing to predict thegeographic and seasonal distribution of *Phlebotomus papatasi* in southwest Asia. *The American Journal of Tropical Medicine and Hygiene*, *54*(5), 530–536. doi:10.4269/ajtmh.1996.54.530 PMID:8644911

CSIRO. (2011). *Sustainability in Australia's Arid Lands*. Retrieved from http://www.csiro.au/Outcomes/Water/Rural-and-regional-water/arid-land-sustainability.aspx

Dai, A. (2012). Increasing drought under global warming in observations and models. *Nature Climate Change*, *3*(1), 52–58. doi:10.1038/nclimate1633

Dale, N. M., & Fuller, H. L. (1980). Effect of Diet Composition on Feed Intake and Growth of Chicks Under Heat Stress. II. Constant vs. Cycling Temperatures. *Poultry Science*, *59*(7), 1434–1441. doi:10.3382/ps.0591434 PMID:7393854

DaMatta, F. M., Grandis, A., Arenque, B. C., & Buckeridge, M. S. (2010). Impacts of climate changes on crop physiology and food quality. *Food Research International*, *43*(7), 1814–1823. doi:10.1016/j.foodres.2009.11.001

Damek-Poprawa, M., & Sawicka-Kapusta, K. (2003). Damage to the liver, kidney, and testis with reference to burden of heavy metals in yellow-necked mice from areas around steelworks and zinc smelters in Poland. *Toxicology*, *186*(1-2), 1–10. doi:10.1016/S0300-483X(02)00595-4 PMID:12604166

Daryanoosh, S. M., Goudarzi, G., Khaniabadi, Y. O., Armin, H., Bassiri, H., & Khaniabadi, F. O. (2016). Effect of exposure to PM10 on cardiovascular diseases hospitalizations in Ahvaz, Khorramabad and Ilam, Iran during 2014. *Iranian Journal of Health, Safety and Environment*, *3*, 428–433.

Dash, S. K., Jenamani, R. K., Kalsi, S. R., & Panda, S. K. (2007). Some evidence of climate change in twentieth-century India. *Climatic Change*, *85*(3-4), 299–321. doi:10.1007/s10584-007-9305-9. doi:10.100710584-007-9305-9

Davidson, E. A. (2009). The contribution of manure and fertilizer nitrogen to atmospheric nitrous oxide since 1860. *Nature Geoscience*, *2*(9), 659–662. doi:10.1038/ngeo608

Davidson, E. A. (2012). Excess nitrogen in the US environment: Trends, risks, and solutions. *ESA Issues Ecology*, *15*, 1–16.

Davidson, E. A., & Janssens, I. A. (2006). Temperature Sensitivity of Soil Carbon Decomposition and Feedbacks to Climate Change. *Nature*, *440*(7081), 165–173. doi:10.1038/nature04514 PMID:16525463

Davies, J., Poulsen, L., Schulte-Herbrüggen, B., Mackinnon, K., Crawhall, N., Henwood, W.D., … Gudka, M. (2012). *Conservaton de la biodiversité des zones arides*. Academic Press.

Dawit, G., Girma, Z., & Simenew, K. (2013). A review on biology, epidemiology andpublic health significance of leishmaniasis. *Journal of Bacteriology & Parasitology*, *4*, 166. doi:10.4172/2155-9597.1000166

Dawson, T. P., Jackson, S. T., House, J. I., Prentice, I. C., & Mace, G. M. (2011). Beyond predictions: Biodiversity conservation in a changing climate. *Science*, *332*(6025), 53–58. doi:10.1126cience.1200303 PMID:21454781

Day, B. A., & Monroe, C. M. (1999). *Environmental Education and Communication for Sustainable World (Handbook for International Practitioners)*. Washington, DC: GreenCom.

De Basilio, V., Vilariño, M., Yahav, S., & Picard, M. (2001). Early Age Thermal Conditioning and a Dual Feeding Program for Male Broilers Challenged by Heat Stress. *Poultry Science*, *80*(1), 29–36. doi:10.1093/ps/80.1.29 PMID:11214333

De Grenade, R. (2013). Date palm as a keystone species in Baja California peninsula, Mexico oases. *Journal of Arid Environments*, *94*, 59–67. doi:10.1016/j.jaridenv.2013.02.008

De Groot, R., Brander, L., Van Der Ploeg, S., Costanza, R., Bernard, F., Braat, L., ... Hussain, S. (2012). Global estimates of the value of ecosystems and their services in monetary units. *Ecosystem Services*, *1*(1), 50–61. doi:10.1016/j.ecoser.2012.07.005

De Han, C. (2001). *Livestock development: Implications for rural poverty, the environment, and global food security*. The World Bank. doi:10.1596/0-8213-4988-0

De Jong, C. (2007). Impounded rivers: Aridification and geomorphological risks. *IAHS-AISH Publication.*

DeAngelo, B. J. (2013). Bounding the role of black carbon in the climate system: A scientific assessment. *Journal of Geophysical Research, D, Atmospheres*, *118*(11), 5380–5552. doi:10.1002/jgrd.50171

DEFCS. (1990). *Projet de création du Parc National d'Iriqui. Rapport inédit.* Direction des Eaux et Forêts et de la Conservation des Sols.

Defense, H. (1996). Health effects of outdoor air pollution. *American Journal of Respiratory and Critical Care Medicine*, 477–498. PMID:8564086

DEFRA. (2005). *Impact of climate change on soil functions. Final Project Report*. London, UK: Research and Development.

Degefu, M. A., Assen, M., & McGahey, D. (2018). *Climate variability and impact in ASSAR's East African region. CARIAA-ASSAR Working Paper*. International Development Research Centre, Ottawa, Canada and UK Aid, London, United Kingdom. Retrieved from www.assar.uct.ac.za

Dehghani, M., Anushiravani, A., Hashemi, H., & Shamsedini, N. (2014). Survey on air pollution and cardiopulmonary mortality in shiraz from 2011 to 2012: An analytical-descriptive study. *International Journal of Preventive Medicine*, *5*, 734–740. PMID:25013693

Dejmek, J., Selevan, S. G., Benes, I., Solanský, I., & Srám, R. J. (1999). Fetal growth and maternal exposure to particulate matter during pregnancy. *Environmental Health Perspectives*, *107*(6), 475–480. doi:10.1289/ehp.99107475 PMID:10339448

Denny, P. (Ed.). (1985). *The ecology and management of African wetland vegetation. Geobotany 6*. Dordrecht, The Netherlands: Dr. W. Junk Publishers. doi:10.1007/978-94-009-5504-2

Denton, F. (2013). *Africa on the rise: can the opportunities from climate change spring the continent to transformative development?* An inaugural speech at the 3rd annual climate change and development, Addis Ababa, Ethiopia.

Denton, F. (2014). Does climate – smart agriculture hold the key to Africa banishing hunger by 2025? ACPC/UNECA.

Dharumarajan, Bishop, Hegde, & Singh. (2017a). Prediction of effect of climate change on soil salinity using recent machine language techniques. *Proceedings of XIII Agricultural science congress*, 28.

Dharumarajan, & Lalitha, Hegde, Somashekar, & Singh. (2017c). Biophysical characterisation of Kangeyam grasslands for land degradation vulnerability analysis. *Range Management and Agroforestry*, *38*(1), 10–16.

Dharumarajan, Lalitha, & Hegde, Janani, Rajavat, Sastry, & Singh. (2018a). Status of desertification in South India- Assessment, Mapping and Change detection analysis. *Current Science*, *115*(2), 331–338.

Dharumarajan, S., Bishop, T. F. A., Hegde, R., & Singh, S. K. (2018b). Desertification vulnerability index—an effective approach to assess desertification processes: A case study in Anantapur District, Andhra Pradesh, India. *Land Degradation & Development*, 1–12. doi:10.1002/ldr.2850

Dharumarajan, S., Lalitha, M., Natarajan, A., Naidu, L. G. K., Balasubramanian, R., Hegde, R., ... Singh, S. K. (2017b). Biophysical and Socio Economic Causes for Increasing Fallow Lands in Tamil Nadu. *Soil Use and Management*, *33*(3), 487–498. doi:10.1111um.12361

Dharumarajan, S., Lalitha, M., & Vasundhara, R., & Hegde, R. (2016). The Major Biophysical Indicators of Desertification in Arid and Semi-arid regions of India. *Agropedology (Nagpur)*, *26*, 189–197.

Dholakia, H. H., Bhadra, D., & Garg, A. (2014). Short term association between ambient air pollution and mortality and modification by temperature in five Indian cities. *Atmospheric Environment*, *99*, 168–174. doi:10.1016/j.atmosenv.2014.09.071

Ding, Q., Chen, X., Hilborn, R., & Chen, Y. (2017). Vulnerability to impacts of climate change on marine fisheries and food security. *Marine Policy*, *83*, 55–61. doi:10.1016/j.marpol.2017.05.011

Dixon, G. R. (2012). Climate change – impact on crop growth and food production, and plant pathogens. *Canadian Journal of Plant Pathology*, *34*(3), 362–379. doi:10.1080/07060661.2012.701233

Dixon, R., & Kahn, D. (2004). Genetic regulation of biological nitrogen fixation. *Nature Reviews. Microbiology*, *2*(8), 621–631. doi:10.1038/nrmicro954 PMID:15263897

Doak, D. F., & Morris, W. F. (2010). Demographic compensation and tipping points in climate-induced range shifts. *Nature*, *467*(7318), 959–962. doi:10.1038/nature09439 PMID:20962844

Dockery, D. W. (2009). Health effects of particulate air pollution. *Annals of Epidemiology*, *19*(4), 257–263. doi:10.1016/j.annepidem.2009.01.018 PMID:19344865

Dockery, D. W., Pope, C. A. III, Xu, X., Spengler, J. D., Ware, J. H., Fay, M. E., ... Speizer, F. A. (1933). An Association between Air Pollution and Mortality in Six U.S. Cities. *The New England Journal of Medicine*, *329*(24), 1753–1759. doi:10.1056/NEJM199312093292401 PMID:8179653

Dockery, D. W., Schwartz, J., & Spengler, J. D. (1992). Air Pollution and Daily Mortality: Associations with Particulates and Acid Aerosols. *Environmental Research*, *59*(2), 362–373. doi:10.1016/S0013-9351(05)80042-8 PMID:1464289

Dockery, D. W., Speizer, F. E., Stram, D. O., Ware, J. H., Spengler, J. D., & Ferris, B. G. Jr. (1989). Effects of Inhalable Particles on Respiratory Health of Children. *The American Review of Respiratory Disease*, *139*(3), 587–594. doi:10.1164/ajrccm/139.3.587 PMID:2923355

Dockery, D. W., & Stone, P. H. (2007). Cardiovascular risks from fine particulate air pollution. *The New England Journal of Medicine*, *356*(5), 511–513. doi:10.1056/NEJMe068274 PMID:17267912

Dockery, D. W., & Trichopoulos, D. (1997). Risk of Lung Cancer from Environmental Exposures to Tobacco Smoke. *Cancer Causes & Control*, *8*(3), 333–345. doi:10.1023/A:1018401120233 PMID:9498897

Dodd, K. T., & Gross, D. R. (1980). Ammonia inhalation toxicity in cats: A study of acute and chronic respiratory dysfunction. *Archives of Environmental Health*, *35*(1), 6–14. doi:10.1080/00039896.1980.10667455 PMID:7362273

DOEC. (1999). *Etat de l'environnement au Maroc, Rapport sur l'état de l'environnement du Maroc (REEM)*. Direction de l'Observation, des Etudes et de la Coordination.

Doha, S. A., & Samy, A. M. (2010). Bionomics of phlebotomine sand flies (Diptera: Psychodidae) in the province of Al-Baha, Saudi Arabia. *Memorias do Instituto Oswaldo Cruz*, *105*(7), 850–856. doi:10.1590/S0074-02762010000700002 PMID:21120352

Dokhan, M. R., Kenawy, M. A., Doha, S. A., El-Hosary, S. S., Shaibi, T., & Annajar, B. B. (2016). Entomological studies of phlebotomine sand flies (Diptera: Psychodidae) in relation to cutaneous leishmaniasis transmission in Al Rabta, North West of Libya. *Acta Tropica*, *154*, 95–101. doi:10.1016/j.actatropica.2015.11.004 PMID:26589378

Domene, L. A. F., & Ayres, R. U. (2001). Nitrogen's role in industrial *systems. Journal of Industrial Ecology*, *5*(1), 77–103. doi:10.1162/108819801753358517

Dominici, F., Peng, R. D., Bell, M. L., Pham, L., McDermott, A., Zeger, S. L., & Samet, J. M. (2006). Fine particulate air pollution and hospital admission for cardiovascular and respiratory disease. *Journal of the American Medical Association*, *295*(10), 1127–1134. doi:10.1001/jama.295.10.1127 PMID:16522832

Donat, M. G., Lowry, A. L., Alexander, L. V., O'Gorman, P. A., & Maher, N. (2016). More extreme precipitation in the world's dry and wet regions. *Nature Climate Change*, *6*(5), 508–513. doi:10.1038/nclimate2941

Dong, W., Cui, B., Liu, Z., & Zhang, K. (2014). Relative effects of human activities and climate change on the river runoff in an arid basin in northwest China. *Hydrological Processes*, *28*(18), 4854–4864. doi:10.1002/hyp.9982

Donham, K. J. (1991). Association of environmental air contaminants with disease and productivity in swine. *American Journal of Veterinary Research*, *52*, 1723–1730. PMID:1767997

Douglass, D. C., Singer, B. S., Kaplan, M. R., Ackert, R. P., Mickelson, D. M., & Caffee, M. W. (2005). Evidence of early Holocene glacial advances in southern South America from cosmogenic surface-exposure dating. *Geology*, *33*(3), 237–240. doi:10.1130/G21144.1. doi:10.1130/G21144.1

Drielsma, M. J., Manion, G., Love, J., Williams, K. J., Harwood, T., & Saremi, H. (2015). 3C modelling for biodiversity under climate change. NSW Office of Environment and Heritage.

Drielsma, M., & Ferrier, S. (2009). Rapid evaluation of metapopulation persistence in highly variegated landscapes. *Biological Conservation*, *142*(3), 529–540. doi:10.1016/j.biocon.2008.11.018

Driouech, F., Déqué, M., & Mokssit, A. (2009). Numerical simulation of the probability distribution function of precipitation over Morocco. *International Journal Climate Dynamics*, *32*(7-8), 1055–1063. doi:10.100700382-008-0430-6

Driouech, F., Déqué, M., & Sánchez-Gómez, E. (2010). Weather regimes—Moroccan precipitation link in a regional climate change simulation. *Global and Planetary Change*, *72*(1-2), 1–10. doi:10.1016/j.gloplacha.2010.03.004

Droogers, P., Immerzeel, W. W., Terink, W., Hoogeveen, J., Bierkens, M. F. P., van Beek, L. P. H., & Debele, B. (2012). Water Resources Trends in Middle East and North Africa Towards 2050. *Hydrology and Earth System Sciences*, *16*(9), 3101–3114. doi:10.5194/hess-16-3101-2012

Drost, M., Ambrose, J. D., Thatcher, M. J., Cantrell, C. K., Wolfsdorf, K. E., Hasler, J. F., & Thatcher, W. W. (1999). Conception rates after artificial insemination or embryo transfer in lactating dairy cows during summer in Florida. *Theriogenology*, *52*(7), 1161–1167. doi:10.1016/S0093-691X(99)00208-3 PMID:10735094

Dubey, K. K., & Fulekar, M. H. (2011). Effect of pesticides on the seed germination of *Cenchrus setigerus* and *Pennisetum pedicellatum* as monocropping and co-cropping system: Implications for rhizospheric bioremediation. *Romanian Biotechnological Letters*, *16*(1), 5908–5918.

Dunnell, K. L., & Travers, S. E. (2011). Shifts in the flowering phenology of the northern Great Plains: Patterns over 100 years. *American Journal of Botany*, *98*(6), 935–945. doi:10.3732/ajb.1000363 PMID:21613073

Duru, U. E. (2008). *Impact of climate change on chemical and biological properties in contaminated soils*. Abstract Presented at the Cambridge University Geotechnical Engineering Group.

Economopoulos, A. P. (1993). *Assessment of Sources of Air Water and Land Pollution. A Guide to Rapid Source Inventory Techniques and Their Use in Formulating Environmental Control Strategies. Part One: Rapid Inventory Techniques in Environmental Pollution. Part Two: Approaches for Consideration in Formulating Environmental Control Strategies (Unpublished document WHO/YEP/93.1)*. Geneva: WHO.

EEA. (2015). *Air Quality in Europe - 2015 Report* (EEA Report No 5/2015). European Environmental Agency. doi:10.2800/62459

Ehrenfeld, D. (2005). The environmental limits to globalization. *Conservation Biology*, *19*(2), 318–326. doi:10.1111/j.1523-1739.2005.00324.x

El Aïch, A., Morand-Fehr, P., Bas, P., Araba, A., & Bourbouze, A. (2008). Meat from young goats raised in Argan Tree Forest (Morocco): Emerging product to valorise. *Options Méditerranéennes*, 205-210.

El Fasskaoui, B. (2009). Fonctions, défis et enjeux de la gestion et du développement durables dans la Réserve de Biosphère de l'Arganeraie (Maroc). *Études caribéennes*, (12). Retrieved from https://journals.openedition.org/etudescaribeennes/3711

El Kamoune, I. (2009). *Application du modèle Stream pour la simulation de l'érosion dans le bassin versant Tleta* (Rif occidental, Maroc). Science du sol, IAV Hassan II.

El Katrani, H. (2014). *Impact des utilisations des terres sur les propriétés hydrologiques du sol au niveau de bassin versant Tleta Rif occidental. Mémoire de 3ème cycle*. ENFI.

El Mrabet, S., Ouahmane, L., El Mousadik, A., Msanda, F., & Abbas, Y. (2014). The effectiveness of arbuscular mycorrhizal inoculation and bio-compost addition for enhancing reforestation with Argania spinosa in Morocco. *Open Journal of Forestry*, *4*(1), 14–23. doi:10.4236/ojf.2014.41003

El Yadari, H. (2012). *Evolution spatio-temporelle des conditions de surface dans le bassin versant Tleta (Rif Occidental). Management des Ressources en sols et en Eaux*. IAV Hassan II.

El-Badry, A. A., Al-Juhani, A. M., Ibrahim, E. K. D., & Al-Zubiany, S. F. (2009). Sand Flies Distribution and Bionomics, in Al-Madinah Al-Munawarah. *Research Journal of Parasitology*, *4*(1), 1–11. doi:10.3923/jp.2009.1.11 PMID:18576887

El-Badry, A., Al-Juhani, A., Ibrahim, E. K., & Al-Zubiany, S. (2008). Distribution of sand flies in El-Nekheil province, in Al-Madinah Al-Munawwarah region, western of Saudi Arabia. *Parasitology Research*, *103*(1), 151–156. doi:10.100700436-008-0942-3 PMID:18350318

El-Beltagy, A. & Madkour, M. (2012). *Impact of climate change on arid lands agriculture*. doi:10.1186/2048-7010-1-3

Endlicher, W. (2000). Mittelmeerländer. In Regionale Klimatologie: Die Alte Welt: Europa, Afrika, Asien. Teubner.

Ephraums, J. J., & Jenkins, G. J. J. (1992). *Climate change 1992*. Cambridge University Press.

Epton, M. J., Dawson, R. D., Brooks, W. M., Kingham, S., Aberkane, T., Cavanagh, J. A., ... Brown, L. (2008). The effect of ambient air pollution on respiratory health of school children: A panel study. *Environmental Health*, *7*(1), 16. doi:10.1186/1476-069X-7-16 PMID:18479529

Escobedo, F. J., Kroeger, T., & Wagner, J. E. (2011). Urban forests and pollution mitigation: Analyzing ecosystem services and disservices. *Environ. Pollut.*, *159*(8-9), 2078–2087. doi:10.1016/j.envpol.2011.01.010 PMID:21316130

Esteban, G. F., & Finlay, B. J. (2010). Conservation work is incomplete without cryptic biodiversity. *Nature*, *463*(7279), 293. doi:10.1038/463293c PMID:20090730

European Centre for Nature Conservation. (2016). Retrieved April 21, 201 from OpenNESS project: http://www.openness-project.eu/

European Commission (EC) Directive. (2008). *50/EC of the European Parliament and of the Council on Ambient Air Quality and Cleaner Air for Europe*. Available online: http://eur-lex.europa.eu/legal-content/

European Commission. (2008a). *Commission Regulation 692/2008 of 18 July 2008 Implementing and Amending Regulation (EC) No 715/2007 of the European Parliament and of the Council on Type-approval of Motor Vehicles with Respect to Emissions from Light Passenger and Commercial Vehicles*. Author.

European Commission. (2008b). *Directive 2008/50/EC. Commission Regulation 692/2008 of 18 July 2008 Implementing and Amending Regulation (EC) No 715/2007 of the European Parliament and of the Council on Type-approval of Motor Vehicles with Respect to Emissions from Light Passenger and Commercial Vehicles*. Author.

European Commission. (2012). *Commission Regulation (EU) No 459/2012 of 29 May 2012 Amending Regulation (EC) No 715/2007 of the European Parliament and of the Council and Commission Regulation (EC) No 692/2008 as Regards Emissions from Light Passenger and Commercial Vehicles (Euro 6)*. Author.

European Environment Agency. (2012). *Climate Change, Impacts and Vulnerability inEurope 2012: An indicator-based Report*. Report no. 12. Copenhagen, Denmark: Author.

European Union. (2000). *Caring for Our Future: Action for Europe's Environment* (3rd.ed.). Luxembourg: EU.

EVI. (2003). *The Demonstration Environmental Vulnerability Index (EVI)*. South Pacific Applied Geoscience Commission, SOPAC Technical Report 356.

EVICC. (2011). *Evaluation de la vulnérabilité et des impacts du changement climatique dans les oasis du Maroc –EVICCa-et structuration de stratégies territoriales d'adaptation. Mission 1.1: Bilan Diagnostic des vulnérabilités climatiques et des capacités d'adaptation en situation actuelle Bassin oasien de Dadès*. Author.

Ewing, R., Bartholomew, K., Winkelman, S., Walters, J., & Chen, D. (2008). *Growing cooler: The evidence on urban development and climate change*. Washington, DC: Urban Land Institute.

Eyotaru, O. (2013, July 23). East African Integration Hindered by Information Gaps – Minister. Daily Monitor Newspaper, p. 1.

Ezra, D., Castillo, U. F., & Strobel, G. A. (2004). Coronamycins, peptide antibiotics produced by a verticillate Streptomyces sp. (MSU-2110) endophytic on *Monstera sp. Microbiology*, *150*(4), 785–793. doi:10.1099/mic.0.26645-0 PMID:15073289

Fadlelmawla, A., Abdel-Halen, M., & Vissers, M. (1999). Preliminary plans for artificial recharge with exclaimed wastewater in Egypt. *9th Biennial Symposium on Artificial recharge of groundwater*, Tempe, AZ.

Fairley, D. (1990). The Relationship of Daily Mortality to Suspended Particulates in Santa Clara County, 1980–1986. *Environmental Health Perspectives*, *89*, 159–168. doi:10.1289/ehp.9089159 PMID:2088743

Fakhro, A., Andrade-Linares, D. R., Von Bargen, S., Bandte, M., Büttner, C., Grosch, R., ... Franken, P. (2010). Impact of *Piriformospora indica*on tomato growth and on interaction with fungal and viral pathogens. *Mycorrhiza*, *20*(3), 191–200. doi:10.100700572-009-0279-5 PMID:19789897

Falcon, W. P., Naylor, R. L., & Asia, S. (2008). Needs for Food Security in 2030 Region. *Science*, *319*, 607–610. doi:10.1126cience.1152339 PMID:18239122

Fang, Y., Naik, V., Horowitz, L. W., & Mauzerall, D. L. (2013). Air pollution and associated human mortality: The role of air pollutant emissions, climate change and methane concentration increases from the preindustrial period to present. *Atmospheric Chemistry and Physics*, *13*(3), 1377–1394. doi:10.5194/acp-13-1377-2013

Fantozzi, F., Monaci, F., Blanusa, T., & Bargagli, R. (2015). Spatio-temporal variations of ozone and nitrogen dioxide concentrations under urban trees and in a nearby open area. *Urban Clim*, *12*, 119–127. doi:10.1016/j.uclim.2015.02.001

FAO. (2003). Retrieved April 7, 2018 from http://www.fao.org/fi/oldsite/FCP/en/SAU/profile.htm

FAO. (2005). *The State of World Fisheries and Aquaculture 2005*. Rome: Food and Agriculture Organisation.

FAO. (2007). *The State of world fisheries and aquaculture – 2006*. Rome, Italy: Food and Agricultural Organization.

FAO. (2009a). *Climate Change Implications for Fisheries and Aquaculture 2009*. Food and Agriculture Organisation.

FAO. (2009b). *Livestock, food security and poverty reduction*. Rome: The State of Food and Agriculture. FAO.

FAO. (2009c). *Livestock in the balance*. Rome: State of Food and Agriculture. FAO.

FAO. (2009c). *The State of Food Insecurity in the World: Economic crises—Impacts and Lessons Learned 8–12*. Food and Agriculture Organization of the United Nations (FAOSTAT). Retrieved March 05, 2015 from http://faostat.fao.org/site/567/default.aspx#ancor

FAO. (2011). *World Livestock 2011-Livestock in food security*. Rome: FAO.

FAO. (2012). *The State of World Fisheries and Aquaculture*. Rome, Italy: FAO.

FAO. (2014). FAO Yearbook: Fishery and Aquaculture Statistics. FAO.

FAO. (2014). *FIGIS*. FishStat (Database). Retrieved from http://data. fao.org/ref/babf3346-ff2d-4e6c-9a40-ef6a50fcd422.html?version=1.0

FAO. (2014). *The State of the World Fisheries and Aquaculture 2014*. Food and Agriculture Organization.

FAO. (2015). *FAO Global Capture Production database updated to 2013 – Summary information*. FAO.

FAO. (2016). *The State of World Fisheries and Aquaculture – 2016 (SOFIA)*. Rome, Italy: FAO.

FAO. (2018). FAO water reports: Drought characteristics and management in North Africa and the Near East. Rome: FAO.

FAOfishstat. (2016). *The State of world fisheries and aquaculture - 2016. Contributing to food security and nutrition for all*. Rome, Italy: Food and Agricultural Organization.

Fattore, E., Paiano, V., Borgini, A., Tittarelli, A., Bertoldi, M., Crosignani, P., & Fanelli, R. (2011). Human health risk in relation to air quality in two municipalities in an industrialized area of Northern Italy. *Environmental Research*, *111*(8), 1321–1327. doi:10.1016/j.envres.2011.06.012 PMID:21764052

Faulde, M., Schrader, J., Heyl, G., Amirih, M., & Hoerauf, A. (2008). Zoonotic cutaneous leishmaniasis outbreak in Mazar-e Sharif, northern Afghanistan: An epidemiological evaluation. *International Journal of Medical Microbiology*, *298*(5-6), 543–550. doi:10.1016/j.ijmm.2007.07.015 PMID:17888728

Felzer, B., Kicklighter, D., Melillo, J., Wang, C., Zhuang, Q., & Prinn, R. (2004). Effects of ozone on net primary production and carbon sequestration in the conterminous United States using a biogeochemistry model. *Tellus*, *56B*(3), 230–248. doi:10.1111/j.1600-0889.2004.00097.x

Fennane, M. & Ibn Tattou, M. (2012). Statistiques et commentaires sur l'inventaire actuel de la flore vasculaire du Maroc. *Bulletin de l'Institut Scientifique, 34*(1), 1-9.

Fertah, M., Belfkira, A., Taourirte, M., & Brouillette, F. (2017). Extraction and characterization of sodium alginate from Moroccan Laminaria digitata brown seaweed. *Arabian Journal of Chemistry*, *10*, S3707–S3714. doi:10.1016/j.arabjc.2014.05.003

Fink, A., Christoph, M. H., Paeth, H., Born, K., Kerschgens, M., & Piecha, K. (2010). *Climate. Impacts of global change on the hydrological cycle in west and northwest Africa*. Heidelberg, Germany: Springer.

Fiore, A. M. (2014). Atmospheric chemistry: No equatorial divide for a cleansing radical. *Nature*, *513*(7517), 176–178. doi:10.1038/513176a PMID:25209794

Fiore, A. M., Jacob, D. J., Field, B. D., Streets, D. G., Fernandes, S. D., & Jang, C. (2002). Linking ozone pollution and climate change: The case for controlling methane. *Geophysical Research Letters*, *29*(19), 1919. doi:10.1029/2002GL015601

Fiore, A. M., Naik, V., & Leibensperger, E. M. (2015). Air quality and climate connections. *Journal of the Air & Waste Management Association*, *65*(6), 645–685. doi:10.1080/10962247.2015.1040526 PMID:25976481

Flegel, T. W. (2006). Detection of major penaeid shrimp viruses in Asia, a historical perspective with emphasis on Thailand. *Aquaculture (Amsterdam, Netherlands)*, *258*(1), 1–33. doi:10.1016/j.aquaculture.2006.05.013

Foden, J., Rogers, S. I., & Jones, A. P. (2008). A critical review of approaches to aquatic environmental assessment. *Marine Pollution Bulletin*, *56*(11), 1825–1833. doi:10.1016/j.marpolbul.2008.08.017 PMID:18840385

Foley, J., DeFries, R., Asner, G., Barford, C., Bonan, G., Carpenter, S., & Snyder, P. (2005). Global consequences of land use. *Science*, *309*(5734), 570–574. doi:10.1126cience.1111772 PMID:16040698

Folke, C., Carpenter, S. R., Walker, B. H., Scheffer, M., Chapin, T., & Rockström, J. (2010). *Resilience thinking: Integrating resilience, adaptability and transformability*. Retrieved from http://www.ecologyandsociety.org/vol15/iss4/art20/

Ford, M. J. (1982). *The changing climate: responses of the natural fauna and flora*. London: Allen and Unwin.

Forster, P., Ramaswamy, V., Artaxo, P., Berntsen, T., Betts, R., Fahey, D. W., . . . Van Dorland, R. (2007). Changes in atmospheric constituents and in radiative forcing. In S. Solomon, D. Qin, M. Manning, Z. Chen, M. Marquis, K. B. Averyt, M. Tignor, & H.L. Miller (Eds.), Climate change 2007: The physical science basis. Cambridge, UK: Cambridge University Press.

Fry, G. (1984). Australia and the South Pacific. *Diplomacy in the marketplace: Australia in world affairs, 90*, 175-77. Filippidou EC, Koukouliata A. Ozone effects on the respiratory system. *Progress in Health Sciences*, *2011*(1), 144–155.

Fuzzi, S., Baltensperger, U., Carslaw, K., Decesari, S., Denier Van Der Gon, H., Facchini, M. C., ... Nemitz, E. (2015). Particulate matter, air quality and climate: Lessons learned and future needs. *Atmospheric Chemistry and Physics*, *15*(14), 8217–8299. doi:10.5194/acp-15-8217-2015

Gaiero, J. R., McCall, C. A., Thompson, K. A., Day, N. J., Best, A. S., & Dunfield, K. E. (2013). Inside the root microbiome: Bacterial root endophytes and plant growth promotion. *American Journal of Botany*, *100*(9), 1738–1750. doi:10.3732/ajb.1200572 PMID:23935113

Gairola, S., Shariff, N. M., Bhatt, A., & Kala, C. P. (2010). Influence of Climate Change on Secondary Chemicals in High Altitude Himalayas: Issues Need Immediate Attention. *Journal of Medicinal Plants Research*, *4*(18), 1825–1829.

Gallagher, J., Baldauf, R., Fuller, C. H., Kumar, P., Gill, L. W., & McNabola, A. (2015). Passive methods for improving air quality in the built environment: A review of porous and solid barriers. *Atmospheric Environment*, *120*, 61–70. doi:10.1016/j.atmosenv.2015.08.075

Gao, Q., Ci, L., & Yu, M. (2002). Modeling wind and water erosion in northern China under climate and land use changes. *Journal of Soil and Water Conservation*, *57*, 46–55.

Gauderman, W. J., Vora, H., McConnell, R., Berhane, K., Gilliland, F., Thomas, D., ... Peters, J. (2007). Effect of exposure to traffic on lung development from 10 to 18 years of age: A cohort study. *Lancet*, *369*(9561), 571–577. doi:10.1016/S0140-6736(07)60037-3 PMID:17307103

Gauss, M., Myhre, G., Isaksen, I. S. A., Grewe, V., Pitari, G., Wild, O., ... Sanderson, M. G. (2003). Radiative forcing in the 21st century due to ozone changes in the troposphere and the lower stratosphere. *Journal of Geophysical Research*, *108*(D9), 4292. doi:10.1029/2002JD002624

Gebbers, R., & Adamchuk, V. I. (2010). Precision Agriculture and Food Security. *Science, 327*(5967), 828-831. Retrieved from http://science.sciencemag.org/content/327/5967/828.abstract

Gebremedhin, B., Ficetola, G. F., Naderi, S., Rezaei, H. R., Maudet, C., Rioux, D., ... Taberlet, P. (2009). Frontiers in identifying conservation units: From neutral markers to adaptive genetic variation. *Animal Conservation*, *12*(2), 107–109. doi:10.1111/j.1469-1795.2009.00255.x

Gedney, N., Cox, P. M., Betts, R. A., Boucher, O., Huntingford, C., & Stott, P. A. (2006). Detection of a direct carbon dioxide effect in continental river runoff records. *Nature*, *439*(7078), 835–838. doi:10.1038/nature04504 PMID:16482155

Geravandi, S., Goudarzi, G., Soltani, F., Salmanzadeh, S., Ghomeishi, A., Zalaghi, E., ... Mohammadi, M. (2016). The cardiovascular and respiratory deaths attributed to sulfur dioxide in Kermanshah. *Majallah-i Danishgah-i Ulum-i Pizishki-i Kirmanshah*, *19*, 319–326.

Germaine, K., Liu, X., Cabellos, G., Hogan, J., Ryan, D., & Dowling, D. N. (2006). Bacterial endophyte-enhanced phyto-remediation of the organochlorine herbicide 2, 4-dichlorophenoxyacetic acid. *FEMS Microbiology Ecology*, *57*(2), 302–310. doi:10.1111/j.1574-6941.2006.00121.x PMID:16867147

Gerstengarbe, F. W., & Werner, P. C. (2007). Der rezente Klimawandel. In Der Klimawandel – Einblicke, Rückblicke und Ausblicke (pp. 34–43). Humboldt University.

Geyer, J., Kiefer, I., Kreft, S., Chavez, V., Salafsky, N., Jeltsch, F., & Ibisch, P. L. (2011). Classification of climate-change-induced stresses on biological diversity. *Conservation Biology*, *25*(4), 708–715. doi:10.1111/j.1523-1739.2011.01676.x PMID:21488958

Ghallabi, L. B., Messouli, M., & Yacoubi, M. (2011). Integrated approaches to the assessment of the impacts of climate and socio-economic change on groundwater resources in the Tensift basin, Morocco. *Int. J. Water Resour. Arid Environ*, *1*, 219–225.

Gharehchahi, E., Mahvi, A. H., Amini, H., Nabizadeh, R., Akhlaghi, A. A., Shamsipour, M., & Yunesian, M. (2013). Health impact assessment of air pollution in Shiraz, Iran: A two- part study. *Iranian Journal of Environmental Health Sciences & Engineering*, *11*, 2–8. PMID:24499576

Ghio, A. J., & Huang, Y. C. (2004). Exposure to concentrated ambient particles (CAPs): A review. *Inhalation Toxicology*, *16*, 53.

Ghio, A. J., & Samet, J. (1999). *Air pollution and health: metals and air pollution particles*. Academic Press. doi:10.1016/B978-012352335-8/50104-6

Ghiyathoddin, M. (2006). *Air pollution—sources, effects and control*. Tehran: Tehran University Press.

Ghosh, R., Rankin, J., Pless-Mulloli, T., & Glinianaia, S. (2007). Does the effect of air pollution on pregnancy outcomes differ by gender? A systematic review. *Environmental Research*, *105*(3), 400–408. doi:10.1016/j.envres.2007.03.009 PMID:17493608

Ghozikali, M. G., Heibati, B., Naddafi, K., Kloog, I., Conti, G. O., Polos, R., & Ferrante, M. (2016). Evaluation of chronic obstructive pulmonary disease (COPD) attributed to atmospheric O_3, NO_2, and SO_2 using Air Q model (2011–2012 year). *Environmental Research*, *144*, 99–105. doi:10.1016/j.envres.2015.10.030 PMID:26599588

Gibelin, A. L., & Déqué, M. (2003). Anthropogenic climate change over the Mediterranean region simulated by a global variable resolution model. *Climate Dynamics*, *20*(4), 327–339. doi:10.100700382-002-0277-1

Giesecke, W. H. (1985). The effect of stress on udder health of dairy cows. *Journal of Veterinary Research (Pulawy)*, *52*, 175–193. PMID:3911132

Gilboa, S. M., Mendola, P., Olshan, A. F., Langlois, P. H., Savitz, D. A., Loomis, D., ... Fixler, D. E. (2005). Relation between ambient air quality and selected birth defects, Seven County Study, Texas, 1997–2000. *American Journal of Epidemiology*, *162*(3), 238–252. doi:10.1093/aje/kwi189 PMID:15987727

Giorgi, F., & Bi, X. (2005). Regional changes in surface climate interannual variability for the 21st century from ensembles of global model simulations. *Geophysical Research Letters*, *32*(13), L13701. doi:10.1029/2005GL023002

Giorgi, F., & Lionello, P. (2008). Climate change projections for the Mediterranean region. *Global and Planetary Change*, *63*(2-3), 90–104. doi:10.1016/j.gloplacha.2007.09.005

Glass, D. (2000). *U.S and International markets for Phytoremediation, D.* Needham, MA: Glass Associates, Inc.

Glaves, P., & Egan, D. (2013). Some Methods and Approached to Valuing Ecosystem Services. Academic Press.

Glick, B. R. (2012). Plant growth-promoting bacteria: mechanisms and applications. *Scientific*, article ID 963401.

Glick, B. R. (2015). *Beneficial Plant-Bacterial Interactions*. Springer.

Glick, P., Stein, B. A., & Edelson, N. A. (2011). *Scanning the conservation horizon: a guide to climate change vulnerability assessment*. Washington, DC: National Wildlife Federation.

Glinianaia, S. V., Rankin, J., Bell, R., Pless-Mulloli, T., & Howel, D. (2004). Particulate air pollution and fetal health: A systematic review of the epidemiologic evidence. *Epidemiology (Cambridge, Mass.)*, *15*(1), 36–45. doi:10.1097/01.ede.0000101023.41844.ac PMID:14712145

Global Footprint Network. (2018). *Footprint and Biocapacity Atlas of Francophonie Member Nations.* Retrieved from https://www.footprintnetwork.org/content/images/article_uploads/Francophonie_English.pdf

Glowka, L., Burhenne-Guilmin, F., Synge, H., McNeely, J. A., & Gundling, L. (1994). *A guide to the convention on biological diversity* (No. 333.95 G946). IUCN, Bonn (Alemania). Environmental Law Centre IUCN, Gland (Suiza). Biodiversity Programme.

Glueck, M. F., & Stockton, C. W. (2001). Reconstruction of the North Atlantic oscillation, 1429–1983. *International Journal of Climatology: A Journal of the Royal Meteorological Society, 21*(12), 1453-1465.

Gogus, U., & Smith, C. (2010). N-3 omega fatty acids: A review of current knowledge. *International Journal of Food Science & Technology*, *45*(3), 417–436. doi:10.1111/j.1365-2621.2009.02151.x

Gollifer, S., & Gorman, S. (2016). *An introductory Workshop Report on Gender Mainstreaming in Higher Education at University of Dodoma*. INASP.

Gossling, S., & Hall, C. M. (2006). An introduction to tourism and global environmental change. In *Tourism and Global Environmental Change* (pp. 1–34). Londres: Routledge.

Goudarzi, G., Geravandi, S., Foruozandeh, H., Babaei, A., Alavi, N., Niri, M., ... Mohammadi, M. (2015a). Cardiovascular and respiratory mortality attributed to ground-level ozone in Ahvaz, Iran. *Environmental Monitoring and Assessment*, *187*(8), 1–9. doi:10.100710661-015-4674-4 PMID:26141926

Goudarzi, G., Geravandi, S., Foruozandeh, H., Babaei, A., Alavi, N., Niri, M., ... Mohammadi, M. (2015b). Total number of deaths and respiratory mortality attributed to particulate matter (PM10) in Ahvaz, Iran during 2009. International. *Journal of Environmental Health Engineering*, *4*, 1–7.

Gouveia, N., Bremner, S. A., & Novaes, H. M. D. (2004). Association between ambient air pollution and birth weight in Sao Paulo, Brazil. *Journal of Epidemiology and Community Health*, *58*(1), 11–17. doi:10.1136/jech.58.1.11 PMID:14684720

Government of Uganda. (1999). *A Simplified Version of the United Nations Convention on the Elimination of all forms of Discrimination against Women (CEDAW)*. Gov. of Uganda.

Government of Uganda. (2013). *Climate Change Policy 2015. Ministry of Water and Environment.* Kampala, Uganda: Government of Uganda.

Government of Uganda. (2017). *National Population and Houwsing Census (2014): Area Specific Profiles for Isingiro Diistrict*. Kampala, Uganda: Government of Uganda.

Gownaris, N. J., Pikitch, E., Aller, J. Y., Kaufman, L., Kolding, J., Lwiza, K. M. M., … Rountos, K. J. (2016). *Reconciling fisheries and water level fluctuations in the world's largest desert lake.* Retrieved from doi:10.1002/eco.1769

Grace, D., Bett, B., Lindahl, J., & Robinson, T. (2015). *Climate and livestock disease: assessing the vulnerability of agricultural systems to livestock pests under climate change scenarios*. CCAFS Working Paper no.116. CGIAR Research Program on Climate Change, Agriculture and Food Security (CCAFS), Copenhagen. Retrieved from https://cgspace.cgiar.org/rest/bitstreams/55094/retrieve

Greaver, T. L., Sullivan, T. J., Herrick, J. D., Barber, M. C., Baron, J. S., Cosby, B. J., ... Novak, K. J. (2012). Ecological effects of nitrogen and sulfur air pollution in the US: What do we know? *Front. Ecol*, *10*(7), 365–372. doi:10.1890/110049

Green, R. E., Harley, M., Miles, L., Scharlemann, J., Watkinson, A., & Watts, O. (2007). Global climate change and biodiversity. *Global Climate Change and Biodiversity.*

Gregory, P. J., Johnson, S. N., Newton, A. C., & Ingram, J. S. I. (2009). Integrating pests and pathogens into the climate change/food security debate. *Journal of Experimental Botany*, *60*(10), 2827–2838. doi:10.1093/jxb/erp080 PMID:19380424

Grime, J. P. (1997). Biodiversity and ecosystem function: The debate deepens. *Science*, *277*(5330), 1260–1261. doi:10.1126cience.277.5330.1260

Groffman, P. M., Baron, J. S., Blett, T., Gold, A. J., Goodman, I., Gunderson, L. H., & Poff, N. L. (2006). Ecological thresholds: The key to successful environmental management or an important concept with no practical application? *Ecosystems (New York, N.Y.)*, *9*(1), 1–13. doi:10.100710021-003-0142-z

Grulke, N. E., Minnich, R. A., Paine, T. D., Seybold, S. J., Chavez, D. J., Fenn, M. E., . . . Dunn, A. (2009). Air pollution increases forest susceptibility to wildfires In A case study in the San Bernardino Mountains in southern California. Academic Press.

Grulke, N. E., Paoletti, E., & Heath, R. L. (2006). Direct measurements of foliar ozone uptake in crop and tree species. *Environ. Pollut.*, *146*, 640–647. doi:10.1016/j.envpol.2006.04.014 PMID:16777298

Gryparis, A., Forsberg, B., Katsouyanni, K., Analitis, A., Touloumi, G., Schwartz, J., ... Dörtbudak, Z. (2004). al. Acute effects of ozone on mortality from the "air pollution and health: A European approach" project. *American Journal of Respiratory and Critical Care Medicine*, *170*(10), 1080–1087. doi:10.1164/rccm.200403-333OC PMID:15282198

Guan, S. H., Sattler, I., Lin, W. H., Guo, D. A., & Grabley, S. (2005). p-Aminoacetophenonic acids produced by a mangrove endophyte: Streptomyces griseus subspecies. *Journal of Natural Products*, *68*(8), 1198–1200. doi:10.1021/np0500777 PMID:16124760

Guemimi, A. (1991). Drainage des sols salés dans le Tafilalet. *Revue Homme terre et eau*, 47-60.

Guennouni, A. (2012). Filière phoénicicole: Des acquis à consolide. *Agriculture du Maghreb*, *62*, 57–59.

Guler, G. O., Kiztanir, B., Aktumsek, A., Citil, O. B., & Ozparlak, H. (2008). Determination of the seasonal changes on total fatty acid composition and ω3/ω6 ratios of carp (Cyprinus carpio L.) muscle lipids in Beysehir Lake (Turkey). *Food Chemistry*, *108*(2), 689–694. doi:10.1016/j.foodchem.2007.10.080 PMID:26059149

Guo, B., Dai, J. R., Ng, S., Huang, Y., Leong, C., Ong, W., & Carte, B. K. (2000). Cytonic acids A and B: Novel tridepside inhibitors of hCMV protease from the endophytic fungus Cytonaema species. *Journal of Natural Products*, *63*(5), 602–604. doi:10.1021/np990467r PMID:10843568

Gyasi, E. A., Kranjac-Berisavljevic, G., Blay, E. T., & Oduro, W. (Eds.). (2004). *Managing agrodiversity the traditional way: Lessons from West Africa in sustainable use of biodiversity and related natural resources*. Tokyo: UN Univ. Press.

Haase, D., Larondelle, N., Andersson, E., Artmann, M., Borgström, S., Breuste, J., ... Kabisch, N. (2014). A quantitative review of urban ecosystem service assessments: Concepts, models, and implementation. *Ambio*, *43*(4), 413–433. doi:10.100713280-014-0504-0 PMID:24740614

Haddouch, M. (1996). Situation actuelle et perspectives de développement du palmier dattier au Maroc. *Options Mediterr*, *28*, 63–79.

Ha, E. H., Hong, Y. C., Lee, B. E., Woo, B. H., Schwartz, J., & Christiani, D. C. (2001). Is air pollution a risk factor for low birth weight in Seoul? *Epidemiology (Cambridge, Mass.)*, *12*(6), 643–648. doi:10.1097/00001648-200111000-00011 PMID:11679791

Hahn, G. L. (1985). Management and housing of farm animals in hot environment. In M. K. Yousef (Ed.), *Stress Physiology of Livestock. Ungulates* (Vol. 2, pp. 151–176). Boca Raton, FL: CRC Press.

Haines-Young, R., & Potschin, M. (2011). Common International Classification of Ecosystem Services (CICES): 2011 Update. Report to the European Environmental Agency, Nottingham, UK.

Haleem, D., Von Wintzingerode, F., Moter, A., Moawad, H., & Göbel, U. B. (2000). Phylogenetic analysis of rhizosphere-associated β-subclass proteobacterial ammonia oxidizers in a municipal wastewater treatment plant based on rhizoremediation technology. *Letters in Applied Microbiology*, *31*(1), 34–38. doi:10.1046/j.1472-765x.2000.00760.x PMID:10886611

Hallmann, A., Benhamou, N., & Kloepper, J. W. (1997). Bacterial endophytes in cotton:mechanisms of entering the plant. *Canadian Journal of Microbiology*, *43*(6), 577–582. doi:10.1139/m97-081

Hall, S. J., Hilborn, R., Andrew, N. L., & Allison, E. H. (2013). Innovations in capture fisheries are an imperative for nutrition security in the developing world. *Proceedings of the National Academy of Sciences of the United States of America*, *110*(21), 8393–8398. doi:10.1073/pnas.1208067110 PMID:23671089

Halpern, B. S., Walbridge, S., Selkoe, K. A., Kappel, C. V., Micheli, F. D., Agrosa, C., … Watson, R. (2008). A Global Map of Human Impact on Marine Ecosystems. *Science, 319*(5865), 948-952. Retrieved from http://science.sciencemag.org/content/319/5865/948.abstract

Halpern, B. S., Longo, C., Hardy, D., McLeod, K. L., Samhouri, J. F., Katona, S. K., ... Zeller, D. (2012). An index to assess the health and benefits of the global ocean. *Nature*, *488*(7413), 615–620. doi:10.1038/nature11397 PMID:22895186

Hamlat, A., Errih, M., & Guidoum, A. (2003). Simulation of Water Resources Management Scenarios in Western Algeria Watersheds Using WEAP Model. *Arabian Journal of Geosciences*, *6*(7), 2225–2236. doi:10.100712517-012-0539-0

Hammond, S. T., Brown, J. H., Burger, J. R., Flanagan, T. P., Fristoe, T. S., Mercado-Silva, N., ... Okie, J. G. (2015). Food Spoilage, Storage, and Transport: Implications for a Sustainable Future. *Bioscience*, *65*(8), 758–768. doi:10.1093/biosci/biv081

Hammouda, N. (2010). Evolution de l'occupation des sols et de l'érosion dans le bassin versant Tleta. Management des Ressources en sols et en Eaux, IAV Hassan II.

Hamra, G. B., Guha, N., Cohen, A., Laden, F., Raaschou-Nielsen, O., Samet, J. M., ... Loomis, D. (2014). Outdoor particulate matter exposure and lung cancer: A systematic review and meta- analysis. *Environmental Health Perspectives*, *122*(9), 906–911. doi:10.1289/ehp/1408092 PMID:24911630

Hanh, V. T., Moore, P., & Emerton, L. (n.d.). Review of laws and policies related to payment for ecosystem services in Viet Nam. *Viet Nam Ctry. Program.*

Hannah, L. (2010). A global conservation system for climate-change adaptation. *Conservation Biology, 24*(1), 70–77. doi:10.1111/j.1523-1739.2009.01405.x PMID:20121843

Hannah, L., Midgley, G., Andelman, S., Araújo, M., Hughes, G., Martinez-Meyer, E., & Williams, P. (2007). Protected area needs in a changing climate. *Frontiers in Ecology and the Environment, 5*(3), 131–138. doi:10.1890/1540-9295(2007)5[131:PANIAC]2.0.CO;2

Hansen, R., Frantzeskaki, N., McPhearson, T., Rall, E., Kabisch, N., Kaczorowska, A., … Pauleit, S. (2015). The uptake of the ecosystem services concept in planning discourses of European and American cities. *Ecosystem Services, 12*, 228–246.

Hansen, C., Luben, T. J., Sacks, J. D., Olshan, A., Jeffay, S., Strader, L., & Perreault, S. D. (2010). The effect of ambient air pollution on sperm quality. *Environmental Health Perspectives, 118*(2), 203–209. doi:10.1289/ehp.0901022 PMID:20123611

Hansen, J., Nazarenko, L., Ruedy, R., Sato, M., Willis, J., Del Genio, A., ... Tausnev, N. (2005). Earth's energy imbalance: Confirmation and implications. *Science, 308*(5727), 1431–1435. doi:10.1126cience.1110252 PMID:15860591

Hansen, J., Ruedy, R., Sato, M., & Lo, K. (2010). Global surface temperature change. *Reviews of Geophysics, 48*(4), RG4004. doi:10.1029/2010RG000345

Hansen, P. J., Drost, M., Rivera, R. M., Paula-Lopes, F. F., Al-Katanani, Y. M., Kringinger, C. E., & Chase, C. C. (2001). Adverse impact of heat stress on embryo production. *Theriogenology, 55*(00), 91–103. doi:10.1016/S0093-691X(00)00448-9 PMID:11198091

Haramis, G. M. (1984). Acid alteration of experimental wetlands: Effects on diversity andabundance of macroinvertebrates and growth, behaviours and survival of juvenile black ducks. Draft research summary, US Food and Wildlife Service, Eastern Energy and Land Use Team.

Hardoim, P. R., Van, O., Verbeek, L. S., & Berg, G. (2015). The hidden world within plants: Ecological and evolutionary considerations for defining functioning of microbial endophytes. *Microbiology and Molecular Biology Reviews, 79*(3), 293–320. doi:10.1128/MMBR.00050-14 PMID:26136581

Hardoim, P. R., Van, O., Verbeek, L. S., & Van, E. (2008). Properties of bacterial endophytes and their proposed role in plant growth. *Trends in Microbiology, 16*(10), 463–471. doi:10.1016/j.tim.2008.07.008 PMID:18789693

Harry, E. G. (1978). Air pollution in farm buildings and methods of control: A review. *Avian Pathology, 7*(4), 441–454. doi:10.1080/03079457808418301 PMID:18770401

Hartley, S. E., & Gange, A. C. (2009). Impacts of plant symbiotic fungi on insect herbivores: Mutualism in a multitrophic context. *Annual Review of Entomology, 54*(1), 323–342. doi:10.1146/annurev.ento.54.110807.090614 PMID:19067635

Hartvigsen, G., Kinzig, A., & Peterson, G. (1998). Complex adaptive systems: Use and analysis of complex adaptive systems in ecosystem science: Overview of special section. *Ecosystems (New York, N.Y.), 1*(5), 427–430. doi:10.1007100219900036

Harvell, C. D., Kim, K., Burkholder, J. M., Colwell, R. R., Epstein, P. R., & Grimes, D. J. (1505). … Vasta, G. R. (1999). Emerging Marine Diseases--Climate Links and Anthropogenic Factors. [abstract]. *Science, 285*(5433), LP-1510. Retrieved from http://science.sciencemag.org/content/285/5433/1505

Hays, J. D., Imbrie, J., & Shackleton, N. J. (1976). Variations in the earth's orbit: Pacemaker of the ice ages. *Science, 194*(4270), 1121–1132. doi:10.1126cience.194.4270.1121 PMID:17790893

HCEFLCD. (2007). *Etude d'aménagement du parc national d'Iriqui. Rapport de synthèse*. HCEFLCD-DREF SO.

HCP. (2010). *Littoral marocain: un poids économique et démographique important*. Haut Commissariat au Plan du Maroc, présentation du 23/09/2010 à l'IRES lors de la journée d'étude "Littoral marocain et changement climatique".

HCP. (2015). *Caractéristiques Démographiques et Socio-Economiques: REGION DRAA-TAFILALET.* Direction Régionale Drâa-Tafilalet.

Heal, G., & Millner, A. (2014). ReflectionsUncertainty and Decision Making in Climate Change Economics. *Review of Environmental Economics and Policy*, *8*(1), 120–137. doi:10.1093/reep/ret023

Health and Environment Alliance. Climate Change. (2009). Retrieved from www.env-health.org/r/93

Heath, Pollard, & Thomas. (1984). *Atlas of Butterflies of Britain and Ireland.* Viking Press.

Heath, J. A., Steenhof, K., & Foster, M. A. (2012). Shorter migration distances associated with higher winter temperatures suggest a mechanism for advancing nesting phenology of American kestrels Falco sparverius. *Journal of Avian Biology*, *43*(4), 376–384. doi:10.1111/j.1600-048X.2012.05595.x

Hegland, S. J., Nielsen, A., Lázaro, A., Bjerknes, A.-L., & Totland, Ø. (2009). How does Climate Warming affect Plant-Pollinator Interactions? *Ecology Letters*, *12*(2), 184–195. doi:10.1111/j.1461-0248.2008.01269.x PubMed. doi:10.1111/j.1461-0248.2008.01269.x PMID:19049509

HEI. (2011). Coordinated studies of short-term exposure to air pollution and daily mortality in two Indian cities. Public Health and Air Pollution in Asia (PAPA), research report 157. Health Effects Institute.

Heidecke, C. (2009). *Economic Analysis of Water Use and Management in the Middle Drâa Valley in Morocco* (Ph.D. Thesis). University of Bonn, Bonn, Germany.

Heidecke, C., & Schmidt, T. (2008). Agricultural Land Use. In Impetus Atlas Morocco: Research Results 2000–2007 (pp. 67–68). University of Bonn.

Heikens, A. (2006). *Arsenic contamination of irrigation water, soil and crops in Bangladesh: Risk implications for sustainable agriculture and food safety in Asia. RAP Publication 2006/20*. Bangkok: Food and Agriculture Organization of the United Nations, Regional Office for Asia and the Pacific.

Heisler, G. M. (1986). Effects of individual trees on the solar radiation climate of small buildings. *Urban Ecology*, *9*(3/4), 337–359. doi:10.1016/0304-4009(86)90008-2

He, K., Song, Y., Daviglus, M. L., Liu, K., Van Horn, L., Dyer, A. R., & Greenland, P. (2004). Accumulated evidence on fish consumption and coronary heart disease mortality: A meta-analysis of cohort studies. *Circulation*, *109*(22), 2705–2711. doi:10.1161/01.CIR.0000132503.19410.6B PMID:15184295

Held, I. M., & Soden, B. J. (2006). Robust Responses of the Hydrological Cycle to Global Warming. *Journal of Climate*, *19*(21), 5686–5699. doi:10.1175/JCLI3990.1

Hellegers, P., Immerzeel, W., & Droogers, P. (2003). Economic Concepts to Address Future Water Supply-Demand Imbalances in Iran, Morocco and Saudi Arabia. *Journal of Hydrology (Amsterdam)*, *502*, 62–67. doi:10.1016/j.jhydrol.2013.08.024

Herczeg, A. L., Dogramaci, S. S., & Leaney, F. W. J. (2001). Origin of dissolved salts in a large, semi-arid groundwater system: Murray Basin, Australia. *Marine & Freshwater Research*, *52*(1), 41–52. doi:10.1071/MF00040

Hernandez-Ochoa, I. M., Asseng, S., Kassie, B. T., Xiong, W., Robertson, R., Pequeno, D. N. L., & Hoogenboom, G. (2018). Climate change impact on Mexico wheat production. *Agricultural and Forest Meteorology*, *263*, 373–387. doi:10.1016/j.agrformet.2018.09.008

Herrera, R. G., Puyol, D. G., MartÍn, E. H., Presa, L. G., & Rodríguez, P. R. (2001). Influence of the North Atlantic oscillation on the Canary Islands precipitation. *Journal of Climate*, *14*(19), 3889–3903. doi:10.1175/1520-0442(2001)014<3889:IO TNAO>2.0.CO;2

Herrero, M., Wirsenius, S., Henderson, B., Rigolot, C., Thornton, P., Havlík, P., ... Gerber, P. J. (2015). Livestock and the Environment: What Have We Learned in the Past Decade? *Annual Review of Environment and Resources*, *40*(1), 177–202. doi:10.1146/annurev-environ-031113-093503

Hertig, E. (2004). *Niederschlags und Temperaturabschätzungen für den Mittelmeerraum unter anthropogen verstärktem Treibhauseffekt*. Wuerzburg: University of Wuerzburg.

Hertig, E., & Jacobeit, J. (2008). Downscaling future climate change: Temperature scenarios for the Mediterranean area. *Global and Planetary Change*, *63*(2-3), 127–131. doi:10.1016/j.gloplacha.2007.09.003

Heywood, V. H., & Watson, R. T. (1995). *Global biodiversity assessment* (Vol. 1140). Cambridge, UK: Cambridge University Press.

Hickley, P., Bailey, R. G., Harper, D. M., Kundu, R., Muchiri, S. M., North, R., & Taylor, A. (2002). The status and future of the Lake Naivasha fishery, Kenya. *Hydrobiologia*, *488*(1/3), 181–190. doi:10.1023/A:1023334715893

Hiebenthal, C., Philipp, E. E. R., Eisenhauer, A., & Wahl, M. (2013). Effects of seawater pCO_2 and temperature on shell growth, shell stability, condition and cellular stress of Western Baltic Sea *Mytilus edulis* (L.) and *Arctica islandica* (L.). *Marine Biology*, *160*(8), 2073–2087. doi:10.100700227-012-2080-9

Hiebenthal, C., Philipp, E., Eisenhauer, A., & Wahl, M. (2012). Interactive effects of temperature and salinity on shell formation and general condition in Baltic Sea *Mytilus edulis* and Arctica islandica. *Aquatic Biology*, *14*(3), 289–298. doi:10.3354/ab00405

Hobbies, J. E., Carpenter, S. R., Grimm, N. B., Gosz, J. R., & Seastedt, T. R. (2003). The US long-term ecological research program. *Bioscience*, *53*(1), 21–32. doi:10.1641/0006-3568(2003)053[0021:TULTER]2.0.CO;2

Hoegh-Guldberg, O., Mumby, P. J., Hooten, A. J., Steneck, R. S., Greenfield, P., & Gomez, E. (1737). … Hatziolos, M. E. (2007). Coral Reefs Under Rapid Climate Change and Ocean Acidification. [abstract]. *Science*, *318*(5857), LP-1742. Retrieved from http://science.sciencemag.org/content/318/5857/1737

Hoffmann, A. A., & Sgrò, C. M. (2011). Climate change and evolutionary adaptation. *Nature*, *470*(7335), 479–485. doi:10.1038/nature09670 PMID:21350480

Hofmann, E., Ford, S., Powell, E., & Klinck, J. (2001). Modeling studies of the effect of climate variability on MSX disease in eastern oyster (Crassostrea virginica) populations. *Hydrobiologia*, *460*(1), 195–212. doi:10.1023/A:1013159329598

Holmes, R. T. (2011). Avian population and community processes in forest ecosystems: Long-term research in the Hubbard Brook Experimental Forest. *Forest Ecology and Management*, *262*(1), 20–32. doi:10.1016/j.foreco.2010.06.021

Holmes, R. T., & Sherry, T. W. (2001). Thirty-year bird population trends in an unfragmented temperate deciduous forest: Importance of habitat change. *The Auk*, *118*(3), 589–609. doi:10.1642/0004-8038(2001)118[0589:TYBPTI]2.0.CO;2

Hooper, D. U., Chapin, F. S., Ewel, J. J., Hector, A., Inchausti, P., Lavorel, S., & Schmid, B. (2005). Effects of biodiversity on ecosystem functioning: A consensus of current knowledge. *Ecological Monographs*, *75*(1), 3–35. doi:10.1890/04-0922

Horrocks, L. A., & Yeo, Y. K. (1999). Health benefits of docosahexaenoic acid (DHA). *Pharmacological Research, 40*(3), 211–225.

Houghton. (2001). *Climate Change 2001: The Scientific Basis*. Cambridge University Press.

Houry, H. (2014). *Etude des stratégies et pratiques adaptatives des exploitations agricoles dans les mosaïques paysagères à agrosystème pluvio-méditéranéen: Cas du bassin versant Tleta », PFE, option: Ingénierie du développement*. ENA-Meknès.

Huang, J., Ji, M., Xie, Y., Wang, S., He, Y., & Ran, J. (2016). Global semi-arid climate change over last 60 years. *Climate Dynamics*, *46*(3–4), 1131–1150. doi:10.100700382-015-2636-8

Huang, J., Minnis, P., Yan, H., Yi, Y., Chen, B., Zhang, L., & Ayers, J. K. (2010). Dust aerosol effect on semi-arid climate over Northwest China detected from A-Train satellite measurements. *Atmospheric Chemistry and Physics*, *10*(14), 6863–6872. doi:10.5194/acp-10-6863-2010

Huang, Y. C., & Ghio, A. J. (2006). Vascular effects of ambient pollutant particles and metals. *Current Vascular Pharmacology*, *4*(3), 199–203. doi:10.2174/157016106777698351 PMID:16842137

Hübener, H., Schmidt, M., Sogalla, M., & Kerschgens, M. (2005). Simulating evapotranspiration in a semi-arid environment. *Theoretical and Applied Climatology*, *80*(2-4), 153–167. doi:10.100700704-004-0097-9

Huete, A. R., Miura, T., & Gao, X. (2003). Land cover conversion and degradation analyses through coupled soil-plant biophysical parameters derived from hyperspectral EO-1 Hyperion. *IEEE Transactions on Geoscience and Remote Sensing*, *41*(6), 1268–1276. doi:10.1109/TGRS.2003.813209

Hu, G., & Jia, L. (2015). Monitoring of evapotranspiration in a semi-arid inland river basin by combining microwave and optical remote sensing observations. *Remote Sensing*, *7*(3), 3056–3087. doi:10.3390/rs70303056

Hughes, T. P., Baird, A. H., Bellwood, D. R., Card, M., Connolly, S. R., Folke, C., & Lough, J. M. (2003). Climate change, human impacts, and the resilience of coral reefs. *Science*, *301*(5635), 929–933. doi:10.1126cience.1085046 PMID:12920289

Hulin, M., Simoni, M., Viegi, G., & Annesi-Maesano, I. (2012). Respiratory health and indoor air pollutants based on quantitative exposure assessments. *The European Respiratory Journal*, *40*(4), 1033–1045. doi:10.1183/09031936.00159011 PMID:22790916

Hungria, M., Campo, R. J., Souza, E. M., & Pedrosa, F. O. (2010). Inoculation with selected strains of *A. brasilense* and *A. lipoferum* improves yields of maize and wheat in Brazil. *Plant and Soil*, *331*(1-2), 413–425. doi:10.100711104-009-0262-0

Hurrell, J. W., & Van Loon, H. (1997). Decadal variations in climate associated with the North Atlantic Oscillation. *Climatic Change*, *36*(3/4), 301–326. doi:10.1023/A:1005314315270

Ibrahim, A. A., & Abdoon, A. M. (2005). Distribution and population dynamics of Phlebotomus sand flies (Diptera: Psychodidae) in an endemic area of cutaneous leishmaniasis in Asir Region, Southwestern Saudi Arabia. *Journal of Entomology*, *2*(Suppl 1), 102–108. doi:10.3923/je.2005.102.108

IHME. (2016). *Poor air quality kills 5.5 million worldwide annually. Institute for health metrics and evaluation*. Available at: http://www.healthdata.org/news- release/poor-air-quality-kills-55-million-worldwide-annually

IIASA/FAO. (2012). *Global Agro-ecological Zones (GAEZ v3.0)*. IIASA and FAO.

Ilahiane, H. (1996). Small-Scale Irrigation in a Multiethnic Oasis Environment: The Case of Zaouit AmelkisVillage. *Southeast Morocco Journal of Political Ecology*, *3*, 89. doi:10.2458/v3i1.20459

ILTER. (1998). *International Long-Term Ecological Research Network. US LTER Network*. Albuquerque, NM: Department of Biology, University of New Mexico.

Imhoff, M. L., & Bounoua, L. (2006). Exploring global patterns of net primary production carbon supply and demand using satellite observations and statistical data. *Journal of Geophysical Research, D, Atmospheres*, *111*(D22), D22S12. doi:10.1029/2006JD007377

Inhamuns, A. J., & Franco, M. R. B. (2008). EPA and DHA quantification in two species of freshwater fish from Central Amazonia. *Food Chemistry*, *107*(2), 587–591. doi:10.1016/j.foodchem.2007.07.032

Intergovernmental Panel on Climate Change (CH). Climate change. (2007). *The physical science basis*. Geneva: Intergovernmental Panel on Climate Change; 2007 (cited 2007 Jan 01). Available from: http://www.ipcc.ch/pdf/assessment-report/ar4/wg1/ar4_wg1_full_report.pdf

International Assessment of Agricultural Knowledge (IAASTD). (2009). *Agriculture at a Crossroads, Global Report Chs 1, 4*. Island Press. Retrieved from http:// www.agassessment.org/reports/IAASTD/EN/Agriculture at a Crossroads_Global Report (English).pdf

IPCC. (2001). *Synthesis report 2001 (Contribution of Working Group I, II, and III to the Third Assessment Report of the Intergovernmental Panel on Climate Change. 2001)*. Cambridge, UK: Cambridge University Press.

IPCC. (2007). *(Intergovernmental Panel on Climate Change). Climate Change: Synthesis Report. Contributions of Working Groups I, II, and III to the Fourth Assessment Report of the Intergovernmental Panel on Climate Change (AR4). Core writing team* (R. K. Pachauri & A. Reisinger, Eds.). Geneva, Switzerland: IPCC.

IPCC. (2007). *Climate change 2007 (Fourth Assessment Report, Intergovernmental Panel on Climate Change)*. Cambridge, UK: Cambridge University Press.

IPCC. (2007). *Climate Change 2007: Impacts, Adaptation*. IPCC.

IPCC. (2007). Synthesis Report. Contribution of Working Groups I, II and III to the Fourth Assessment Report of the Intergovernmental Panel on Climate Change, Geneva, Switzerland.

IPCC. (2007). The Fourth Assessment Report: climate change (AR4). The AR4 Synthesis Report. In *The Working Group I Report: The Physical Science Basis (ISBN 9780521 88009–1 Hardback; 978 0521 70596-7 Paperback), the Working Group II Report Impacts: Adaptation and Vulnerability, the Working Group III Report: Mitigation of Climate Change*. Cambridge University Press.

IPCC. (2013). Climate change 2013: The physical science. In *Contribution of Working Group l to the Fifth Assessment Report of the Intergovernmental Panel on Climate Change*. Cambridge, UK: Cambridge University Press.

IPCC. (2014). *Climate change 2014: impacts, adaptation, and vulnerability. (Contribution of Working Group II to the Fifth Assessment Report of the Intergovernmental Panel on Climate Change)*. Cambridge, UK: Cambridge University Press.

IPCC. (2014). *Climate Change 2014: Synthesis Report. Contribution of Working Groups I, II and III to the Fifth Assessment Report of the Intergovernmental Panel on Climate Change. Climate Change 2014: Synthesis Report*. Contribution of Working Groups I, II and III to the Fifth Assessment Report of the Intergovernmental Panel on Climate Change. doi:10.1017/CBO9781107415324

IPCC. (2014). *IPCC Synthesis of Scientific-Technical Information Relevant to Interpreting Article 2 of the UN Framework Convention on Climate Change*. IPCC.

IRATE. (2006). *Atlas*. Région Meknès-Tafilalet, Inspection Régionale de l'Aménagement du Communication Territoire et de l'Environnement, Edition.

IRATE. (2006). *L'Atlas*. Région Meknes-Tafilalet, Inspection Régionale de l'Aménagement du Territoire et de l'Environnement.

Isakov, V., Touma, J. S., & Khlystov, A. (2007). A Method of Assessing Air Toxics Concentrations in Urban Areas Using Mobile Platform Measurements. *Journal of the Air & Waste Management Association*, *57*(11), 1286–1295. doi:10.3155/1047-3289.57.11.1286 PMID:18069452

Isingiro District Local Government. (2017). *Five Year District Local Government Development*. Author.

IUCN. (2010). *Building resilience to climate change: Ecosystem-based adaptation and lessons from the field*. Gland, Switzerland: IUCN.

Iverson, L. R., Prasad, A. M., Matthews, S. N., & Peters, M. (2008). Estimating potential habitat for 134 eastern US tree species under six climate scenarios. *Forest Ecology and Management*, *254*(3), 390–406. doi:10.1016/j.foreco.2007.07.023

Jackson, J. B. C., Kirby, M. X., Berger, W. H., Bjorndal, K. A., Botsford, L. W., Bourque, B. J., ... Warner, R. R. (2001). Historical Overfishing and the Recent Colllapse of Coastal Ecosystems. *Science*, *293*(5530), 629–638. doi:10.1126cience.1059199 PMID:11474098

Jacob, D. J., & Winner, D. A. (2009). Effect of climate change on air quality. *Atmospheric Environment*, *43*(1), 51–63. doi:10.1016/j.atmosenv.2008.09.051

Jacobeit, J. (1994). Empirical estimations of Mediterranean rainfall changes in the transitional seasons due to enhanced greenhouse warming. In R. Brazdil & M. Kolnar (Eds.), *Contemporary Climatology (pp. 266–71)*. Academic Press.

Jacobsen, C. S. (1997). Plant protection and rhizosphere colonization of barley by seed inoculated herbicide degrading Burkholderia (*Pseudomonas*) cepacia DBO1 (pRO101) in 2, 4-D contaminated soil. *Plant and Soil*, *189*(1), 139–144. doi:10.1023/A:1004296615446

Jacobson, M. Z. (2007). Effects of ethanol (E85) versus gasoline vehicles on cancer and mortality in the United States. *Environmental Science & Technology*, *41*(11), 4150–4157. doi:10.1021/es062085v PMID:17612204

Jakubiak-Lasocka, J., Lasocki, J., & Badyda, A. J. (2015). The influence of particulate matter on respiratorymorbidity andmortality in children and infants. *Advances in Experimental Medicine and Biology*, *849*, 39–48. doi:10.1007/5584_2014_93 PMID:25381559

Jalgaonwala, R. E., &Mahajan, R. T. (2011). Isolation and characterization of endophytic bacteria from roots of *Pongamia Glabra Vent. International Journal of Pharmaceutical and Biological Science*, *1*, 280-287.

Jarsj, Ö. J., Asokan, S. M., Shibuo, Y., & Destouni, G. (2008). Water Scarcity In The Aral Sea Drainage Basin: Contributions Of Agricultural Irrigation Anda Changing Climate. In Environmental Problems of Central Asia and their Economic, Social and Security Impacts (pp. 99-108). Springer.

Jedrychowski, W., Bendkowska, I., Flak, E., Penar, A., Jacek, R., Kaim, I., ... Perera, F. P. (2004). Estimated risk for altered fetal growth resulting from exposure to fine particles during pregnancy: An epidemiologic prospective cohort study in Poland. *Environmental Health Perspectives*, *112*(14), 1398–1402. doi:10.1289/ehp.7065 PMID:15471732

Jeong, S. (2013). The impact of air pollution on human health in Suwon City. *Asian J Atmos Environ*, *7*(4), 227–233. doi:10.5572/ajae.2013.7.4.227

Jerrett, M., Burnett, R. T., Pope, C. A., Thurston, G., & Krewski, D. (2009). Long-term ozone exposure and mortality. *The New England Journal of Medicine*, *360*(11), 1085–1095. doi:10.1056/NEJMoa0803894 PMID:19279340

Jetz, W., Wilcove, D. S., & Dobson, A. P. (2007). Projected impacts of climate and land-use change on the global diversity of birds. *PLoS Biology*, *5*(6), e157. doi:10.1371/journal.pbio.0050157 PMID:17550306

Johannsen, I. M., Hengst, J. C., Goll, A., Höllermann, B., & Diekküger, B. (2016). Future ofWater Supply and Demand in the Middle Drâa Valley, Morocco, under Climate and Land Use Change. *Water (Basel)*, *8*(8), 313. doi:10.3390/w8080313

John, M., Pannell, D., & Kingwell, R. (2005). Climate change and the economics of farm management in the face of land degradation: Dryland salinity in western Australia. *Canadian Journal of Agricultural Economics-Revue Canadienne D Agroeconomie, 53*(4), 443–459. doi:10.1111/j.1744-7976.2005.00029.x

Johnson, J. C., Christian, R. R., Brunt, J. W., Hickman, C. R., & Waide, R. B. (2010). Evolution of collaboration within the US long-term ecological research network. *Bioscience*, *60*(11), 931–940. doi:10.1525/bio.2010.60.11.9

Johnson, J. E., & Welch, D. J. (2010). Marine fisheries management in a changing climate: A review of vulnerability and future options. *Reviews in Fisheries Science*, *18*(1), 106–124. doi:10.1080/10641260903434557

Johnson, J. M., Alex, T., & Oelmu, R. (2014). *Piriformospora indica*: The versatile and multifunctional root endophytic fungus for enhanced yield and tolerance to biotic and abiotic stress in crop plants. *Journal of Tropical Agriculture*, *52*, 103–122.

Jones, P. G., & Thornton, P. K. (2003). The potential impacts of climate change on maize production in Africa and Latin America in 2055. *Global Environmental Change*, *13*(1), 51–59. doi:10.1016/S0959-3780(02)00090-0

Jones, P. G., & Thornton, P. K. (2009). Croppers to livestock keepers: Livelihood transitions to 2050 in Africa due to climate change. *Environmental Science & Policy*, *12*(4), 427–437. doi:10.1016/j.envsci.2008.08.006

Jong, D. C., Machauer, R., Reichert, B., Cappy, S., Viger, R., & Leavesley, G. (2004). An integrated geomorphological and hydrogeological MMS modeling framework for a semi-arid mountain basin inthe High Atlas, southern Morocco. In *International Congress: Complexity and Integrated Resources Management*. International Environmental Modelling and Software Society.

Jorge-Romero, G., Lercari, D., Ortega, L., & Defeo, O. (2019). Long-term ecological footprints of a man-made freshwater discharge onto a sandy beach ecosystem. *Ecological Indicators*, *96*, 412–420. doi:10.1016/j.ecolind.2018.09.024

Jouzel, J., Masson-Delmotte, V., Cattani, O., Dreyfus, G., Falourd, S., Hoffmann, G., ... Wolff, E. W. (2007). Orbital and millennial antarctic climate variability over the past 800,000 years. *Science*, *317*(5839), 793–796. doi:10.1126cience.1141038 PMID:17615306

Jurewicz, J., Hanke, W., Radwan, M., & Bonde, P. J. (2009). Environmental factors and semen quality. *International Journal of Occupational Medicine and Environmental Health*, *22*(4), 1–25. doi:10.2478/v10001-009-0036-1 PMID:20053623

Kabiri, L. (2003). *Les Ressources en eaux dans les Oasis de Sud Marocain: cas de Tafilalt*. Errachidia, Maroc: Ghèris et de Ferkla.

Kabiri, L. (2004). *Contribution à la connaissance, la préservation et la valorisation des Oasis du Sud marocain: cas de Tafilalt*. Thèse d'habilitation universitaire, Facultés des Sciences et Techniques, Errachidia, Université My Ismaïl.

Kabiri, L. (2004). *Contribution à la connaissance, la préservation et la valorisation des Oasis du Sud marocain: cas de Tafilalt*. Thèse d'habilitation universitaire, Facultés des Sciences et Techniques, Errachidia, Université My Ismaïl, Maroc.

Kagawa, J. (1985). Evaluation of biological significance of nitrogen oxides exposure. *The Tokai Journal of Experimental and Clinical Medicine*, *10*, 348.

Kala. (2014). Rhizoremediation: A Promising Rhizosphere Technology. *IOSR-JESTFT, 8*(8), 23-27.

Kaly, U., Pratt, C., & Mitchell, J. (2004). The demonstration environmental vulnerability index (EVI). Pacific Islands Applied Geoscience Commission (SOPAC).

Kaly, U., Pratt, C., Mitchell, J., & Howorth, R. (2003). The demonstration environmental vulnerability index (EVI). *Pacific Islands Applied Geoscience Commission (SOPAC)*. Retrieved from gsd.spc.int/sopac/evi/Files/EVI%202004%20 Technical%20Report.pdf

Kaly, U., Pratt, C., & Mitchell, J. (2005). *Building Resilience in SIDS: The Environmental Vulnerability Index. Final Report. SOPAC*. UNEP.

Kane, K. H. (2011). Effects of endophyte infection on drought stress tolerance of *Lolium perenne* accessions from the Mediterranean region. *Environmental and Experimental Botany*, *71*(3), 337–344.

Kan, H. D., London, S. J., Chen, G. H., Zhang, Y. H., Song, G. X., Zhao, N. Q., ... Chen, B. H. (2008). Season, sex, age, and education as modifiers of the effects of outdoor air pollution on daily mortality in Shanghai, China: The public health and air pollution in Asia (PAPA) study. *Environmental Health Perspectives*, *116*(9), 1183–1188. doi:10.1289/ehp.10851 PMID:18795161

Kannadan, S., & Rudgers, J. A. (2008). Endophyte symbiosis benefits a rare grass under low water availability. *Functional Ecology*, *22*(4), 706–713. doi:10.1111/j.1365-2435.2008.01395.x

Kapelle, M., Margret, M. I., Vuiren, V., & Baas, B. (1999). Effects of Climate Change on Biodiversity; A Review & Identification of Key Research Issues. *Biodiversity and Conservation*, *8*, 1393–1394.

Kappelle, M., Van Vuuren, M. M., & Baas, P. (1999). Effects of climate change on biodiversity: A review and identification of key research issues. *Biodiversity and Conservation*, *8*(10), 1383–1397. doi:10.1023/A:1008934324223. doi:10.1023/A:1008934324223

Karasti, H., & Baker, K. S. (2008). Digital data practices and the long term ecological research program growing global. *International Journal of Digital Curation*, *3*(2), 42–58. doi:10.2218/ijdc.v3i2.57

Karmaoui, A. (2016). Economic Value of Regulating and Cultural Ecosystem Services of Oasean Biome: Case Study in Middle Draa Valley. In Economic Modeling, Analysis, and Policy for Sustainability (pp. 307-326). IGI Global.

Karmaoui, A. (2018). Environmental Vulnerability to Climate Change in Mediterranean Basin: Socio-Ecological Interactions Between North and South. In *Hydrology and Water Resource Management: Breakthroughs in Research and Practice* (pp. 61-96). IGI Global.

Karmaoui, A., & Zerouali, S. (2018). Modeling the vulnerability to Zoonotic Cutaneous Leishmaniasis at the local scale. *Asian Journal of Applied Sciences*. Retrieved from https://scialert.net/abstract/?doi=ajaps.0000.91815.91815

Karmaoui, A., Ifaadassan, I., Messouli, M., & Yacoubi Khebiza, M. (2015a). Characterization of Common Environmental Indicators of the Moroccan Oasean Biome, Pilot Study in the Reserve Biosphere of Oases in Southern Morocco. *Advances in Research*. Doi:10.9734/AIR/2015/19369

Karmaoui, A., Messouli, M., Yacoubi, Khebiza, M. & Ifaadassan, I. (2014a). Environmental Vulnerability to Climate Change and Anthropogenic Impacts in Dryland (Pilot Study: Middle Draa Valley, South Morocco). *J Earth Sci Clim Change, S11*, 2. doi:10.4172/2157-7617.S11-002

Karmaoui, A. (2016). Ecosystem Services of the Argan Forest, the Current State and Trends. *Advances in Research*, *8*(1), 1–13. doi:10.9734/AIR/2016/21353

Karmaoui, A. (2018). The cutaneous leishmaniasis vulnerability index (CLVI). *Acta Ecologica Sinica*, *38*(4), 288–295. doi:10.1016/j.chnaes.2018.01.001

Karmaoui, A., Balica, S. F., & Messouli, M. (2016b). Analysis of applicability of flood vulnerability index in Pre-Saharan region, Morocco. *Natural Hazards and Earth System Sciences, 2016*. doi:10.5194/nhess-2016-96

Karmaoui, A., Ifaadassan, I., Babqiqi, A., Messouli, M., & Khebiza, M. Y. (2016). Analysis of the Water Supply-demand Relationship in the Middle Draa Valley, Morocco, under Climate Change and Socio-economic Scenarios. *Journal of Scientific Research & Reports*, *9*(4), 1–10. doi:10.9734/JSRR/2016/21536

Karmaoui, A., Ifaadassan, I., Messouli, M., & Khebiza, M. Y. (2015). Sustainability of the Moroccan oasean system (Case study: Middle Draa Valley). *Global J TechnolOptim*, *6*, 170. doi:10.4172/2229-8711.1000170

Karmaoui, A., Ifaadassan, I., Messouli, M., & Yacoubi Khebiza, M. (2015b). Monetarization of Ecosystem Services of Oasean Biome (Case study: Provisioning services of Middle Draa Valley Oases). *British Journal of Applied Science and Technology*, *11*(6), 1–18. doi:10.9734/BJAST/2015/19689

Karmaoui, A., Messouli, M., Ifaadassan, I., & Khebiza, M. Y. (2014b). A multidisciplinary approach to assess the environmental vulnerability at local scale in context of climate change (Pilot study in upper draa valley, South Morocco). *Glob. J. Technol. Optim*, *6*, 1–11. doi:10.4172/2229-8711.1000167

Karmaoui, A., Messouli, M., Khebiza, Y. M., & Ifaadassan, I. (2014). Environmental Vulnerability to climate change and anthropogenic impacts in dryland,(Pilot study: Middle Draa Valley, South Morocco). *Journal of Earth Science & Climatic Change*, (S11), 1.

Karmaoui, A., Messouli, M., & Yacoubi, M. (2015b). Vulnerability of Ecosystem Services to Climate Change and Anthropogenic Impacts in South East of Morocco (Case study: Drying up of Iriki lake). *The International Journal of Climate Change: Impacts and Responses.*, *7*(3), 1835–7156.

Karmaoui, A., Minucci, G., Messouli, M., Khebiza, M. Y., Ifaadassan, I., & Babqiqi, A. (2018). Climate Change Impacts on Water Supply System of the Middle Draa Valley in South Morocco. In *Climate Change, Food Security and Natural Resource Management* (pp. 163–178). Cham: Springer.

Karmaoui, A., & Moumane, A. (2016). Changes in the Environmental Vulnerability of Oasean System (desert oasis), Pilot Study in Middle Draa Valley, Morocco. *Expert Opin Environ Biol*, *5*(3), 3. doi:10.4172/2325-9655.1000135

Karmaoui, A., & Zerouali, S. (2018). (in press). Modeling the vulnerability to Zoonotic Cutaneous Leishmaniasis at the local scale. *Asian Journal of Applied Sciences*.

Kasai, K., Usami, T., Lee, J., Ishikawa, S. I., & Oikawa, T. (2000). Responses of Ectomycorrhizal Colonization and Morphotype Assemblage of Quercusmyrsinaefolia Seedlings to Elevated Air Temperature and Elevated Atmospheric CO_2. *Microbes and Environments*, *15*(4), 197–207. doi:10.1264/jsme2.2000.197

Katsouyanni, K., Zmirou, D., Spix, C., Sunyer, J., Schouten, J., Pönkä, A., & … . (1995). Short-term effects of air pollution on health: a European approach using epidemiological time-series data—the APHEA project: background, objectives, design. *The European Respiratory Journal*, *8*, 1030–1038. PMID:7589368

Keddal, H. (2006). *National study on date palm irrigation and associated crops in the kingdom of morocco*. FAO.

Kedima. (2014). *Rapport sur la situation de la nappe à Guerdane*. Rapport Interne. Providence verte. Rabat.

Kelley, S. L., Aitchison, E. W., Deshpande, M., Schnoor, J. L., & Alvarez, P. J. (2001). Biodegradation of 1, 4-dioxane in planted and unplanted soil: Effect of bioaugmentation with *Amycolata* sp. CB1190. *Water Research*, *35*(16), 3791–3800. doi:10.1016/S0043-1354(01)00129-4 PMID:12230161

Khaefi, M., Hassani, G., Yari, A. R., Soltani, F., Dobaradaran, S., Moogahi, S., ... Khaniabadi, Y. O. (2017). Association of particulatematter impact on prevalence of chronic obstructive pulmonary disease in Ahvaz, Southwest Iran during 2009–2013. *Aerosol and Air Quality Research*, *17*(1), 203–207. doi:10.4209/aaqr.2015.11.0628

Khan, A. L., Waqas, M., Khan, A. R., Hussain, J., Kang, S. M., Gilani, S. A., ... Lee, I.-J. (2013). Fungal endophyte-Penicilliumjanthinellum LK5 improves growth of ABA-deficient tomato under salinity. *World Journal of Microbiology & Biotechnology*, *29*(11), 2133–2144. doi:10.100711274-013-1378-1 PMID:23842755

Khardi, A. (2003). Aperçu des problèmes d'environnement pour l'agriculture oasienne. *Revue Homme terre et eau*, 127.

Khomsi, K., Mahe, G., Tramblay, Y., Sinan, M., & Snoussi, M. (2015). Trends in rainfall and temperature extremes in Morocco. *Natural Hazards and Earth System Sciences*, *3*, 1175–1201. doi:10.5194/nhessd-3-1175-2015

Khoshoo, T. N. (1993). Himalayan biodiversity conservation: an overview. *Himalayanbiodiversity: conservation strategies*, 5-35.

Khoury, C. K., Bjorkman, A. D., Dempewolf, H., Ramirez-Villegas, J., Guarino, L., Jarvis, A., ... Struik, P. C. (2014). Increasing homogeneity in global food supplies and the implications for food security. *Proceedings of the National Academy of Sciences*, *111*(11), 4001-4006. Retrieved from http://www.pnas.org/content/111/11/4001.abstract

Kimbrough, R. D., Carter, C. D., Liddle, J. A., Cline, R. E., & Phillips, P. E. (1977). Epidemiology and pathology of a tetrachlorodibenzodioxin poisoning episode. *Archives of Environmental Health*, *32*(2), 77–86. doi:10.1080/00039896.1977.10667259 PMID:557961

Kim, E. S. (2006). Development, potentials, and challenges of the International Long-Term Ecological Research (ILTER) Network. *Ecological Research*, *21*(6), 788–793. doi:10.100711284-006-0049-1

Kingdom of Morocco. (2013). *Green Morocco Plan*. Retrieved from http://www.maroc.ma/en/content/green-morocoo-plan

Kingsley, M. T., Fredrickson, J. K., Metting, F. B., & Seidler, R. J. (1994). *Environmental restoration using plant-microbe bioaugmentation. In Bioremediation of chlorinated and polyaromatic hydrocarbon compounds* (pp. 287–292). Boca Raton, FL: Lewis Publishers.

Kinney, P. L. (2008). Climate change, air quality, and human health. *American Journal of Preventive Medicine*, *35*(5), 459–467. doi:10.1016/j.amepre.2008.08.025 PMID:18929972

Kiruhura District Local Government. (2016). *Five Year District Local Government Development*. Author.

Kitzes, J., & Harte, J. (2014). Beyond the species–area relationship: Improving macroecological extinction estimates. *Methods in Ecology and Evolution*, *5*(1), 1–8. doi:10.1111/2041-210X.12130

Klose, A. (2009). *Soil Characteristics and Soil Erosion by Water in a Semi-arid Catchment (Wadi Drâa, South Morocco) under the Pressure of Global Change* (Dissertation). University of Bonn, Bonn, Germany.

Klose, S. (2012). *Regional Hydrogeology and Groundwater Budget Modeling in the Arid Middle Drâa Catchment* (Dissertation). University of Bonn, Bonn, Germany.

Knapp, S., Stadler, J., Harpke, A., & Klotz, S. (2016). Dispersal traits as indicators of vegetation dynamics in long-term old-field succession. *Ecological Indicators*, *65*, 44–54. doi:10.1016/j.ecolind.2015.10.003

Knight, A. T., Cowling, R. M., Rouget, M., Balmford, A., Lombard, A. T., & Campbell, B. M. (2008). Knowing but not doing: Selecting priority conservation areas and the research-implementation gap. *Conservation Biology*, *22*(3), 610–617. doi:10.1111/j.1523-1739.2008.00914.x PMID:18477033

Knippertz, P., Ulbrich, U., Marques, F., & Corte-Real, J. (2003b). Decadal changes in the link between El Niño and springtime North Atlantic Oscillation and European–North African rainfall. *International Journal of Climatology: A Journal of the Royal Meteorological Society, 23*(11), 1293-1311.

Knippertz, P., Christoph, M., & Speth, P. (2003). Long-term precipitation variability in Morocco and the link to the large-scale circulation in recent and future climates. *Meteorology and Atmospheric Physics, 83*(1-2), 67–88. doi:10.100700703-002-0561-y

Knippertz, P., Fink, A. H., Reiner, A., & Speth, P. (2003a). Three late summer/early autumn cases of tropicalextra-tropical interactions causing precipitation in Northwest Africa. *Monthly Weather Review, 131*(1), 116–135. doi:10.1175/1520-0493(2003)131<0116:TLSEAC>2.0.CO;2

Kobayashi, D. Y., & Palumbo, J. D. (2000). Bacterial endophytes and their effectson plants and uses in agriculture. In C. W. Bacon & J. F. White (Eds.), *Microbialendophytes* (pp. 199–236). New York: Marcel Dekker.

Kofi, A. D. (2014). CCDA- Live. New initiative will help finance efforts to harness information to fight climate change. Daily news from 4th CCDA-8-10 October 2014. *Friday, 10*, 1.

Kohli, R. K., Batish, D. R., & Singh, H. P. (2005). Need to Establish Long-Term Ecological Research Network in India. Bulletin of the National Institute of Ecology, 15, 271-276.

Kohlmaier, G. H., Siré, E. -O, Janecek, A., Keeling, C. D., Piper, S. C., & Revelle, R. (1989). Modelling the seasonal contribution of a CO2 fertilization effect of the terrestrial vegetation to the amplitude increase in atmospheric CO2 at Mauna Loa Observatory. *Tellus B, 41 B*(5), 487–510. doi:10.1111/j.1600-0889.1989.tb00137.x

Kolding, J., Bundy, A., van Zwieten, P. A. M., & Plank, M. J. (2016b). Fisheries, the inverted food pyramid. *ICES Journal of Marine Science, 73*(6), 1697–1713. doi:10.1093/icesjms/fsv225

Kolding, J., van Zwieten, P. A. M., Marttin, F., & Poulain, F. (2016a). Fisheries in the drylands of Sub-Saharan Africa "Fish come with the Rains". *FAO Fisheries and Aquaculture Circular*, C1118.

Kooijman, S. A. L. M. (2010). *Dynamic Energy Budget Theory for Metabolic Organisation*. Cambridge, UK: Cambridge University Press.

Koposzki, H. (1992). Effects of acid and nitrogen deposition on the mesofauna, especially the collembolan. In *Proceedings of Air Pollution and Interactions between Organisms in Forest Ecosystems, 15th IUFRO International Meeting of Specialists on Air Pollution Effects on Forest Ecosystems* (p. 46). Academic Press.

Korade, D. L., & Fulekar, M. H. (2009). Rhizosphere remediation of chlorpyrifos in mycorrhizospheric soil using ryegrass. *Journal of Hazardous Materials, 172*(2), 1344–1350. doi:10.1016/j.jhazmat.2009.08.002 PMID:19720454

Korek, M. J., Bellander, T. D., Lind, T., Bottai, M., Eneroth, K. M., Caracciolo, B., ... Penell, J. C. (2015). Traffic related air pollution exposure and incidence of stroke in four cohorts from Stockholm. *Journal of Exposure Science & Environmental Epidemiology, 25*(5), 517–523. doi:10.1038/jes.2015.22 PMID:25827311

Kothawale, D. R., Munot, A. A., & Kumar, K. K. (2010). Surface air temperature variability over India during 1901–2007, and its association with ENSO. *Climate Research, 42*(2), 89–104. doi:10.3354/cr00857. doi:10.3354/cr00857

Kourouma. (2017). *Evaluation économique du service de stabilisation des sols et de contrôle de l'érosion d'un agro-système pluvial méditerranéen* (Maroc). Mémoire de recherche Supagro de Montpellier. Retrieved from http://www.fao.org/nr/climpag/pub/FAO_WorldBank_Study_CC_Morocco_2008.pdf

Kuemmerlen, M., Stoll, S., Sundermann, A., & Haase, P. (2016). Long-term monitoring data meet freshwater species distribution models: Lessons from an LTER-site. *Ecological Indicators, 65*, 122–132. doi:10.1016/j.ecolind.2015.08.008

Kuhn, A., Heidecke, C., Roth, A., Goldbach, H., Burkhardt, J., Linstädter, A., ... Gaiser, T. (2010). *II-5.1 Importance of resource management for livelihood security under Climate Change in Southern Morocco*. Academic Press.

Kumar, K. (2009). Impact of climate change on India's monsoon climate and development of high resolution climate change scenarios for India. *MoEF*.

Kumar, M., Yadav, V., Tuteja, N., & Johri, A. K. (2009). Antioxidant enzyme activities in maize plants colonized with *Piriformosporaindica. Microbiology, 155*(3), 780–790. doi:10.1099/mic.0.019869-0 PMID:19246749

Kumar, V., & Chopra, A. K. (2009). Impact of Climate Change on Biodiversity in India with Special reference to Himalayan Region- An Overview. *Journal of Applied and Natural Science, 1*(1), 117–122. doi:10.31018/jans.v1i1.48. doi:10.31018/jans.v1i1.48

Kuo, C. Y., Wong, R. H., Lin, J. Y., Lai, J. C., & Lee, H. (2006). Accumulation ofchromium and nickel metals in lung tumors from lung cancer patients in Taiwan. *Journal of Toxicology and Environmental Health, 69*(14), 1337–1344. doi:10.1080/15287390500360398 PMID:16760140

Kurpius, M. R., & Goldstein, A. H. (2003). Gas-phase chemistry dominates O3 loss to a forest, implying a source of aerosols and hydroxyl radicals to the atmosphere. *Geophysical Research Letters, 30*(7), 1371. doi:10.1029/2002GL016785

Kwon, T. S. (2014). Empirical test of the influence of global warming and forest disturbance on ant fauna at the Gwangneung Forest Long Term Ecological Research site, South Korea. *Journal of Asia-Pacific Biodiversity, 7*(3), 252–257. doi:10.1016/j.japb.2014.06.004

Lacasaña, M., Esplugues, A., & Ballester, F. (2005). Exposure to ambient air pollution and prenatal and early childhood health effects. *European Journal of Epidemiology, 20*(2), 183–199. doi:10.100710654-004-3005-9 PMID:15792286

Lane, R. P., Abdel-Hafez, S., & Kamhawi, S. (1988). The distribution of phlebotomine sandflies in the principal ecological zones of Jordan. *Medical and Veterinary Entomology, 2*(3), 237–246. doi:10.1111/j.1365-2915.1988.tb00190.x PMID:2980180

Lan, K.-W., Lee, M.-A., Zhang, C. I., Wang, P.-Y., Wu, L.-J., & Lee, K.-T. (2014). Effects of climate variability and climate change on the fishing conditions for grey mullet (Mugil cephalus L.) in the Taiwan Strait. *Climatic Change, 126*(1–2), 189–202. doi:10.100710584-014-1208-y

Laouina, A. (2006). Prospective « Maroc 2030 ». Gestion durable des ressources naturelles et de la biodiversité au Maroc. *Haut commissariat au plan*.

Larsen, B. B., Miller, E. C., Rhodes, M. K., & Wiens, J. J. (2017). Inordinate fondness multiplied and redistributed: The number of species on earth and the new pie of life. *The Quarterly Review of Biology, 92*(3), 229–265. doi:10.1086/693564. doi:10.1086/693564

Last, P. R., White, W. T., Gledhill, D. C., Hobday, A. J., Brown, R., Edgar, G. J., & Pecl, G. (2010). Long-term shifts in abundance and distribution of a temperate fish fauna: A response to climate change and fishing practices. *Global Ecology and Biogeography, 20*(1), 58–72. doi:10.1111/j.1466-8238.2010.00575.x

Lavergne, S., Mouquet, N., Thuiller, W., & Ronce, O. (2010). Biodiversity and climate change: Integrating evolutionary and ecological responses of species and communities. *Annual Review of Ecology Evolution and Systematics, 41*(1), 321–350. doi:10.1146/annurev-ecolsys-102209-144628

Law, K. (2010). More ozone over North America. *Nature, 463*, 307–308.

Leadley, P. (2010). *Biodiversity scenarios: projections of 21st century change in biodiversity, and associated ecosystem services: a technical report for the global biodiversity outlook 3 (No. 50)*. UNEP/Earthprint.

Lean, J. L., & Rind, D. H. (2008). *How natural and anthropogenic influences alter global and regional surface temperatures: 1889 to 2006*. doi:10.1029/2008GL034864

Lean, J., Rottman, G., Harder, J., & Kopp, G. (2005). SORCE contributions to new understanding of global change and solar variability. *Solar Physics*, *230*(1-2), 27–53. doi:10.100711207-005-1527-2

Lee, C., Schlemme, C., Murray, J., & Unsworth, R. (2015). The cost of climate change: Ecosystem services and wildland fires. *Ecological Economics*, *116*, 261–269. doi:10.1016/j.ecolecon.2015.04.020

Lee, J. J., Phillips, D. L., & Benson, V. W. (1999). Soil erosion and climate change: Assessingpotential impacts and adaptation practices. *Journal of Soil and Water Conservation*, *54*, 529–536.

Lee, J. J., Phillips, D. L., & Dodson, R. F. (1996). Sensitivity of the US corn belt to climatechangeandelevatedCO2: II. Soil erosion and organic carbon. *Agricultural Systems*, *52*(4), 503–552. doi:10.1016/S0308-521X(96)00015-7

Leone, F., De richemond, N., & Vinet, F. (2010). Aléas naturels et gestion des risques. Édit. PUF, collection L. Géographie.

Lévite, H., Sally, H., &Cour, J. (2003). Testing water demand management scenarios in a water-stressed basin in South Africa: application of the WEAP model. *Physics and Chemistry of the Earth, 28*(20-27), 779-786.

Leye, M. L. (2015). *Etude de la perception de l'évolution paysagère des écosystèmes fragiles et du rôle des acteurs institutionnels dans l'adaptation des agriculteurs aux changements climatiques: Cas du Bassin versant de Tleta (Tanger-Tetouan). Ingénierie du développement*. ENA-Meknès.

Liang, W., Wei, H., & Kuo, H. (2009). Association between daily mortality from respiratory and cardiovascular diseases and air pollution in Taiwan. *Environmental Research*, *109*(1), 51–58. doi:10.1016/j.envres.2008.10.002 PMID:19027107

Liddicoat, C., Hayman, P., Alexander, B., Rowland, J., Maschmedt, D., Young, M.-A., . . . Sweeney, S. (2012). Climate change, wheat production and erosion risk in South Australia's cropping zone: linking crop simulation modelling to soil landscape mapping. Government of South Australia, Department of Environment, Water and Natural Resource.

Li, H. Y., Wei, D. Q., Shen, M., & Zhou, Z.-P. (2012). Endophytes and their role in phytoremediation. *Fungal Diversity*, *54*(1), 11–18. doi:10.100713225-012-0165-x

Lilleskov, E. A., Liu, L., Lynch, J., Nadelhoffer, K., Perakis, S. S., Robin-Abbott, M. J., ... Dennis, R. L. (2011). Effects of nitrogen deposition and empirical nitrogen critical loads for ecoregions of the United States. *Ecological Applications*, *21*(8), 3049–3082. doi:10.1890/10-2341.1

Lim, S. S., Vos, T., Flaxman, A. D., Danaei, G., Shibuya, K., Adair-Rohani, H., ... Ezzati, M. (2012). A comparative risk assessment of burden of disease and injury attributable to 67 risk factors and risk factor clusters in 21 regions, 1990–2010: A systematic analysis for the global burden of disease study 2010. *Lancet*, *380*(9859), 2224–2260. doi:10.1016/S0140-6736(12)61766-8 PMID:23245609

Linderman, R. G. (1994). Role of AM fungi in biocontrol. In F. L. Pfleger & R. G. Linderman (Eds.), *Mycorrhizae and Plant Health* (pp. 1–25). St. Paul, MN: APS Press.

Lin, J.-T., Patten, K. O., Liang, X.-Z., & Wuebbles, D. J. (2008). Effects of intercontinental transport on surface ozone over the United States at the present and future. *Geophysical Research Letters*, *35*(2), L02805. doi:10.1029/2007GL031415

Lionello, P., Malanotte-Rizzoli, P., & Boscolo, R. (Eds.). (2006). *Mediterranean climate variability* (Vol. 4). Elsevier.

Lioubimtseva, E., & Cole, R. (2006). Uncertainties of climate change in aridenvironments of Central Asia. *Reviews in Fisheries Science*, *14*(1-2), 29–49. doi:10.1080/10641260500340603

Lioy, P. J. (1991). *Human Exposure Assessment for Airborne Pollutants -Advances and Opportunities*. Washington, DC: National Academy of Sciences.

Lipper, L., Thornton, P., Campbell, B. M., Baedeker, T., Braimoh, A., Bwalya, M., ... Torquebiau, E. F. (2014). Climate-smart agriculture for food security. *Nature Climate Change*, *4*(12), 1068–1072. doi:10.1038/nclimate2437

Li, R., Wang, C., & Wu, D. (2018). Changes in precipitation recycling over arid regions in the Northern Hemisphere. *Theoretical and Applied Climatology*, *131*(1–2), 489–502. doi:10.100700704-016-1978-4

Li, S., Williams, G., Jalaludin, B., & Baker, P. (2012). Panel studies of air pollution on children's lung function and respiratory symptoms: A literature review. *The Journal of Asthma*, *49*(9), 895–910. doi:10.3109/02770903.2012.724129 PMID:23016510

Lischeid, G., Kalettka, T., Merz, C., & Steidl, J. (2016). Monitoring the phase space of ecosystems: Concept and examples from the Quillow catchment, Uckermark. *Ecological Indicators*, *65*, 55–65. doi:10.1016/j.ecolind.2015.10.067

Liu, C., Hofstra, N., & Franz, E. (2013). Impacts of climate change on the microbial safety of pre-harvest leafy green vegetables as indicated by Escherichia coli O157 and Salmonella spp. *International Journal of Food Microbiology*, *163*(2), 119–128. doi:10.1016/j.ijfoodmicro.2013.02.026 PMID:23558195

Liu, C., & Xia, J. (2004). Water problems and hydrological research in the Yellow River and the Huai and Hai River basins of China. *Hydrological Processes*, *18*(12), 2197–2210. doi:10.1002/hyp.5524

Liu, J., Li, S., Ouyang, Z., Tam, C., & Chen, X. (2008). Ecological and socioeconomic effects of China's policies for ecosystem services. *Proceedings of the National Academy of Sciences of the United States of America*, *105*(28), 9477–9482. doi:10.1073/pnas.0706436105 PMID:18621700

Liu, Y., Stanturf, J., & Goodrick, S. (2010). Trends in global wildfire potential in a changing climate. *Forest Ecology and Management*, *259*(4), 685–697. doi:10.1016/j.foreco.2009.09.002

Lobell, D. B., Burke, M. B., Tebaldi, C., Mastrandrea, M. D., Falcon, W. P., & Naylor, R. L. (2008). Prioritizing Climate Change Adaptation Needs for Food Security in 2030. *Science, 319*(5863), 607-610. Retrieved from http://science.sciencemag.org/content/319/5863/607.abstract

Lodewyckx, C., Taghavi, S., Mergeay, M., Vangronsveld, J., Clijsters, H., & Van der Lelie, D. (2001). The effect of recombinant heavy metal resistant endophytic bacteria in heavy metal uptake by their host plant. *International Journal of Phytoremediation*, *3*(2), 173–187. doi:10.1080/15226510108500055

Lodewyckx, C., Vangronsveld, J., Porteous, F., Moore, E. R. B., Taghavi, S., Mezgeay, M., & Van der Lelie, D. (2002). Endophytic bacteria and their potential applications. *Critical Reviews in Plant Sciences*, *21*(6), 583–606. doi:10.1080/0735-260291044377

Loghman-Adham, M. (1997). Renal effects of environmental and occupational lead exposure. Environ. Health Perspect. 105, 928.Jarup, L., 2003. Hazards of heavy metal contamination. *British Medical Bulletin*, *68*, 167.

Lopes, C. (2013). *Can Climate Change Spring Africa's Transformative Development? A Statement by the UN Undersecretary-General and ECA Executive Secretary on 21 October 2013,on the 3rd Annual Conference on Climate Change and Development in Africa from 21-23 October 2013 at United Nations conference centre*. Addis Ababa, Ethiopia: ECA, ACPC.

Lowe-McConnell, R. H. (1979). Ecological aspects of seasonality in fishes of tropical waters. *Symp Zool Soc Lond London*, *44*, 219–241.

Lozano, F. D., & Schwartz, M. W. (2005). Comparative taxonomic structure of the floras of two Mediterranean-climate regions: Iberia and California. *Diversity & Distributions*, *11*(5), 399–408. doi:10.1111/j.1366-9516.2005.00168.x

Lu, D., & Weng, Q. (2007). A survey of image classification methods and techniques for improving classification performance. *International Journal of Remote Sensing*, *28*(5), 823–870. doi:10.1080/01431160600746456

Ludwig, F., & Asseng, S. (2006). Climate change impacts on wheat production in a Mediterranean environment in Western Australia. *Agricultural Systems*, *90*(1), 159–179. doi:10.1016/j.agsy.2005.12.002

Lundqvist, L. (2016). Planning for climate change adaptation in a multi-level context: The Gothenburg metropolitan area. *European Planning Studies*, *24*(1), 1–20. doi:10.1080/09654313.2015.1056774

Lutgens, F. K., Edward, J. T., & Tasa, D. G. (1998). *The Atmosphere: An Introduction to Meteorology (Mastering Meteorology Series)* (13th ed.). Prentice-Hall, Incorporated.

Lu, Z., Wei, Y., Xiao, H., Zou, S., Ren, J., & Lyle, C. (2015). Trade-offs between midstream agricultural production and downstream ecological sustainability in the Heihe River basin in the past half century. *Agricultural Water Management*, *152*, 233–242. doi:10.1016/j.agwat.2015.01.022

Lv, S., Tian, L. G., Liu, Q., Qian, M. B., Fu, Q., Steinmann, P., ... Zhou, X. N. (2013). Water-related parasitic diseases in China. *International Journal of Environmental Research and Public Health*, *10*(5), 1977–2016. doi:10.3390/ijerph10051977 PMID:23685826

Lybbert, T. J., Aboudrare, A., Chaloud, D., Magnan, N., & Nash, M. (2011). Booming markets for Moroccan argan oil appear to benefit some rural households while threatening the endemic argan forest. *Proceedings of the National Academy of Sciences of the United States of America*, *108*(34), 13963–13968. doi:10.1073/pnas.1106382108 PMID:21873185

Lyons, P. C., Evans, J. J., & Bacon, C. W. (1990). Effects of fungal endophyte *Acremonium coenophialum* on nitrogen accumulation and metabolism in tall fescue. *Plant Physiology*, *92*(3), 726–732. doi:10.1104/pp.92.3.726 PMID:16667341

Mace, B. G. M. (2014). Changes in the perception and goals of nature conservation require a solid scientific basis. *Science, 345*(6204).

Mace, G. M. (2014). Whose conservation? *Science, 345*(6204), 1558-1560. Retrieved from http://science.sciencemag.org/content/345/6204/1558.abstract

Mace, G. M., Norris, K., & Fitter, A. H. (2012). Biodiversity and ecosystem services: A multilayered relationship. *Trends in Ecology & Evolution*, *27*(1), 19–26. doi:10.1016/j.tree.2011.08.006 PMID:21943703

Mace, G. M., & Purvis, A. (2008). Evolutionary biology and practical conservation: Bridging a widening gap. *Molecular Ecology*, *17*(1), 9–19. doi:10.1111/j.1365-294X.2007.03455.x PMID:17696991

Mackenzie, C. L., Ormondroyd, G. A., Curling, S. F., Ball, R. J., Whiteley, N. M., & Malham, S. K. (2014). Ocean Warming, More than Acidification, Reduces Shell Strength in a Commercial Shellfish Species during Food Limitation. *PLoS One*, *9*(1), e86764. doi:10.1371/journal.pone.0086764 PMID:24489785

Maclean, I. M., & Wilson, R. J. (2011). Recent ecological responses to climate change support predictions of high extinction risk. *Proceedings of the National Academy of Sciences of the United States of America*, *108*(30), 12337–12342. doi:10.1073/pnas.1017352108 PMID:21746924

MacMynowski, D. P., & Root, T. L. (2007). Climate and the complexity of migratory phenology: Sexes, migratory distance, and arrival distributions. *International Journal of Biometeorology*, *51*(5), 361–373. doi:10.100700484-006-0084-1 PMID:17245563

Macpherson, G. L., & Sophocleous, M. (2004). Fast ground-water mixing and basal recharge in an unconfined, alluvial aquifer, Konza LTER Site, Northeastern Kansas. *Journal of Hydrology (Amsterdam)*, *286*(1-4), 271–299. doi:10.1016/j.jhydrol.2003.09.016

Maczka, K., Matczak, P., Pietrzyk-Kaszyńska, A., Rechciński, M., Olszańska, A., Cent, J., & Grodzińska-Jurczak, M. (2016). Application of the ecosystem services concept in environmental policy—A systematic empirical analysis of national level policy documents in Poland. *Ecological Economics*, *128*, 169–176. doi:10.1016/j.ecolecon.2016.04.023

Madden, E. F., & Fowler, B. A. (2000). Mechanisms of nephrotoxicity from metal combinations: A review. *Drug and Chemical Toxicology*, *23*(1), 1–12. doi:10.1081/DCT-100100098 PMID:10711385

Maisonet, M., Bush, T. J., Correa, A., & Jaakkola, J. J. (2001). Relation between ambient air pollution and low birth weight in the northeastern United States. *Environmental Health Perspectives*, *109*(suppl. 3), 351–356. PMID:11427384

Majhi, A. (2018). Review article role of integrated duck cum fish farming and its economic efficiency : a study in Purulia district, West Bengal. Academic Press.

Ma, L., Wu, J., Liu, W., & Abuduwaili, J. (2014). Distinguishing between anthropogenic and climatic impacts on lake size: A modeling approach using data from Ebinur Lake in arid northwest China. *Journal of Limnology*, *73*(2). doi:10.4081/jlimnol.2014.852

Malcolm, G. M., Lopez, J. C., Koide, R. T., & Eissenstat, D. M. (2008). Acclimation to temperature and temperature sensitivity of metabolism by ectomycorrhizal fungi. *Global Change Biology*, *14*(5), 1169–1180. doi:10.1111/j.1365-2486.2008.01555.x

Malcolm, J. R., Liu, C., Neilson, R. P., Hansen, L., & Hannah, L. E. E. (2006). Global warming and extinctions of endemic species from biodiversity hotspots. *Conservation Biology*, *20*(2), 538–548. doi:10.1111/j.1523-1739.2006.00364.x PMID:16903114

Maltezou, H. C. (2008). Visceral leishmaniasis: Advances in treatment. *Recent Patents on Anti-infective Drug Discovery*, *3*(3), 192–198. doi:10.2174/157489108786242341 PMID:18991801

Mandal, P. K. (2005). Dioxin: A review of its environmental effects and its aryl hydrocarbon receptor biology. *Journal of Comparative Physiology*, *175*(4), 221–230. doi:10.100700360-005-0483-3 PMID:15900503

Mandyam, K., & Jumpponen, A. (2005). Abundance and possible functions of the root-colonising dark septate endophytic fungi. Studies Mycology, 53, 173-189.

Mannes, T., Jalaludin, B., Morgan, G., Lincoln, D., Sheppeard, V., & Corbett, S. (2005). Impact of ambient air pollution on birth weight in Sydney, Australia. *Occupational and Environmental Medicine*, *62*(8), 524–530. doi:10.1136/oem.2004.014282 PMID:16046604

Manvendra, K., Dubey, P. C., Zender, C. S., & Folland, C. K. (2010). Global warming and the next ice age. *Bulletin of the American Meteorological Society*, *89*(12), 1905–1909.

Manzo, N. D., Slade, R., & Dye, J. A. (2010). Interrelationship of lung inflammation and air pollutant exposure on cellular oxidative stress and epithelial injury. *Proceedings 27th Symposium of the Veterinary Comparative Respiratory Society*.

Marcogliese, D. J. (2016). The distribution and abundance of parasites in aquatic ecosystems in a changing climate: More than just temperature. *Integrative and Comparative Biology*, *56*(4), 611–619. doi:10.1093/icb/icw036 PMID:27252206

Margat, J. (1962). *Mémoire explicative de la carte hydrogéologique au 1/50000 de la plaine de Tafilalet*. Academic Press.

Marín, P., Magan, N., Vázquez, C., & González-Jaén, M. T. (2010). Differential effect of environmental conditions on the growth and regulation of the fumonisin biosynthetic gene FUM1 in the maize pathogens and fumonisin producers Fusarium verticillioides and Fusarium proliferatum. *FEMS Microbiology Ecology*, *73*(2), 303–311. doi:10.1111/j.1574-6941.2010.00894.x PMID:20491926

Marques, A., Nunes, M. L., Moore, S. K., & Strom, M. S. (2010). Climate change and seafood safety: Human health implications. *Food Research International*, *43*(7), 1766–1779. doi:10.1016/j.foodres.2010.02.010

Marquez, L. M., Redman, R. S., Rodriguez, R. J., & Roossinck, M. J. (2007). A virus in a fungus in a plant: Three-way symbiosis required for thermal tolerance. *Science*, *315*(5811), 513–516. doi:10.1126cience.1136237 PMID:17255511

Mason, C. F., & MacDonald, S. M. (1987). Acidification and otter (Lutra lutra) distribution on a British river. *Mammalia*, *51*(1), 81–87.

Ma, Y., Oliveira, R. S., Freitas, H., & Zhang, C. (2016). Biochemical and Molecular Mechanisms of Plant-Microbe-Metal Interactions: Relevance for Phytoremediation. *Frontiers in Plant Science*, *7*, 918. doi:10.3389/fpls.2016.00918 PMID:27446148

Ma, Y., Oliveira, R. S., Nai, F., Rajkumar, M., Luo, Y., Rocha, I., & Freitas, H. (2015). The hyperaccumulator Sedum plumbizincicola harbors metal-resistant endophytic bacteria that improve its phytoextraction capacity in multi-metal contaminated soil. *Journal of Environmental Management*, *156*, 62–69. doi:10.1016/j.jenvman.2015.03.024 PMID:25796039

Ma, Y., Rajkumar, M., Luo, Y. M., & Freitas, H. (2011). Inoculation of endophytic bacteria on host and non-host plants – effects on plant growth and Ni uptake. *Journal of Hazardous Materials*, *196*, 230–237. doi:10.1016/j.jhazmat.2011.08.034 PMID:21872991

Ma, Y., Rajkumar, M., Luo, Y., & Freitas, H. (2013). Phytoextraction of heavy metal polluted soils using *Sedum plum bizin cicola* inoculated with metal mobilizing *Phyllobacterium myrsinacearum* RC6b. *Chemosphere*, *93*(7), 1386–1392. doi:10.1016/j.chemosphere.2013.06.077 PMID:23890964

Mayer, H. (1999). Air pollution in cities. *Atmospheric Environment*, *33*(24–25), 4029–4037. doi:10.1016/S1352-2310(99)00144-2

McCarthy, J. J., Canziani, O. F., Leary, N. A., Dokken, D. J., & White, K. S. (Eds.). (2001). *Climate change 2001: impacts, adaptation, and vulnerability: contribution of Working Group II to the third assessment report of the Intergovernmental Panel on Climate Change* (Vol. 2). Cambridge University Press.

McCarthy, M. C., Hafner, H. R., Chinkin, L. R., & Charrier, J. G. (2007). Temporal Variability of Selected Air Toxics in the United States. *Atmospheric Environment*, *41*(34), 7180–7194. doi:10.1016/j.atmosenv.2007.05.037

Mcclanahan, T., Allison, E. H., & Cinner, J. E. (2015). Managing fisheries for human and food security. *Fish and Fisheries*, *16*(1), 78–103. doi:10.1111/faf.12045

McDaniel, C. D., Bramwell, R. K., Wilson, J. L., & Howarth Birkett, J. R. (1995). Fertility of Male and Female Broiler Breeders Following Exposure to Elevated Ambient Temperatures1. *Poultry Science*, *74*(6), 1029–1038. doi:10.3382/ps.0741029 PMID:7644414

McGrath, J. M., Betzelberger, A. M., Wang, S., Shook, E., Zhu, X. G., Long, S. P., & Ainsworth, E. A. (2015). An analysis of ozone damage to historical maize and soybean yields in the United States. *Proceedings of the National Academy of Sciences of the United States of America*, *112*(46), 14390–14395. doi:10.1073/pnas.1509777112 PMID:26578785

McGuinness, M., & Dowling, D. (2009). Plant-Associated Bacterial Degradation of Toxic Organic Compounds in Soil. *International Journal of Environmental Research and Public Health*, *62*, 226–2247. PMID:19742157

MEA. (2005). *Dryland Systems. Ecosystems and human well-being: current state and trends.*

Meckler, M. (1991, November). Indoor air quality from commissioning through building operation. *ASHRAE Journal*, 42–48.

Meddich, A., Ait El Mokhtar, M., Bourzik, W., Mitsui, T., Baslam, M., & Hafidi, M. (2018). Optimizing growth and tolerance of date palm (Phoenix dactylifera L.) to drought, salinity, and vascular fusarium-induced wilt (Fusarium oxysporum) by application of arbuscular mycorrhizal fungi (AMF). In *Root Biology* (pp. 239–258). Cham: Springer. doi:10.1007/978-3-319-75910-4_9

Meddich, A., Ait El Mokhtar, M., Wahbi, S., & Boumezzough, A. (2017). Évaluation des potentialités mycorhizogènes en lien avec les paramètres physico-chimiques des sols de palmeraies du Maroc (Marrakech et Tafilalet). *Cahiers Agricultures*, *26*(4), 45012. doi:10.1051/cagri/2017044

Meddi, M. M., Assani, A. A., & Meddi, H. (2010). Temporal variability of annual rainfall in the Macta and Tafna catchments, Northwestern Algeria. *Water Resources Management*, *24*(14), 3817–3833. doi:10.100711269-010-9635-7

Medina, A., Rodriguez, A., & Magan, N. (2015). Changes in environmental factors driven by climate change: effects on the ecophysiology of mycotoxigenic fungi. In *Climate Change and Mycotoxins* (pp. 71–90). De Gruyter. doi:10.1515/9783110333619-006

Meehl, G. A., Stocker, T. F., Collins, W. D., Friedlingstein, P., Gaye, T., Gregory, J. M., & Raper, S. C. (2007). *Global climate projections*. Academic Press.

Meesenburg, H., Ahrends, B., Fleck, S., Wagner, M., Fortmann, H., Scheler, B., ... Meiwes, K. J. (2016). Long-term changes of ecosystem services at Solling, Germany: Recovery from acidification, but increasing nitrogen saturation? *Ecological Indicators*, *65*, 103–112. doi:10.1016/j.ecolind.2015.12.013

Mendes, A., Pereira, C., Mendes, D., Aguiar, L., Neves, P., Silva, S., ... Teixeira, J. P. (2013). Indoor air quality and thermal comfort- Results of a pilot study in elderly care centers in Portugal. *Journal of Toxicology and Environmental Health. Part A.*, *76*(4-5), 333–343. doi:10.1080/15287394.2013.757213 PMID:23514075

Mensah, A., Anderson, K., & Nelson, W. (2016). *Review of adaptation related policies in Ghana* (No. 107642). DECCMA working paper, deltas, vulnerability and climate change: migration and adaptation, IDRC project.

Messouli, M., Ben Salem, A., Ghallabi, B., & Yacoubi-Khebiza, M. (2009). Ecohydrology and groundwater resources management under global changes: A pilot study in presaharan basins of southern Morocco. Options Méditerranéennes, 88, 255-264.

Messouli, M., Rochdane, S., El Alami El Filali, A., & Bouarab, L. (2009). Groundwater–surface water interactions and the ecohydrology of arid regions: evidence from the Tafilalt oasis of southern Morocco Groundwater and Climate in Africa. *Proceedings of the Kampala Conference, 334*, 61-67.

Messouli, M., Bensalem, A., & Rochdane, S. (2013). *Développement d'un concept d'Indice de Vulnérabilité au Changement Climatique et Environnementale au Niveau Régional. Département du développement durable (MNSSD).* Région Moyen-Orient et Afrique du Nord. Banque Mondiale.

Metternicht, G. (2007). Remote sensing. In K. Kemp (Ed.), Encyclopedia of geographic information science. Sage Publ.

Metternicht, G. (2003). Categorical fuzziness: A comparison between crisp and fuzzy class boundary modelling for mapping salt-affected soils using Landsat TM data and a classification based on anion ratios. *Ecological Modelling*, *168*(3), 371–389. doi:10.1016/S0304-3800(03)00147-9

Metternicht, G., Zinck, J. A., Blanco, P. D., & Del Valle, H. F. (2010). Remote Sensing of Land Degradation: Experiences from Latin America and the Caribbean. *Journal of Environmental Quality*, *39*(1), 42–61. doi:10.2134/jeq2009.0127 PMID:20048293

Meyers, L. A., & Bull, J. J. (2002). Fighting change with change: Adaptive variation in an uncertain world. *Trends in Ecology & Evolution*, *17*(12), 551–557. doi:10.1016/S0169-5347(02)02633-2

Michener, W. K., Brunt, J. W., Helly, J. J., Kirchner, T. B., & Stafford, S. G. (1997). Nongeospatial metadata for the ecological sciences. *Ecological Applications*, *7*(1), 330–342. doi:10.1890/1051-0761(1997)007[0330:NMFTES]2.0.CO;2

Michener, W. K., Porter, J., Servilla, M., & Vanderbilt, K. (2011). Long term ecological research and information management. *Ecological Informatics*, *6*(1), 13–24. doi:10.1016/j.ecoinf.2010.11.005

Miethke, M., & Marahiel, M. A. (2007). Siderophore-based iron acquisition and pathogen control. *Microbiology and Molecular Biology Reviews*, *71*(3), 413–445. doi:10.1128/MMBR.00012-07 PMID:17804665

Mieza, M. S., Kovac, F., & Harán, R. (2004). Estimación de superficies inundadasmedianteimágenes. Proc. Simposio-Latinoamericano de PercepciónRemota.

Mili, A. F., Saghrouni, F., BenSaid, Z., Saadi-BenAoun, Y., Guizani, I., & BenSaid, M. (2012). Retrospective Analysis of Leishmaniasis in Central Tunisia: An Update on Emerging Epidemiological Trends. In Current Topics in Tropical Medicine. InTech.

Millennium Ecosystem Assessment. (2005). *Ecosystems and Human Well- Being: Synthesis*. Washington, DC: Island Press.

Millennium Ecosystem Assessment. (2005). *Ecosystems and human well-being: synthesis*. Washington, DC: Island Press.

Miller, C. M., Miller, R. V., Garton-Kenny, D., Redgrave, B., Sears, J., Condron, M. M., ... Strobel, G. A. (1998). Ecomycins unique antimycotics from *Pseudomonas viridi flava*. *Journal of Applied Microbiology*, *84*(6), 937–944. doi:10.1046/j.1365-2672.1998.00415.x PMID:9717277

Miller-Rushing, A. J., Høye, T. T., Inouye, D. W., & Post, E. (2010). The effects of phenological mismatches on demography. *Philosophical Transactions of the Royal Society of London. Series B, Biological Sciences*, *365*(1555), 3177–3186. doi:10.1098/rstb.2010.0148 PMID:20819811

Ministry of Energy, Mines, Water and Environment (Secrétariat d'État chargé de l'eau et de l'environpressure). (2009). *Adaptation aux changements climatiques au Maroc: pour des Oasis résilientes*. Author.

Ministry of Environment of Morocco, (2001). *Rapport de l'Etat de l'environnement du Maroc, 292*. Author.

Ministry of Water, Lands and Environment. (2010). Wetland Sector Strategic Plan 2001-2010, Kampala. Government of Uganda.

Minucci, G., & Karmaoui, A. (2017). Exploring the Water-Food-Energy and Climate Nexus: Insights from the Moroccan Draa Valley. In *Peri-Urban Areas and Food-Energy-Water Nexus* (pp. 89–97). Cham: Springer. doi:10.1007/978-3-319-41022-7_11

Miranda, M. L., Maxson, P., & Edwards, S. (2009). Environmental contributions to disparities in pregnancy outcomes. *Epidemiologic Reviews*, *31*(1), 67–83. doi:10.1093/epirev/mxp011 PMID:19846592

Mires, D. (2000). Development of inland aquaculture in arid climates: Water utilization strategies applied in Israel. *Fisheries Management and Ecology*, *7*(1–2), 189–195. doi:10.1046/j.1365-2400.2000.00186.x

Mirtl, M., Borer, E. T., Djukic, I., Forsius, M., Haubold, H., Hugo, W., ... Orenstein, D. E. (2018). Genesis, goals and achievements of long-term ecological research at the global scale: A critical review of ILTER and future directions. *The Science of the Total Environment*, *626*, 1439–1462. doi:10.1016/j.scitotenv.2017.12.001 PMID:29898550

Mitchell, M. A., & Kettlewell, P. J. (1998). Physiological stress and welfare of broiler chickens in transit: Solutions not problems! *Poultry Science*, *77*(12), 1803–1814. doi:10.1093/ps/77.12.1803 PMID:9872583

Mohanraj, R., & Azeez, P. A. (2005). Urban development and particulate air pollution in Coimbatore city, India. *The International Journal of Environmental Studies*, *62*(1), 69–78. doi:10.1080/0020723042000261713

Mohnen, V., Goldstein, W., & Wang, W. C. (1993). Tropospheric ozone and climate change. *Air & Waste*, *43*(10), 1332–1334. doi:10.1080/1073161X.1993.10467207

Mollenhauer, H., Kasner, M., Haase, P., Peterseil, J., Wohner, C., Frenzel, M., ... Zacharias, S. (2018). Long-term environmental monitoring infrastructures in Europe: Observations, measurements, scales, and socio-ecological representativeness. *The Science of the Total Environment*, *624*, 968–978. doi:10.1016/j.scitotenv.2017.12.095 PMID:29275260

Montoya, J. M., & Raffaelli, D. (2010). *Climate change, biotic interactions and ecosystem services*. Academic Press.

Mooney, H. A., Ehrlich, P. R., & Daily, G. (1997). *Ecosystem Services: A Fragmentary History*. Washington, DC: Island Press.

Moroccan Ministry of Health. (2015). *Santé en Chiffre*. Rabat: Moroccan Ministry of Health.

Movia, C. (1980). Inventario de la erosiónen la Patagonia Argentina basadoenimágenes Landsat y fotografíaaérea. *Proc. Symposium Arid Land Resource Inventories: Cost-efficient methods*.

Muiruri, H. K., & Harrison, P. C. (1991). Effect of Roost Temperature on Performance of Chickens in Hot Ambient Environments. *Poultry Science*, *70*(11), 2253–2258. doi:10.3382/ps.0702253 PMID:1754541

Müller, F., Baessler, C., Frenzel, M., Klotz, S., & Schubert, H. (2010). Long-term ecosystem research between theory and application–an introduction. In *Long-Term Ecological Research* (pp. 3–7). Dordrecht: Springer. doi:10.1007/978-90-481-8782-9_1

Munday, P. L., Jones, G. P., Pratchett, M. S., & Williams, A. J. (2008). Climate change and the future for coral reef fishes. *Fish and Fisheries*, *9*(3), 261–285. doi:10.1111/j.1467-2979.2008.00281.x

Murray-Tortarolo, G., Friedlingstein, P., Sitch, S., Seneviratne, S. I., Fletcher, I., Mueller, B., ... Zeng, N. (2016). The dry season intensity as a key driver of NPP trends. *Geophysical Research Letters*, *43*(6), 2632–2639. doi:10.1002/2016GL068240

Mustafa, M. B., Hussein, S. M., Ibrahim, E. A., Al-Seghayer, S. M., Al Amri, S. A., & Gradoni, L. (1994). Phlebotomus papatasi (Scopoli), vector of zoonotic cutaneous leishmaniasis in Riyadh province, Saudi Arabia. *Transactions of the Royal Society of Tropical Medicine and Hygiene*, *88*(1), 40. doi:10.1016/0035-9203(94)90489-8 PMID:8153996

Naeem, S., Bunker, D. E., Hector, A., Loreau, M., & Perrings, C. (Eds.). (2009). *Biodiversity, ecosystem functioning, and human wellbeing: an ecological and economic perspective*. Oxford University Press. doi:10.1093/acprof:oso/9780199547951.001.0001

Nakano, S. I., & Tetsukazu, Y. (2012). *The Biodiversity observation network in the Asia-Pacific region* (T. Nakashizuka, Ed.). Springer. doi:10.1007/978-4-431-54032-8

Nakhli, F. (2011). Ressources en eau/Souss: menaces sur la nappe phréatique. *L'économiste,* 3547.

Nam, S., Phommakone, S., Vuthy, L., Samphawamana, T., Son, N. H., Khumsri, M., ... Starr, P. (2015). Lower Mekong fisheries estimated to be worth around $17 billion a year. *Catch Cult.*, *21*, 4–7.

Naruševičius, V., & Matiukas, G. (2011). Introducing the concept of ecosystem services: Inventory and economic valuation on local scale in Lithuania. *Sustainable Development Strategy and Practise*, *1*(5), 120–136.

National Environment Management Agency. (2012). *State of the Environment Report for Uganda 2002. Ministry of Water, Lands and Environment*. Kampala: Government of Uganda.

National Environment Management Authority. (2000/2001). *State of Environment Report 2000/2001. Government of Uganda, Uganda*. Kampala: NEMA.

National Environment Management Authority. (2003). *Guidelines for District State of Environment Report (DSOER). Preparation in Uganda* [Final Report]. Kampala: NEMA.

National Research Council. (1982). *Solar variability, weather, and climate*. Washington, DC: National Academy Press.

Nawrot, T., Plusquin, M., Hogervorst, J., Roels, H. A., Celis, H., Thijs, L., ... Staessen, J. A. (2006). Environmental exposure to cadmium and risk of cancer: A prospective population-based study. *The Lancet Oncology*, *7*(2), 119–126. doi:10.1016/S1470-2045(06)70545-9 PMID:16455475

Nayak, P. K., Nayak, A. K., Panda, B. B., Lal, B., Gautam, P., Poonam, A., … Jambhulkar, N. N. (2018). Ecological mechanism and diversity in rice based integrated farming system. *Ecological Indicators, 91*, 359–375.

Neas, L., Williams, R. W., & Devlin, R. B. (2004). Particulate matter exposure in cars is associated with cardiovascular effects in healthy young men. *American Journal of Respiratory and Critical Care Medicine*, *169*(8), 934–940. doi:10.1164/rccm.200310-1463OC PMID:14962820

Neiland, A. E., & Béné, C. (2006). Tropical River Fisheries Valuation: A Global Synthesis and Critical Review. Colombo, Sri Lanka: Academic Press.

Neisi, A., Goudarzi, G., Babaei, A., Vosoughi, M., Hashemzadeh, H., Naimabadi, A., ... Hashemzadeh, B. (2016). Study of heavy metal levels in indoor dust and their health risk assessment in children of Ahvaz City, Iran. *Toxin Reviews*, *35*(1-2), 16–23. doi:10.1080/15569543.2016.1181656

Nemery, B., Hoet, P., & Nemmar, A. (2001). The Meuse Valley fog of 1930: An air pollution disaster. *Lancet*, *357*(9257), 704–708. doi:10.1016/S0140-6736(00)04135-0 PMID:11247570

Newman, L. A., & Reynolds, C. M. (2005). Bacteria and phytoremediation: New uses for endophytic bacteria in plants. *Trends in Biotechnology*, *23*(1), 6–8. doi:10.1016/j.tibtech.2004.11.010 PMID:15629849

Nicholls, R. J., Wong, P. P., Burkett, V. R., Codignotto, J. O., Hay, J. E., McLean, R. F., . . . Woodroffe, C. D. (2007). Coastal systems and low-lying areas. In M.L. Parry, O.F. Canziani, J.P. Palutikof, P.J. Linden, & C.E. Hanson (Eds), Climate Change 2007: impacts, adaptation and vulnerability (pp. 315–356). Contribution of working group II to the Fourth Assessment Report of the Intergovernmental Panel on Climate Change. Cambridge, UK: Cambridge University Press.

Nichols, T. D., Wolf, D. C., Rogers, H. B., Beyrouty, C. A., & Reynolds, C. M. (1997). Rhizosphere microbial populations in contaminated soils. *Water, Air, and Soil Pollution*, *95*(1-4), 165–178. doi:10.1007/BF02406163

Nisa, H., Kamili, A. N., Nawchoo, I. A., Shafi, S., Shameem, N., & Bandh, S. A. (2015). Fungal endophytes as prolific source of phytochemicals and other bioactive natural products: A review. *Microbial Pathogenesis*, *82*, 50–59. doi:10.1016/j.micpath.2015.04.001 PMID:25865953

NOAASTAT. (2019). Retrieved April 06, 2018, from https://www.st.nmfs.noaa.gov/commercial-fisheries/commercial-landings/annual-landings/index

Noss, R. F. (1990). Indicators for monitoring biodiversity: A hierarchical approach. *Conservation Biology*, *4*(4), 355–364. doi:10.1111/j.1523-1739.1990.tb00309.x

Nourmoradi, H., Goudarzi, G., Daryanoosh, S.M., Omidi-Khaniabadi, F., & Jourvand, M., & Omidi-Khaniabadi, Y. (2015). Health impacts of particulate matter in air by Air Q model in Khorramabad City, Iran. *J Bas Res Med Sci.*, *2*, 44–52.

Nowak, D. J. (2000). Tree species selection, design, and management to improve air quality. *Construction Technology: Annual Meeting Proceedings of the American Society of Landscape Architects.*

Nowak, D. J., Civerolo, K. L., Rao, S. T., Sistla, G., Luley, C. J., & Crane, D. E. (2000). A modeling study of the impact of urban trees on ozone. *Atmospheric Environment, 34*, 1601–1613. doi:10.1016/S1352-2310(99)00394-5

Nowak, D. J., Crane, D. E., & Stevens, J. C. (2006). Air pollution removal by urban trees and shrubs in the United States. *Urban Forestry & Urban Greening*, *4*(3-4), 115–123. doi:10.1016/j.ufug.2006.01.007

Nowak, D. J., Hirabayashi, S., Bodine, A., & Hoehn, R. (2013). Modeled PM2.5 removal by trees in ten U.S. cities and associated health effects. *Environmental Pollution*, *178*, 395–402. doi:10.1016/j.envpol.2013.03.050 PMID:23624337

OECD-FAO (OECD/Food and Agriculture Organization of the United Nations). (2014). *OECD- FAO Agricultural Outlook 2014*. OECD Publishing. doi:10.1787/agr_outlook-2014-

Oehninger, S. (2001). Strategies for the infertile man. *Seminars in Reproductive Medicine*, *19*(03), 231–237. doi:10.1055-2001-18042 PMID:11679904

Okaka, W. (2002). An Environmental Education Programme: Uganda Polytechnic Kyambogo.Applied environmental education and communication. Taylor & Francis.

Okaka, W. (2007). *The Role of Media Communications in Developing Tourism Policy and Cross-Cultural Communication for Peace, Security for Sustainable Tourism Industry in Africa.* A Paper at the 4th International Institute of Peace through Tourism (IIPT) African Conference on Peace through Tourism at Educators' Forum, Uganda, Kampala.

Okaka, W. (2010). Developing Regional Communications Campaigns Strategy For Environment and Natural Resources Management Policy Awareness For The East African Community. *Research Journal of Environmental and Earth Sciences*, *2*(2), 106–111.

Okaka, W. (2013). *The dawn of community radio in Bangladesh*. Lambert Academic publishers.

Olivier, J. G. J., Janssens-Maenhout, G., Peters, J. A. H. W., & Wilson, J. (2011). *Long- term trend in global CO_2 emissions 2011 report*. The Hague: PBL Netherlands.

Ollinger, S. V., Aber, J. D., & Reich, P. B. (1997). Simulating ozone effects on forest productivity: Interactions among leaf-, canopy-, and stand-level processes. *Ecological Applications*, *7*(4), 1237–1251. doi:10.1890/1051-0761(1997)007[1237:SOEOFP]2.0.CO;2

Olson, I. C., Kozdon, R., Valley, J. W., & Gilbert, P. U. P. A. (2012). Mollusk Shell Nacre Ultrastructure Correlates with Environmental Temperature and Pressure. *Journal of the American Chemical Society*, *134*(17), 7351–7358. doi:10.1021/ja210808s PMID:22313180

Olsson, D., Mogren, I., & Forsberg, B. (2013). Air pollution exposure in early pregnancy and adverse pregnancy outcomes: A register-based cohort study. *BMJ Open*, *3*(2), e001955. doi:10.1136/bmjopen-2012-001955 PMID:23386578

OPERAs. (2016). *OPERAs- Ecosystem Science for Policy & Practice*. Retrieved April 21, 2018, from http://www.operas-project.eu/

Oppenheimer, M., Campos, M., Warren, R., Birkmann, J., Luber, G., O'Neill, B., Takahashi, K. (2014). Emergent risks and key vulnerabilities. In *Climate Change 2014: Impacts, Adaptation, and Vulnerability. Part A: Global and Sectoral Aspects. Contribution of Working Group II to the Fifth Assessment Report of the Intergovernmental Panel on Climate Change*. Cambridge University Press. Retrieved from http://www.ipcc.ch/report/ar5/wg2/

OPPLA. (2015). Retrieved May 27, 2018, from http://oppla.eu/

Orr, S., Pittock, J., Chapagain, A., & Dumaresq, D. (2012). Dams on the Mekong River: Lost fish protein and the implications for land and water resources. *Global Environmental Change*, *22*(4), 925–932. doi:10.1016/j.gloenvcha.2012.06.002

Oshaghi, M. A., Rasolian, M., Shirzadi, M. R., Mohtarami, F., & Doosti, S. (2010). First report on isolation of Leishmania tropica from sandflies of a classical urban Cutaneous leishmaniasis focus in southern Iran. *Experimental Parasitology*, *126*(4), 445–450. doi:10.1016/j.exppara.2010.05.020 PMID:20570590

Owen, N. L., & Hundley, N. (2004). Endophytes–the chemical synthesizers inside plants. *Science progress*, *87*(2), 79–99. doi:10.3184/003685004783238553 PMID:15782772

Ozgul, A., Childs, D. Z., Oli, M. K., Armitage, K. B., Blumstein, D. T., Olson, L. E., & Coulson, T. (2010). Coupled dynamics of body mass and population growth in response to environmental change. *Nature*, *466*(7305), 482–485. doi:10.1038/nature09210 PMID:20651690

PACC. (2011). *Oasis et Zones Arides Résilientes Territoires Engagés*. Retrieved from www.oasisadaptation.com

Packard, S., & Mutel, C. F. (Eds.). (1997). *The Tallgrass Restoration Handbook*. Washington, DC: Island.

Paeth, H., Born, K., Girmes, R., Podzun, R., & Jacob, D. (2009). Regional climate change in tropical and northern Africa due to greenhouse forcing and land use changes. *Journal of Climate*, *22*(1), 114–132. doi:10.1175/2008JCLI2390.1

Pai, S. G., Riley, M. B., & Camper, N. D. (2001). Microbial degradation of mefenoxam in rhizosphere of Zinnia angustifolia. *Chemosphere*, *44*(4), 577–582. doi:10.1016/S0045-6535(00)00368-4 PMID:11482644

Palutikof, J. P., & Wigley, T. M. L. (1996). Developing climate change scenarios for the Mediterranean region. In Climatic Change and the Mediterranean (pp. 27–56). Academic Press.

Panitz, C., Sperl, M. F., Hennig, J., Klucken, T., Hermann, C., & Mueller, E. M. (2018). Fearfulness, neuroticism/anxiety, and COMT Val158Met in long-term fear conditioning and extinction. *Neurobiology of Learning and Memory*, *155*, 7–20. doi:10.1016/j.nlm.2018.06.001 PMID:29883709

Pardo, L. H., Fenn, M. E., Goodale, C. L., Geiser, L. H., Driscoll, C. T., Allen, E. B., ... Haines, A. (2014). Climate change: Challenges and opportunities for global health. *Journal of the American Medical Association*, *312*(15), 1565–1580. doi:10.1001/jama.2014.13186 PMID:25244362

Parmesan, C. (1996). Climate and species' range. *Nature*, *382*(6594), 765–766. doi:10.1038/382765a0. doi:10.1038/382765a0

Parmesan, C. (2006). Ecological and evolutionary responses to recent climate change. *Annual Review of Ecology Evolution and Systematics*, *37*(1), 637–669. doi:10.1146/annurev.ecolsys.37.091305.110100

Parmesan, C. (2007). Influences of species, latitudes and methodologies on estimates of phenological response to global warming. *Global Change Biology*, *13*(9), 1860–1872. doi:10.1111/j.1365-2486.2007.01404.x

Parmesan, C., & Yohe, G. (2003). A globally coherent fingerprint of climate change impacts across natural systems. *Nature*, *421*(6918), 37–42. doi:10.1038/nature01286 PMID:12511946

Parton, W. J., Scurlock, J. M. O., Ojima, D. S., Schimel, D. S., Hall, D. O., & Members, S. G. (1995). Impact of Climate Change on Grassland Production and Soil Carbon Worldwide. *Global Change Biology*, *1*(1), 13–22. doi:10.1111/j.1365-2486.1995.tb00002.x

Paruelo, J. M., Jobbagy, E., & Sala, O. (2001). Current distribution of ecosystem functional types in temperate South America. *Ecosystems (New York, N.Y.)*, *4*(7), 683–698. doi:10.100710021-001-0037-9

Patricola, C. M., & Cook, K. H. (2010). Northern African climate at the end of the twenty-first century: An integrated application of regional and global climate models. *Climate Dynamics*, *35*(1), 193–212. doi:10.100700382-009-0623-7

Paul, I., Mandal, L., & Datta, S. (2018). *Feasibility study of the small-scale poultry-cum-fish farming in homestead ponds for empowerment of rural women: A case study in the new alluvial zone of west.* Academic Press.

Pauling, A., Luterbacher, J., Casty, C., & Wanner, H. (2006). Five hundred years of gridded high-resolution precipitation reconstructions over Europe and the connection to large-scale circulation. *Climate Dynamics*, *26*(4), 387–405. doi:10.100700382-005-0090-8

Pawson, S. M., Brin, A., Brockerhalf, E. G., Lamb, D., Payn, T. W., Paqquet, A., & Parrotta, J. A. (2013). Plantation Forests, Climate Change & Biodiversity. *Biodiversity and Conservation*, *22*(5), 1203–1227. doi:10.1007/s10531-013-0458-8. doi:10.100710531-013-0458-8

PDES. (2004). Le plan de développement économique et social 2000-2004. Région de Tanger-Tétouan, 649-683.

Peel, M. C., Finlayson, B. L., & McMahon, T. A. (2007). Updated world map of the Koppen-Geiger climate classification. *Hydrology and Earth System Sciences*, *11*(5), 1633–1644. doi:10.5194/hess-11-1633-2007

Pereira, H. M., Leadley, P. W., Proença, V., Alkemade, R., Scharlemann, J. P., Fernandez-Manjarrés, J. F., ... Chini, L. (2010). Scenarios for global biodiversity in the 21st century. *Science*, *330*(6010), 1496–1501. doi:10.1126cience.1196624 PMID:20978282

Pereira, L. S. (2003). Irrigation Demand Management to Cope with Drought and Water Scarcity BT - Tools for Drought Mitigation in Mediterranean Regions. Dordrecht: Springer Netherlands; doi:10.1007/978-94-010-0129-8_2

Peters, A., von Klot, S., Heier, M., Trentinaglia, I., Hörmann, A., Wichmann, H. E., & Löwel, H. (2004). Exposure to traffic and the onset of myocardial infarction. *The New England Journal of Medicine*, *351*(17), 1721–1730. doi:10.1056/NEJMoa040203 PMID:15496621

Peterson, T. C., & Manton, M. J. (2008). Monitoring changes in climate extremes: A tale of international collaboration. *Bulletin of the American Meteorological Society*, *89*(9), 1266–1271. doi:10.1175/2008BAMS2501.1

Petrini, O. (1991). Fungalendophytes of tree leaves. In J. H. Andrews & S. S. Hirano (Eds.), *Microbial Ecology of Leaves* (pp. 179–197). Berlin: Springer. doi:10.1007/978-1-4612-3168-4_9

Phillips, S. J., Dudík, M., Elith, J., Graham, C. H., Lehmann, A., Leathwick, J., & Ferrier, S. (2009). Sample selection bias and presence-only distribution models: Implications for background and pseudo-absence data. *Ecological Applications*, *19*(1), 181–197. doi:10.1890/07-2153.1 PMID:19323182

Piao, S., Friedlingstein, P., Ciais, P., de Noblet-Ducoudré, N., Labat, D., & Zaehle, S. (2007). Changes in climate and land use have a larger direct impact than rising CO2 on global river runoff trends. *Proceedings of the National Academy of Sciences, 104*(39), 15242 LP-15247. Retrieved from http://www.pnas.org/content/104/39/15242.abstract

Pickrell, J. (1991). Hazards in confinement housing - gases and dusts in confined animal houses for swine, poultry, horses and humans. *Veterinary and Human Toxicology*, *33*, 32–39. PMID:2017863

Pieterse, C. M., Zamioudis, C., Berendsen, R. L., Weller, D. M., Van Wees, S. C., & Bakker, P. A. (2014). Induced systemic resistance by beneficial microbes. *Annual Review of Phytopathology*, 52. PMID:24906124

Pinder, R.W., Bettez, N.D., Bonan, G.B., Greaver, T. L., Widier, W.R., Schlesinger,W., & Davidson, E. A. (2013). Impacts of human alteration of the nitrogen cycle in the US on radiative forcing. *Biogeochemistry, 114*, 25–40.

Pirovani, D. B., Pezzopane, J. E. M., Xavier, A. C., Pezzopane, J. R. M., de Jesus Júnior, W. C., Machuca, M. A. H., ... dos Santos, A. R. (2018). Climate change impacts on the aptitude area of forest species. *Ecological Indicators*, *95*, 405–416. doi:10.1016/j.ecolind.2018.08.002

Pitt, N. R., Poloczanska, E. S., & Hobday, A. J. (2010). Climate-driven range changes in Tasmanian intertidal fauna. *Marine & Freshwater Research*, *61*(9), 963–970. doi:10.1071/MF09225

Plan I for 2010/2011-2015. (2016). Kampala, Uganda: Isingiro District Local Government.

Plan II for 2015/2016-2019/2020. Uganda, Kampala: Government of Uganda Isingiro District Local Government. (2017). Five Year District Local Government Development.

Plan II for 2015/2016-2019/2020. Uganda, Kampala: Government of Uganda Isingiro District Planning Unit. (2014). Five Year District Local Government Development.

Plan, I. I. 2010/2011-2015/2016. Uganda, Kampala: Government of Uganda.

Platt, S. (2008). Current Development of the Population in the Provinces of Ouarzazate and Zagora. In Impetus Atlas Morocco: Research Results 2000–2007 (pp. 59–60). University of Bonn.

Płażek, A., Hura, K., Rapacz, M., & Żur, I. (2001). The influence of ozone fumigation on metabolic efficiency and plant resistance to fungal pathogens. *Journal of Applied Botany and Food Quality*, *75*, •••.

PNUD. (2012). *Rapport PNUD Maroc en action*. Programme des Nations Unies pour le développement au Maroc.

PNUED. (2013). Ecosystem Assessment of Tafilalet Oasis. *Synthesis*, 1-24.

Pope, C. A. III. (1989). Respiratory Disease Associated with Community Air Pollution and a Steel Mill, Utah Valley. *American Journal of Public Health*, *79*(5), 623–628. doi:10.2105/AJPH.79.5.623 PMID:2495741

Pope, C. A. III. (1991). Respiratory Hospital Admissions Associated with PM10 Pollution in Utah, Salt Lake, and Cache Valleys. *Archives of Environmental Health*, *46*(2), 90–97. doi:10.1080/00039896.1991.9937434 PMID:2006899

Pope, C. A. III, Burnett, R. T., Thurston, G. D., Thun, M. J., Calle, E. E., Krewski, D., & Godleski, J. J. (2004). Cardiovascular mortality and long-term exposure to particulate air pollution: Epidemiological evidence of general pathophysiological pathways of disease. *Circulation*, *109*(1), 71–77. doi:10.1161/01.CIR.0000108927.80044.7F PMID:14676145

Pope, C. A. III, Burnett, R. T., Turner, M. C., Cohen, A., Krewski, D., Jerrett, M., ... Thun, M. J. (2011). Lung cancer and cardiovascular disease mortality associated with ambient air pollution and cigarette smoke: Shape of the exposure-response relationships. *Environmental Health Perspectives*, *119*(11), 1616–1621. doi:10.1289/ehp.1103639 PMID:21768054

Pope, C. A. III, & Dockery, D. W. (1992). Acute Health Effects of PM10 Pollution on Symptomatic and Asymptomatic Children. *The American Review of Respiratory Disease*, *145*(5), 1123–1128. doi:10.1164/ajrccm/145.5.1123 PMID:1586057

Pope, C. A. III, Dockery, D. W., Spengler, J. D., & Raizenne, M. E. (1991). Respiratory Health and PM10 Pollution: A Daily Time Series Analysis. *The American Review of Respiratory Disease*, *144*(3_pt_1), 668–674. doi:10.1164/ajrccm/144.3_Pt_1.668 PMID:1892309

Pope, C. A. III, & Kanner, R. E. (1993). Acute Effects of PM10 Pollution on Pulmonary Function of Smokers with Mild to Moderate Chronic Obstructive Pulmonary Disease. *The American Review of Respiratory Disease*, *147*(6_pt_1), 1336–1340. doi:10.1164/ajrccm/147.6_Pt_1.1336 PMID:8503541

Pope, C. A. III, Schwartz, J., & Ransom, M. R. (1992). Daily Mortality and PM10 Pollution in Utah Valley. *Archives of Environmental Health*, *47*(3), 211–217. doi:10.1080/00039896.1992.9938351 PMID:1596104

Pope, C. A. III, Thun, M. J., Namboodiri, M. M., Dockery, D. W., Evans, J. S., Speizer, F. E., & Heath, C. W. Jr. (1995). Particulate air pollution as a predictor of mortality in a prospective study of U.S. adults. *American Journal of Respiratory and Critical Care Medicine*, *151*(3_pt_1), 669–674. doi:10.1164/ajrccm/151.3_Pt_1.669 PMID:7881654

Porter, J. R., Xie, L., Challinor, A. J., Cochrane, K., Howden, S. M., & Iqbal, M. M. (2014). Food security and food production systems. In Climate Change 2014: Impacts, Adaptation, and Vulnerability. Part A. Global and Sectoral Aspects. Contribution of Working Group II to The Fifth Assessment Report of the Intergovernmental Panel on Climate Change. Academic Press.

Pörtner, H. O., & Knust, R. (2007). Climate Change Affects Marine Fishes Through the Oxygen Limitation of Thermal Tolerance. *Science, 315*(5808), 95 LP-97. Retrieved from http://science.sciencemag.org/content/315/5808/95.abstract

Pörtner, H. O., & Knust, R. (2007). Climate change affects marine fishes through the oxygen limitation of thermal tolerance. *Science*, *315*(5808), 95–97. doi:10.1126cience.1135471 PMID:17204649

Potts, S.G., Biesmeijer, J.C., Kremen, C., Neumann, P., Schweizer, O., &Kunin, W.E. (2015). Global Pollinator Declines: Trends, Impacts & Drivers. *Trends EcolEvol.*, 345-353.

Pressey, R. L., Cabeza, M., Watts, M. E., Cowling, R. M., & Wilson, K. A. (2007). Conservation planning in a changing world. *Trends in Ecology & Evolution*, *22*(11), 583–592. doi:10.1016/j.tree.2007.10.001 PMID:17981360

Prober, S. M., & Dunlop, M. (2011). Climate change: A cause for new biodiversity conservation objectives but let's not throw the baby out with the bathwater. *Ecological Management & Restoration*, *12*(1), 2–3. doi:10.1111/j.1442-8903.2011.00563.x

Rabiey, M., Ullah, I., & Shaw, M. W. (2015). The endophytic fungus Piriformosporaindica protects wheat from fusarium crown rot disease in simulated UK autumn conditions. *Plant Pathology*, *64*(5), 1029–1040. doi:10.1111/ppa.12335

RADDO. (2011). Lutte contre la désertification: les oasis au péril du changement climatique. *Le réseau d'associations oasiennes du Maghreb (RADDO)*. Retrieved from www.raddo.org

Raghukumar, C. (2008). Marine fungal biotechnology: An ecologicalperspective. *Fungal Diversity*, *31*, 5–19.

Rajkumar, M., Ae, N., & Freitas, H. (2010). Endophytic bacteria and their potential to enhance heavy metal phytoextraction. *Chemosphere*, *77*(2), 153–156. doi:10.1016/j.chemosphere.2009.06.047 PMID:19647283

Rani Parker, A. (1993). *Another Point of View: A Manual on Gender Analysis Training for Grassroots Workers*. New York: Ink. Publishers.

Rankou, H., Culham, A., Sghir Taleb, M., Ouhammou, A., Martin, G., & Jury, S. L. (2015). Conservation assessments and Red Listing of the endemic Moroccan flora (monocotyledons). *Botanical Journal of the Linnean Society*, *177*(4), 504–575. doi:10.1111/boj.12258

Ransom, M. R., & Pope, C. A. III. (1992). Elementary School Absences and PM10 Pollution in Utah Valley. *Environmental Research*, *58*(1-2), 204–219. doi:10.1016/S0013-9351(05)80216-6 PMID:1511674

Rastogi, S. K., Gupta, B. N., Husain, T., Chandra, H., Mathur, N., Pangtey, B. S., ... Ravenscroft, J. (2006). Effects of pollution on human growth and development: An introduction. *Journal of Physiological Anthropology*, *25*(1), 103–112. doi:10.2114/jpa2.25.103 PMID:16617215

Raven, P. H. (1973). The evolution of Mediterranean floras. In *Mediterranean type ecosystems* (pp. 213–224). Berlin: Springer. doi:10.1007/978-3-642-65520-3_14

Ravi Prasad Rao, B., Salamma, S., & Durga Prasad, K. (2017). From Climate Change and Biodiversity Impacts. *Adaptation and Mitigation-an Overview.*

Ravindranath, N. H., & Sukumar, R. (1998). Climate change and tropical forests in India. In Potential Impacts of Climate Change on Tropical Forest Ecosystems (pp. 423–441). Dordrecht: Springer. doi:10.1007/978-94-017-2730-3_21 doi:10.1007/978-94-017-2730-3_21

Ravindranath, Sukumar, & Saxena. (2006). Impact of climate change on forests in India. *Current Science*, *90*(3), 354–361.

Ray, D. K., Gerber, J. S., MacDonald, G. K., & West, P. C. (2015). Climate variation explains a third of global crop yield variability. *Nature Communications*, *6*(1), 5989. doi:10.1038/ncomms6989 PMID:25609225

Ray, K. S., & Srivastava, A. K. (2000). Is there any change in extreme events like heavy rainfall? *Current Science*, *79*(2), 155–158.

RBOSM. (2008). *Plan cadre de gestion de la Réserve de Biosphère des Oasis du Sud Marocain (RBOSM)*. Version provisoire.

Ready, P. D. (2010). Leishmaniasis emergence in Europe. *Eurosurveillance*, *15*(10), 1–11. PMID:20403308

Redman, R. S., Sheehan, K. B., & Stout, R. G. (2002). Thermo-tolerance generated by plant/fungal symbiosis. *Science*, *298*(5598), 1581. doi:10.1126cience.1072191 PMID:12446900

Rehfeldt, G. E., Crookston, N. L., Sáenz-Romero, C., & Campbell, E. M. (2012). North American vegetation model for land-use planning in a changing climate: A solution to large classification problems. *Ecological Applications*, *22*(1), 119–141. doi:10.1890/11-0495.1 PMID:22471079

Reich, P. B., & Amundson, R. G. (Eds.). (n.d.). Ambient levels of ozone reduce net photosynthesis in tree and crop species. Science, 230, 566–570.

Reinap, A., Wiman, B. L. B., Svenningsson, B., & Gunnarsson, S. (2009). Oak leaves as aerosol collectors: Relationships with wind velocity and particle size distribution. Experimental results and their implications. *Trees (Berlin)*, *23*(6), 1263–1274. doi:10.100700468-009-0366-4

Renevot, G., Bouaziz, A., & Raki, T. R. (2010). Pratiques d'irrigation du palmier dattier dans les systèmes oasiens du Tafilalet, Maroc. *Revue Homme terre et eau, 146*.

Resagro. (2016). Le mensuel des décideurs. *Le palmier dattier face aux changements climatiques*, *71*, 27-29.

Reza, F. M., & Mansour, N. (2006). Entomological studies of Phlebotomus papatasi and P. sergenti (Diptera: Psychodidae) as vectors of cutaneous leishmaniasis in Shiraz, Iran. *The Southeast Asian Journal of Tropical Medicine and Public Health*, *37*, 115–117. PMID:17547064

RGPH. (2004). *Le Recensement Général de la Population et de l'Habitat 2004*. HCP Maroc.

Ribot, J. C., Magalhães, A. R., & Panagides, S. (2005). *Climate Variability, Climate Change and Social Vulnerability in the Semi-arid Tropics*. Cambridge University Press.

Rice, J. C., & Garcia, S. M. (2011). Fisheries, food security, climate change, and biodiversity: Characteristics of the sector and perspectives on emerging issues. *ICES Journal of Marine Science*, *68*(6), 1343–1353. doi:10.1093/icesjms/fsr041

Richardson, A. J., & Poloczanska, E. S. (2008). Under-resourced, under threat. *Science-New York Then Washington*, *320*(5881), 1294–1295. doi:10.1126cience.1156129 PMID:18535230

Riebau, A. R., & Andersen, C. (1985). Wildland Fires and Air Pollution. *Developments in Environmental Science*, *8*, 365–403.

Rignall, K., & Kusunose, Y. (2018). Governing livelihood and land use transitions: The role of customary tenure in southeastern Morocco. *Land Use Policy*, *78*, 91–103. doi:10.1016/j.landusepol.2018.03.035

Riou, Ch. (1990). Bioclimatologie des oasis. *Option Méditerranéennes, série A*, *11*, 207–213.

Rioux, J. A., Rispail, P., Lanotte, G., & Lepart, J. (1984). Relations Phlébotomes-bioclimats en écologie des leishmanioses Corollaires épidémiologiques. L'exemple du Maroc. *Bulletin de la Société Botanique de France. Actualités Botaniques*, *131*(2-4), 549–557.

Risbey, J. S. (1998). Sensitivities of water supply planning decisions to streamflow and climate scenario uncertainties. *Water Policy*, *1*(3), 321–340. doi:10.1016/S1366-7017(98)00023-3

Rissik, D. (2014). *Peer Review Report: Assessment of the Comprehensiveness of the Potential Environmental Impacts, Threats and Risks Identified in the Tamar Lake Inc.* Reports.

Ritz, B., Yu, F., Chapa, G., & Fruin, S. (2000). Effect of air pollution on preterm birth among children born in southern California between 1989 and 1993. *Epidemiology (Cambridge, Mass.)*, *11*(5), 502–511. doi:10.1097/00001648-200009000-00004 PMID:10955401

Robertson, G. P. (2008). Long-term ecological research: Re-inventing network science. *Frontiers in Ecology and the Environment*, *6*(5), 281–281. doi:10.1890/1540-9295(2008)6[281:LERRNS]2.0.CO;2

Rochdane, S., Bounoua, L., Zhang, P., Imhoff, M. L., Messouli, M., & Yacoubi-Khebiza, M. (2014). Combining satellite data and models to assess vulnerability to climate change and its impact on food security in Morocco. *Sustainability*, *6*(4), 1729–1746. doi:10.3390u6041729

Rodriguez, R. J., Henson, J., Van Volkenburgh, E., Hoy, M., Wright, L., Beckwith, F., ... Redman, R. S. (2008a). Stress tolerance in plants via habitat-adapted symbiosis. *The ISME Journal*, *2*(4), 404–416. doi:10.1038/ismej.2007.106 PMID:18256707

Roessig, J. M., Woodle, C. M., Cech, J. J., & Hansen, L. J. (2004). Effects of global climate change on marine and estuarine fish and fisheries. *Reviews in Fish Biology and Fisheries*, *14*(2), 251–275. doi:10.100711160-004-6749-0

Rogers, D. J. (1996). Changes in disease vector distributions. In M. Hulme (Ed.), *Climate Change and Southern Africa: An Exploration of Some Potential Impacts and Implications in the SADC Region. Climatic Research Unit.* Norwich: University of East Anglia.

Romagny, B., & Boujrouf, S. (2010). The rush argan oil, chronicles the heritage value of a Moroccan land. *Hesperis Tamuda*, *45*, 151–166.

Ronchel, M. C., & Ramos, J. L. (2001). Dual system to reinforce biological containment of recombinant bacteria designed for rhizoremediation. *Applied and Environmental Microbiology*, *67*(6), 2649–2656. doi:10.1128/AEM.67.6.2649-2656.2001 PMID:11375176

Rosegrant, M. M. (2009). Changement climatique et sécurité alimentaire au Maroc. Rencontre internationale « Le changement climatique: enjeux et perspectives d'adaptation pour le Maroc », Institut Royal des Etudes Stratégiques (IRES), Rabat, Maroc.

Rosegrant, M. W., & Cline, S. A. (2003). Global Food Security: Challenges and Policies. *Science, 302*(5652), 1917-1919. Retrieved from http://science.sciencemag.org/content/302/5652/1917.abstract

Rosenzweig, C., Casassa, G., Karoly, D. J., Imeson, A., Liu, C., Menzel, A., . . . Hanson, C. E. (2007). Assessment of observed changes and responses in natural and managed systems. In Climate Change 2007: Impacts, Adaptation and Vulnerability. Contribution of Working Group II to the Fourth Assessment Report of the Intergovernmental Panel on Climate Change. Cambridge University Press.

Rothballer, M., Eckert, B., Schmid, M., Fekete, A., Schloter, M., Lehner, A., ... Hartmann, A. (2008). Endophytic root colonization of gramineous plants by Herbaspirillumfrisingense. *FEMS Microbiology Ecology*, *66*(1), 85–95. doi:10.1111/j.1574-6941.2008.00582.x PMID:18761671

Roupsard, P. (2013). *Etude Phénoménologique Du dépôt Sec d'aérosols En Milieu Urbain* (Ph.D Thesis). University of Rouen.

Roupsard, P., Amielh, M., Maro, D., Coppalle, A., Branger, H., Connan, O., ... Talbaut, M. (2013). Measurement in a wind tunnel of dry deposition velocities of submicron aerosol with associated turbulence onto rough and smooth urban surfaces. *Journal of Aerosol Science*, *55*, 12–24. doi:10.1016/j.jaerosci.2012.07.006

Rowell, D. P. (2005). A scenario of European climate change for the late twenty-first century: Seasonal means and interannual variability. *Climate Dynamics*, *25*(7-8), 837–849. doi:10.100700382-005-0068-6

Royaume du Maroc. (2010). *Seconde communication nationale à la Convention-Cadre des Nations Unies sur les Changements Climatiques*. Rabat: Ministère de l'Environnement.

Roy, P. S. (1993). *Remote sensing for forest ecosystem analysis and management*. New Delhi, India: Environmental Studies in India. IBH.

Rubin, J. I., Kean, A. J., Harley, R. A., Millet, D. B., & Goldstein, A. H. (2006). Temperature dependence of volatile organic compound evaporative emissions from motor vehicles. *Journal of Geophysical Research*, *111*(D3), D03305. doi:10.1029/2005JD006458

Ruddell, B. L., Zaslavsky, I., Valentine, D., Beran, B., Piasecki, M., Fu, Q., & Kumar, P. (2014). Sustainable long term scientific data publication: Lessons learned from a prototype Observatory Information System for the Illinois River Basin. *Environmental Modelling & Software*, *54*, 73–87. doi:10.1016/j.envsoft.2013.12.015

Ryan, R. P., Germaine, K., Franks, A., Ryan, D. J., & Dowling, D. N. (2008). Bacterial endophytes: Recent developments and applications. *FEMS Microbiology Letters*, *278*(1), 1–9. doi:10.1111/j.1574-6968.2007.00918.x PMID:18034833

Saaidi, M. (1992). Comportement au champ de 32 cultivars de palmier dattier vis-à-vis du bayoud: 25 années d'observations. *Agronomie*, *12*(5), 359–370. doi:10.1051/agro:19920502

Sabine, C. L. (2004). The oceanic sink for anthropogenic CO_2. *Science*, *305*(5682), 367–371. doi:10.1126cience.1097403 PMID:15256665

SAC. (2016). Desertification and Land Degradation Atlas of India (Based on IRS AWiFS data of 2011-13 and 2003-05). Space Applications Centre, ISRO.

Saebo, A., Hanslin, H. M., Baraldi, R., Rapparini, F., Gawronska, H., & Gawronski, S. W. (2013). Characterization of urban trees and shrubs for particulate deposition, carbon sequestration and BVOC emissions. *Proceedings of International Symposium on Woody Ornamentals of the Temperate Zone*, 509-517. 10.17660/ActaHortic.2013.990.66

Saf'janova, V. M. (1985). Leishmaniasis in USSR. In *Leishmaniasis* (pp. 427–435). Amsterdam: Elsevier.

Safriel, U., & Adeel, Z. (2008). Development paths of drylands: Thresholds and sustainability. *Sustainability Science*, *3*(1), 117–123. doi:10.100711625-007-0038-5

Salamin, N., Wüest, R. O., Lavergne, S., Thuiller, W., & Pearman, P. B. (2010). Assessing rapid evolution in a changing environment. *Trends in Ecology & Evolution, 25*(12), 692–698. doi:10.1016/j.tree.2010.09.009 PMID:20961648

Salam, M. T., Millstein, J., Li, Y. F., Lurmann, F. W., Margolis, H. G., & Gilliland, F. D. (2005). Birth outcomes and prenatal exposure to ozone, carbon monoxide, and particulate matter: Results from the Children's Health Study. *Environmental Health Perspectives, 113*(11), 1638–1644. doi:10.1289/ehp.8111 PMID:16263524

Sandercock, D. A., Hunter, R. R., Nute, G. R., Mitchell, M. A., & Hocking, P. M. (2018, April). Acute Heat Stress-Induced Alterations in Blood Acid-Base Status and\r Skeletal Muscle Membrane Integrity in Broiler Chickens at Two Ages:\r Implications for Meat Quality. *Environmental Toxicology and Chemistry,* 418–425.

Sandmann, E. R. I. C., & Loos, M. A. (1984). Enumeration of 2, 4-D-degrading microorganisms in soils and crop plant rhizospheres using indicator media; high populations associated with sugarcane (*Saccharum officinarum). Chemosphere, 13*(9), 1073–1084. doi:10.1016/0045-6535(84)90066-3

Sapkota, A., Chelikowski, A. P., Nachman, K. E., Cohen, A. J., & Ritz, B. (2010). Exposure to particulate matter and adverse birth outcomes: A comprehensive review and meta analysis. *Air Quality, Atmosphere & Health, 5*(4), 359–381.

Sari, B., Limoncu, M. E., Balcioglu, I. C., Aldemir, A., Tasci, G. T., Kiliç, Y., ... Olgen, M. K. (2015). Seroepidemiological and entomological survey in a new focus of zoonotic visceral leishmaniasis in Kars province, Northeastern Turkey. *Veterinary Parasitology, 209*(3), 179–187. doi:10.1016/j.vetpar.2015.02.021 PMID:25769470

Sarmiento, J. L., Slater, R., Barber, R., Bopp, L., Doney, S. C., Hirst, A. C., ... Stouffer, R. (2004). Response of ocean ecosystems to climate warming. *Global Biogeochemical Cycles, 18*(3), n/a. doi:10.1029/2003GB002134

Sathaye, J., Shukla, P. R., & Ravindranath, N. H. (2006). Climate change, Sustainable development, and India: Global and National Concerns. *Current Science, 19*(3), 314–325.

SBSTTA. (1999). Biological diversity of drylands, arid, semi-arid, savannah, grassland and mediterranean ecosystems. Draft recommendations to COP5. Montreal, Canada: Author.

Scherer, M., & Unshelm, J. (1995). Untersuchungen zur Belastung von Mastschweinen sowie der Umgebung von Mastschweinstallen durch atembaren Feinstaub, stallspezifische Bakterien und Ammoniak. *Zentralblatt fur Hygiene und Umweltmedizin, 196*, 399–415. PMID:7727021

Schernthanner, H. (2007). Application of the fuzzy logic method for landslide susceptibility mapping, as feasible-ready to use approach, in the context of development, Rio Blanco, Nicaragua. In P. Zeil & S. Kienberger (Eds.), *Geoinformation for development: Bridging the divide through partnerships* (pp. 114–122). Heidelberg, Germany: Hüthig GmbH & Co. KG.

Schilling, J., Freier, K. P., Hertig, E., & Scheffran, J. (2012). Climate change, vulnerability and adaptation in North Africa with focus on Morocco. *Agriculture, Ecosystems & Environment, 156*, 12–26. doi:10.1016/j.agee.2012.04.021

Schimmelpfennig, D. (1996). Uncertainty in economic models of climate change impacts. *Climatic Change, 33*(2), 213–234. doi:10.1007/BF00140247

Schlesinger, R. B. (1979). Natural removal mechanisms for chemical pollutants in the environment. *Bioscience, 29*(2), 95–101. doi:10.2307/1307745

Schleyer, C., Görg, C., Hauck, J., & Winkler, K. J. (2015). Opportunities and challenges for mainstreaming the ecosystem services concept in the multi-level policy-making within the EU. *Ecosystem Services, 16*, 174–181. doi:10.1016/j.ecoser.2015.10.014

Schloss, C. A., Nuñez, T. A., & Lawler, J. J. (2012). Dispersal will limit ability of mammals to track climate change in the Western Hemisphere. *Proceedings of the National Academy of Sciences of the United States of America, 109*(22), 8606–8611. doi:10.1073/pnas.1116791109 PMID:22586104

Schlyter, P., Stjernquist, I., Bärring, L., Jönsson, A. M., & Nilsson, C. (2006). Assessment of the impacts of climate change and weather extremes on boreal forests in northern Europe, focusing on Norway spruce. *Climate Research, 31*(1), 75–84. doi:10.3354/cr031075. doi:10.3354/cr031075

Schmidhuber, J., & Tubiello, F. N. (2007). Global food security under climate change. *Proceedings of the National Academy of Sciences, 104*(50), 19703-19708. Retrieved from http://www.pnas.org/content/104/50/19703.abstract

Schmidt, M. (2003). *Development of a fuzzy expert system for detailed land cover mapping in the Draa catchment (Morocco) using high resolution satellite images.* Retrieved from https://www.researchgate.net/profile/Michael_Schmidt15/publication/34686505_Development_of_a_fuzzy_expert_system_for_detailed_land_cover_mapping_in_the_Dra_catchment_Morocco_using_high_resolution_satellite_images/links/5779f71908aeb9427e2c03cb/Development-of-a-fuzzy-expert-system-for-detailed-land-cover-mapping-in-the-Dra-catchment-Morocco-using-high-resolution-satellite-images.pdf

Schnoor, J. L., Aitchison, E. W., Kelley, S. L., & Alvarez, P. J. J. (1998). Phytoremediation and bioaugmentation of 1, 4-dioxane. In *Abstr. Adv. Biol. Syst. Rem. Conf* (pp. 87-91). Academic Press.

Schroder, M., Linkert, K. H., Pedersen, S., Takai, H., Johnsen, J. O., Groot Koerkamp, P. W. G., ... Linke, S. (1998). Emissions of Aerial Pollutants in Livestock Buildings in Northern Europe: Overview of a Multinational Project. *Journal of Agricultural Engineering Research, 70*(1), 3–9. doi:10.1006/jaer.1998.0278

Schucht, S., Colette, A., Rao, S., Holland, M., Schöpp, W., Kolp, P., ... Brignon, J. M. (2015). Moving towards ambitious climate policies: Monetised health benefits from improved air quality could offset mitigation costs in Europe. *Environmental Science & Policy, 50*, 252–269. doi:10.1016/j.envsci.2015.03.001

Schulz, O. (2006) *Analyse schneehydrologischer Prozesse und Schneekartierung im Einzugsgebiet des Oued M'Goun, Zentraler Hoher Atlas (Marokko)* (Dissertation). Mathematisch-Naturwissenschaftliche Fakultät, Universität Bonn.

Schwartz, N., Gätke, P., & Baran, E. (2016). Adaptation to environmental change among fishing-dependent households in Cambodia. Phnom Penh, Cambodia: Inland Fisheries Research and Development Institute (Fisheries Administration) and WorldFish.

Schwartz, J. (1991). Particulate Air Pollution and Daily Mortality in Detroit. *Environmental Research, 56*(2), 204–213. doi:10.1016/S0013-9351(05)80009-X PMID:1769365

Schwartz, J. (1993). Air Pollution and Daily Mortality in Birmingham, Alabama. *American Journal of Epidemiology, 137*(10), 1136–1147. doi:10.1093/oxfordjournals.aje.a116617 PMID:8317443

Schwartz, J., & Dockery, D. W. (1992). Increased Mortality in Philadelphia Associated with Daily Air Pollution Concentrations. *The American Review of Respiratory Disease, 145*(3), 600–604. doi:10.1164/ajrccm/145.3.600 PMID:1546841

Schwartz, J., & Marcus, A. (1990). Mortality and Air Pollution in London: A Time Series Analysis. *American Journal of Epidemiology, 131*(1), 185–194. doi:10.1093/oxfordjournals.aje.a115473 PMID:2403468

Schwartz, J., Slater, D., Larson, T., Pierson, W., & Koenig, J. (1993). Particulate air pollution and hospital emergency room visits for asthma in Seattle. *The American Review of Respiratory Disease, 174*(4), 826–831. doi:10.1164/ajrccm/147.4.826 PMID:8466116

Schwinning, S., Sala, O. E., Loik, M. E., & Ehleringer, J. R. (2004). Thresholds, memory, and seasonality: Understanding pulse dynamics in arid/semi-arid ecosystems. *Oecologia, 141*(2), 191–193. doi:10.100700442-004-1683-3 PMID:15300489

Scott, D. (2006). Climate change and sustainable tourism in the 21st century. In Tourism Research: Policy, Planning, and Prospects (pp.175-248). Waterloo: Série Publications du Département de géographie, Université de Waterloo.

Seager, R., Naik, N., & Vecchi, G. A. (2010). Thermodynamic and Dynamic Mechanisms for Large-Scale Changes in the Hydrological Cycle in Response to Global Warming. *Journal of Climate*, *23*(17), 4651–4668. doi:10.1175/2010JCLI3655.1

SECE. (1997). *Etude Nationale sur la biodiversité, Rapport National sur la biodiversité*. Présenté à la quatrième conférence des parties contractantes à la Convention sur la biodiversité.

Secretarial Department. (2009). *Ministry of Agriculture*. Retrieved from http://www.eauxetforets.gov.ma/

Secretariat, C. B. D. (2010). *Global biodiversity outlook 3*. In Secretariat of the Convention on Biological Diversity.

Sedra, M. H. (2012). Bayoud et charançon rouge: Les ennemis les plus dangereux du palmier dattier. *Agriculture du Maghreb*, *62*, 62–65.

SEI. (2005). *WEAP water evaluation and planning system, Tutorial*. Stockholm Environmental Institute, Boston Center, Tellus Institute.

Seibert, K., Fuehr, F., & Cheng, H. H. (1981). *Theory and Practical Use of Soil-Applied Herbicides Symposium*. European Weed Resource Society.

Sekercioglu, C. H., Schneider, S. H., Fay, J. P., & Loarie, S. R. (2008). Climate change, elevational range shifts, and bird extinctions. *Conservation Biology*, *22*(1), 140–150. doi:10.1111/j.1523-1739.2007.00852.x PMID:18254859

Sellers, P. J., Tucker, C. J., Collatz, G. J., Los, S. O., Justice, C. O., Dazlich, D. A., & Randall, D. A. (1996). A revised land surface parameterization (SiB2) for atmospheric GCMs. Part II: The generation of global fields of terrestrial biophysical parameters from satellite data. *Journal of Climate*, *9*(4), 706–737. doi:10.1175/1520-0442(1996)009<0706:ARLSPF>2.0.CO;2

Sewagudde, S. M. (2009). Lake Victoria's water budget and the potential effects of climate change in the 21st century. *African Journal of Tropical Hydrobiology and Fisheries*, *12*, 22–30.

Sgro, C. M., Lowe, A. J., & Hoffmann, A. A. (2011). Building evolutionary resilience for conserving biodiversity under climate change. *Evolutionary Applications*, *4*(2), 326–337. doi:10.1111/j.1752-4571.2010.00157.x PMID:25567976

Sharma, H. K., Dandotiya, B., & Jadon, N. (2017). Exposure of Air Pollution and Its Health Effects in Traffic Police Persons of Gwalior City, India. *Environmental Claims Journal*, *29*(4), 305–315. doi:10.1080/10406026.2017.1390357

Sharma, R. K., & Agrawal, M. (2005). Biological effects of heavy metals: An overview. *Journal of Environmental Biology*, *26*, 301. PMID:16334259

Shentu, X., Zhan, X., Ma, Z., Yu, X., & Zhang, C. (2014). Antifungal activity of metabolites of the endophytic fungus *Trichoderma brevicompactum* from garlic. *Brazilian Journal of Microbiology*, *45*(1), 248–254. doi:10.1590/S1517-83822014005000036 PMID:24948941

Shibata, H. (2016). Impact of winter climate change on nitrogen biogeochemistry in forest ecosystems: A synthesis from Japanese case studies. *Ecological Indicators*, *65*, 4–9. doi:10.1016/j.ecolind.2015.10.063

Shine, K. P., Stevenson, D. S., Sudo, K., Szopa, S., & Zeng, G. (2006). Radiative forcing since preindustrial times due to ozone change in the troposphere and the lower stratosphere. *Atmospheric Chemistry and Physics*, *6*(3), 575–599. doi:10.5194/acp-6-575-2006

Sicard, P., Lesne, O., Alexandre, N., Mangin, A., & Collomp, R. (2011). Air quality trends and potential health effects-development of an aggregate risk index. *Atmospheric Environment*, *45*(5), 1145–1153. doi:10.1016/j.atmosenv.2010.12.052

Sicard, P., Mangin, A., Hebel, P., & Malléa, P. (2010). Detection and estimation trends linked to air quality and mortality on French Riviera over the 1990-2005 period. *The Science of the Total Environment*, *408*(8), 1943–1950. doi:10.1016/j.scitotenv.2010.01.024 PMID:20138648

Sicard, P., Serra, R., & Rossello, P. (2016). Spatiotemporal trends of surface ozone concentrations and metrics in France. *Environmental Research*, *149*, 122–144. doi:10.1016/j.envres.2016.05.014 PMID:27200478

Siciliano, S., Fortin, N., & Himoc, N. (2001). Selection of specific endophytic bacterial genotypes by plants in response to soil contamination. *Applied and Environmental Microbiology*, *67*(6), 2469–2475. doi:10.1128/AEM.67.6.2469-2475.2001 PMID:11375152

SIDA. (2015). *Gender Amalysis – Principles and Elements. Tool Box 1 (March 2015)*. SIDA.

Sigman, D. M., & Boyle, E. A. (2000). Glacial/interglacial variations in atmospheric carbon dioxide. *Nature*, *407*(6806), 859–869. doi:10.1038/35038000 PubMed. doi:10.1038/35038000 PMID:11057657

Sillman, S., & Samson, P. J. (1995). Impact of temperature on oxidant photochemistry in urban, polluted rural and remote environments. *Journal of Geophysical Research, D, Atmospheres*, *100*(D6), 11497–11508. doi:10.1029/94JD02146

Silva, R. A., West, J. J., Zhang, Y., Anenberg, S. C., Lamarque, J. F., Shindell, D. T., & Horowitz, L. W. (2013). Global premature mortality due to anthropogenic outdoor air pollution and the contribution of past climate change. *Environmental Research Letters*, *8*(3), 034005. doi:10.1088/1748-9326/8/3/034005

Simonneaux, V., Cheggour, A., Deschamps, C., Mouillot, F., Cerdan, O., & Le Bissonnais, Y. (2015). Land use and climate change effects on soil erosion in a semi-arid mountainous watershed (High Atlas, Morocco). *Journal of Arid Environments*, *122*, 64–75. doi:10.1016/j.jaridenv.2015.06.002

Sinervo, B., Mendez-De-La-Cruz, F., Miles, D. B., Heulin, B., Bastiaans, E., Villagrán-Santa Cruz, M., ... Gadsden, H. (2010). Erosion of lizard diversity by climate change and altered thermal niches. *Science*, *328*(5980), 894–899. doi:10.1126cience.1184695 PMID:20466932

Sitch, S., Cox, P. M., Collins, W. J., & Huntingford, C. (2007). Indirect radiative forcing of climatechange through ozone effects on the land-carbon sink. *Nature*, *448*(7155), 791–794. doi:10.1038/nature06059 PMID:17653194

Slama, R., Darrow, L., Parker, J., Woodruff, T. J., Strickland, M., Nieuwenhuijsen, M., ... Ritz, B. (2008). Meeting report: Atmospheric pollution and human reproduction. *Environmental Health Perspectives*, *116*(6), 791–798. doi:10.1289/ehp.11074 PMID:18560536

Smil, V. (2002). Nitrogen and food production: Proteins for human diets. *Ambio*, *31*(2), 126–131. doi:10.1579/0044-7447-31.2.126 PMID:12078001

Smith, B., & Wandel, J. (2006). Adaptation, adaptive capacity and vulnerability. *Global Environmental Change*, *16*(3), 282–292. doi:10.1016/j.gloenvcha.2006.03.008

Smith, N. (2010). *Uneven development: Nature, capital, and the production of space*. University of Georgia Press.

Snyder, M. A., Sloan, L. C., Diffenbaugh, N. S., & Bell, J. L. (2003). Future climate change and upwelling in the California Current. *Geophysical Research Letters*, *30*(15), 1–4. doi:10.1029/2003GL017647

Soden, B. J., & Held, I. M. (2006). An assessment of climate feedbacks in coupled ocean-atmosphere models. *Journal of Climate*, *19*(14), 3354–3360. doi:10.1175/JCLI3799.1

Soleimani, M., Afyuni, M., Hajabbasi, M. A., Nourbakhsh, F., Sabzalian, M. R., & Christensen, J. H. (2010). Phytoremediation ofan aged petro-leum contaminated soil using endophyt infected and non-infected grasses. *Chemosphere, 81*(9), 1084–1090. doi:10.1016/j.chemosphere.2010.09.034 PMID:20961596

Solomon, S., Qin, D., Manning, M., Chen, Z., Marquis, M., Averyt, K. B., & Miller, H. L. (2007). IPCC, 2007: Climate change 2007: The physical science basis. Contribution of Working Group I to the fourth assessment report of the Intergovernmental Panel on Climate Change.

Solomon, S., Qin, D., Manning, M., Marquis, M., Averyt, K., & Tignor, M. l. (2007). Technical summary. In *Climate change. The physical science basis* (pp. 19–940). New York, NY: Cambridge University Press.

Soltani, Z., Fakoorziba, M. R., Moemenbellah-Fard, M. D., Kalantari, M., Akbarpour, M., Faramarzi, H., & Azizi, K. (2015). Phlebotomus papatasi (Diptera: Psychodidae) as the Vector of Leishmania Major in Kharameh District, Southern Iran. *Journal of Health Sciences and Surveillance System, 3*(4), 160-164.

Soltwedel, T., Bauerfeind, E., Bergmann, M., Bracher, A., Budaeva, N., Busch, K., ... Jacob, M. (2016). Natural variability or anthropogenically-induced variation? Insights from 15 years of multidisciplinary observations at the arctic marine LTER site HAUSGARTEN. *Ecological Indicators, 65*, 89–102. doi:10.1016/j.ecolind.2015.10.001

Sopac. (n.d.). *Index.* Retrieved from http://www.sopac.org/index

Souza, R., Ambrosini, A., Luciane, M. P., & Passaglia, T. (2015). Plant growth-promoting bacteria as inoculants in agricultural soils. *Genetics and Molecular Biology, 38*(4), 401–419. doi:10.1590/S1415-475738420150053 PMID:26537605

Sowers, J., Vengosh, A., & Weinthal, E. (2010). Climate change, water resources, and the politics of adaptation in the Middle East and North Africa. *Climatic Change, 104*(3–4), 599–627. doi:10.100710584-010-9835-4

Spaepen, S., Vanderleyden, J., & Remans, R. (2007). Indole-3-acetic acid in microbial and microorganismplant signaling. *FEMS Microbiology Reviews, 31*(4), 425–448. doi:10.1111/j.1574-6976.2007.00072.x PMID:17509086

Speth, P., Christoph, M., & Diekkrüger, B. (Eds.). (2010). *Impacts of global change on the hydrological cycle in West and Northwest Africa.* Springer Science & Business Media. doi:10.1007/978-3-642-12957-5

Sprott, P. (1998). *The International Long-Term Ecological Research Network.* US LTER Network Office Albuquerque New Mexico.

Sraïri, M. T., M'ghar, F. A., Benidir, M., & Bengoumi, M. (2017). Analyse typologique de la diversité et des performances de l'élevage oasien. *Cahiers Agricultures, 26*(1), 15005. doi:10.1051/cagri/2017002

Sram, R. J. (1999). Impact of air pollution on reproductive health. *Environmental Health Perspectives, 107*(11), 542–543. doi:10.1289/ehp.99107a542 PMID:10627171

Sriprang, R., Hayashi, M., Yamashita, M., Ono, H., Saeki, K., & Murooka, Y. (2002). A novel bioremediation system for heavy metals using the symbiosis between leguminous plant and genetically engineered rhizobia. *Journal of Biotechnology, 99*(3), 279–293. doi:10.1016/S0168-1656(02)00219-5 PMID:12385715

Stainforth, D. A., Aina, T., Christensen, C., Collins, M., Faull, N., Frame, D. J., ... Allen, M. R. (2005). Uncertainty in predictions of the climate response to rising levels of greenhouse gases. *Nature, 433*(7024), 403–406. doi:10.1038/nature03301 PMID:15674288

Staudinger, M. D., Grimm, N.B., Staud, T. A., Carter, S.L., & Chapin, F.S. (2012). Impacts of climate change on biodiversity, ecosystems, and ecosystem services. United States Global Change Research Program, Washington, DC.

Staudinger, M. D., Grimm, N. B., Staudt, A., Carter, S. L., & Chapin, F. S. (2012). *Impacts of climate change on biodiversity, ecosystems, and ecosystem services*. Washington, DC: United States Global Change Research Program.

Steffens, W., & Wirth, M. (2005). Freshwater fish - an important source of n-3 polyunsaturated fatty acids: a review. *Archives of Polish Fisheries/Archiwum Rybactwa Polskiego, 13*(1), 5–16. Retrieved from http://search.proquest.com/docview/17136587?accountid=14643%5Cnhttp://mlbsfx.sibi.usp.br:3410/sfxlcl41?url_ver=Z39.88-2004&rft_val_fmt=info:ofi/fmt:kev:mtx:journal&genre=article&sid=ProQ:ProQ:asfaaquaculture&atitle=Freshwater+fish+-+an+important+source+of

Steinfeld, H., De Haan, C., & Blackburn, H. (1997). *Livestock–Environment Interactions: Issues and options*. Brussels: European Commission Directorate-General for Development.

Stępniewska, Z., & Kuźniar, A. (2013). Endophytic microorganisms—Promising applications in bioremediation of greenhouse gases. *Applied Microbiology and Biotechnology*, *97*(22), 9589–9596. doi:10.100700253-013-5235-9 PMID:24048641

Stone, J. K., Bacon, C. W., & White, J. F. (2000). An overview of endophytic microbes: Endophytism defined. In Microbial endophytes. New York: Dekker.

Stotz, O. (2018). The perception of homeownership utility: Short-term and long-term effects. *Journal of Housing Economics*. doi:10.1016/j.jhe.2018.11.003

St-Pierre, N. R., Cobanov, B., & Schnitkey, G. (2003). Economic Losses from Heat Stress by US Livestock Industries[1]. *Journal of Dairy Science*, *86*, E52–E77. doi:10.3168/jds.S0022-0302(03)74040-5

Strobel, G. A., & Long, D. M. (1998). Endophytic microbes embody pharmaceutical potential. *ASM News*, *64*, 263–268.

Stromberg, J. C., Tiller, R., & Richter, B. (1996). Effects of Groundwater Decline on Riparian Vegetation of Semiarid Regions: The San Pedro, Arizona. *Ecological Applications*, *6*(1), 113–131. doi:10.2307/2269558

Sturz, A., & Kimpinski, J. (2004). Endo root bacteria derived from marigolds (*Tagetes spp.*) can decrease soil population densities of root-lesion nematodes in the potato root zone. *Plant and Soil*, *262*(1/2), 241–249. doi:10.1023/B:PLSO.0000037046.86670.a3

Suárez, A., Watson, R. T., & Dokken, D. J. (2002). *Climate change and biodiversity*. Academic Press.

Sulman, B., Desai, A., & Mladenoff, D. (2013). Modeling soil and biomass carbon responses to declining water table in a wetland-rich landscape. *Ecosystems (New York, N.Y.)*, *16*(3), 491–507. doi:10.100710021-012-9624-1

Sumaila, U. R., Cheung, W. W. L., Lam, V. W. Y., Pauly, D., & Herrick, S. (2011). Climate change impacts on the biophysics and economics of world fisheries. *Nature Climate Change*, *1*(9), 449–456. doi:10.1038/nclimate1301

Sun, G., McLaughlin, S. B., Porter, J. H., Uddling, J., Mulholland, P. J., Adams, M. B., & Pederson, N. (2012). Interactive influences of ozone and climate on streamflow of forested watersheds. *Global Change Biology*, *18*(11), 3395–3409. doi:10.1111/j.1365-2486.2012.02787.x

Suttle, K. B., Thomsen, M. A., & Power, M. E. (2007). Species interactions reverse grassland responses to changing climate. *Science*, *315*(5812), 640–642. doi:10.1126cience.1136401 PMID:17272720

Swanson, D. L., & Palmer, J. S. (2009). Spring migration phenology of birds in the Northern Prairie region is correlated with local climate change. *Journal of Field Ornithology*, *80*(4), 351–363. doi:10.1111/j.1557-9263.2009.00241.x

Swarnalatha, Y., Saha, B., & Choudary, L. (2015). Bioactive compound analysis and antioxidant activity of endophytic bacterial extract from *Adhatho dabeddomei*. *Asian Journal of Pharmaceutical and Clinical Research*, *8*(1), 70–72.

Syroka, J., & Toumi, R. (2004). On the withdrawal of the Indian summer monsoon. *Quarterly Journal of the Royal Meteorological Society, 130*(598), 989-1008.

Tabet-Aoul, M. (2008). Impacts du changement climatique sur les agricultures et les ressources hydriques au Maghreb. *Les Notes d'Alerte du CIHEAM*, (48).

Tabet-Aoul, M. (2008). Impacts du changement climatique sur les agricultures et les ressources hydriques au Maghreb. *Les notes d'alerte du CIHEAM.*

Tager, I. B., Balmes, J., Lurmann, F., Ngo, L., Alcorn, S., & Kunzli, N. (2005). Chronic exposure to ambient ozone and lung function in young adults. *Epidemiology (Cambridge, Mass.), 16*, 751. doi:10.1097/01.ede.0000183166.68809.b0 PMID:16222164

Tai, A. P., Martin, M. V., & Heald, C. L. (2014). Threat to future global food security from climate change and ozone air pollution. *Nature Climate Change, 4*(9), 817–821. doi:10.1038/nclimate2317

Talmoudi, K., Bellali, H., Ben-Alaya, N., Saez, M., Malouche, D., & Chahed, M. K. (2017). Modeling zoonotic cutaneous leishmaniasis incidence in central Tunisia from 2009-2015: Forecasting models using climate variables as predictors. *PLoS Neglected Tropical Diseases, 11*(8), e0005844. doi:10.1371/journal.pntd.0005844 PMID:28841642

Tarallo, V. D., Dantas-Torres, F., Lia, R. P., & Otranto, D. (2010). Phlebotomine sand fly population dynamics in a leishmaniasis endemic peri-urban area in southern Italy. *Acta Tropica, 116*(3), 227–234. doi:10.1016/j.actatropica.2010.08.013 PMID:20816927

Tarrier, M., & Benzyane, M. (2003). *Moroccan argan tree is dying: problematic and bio-indication.* Retrieved from http://www.jle.com/fr/revues/sec/e-docs/larganeraie_marocaine_se_meurt_problematique_et_bioindication__262867/article.phtml?cle_doc=000402D3

Taylor, S. G. (2008). Climate warming causes phenological shift in Pink Salmon, Oncorhynchus gorbuscha, behavior at Auke Creek, Alaska. *Global Change Biology, 14*(2), 229–235. doi:10.1111/j.1365-2486.2007.01494.x

Tedersoo, L., Pärtel, K., Jairus, T., Gates, G., Põldmaa, K., & Tamm, H. (2009). Ascomycetes associated with ectomycorrhizas: Moleculardiversity and ecology with particularreference to the *Helotiales. Environmental Microbiology, 11*(12), 3166–3178. doi:10.1111/j.1462-2920.2009.02020.x PMID:19671076

Tedesco, M., Willis, I. C., Hoffman, M. J., Banwell, A. F., Alexander, P. & Arnold, N. S. (2013). *Ice dynamic response to two modes of surface lake drainage on the Greenland ice sheet.* doi:10.1088/1748-9326/8/3/034007

TEEB Foundations, . (2010). *The Economics of Ecosystems and Biodiversity: Ecological and Economic Foundations.* London: Earthscan.

Teh, L. C. L., & Sumaila, U. R. (2013). Contribution of marine fisheries to worldwide employment. *Fish and Fisheries, 14*(1), 77–88. doi:10.1111/j.1467-2979.2011.00450.x

Teoh, S. H. S., Symes, W. S., Sun, H., Pienkowski, T., & Carrasco, L. R. (2019). A global meta-analysis of the economic values of provisioning and cultural ecosystem services. *The Science of the Total Environment, 649*, 1293–1298. doi:10.1016/j.scitotenv.2018.08.422 PMID:30308899

The United Nations Statistics Division (UNSD). (2018). *Global Gender Statistics Programme.* New York: UN. Available at: http://unstats.un.org/unsd/genderstatmanual/Glossary.ashx

Thiering, E., & Heinrich, J. (2015). Epidemiology of air pollution and diabetes. *Trends in Endocrinology and Metabolism, 26*(7), 384–394. doi:10.1016/j.tem.2015.05.002 PMID:26068457

Thomas, C. D., Cameron, A., Green, R. E., Bakkenes, M., Beaumont, L. J., Collingham, Y. C., & Hughes, L. (2004). Extinction risk from climate change. *Nature*, *427*(6970), 145–148. doi:10.1038/nature02121 PMID:14712274

Thompson, J. N. (2005). *The geographic mosaic of coevolution*. University of Chicago Press.

Thornton, P. K., & Gerber, P. J. (2010). Climate change and the growth of the livestock sector in developing countries. *Mitigation and Adaptation Strategies for Global Change*, *15*(2), 169–184. doi:10.100711027-009-9210-9

Thron, R. W. (1996). Direct and indirect exposure to air pollution. *Otolaryngology - Head and Neck Surgery*, *114*(2), 281–285. doi:10.1016/S0194-5998(96)70184-5 PMID:8637751

Thuiller, W., Lavorel, S., Araújo, M. B., Sykes, M. T., & Prentice, I. C. (2005). Climate change threats to plant diversity in Europe. *Proceedings of the National Academy of Sciences of the United States of America*, *102*(23), 8245–8250. doi:10.1073/pnas.0409902102 PMID:15919825

Thurston, G. D. (2008). Outdoor air pollution: sources, atmospheric transport, and human health effects. In H. Kris (Ed.), *International Encyclopedia of Public Health* (pp. 700–712). Oxford, UK: Academic Press. doi:10.1016/B978-012373960-5.00275-6

Todd, B. D., Scott, D. E., Pechmann, J. H., & Gibbons, J. W. (2011). Climate change correlates with rapid delays and advancements in reproductive timing in an amphibian community. *Proceedings of the Royal Society of London B: Biological Sciences, 278*(1715), 2191-2197. 10.1098/rspb.2010.1768

Tol, R. S. J., Downing, T. E., Kulik, O. J., & Smith, J. B. (2009). Distributional aspects of climate change impacts. *Global Environmental Change*, *14*(3), 259–272. doi:10.1016/j.gloenvcha.2004.04.007

Toprak, S., & Ozer, N. (2007). Distribution of sand fly (Diptera: Psychodidae) species and efficiency of capturing methods in Sanliurfa Province, *Turkey. Journal of Medical Entomology*, *44*(1), 23–28. doi:10.1093/jmedent/41.5.23 PMID:17294917

Touchan, R., Anchukaitis, K. J., Meko, D. M., Sabir, M., Attalah, S., & Aloui, A. (2011). Spatiotemporal drought variability in northwestern Africa over the last nine centuries. *Climate Dynamics*, *37*(1-2), 237–252. doi:10.100700382-010-0804-4

Toutain, G. (1979). Eléments d'agronomie saharienne: de la recherche au développement. *INRA-GRET*, 276.

Trenberth, K. E., Jones, P. D., Ambenje, P., & Bojariu, R. (2007). Observations: surface and atmospheric climate change. In Climate Change 2007: The Physical Science Basis. Contribution of Working Group I to the Fourth Assessment Report of the Intergovernmental Panel on Climate Change. Cambridge, UK: Cambridge University Press.

Trenberth, K. E., Dai, A., van der Schrier, G., Jones, P. D., Barichivich, J., Briffa, K. R., & Sheffield, J. (2013). Global warming and changes in drought. *Nature Climate Change*, *4*(1), 17–22. doi:10.1038/nclimate2067

Tribak, A. (2002). The Environmental Impact Of Management Of Water On The irrigation On Forest Ecosystems: Case Of The argan tree. *H.T.E. Review*, 124.

Turner, M. C., Jerrett, M., Pope, C. A. III, Krewski, D., Gapstur, S. M., Diver, W. R., ... Burnett, R. T. (2016). Long-term ozone exposure and mortality in a large prospective study. *American Journal of Respiratory and Critical Care Medicine*, *193*(10), 1134–1142. doi:10.1164/rccm.201508-1633OC PMID:26680605

U.S. Geological Survey. (2009). *Volcanic gases and their effects*. Volcano Hazards Program.

UICN. (2005). *La France et la biodiversité. Enjeux et responsabilités.* Réalisation. Comité français UICN. Retrieved from www.karibu.fr

UNDP. (2005). *Uganda human development report 2005. Linking environment to development: A deliberate choice.* Kampala, Uganda: UNDP.

UNECA/ACPC. (2014). *Fourth conference on climate change and development in Africa summary and recommendations*. Marrakech, Morocco: UNECA/ACPC.

UNEP. (2012). *The Fifth Global Environmental Outlook Report*. LAND.

UNESCO. (2007). Études de cas sur le changement climatique et le patrimoine mondial. Paris, France: UNESCO, Centre du patrimoine mondial.

UNESCO. (2010). Changement climatique: Enjeux et perspectives au Maghreb 2010MO0615. UNESCO.

UNFCCC. (2008). *Report of the Conference of the Parties on its thirteenth session, held in Bali from 3 to 15 December 2007* (Document UNFCCC/CP/2007/6/Add1). UNFCCC.

United Nations Convention to Combat Desertification (UNCCD). (2007). *Desertification, exacerbated by climate change, represents one of the greatest environmental challenges of our times*. Retrieved from www.unccd.int/documents/Desertificationandclimatechange.pdf

Università, R. V, Cuore, S., Agriculture, F., Istituto, L. A., Nazionale, C., Nazionale, C., … Elena, V. R. (2012). *Scientific report submitted to EFSA*. Prepared by Battilani P ./ Università Cattolica del Sacro Cuore, Faculty of Agriculture.

US EPA. (2008). Integrated Science Assessment (ISA) for Oxides of Nitrogen and Sulfur— Environmental Criteria, final report. Washington, DC: Author.

Uysal, N., & Schapira, R. M. (2003). Effects of ozone on lung function and lung diseases. *Current Opinion in Pulmonary Medicine*, *9*, 144.

Vamvakas, S., Bittner, D., & Koster, U. (1993). Enhanced expression of the protooncogenes c-myc and c-fos in normal and malignant renal growth. *Toxicology Letters*, *67*(1-3), 161–172. doi:10.1016/0378-4274(93)90053-Z PMID:8451757

Van Dingenen, R., Dentener, F., Raes, F., Krol, M., Emberson, L., & Cofala, J. (2009). The global impact of ozone on agricultural crop yields under current and future air quality legislation. *Atmospheric Environment*, *43*(3), 604–618. doi:10.1016/j.atmosenv.2008.10.033

Van Grinsven, H. J. M., Holland, M., Jacobsen, B. H., Klimont, Z., Sutton, M. A., & Willems, W. J. (2013). Costs and benefits of nitrogen for Europe and implications for mitigation. *Environmental Science & Technology*, *47*(8), 3571–3579. doi:10.1021/es303804g PMID:23473305

VanBuskirk, K. M., O'Neill, M. T., De La Vega, P., Maier, A. G., Krzych, U., Williams, J., ... Kneteman, N. M. (2009). Preerythrocytic, live-attenuated Plasmodium falciparum vaccine candidates by design. *Proceedings of the National Academy of Sciences of the United States of America*, *106*(31), 13004–13009. doi:10.1073/pnas.0906387106 PMID:19625622

Váry, Z., Mullins, E., Mcelwain, J. C., & Doohan, F. M. (2015). The severity of wheat diseases increases when plants and pathogens are acclimatized to elevated carbon dioxide. *Global Change Biology*, *21*(7), 2661–2669. doi:10.1111/gcb.12899 PMID:25899718

Vaughan, M. M., Huffaker, A., Schmelz, E. A., Dafoe, N. J., Christensen, S., Sims, J., ... Teal, P. E. A. (2014). Effects of elevated [CO_2] on maize defence against mycotoxigenic *Fusarium verticillioides*. *Plant, Cell & Environment*, *37*(12), 2691–2706. doi:10.1111/pce.12337 PMID:24689748

Venners, S. A., Wang, B. Y., Peng, Z. G., Xu, Y., Wang, L. H., & Xu, X. P. (2003). Particulate matter, sulfur dioxide, and daily mortality in Chongqing, China. *Environmental Health Perspectives*, *111*(4), 562–567. doi:10.1289/ehp.5664 PMID:12676616

Verma, S., Varma, A., Rexer, K. H., Hassel, A., Kost, G., Sarbhoy, A., ... Franken, P. (1998). *Piriformo sporaindica*, gen. et sp. nov., a new root-colonizing fungus. *Mycologia*, *90*(5), 896–903. doi:10.1080/00275514.1998.12026983

Vermylen, J., Nemmar, A., Nemery, B., & Hoylaerts, M. F. (2005). Ambient air pollution and acute myocardial infarction. *Journal of Thrombosis and Haemostasis*, *3*(9), 1955–1961. doi:10.1111/j.1538-7836.2005.01471.x PMID:16102102

Veysi, A., Vatandoost, H., Yaghoobi-Ershadi, M. R., Arandian, M. H., Jafari, R., & Hosseini, M. (2012). Comparative study on the effectiveness of Coumavec® and zinc phosphide in controlling zoonotic cutaneous leishmaniasis in a hyperendemic focus in central Iran. *Journal of Arthropod-Borne Diseases*, *6*(1), 18. PMID:23293775

Vicuna, S. (2006). *Predictions of Climate Change Impacts on California Water Resources Using CALSIM II: A Technical Note*. Berkeley, CA: California Climate Change Center.

Vidal, M. (2005). Ajustement entre des systèmes irrigués et des systèmes de culture diversifiés. Fonctionnement de deux oasis du Tafilalet Mémoire de Master.

Vigil, K. M. (2003). *Clean water: an introduction to water quality and pollution control*. Corvallis, OR: Oregon State University Press.

Vingarzan, R. (2004). A review of surface ozone background levels and trends. *Atmospheric Environment*, *38*, 3431–3442.

Volf, P., Ozbel, Y., Akkafa, F., Svobodova, M., Votýpka, J., & Chang, K. P. (2002). Sand flies (Diptera: Phlebotominae) in Sanliurfa, Turkey: relationship of Phlebotomus sergenti with the epidemic of anthroponotic cutaneous leishmaniasis. *Journal of Medical Entomology*, *39*(1), 12–15. doi:10.1603/0022-2585-39.1.12 PMID:11931243

Vörösmarty, C. J., Green, P., Salisbury, J., & Lammers, R. B. (2000). Global Water Resources: Vulnerability from Climate Change and Population Growth. *Science, 289*(5477), 284-288. Retrieved from http://science.sciencemag.org/content/289/5477/284.abstract

Wafula, D. (2012). Where do policy-makers get information? Science Communication Training Workshop for Bio- Innovate Projects, Kigali, Rwanda.

Wagenaar, M. M., & Clardy, J. (2001). Dicerandrols, new antibiotics and cytotoxic dimmers produced by the fungus *Phomopsis longicolla* isolated from an endangered mint. *Journal of Natural Products*, *64*(8), 1006–1009. doi:10.1021/np010020u PMID:11520215

Waide, R. (1998). *The International Long-Term Ecological Research Network*. US LTER Network, National Science Foundation.

Walker, B. H., Holling, C. S., Carpenter, S. R., & Kinzig, A. (2004). Resilience, adaptability and transformability in social–ecological systems. *Ecology and Society*, *9*(2), 5. doi:10.5751/ES-00650-090205

Walther, G. R., Post, E., Convey, P., Menzel, A., Parmesan, C., Beebee, T. J., & Bairlein, F. (2002). Ecological responses to recent climate change. *Nature*, *416*(6879), 389–395. doi:10.1038/416389a PMID:11919621

Wang, D., Zhou, B., Fu, Q., Zhao, Q., Zhang, Q., Chen, J., ... Li, J. (2016). Intense secondary aerosol formation due to strong atmospheric photochemical reactions in summer: Observations at a rural site in eastern Yangtze River Delta of China. *The Science of the Total Environment*, *571*, 1454–1466. doi:10.1016/j.scitotenv.2016.06.212 PMID:27418517

Wang, H., Jacob, D. J., Le Sager, P., Streets, D. G., Park, R. J., Gilliland, A. B., & van Donkelaar, A. (2009). Surface ozone background in the United States: Canadian and Mexican influences. *Atmospheric Environment*, *43*(6), 1310–1319. doi:10.1016/j.atmosenv.2008.11.036

Wang, S., Feng, X., Zeng, X., Ma, Y., & Shang, K. (2009). A study on variations of concentrations of particulate matter with different sizes in Lanzhou, China. *Atmospheric Environment*, *43*(17), 2823–2828. doi:10.1016/j.atmosenv.2009.02.021

Wang, X., Ding, H., Ryan, L., & Xu, X. (1997). Association between air pollution and low birth weight: A community-based study. *Environmental Health Perspectives*, *105*(5), 514–520. doi:10.1289/ehp.97105514 PMID:9222137

Wang, Y. J., Miller, L., Perren, M., & Addis, P. B. (1990). Omega-3 Fatty Acids in Lake Superior Fish. *Journal of Food Science*, *55*(1), 71–73. doi:10.1111/j.1365-2621.1990.tb06018.x

Wanner, H., Brönnimann, S., Casty, C., Gyalistras, D., Luterbacher, J., Schmutz, C., ... Xoplaki, E. (2001). North Atlantic Oscillation–concepts and studies. *Surveys in Geophysics*, *22*(4), 321–381. doi:10.1023/A:1014217317898

Ward, M. N., Lamb, P. J., Portis, D. H., El Hamly, M., & Sebbari, R. (1999). Climate variability in Northern Africa: Understanding droughts in the Sahel and the Maghreb. In *Beyond el Niño* (pp. 119–140). Berlin: Springer. doi:10.1007/978-3-642-58369-8_6

WB/MAPM, FAO, INRA, & DMN. (2008). Workshop on WB/Morocco/FAO Climate Change Study. Rabat: Author.

Weber, B. (2004). *Untersuchungen zum Bodenwasserhaushalt und Modellierung der Bodenwasser-flüsse entlang eines HöhenundAriditätsgradienten (SE Marokko)* (Dissertation). Mathematisch-Naturwissenschaftliche Fakultät. Universität Bonn.

Wegmann, M., Fehrenbach, A., Heimann, S., Fehrenbach, H., Renz, H., Garn, H., & Herz, U. (2005). NO2-induced airway inflammation is associated with progressive airflow limitation and development of emphysema-like lesions in C57bl/6 mice. *Experimental and Toxicologic Pathology*, *56*(6), 341–350. doi:10.1016/j.etp.2004.12.004 PMID:15945273

Welcomme, R. L. (2011). An overview of global catch statistics for inland fish. *ICES Journal of Marine Science*, *68*(8), 1751–1756. doi:10.1093/icesjms/fsr035

Welcomme, R. L., Cowx, I. G., Coates, D., Bene, C., Funge-Smith, S., Halls, A., & Lorenzen, K. (2010). Inland capture fisheries. *Philosophical Transactions of the Royal Society of London. Series B, Biological Sciences*, *365*(1554), 2881–2896. doi:10.1098/rstb.2010.0168 PMID:20713391

Westoby, M. (1991). On long-term ecological research in Australia. In P. G. Risser (Ed.), *Long-Term Ecological Research* (pp. 191–209). New York: John Wiley and Sons.

Weuve, J., Puett, R., Schwartz, J., Yanosky, J., Laden, F., & Grodstein, F. (2012). Exposure to particulate air pollution and cognitive decline in older women. *Archives of Internal Medicine*, *172*(3), 219–227. doi:10.1001/archinternmed.2011.683 PMID:22332151

Weyens, N., Croes, S., Dupae, J., Newman, L., Van der Lelie, D., Carleer, R., & Vangronsveld, J. (2010). Endophytic bacteria improve phytoremediation of Ni and TCE co-contamination. *Environmental Pollution*, *158*(7), 2422–2427. doi:10.1016/j.envpol.2010.04.004 PMID:20462680

Weyens, N., Van der Lelie, D., Taghavi, S., Newman, L., & Vangronsveld, J. (2009). Exploiting plant-microbe partnerships to improve biomass production and remediation. *Trends in Biotechnology*, *27*(10), 591–598. doi:10.1016/j.tibtech.2009.07.006 PMID:19683353

Wheeler, S. M. (2008). State and municipal climate change plans: The first generation. *Journal of the American Planning Association*, *74*(4), 481–496. doi:10.1080/01944360802377973

White, N., Sutherst, R. W., Hall, N., & Whish-Wilson, P. (2003). The vulnerability of the Australian beef industry to impacts of the cattle tick (Boophilus microplus) under climate change. *Climatic Change*, *61*(1–2), 157–190. doi:10.1023/A:1026354712890

WHO Expert Committee on the Control of the Leishmaniases, Meeting, & World Health Organization. (2010). Control of the Leishmaniases: Report of a Meeting of the WHO Expert Committee on the Control of Leishmaniases, Geneva, 22-26 March 2010 (Vol. 949). World Health Organization.

WHO Regional Office for Europe (DK). (2013). *Review of evidence on health aspects of air pollution: REVIHAAP Project*. Copenhagen: WHO Regional Office for Europe.

WHO Regional Office for Europe (DK). (2014). *WHO Expert Meeting: methods and tools for asessing the health risks of air pollution at local, national and international level*. Meeting report. Bonn, Germany, 12-13 May 2014. Copenhagen: WHO Regional Office for Europe.

WHO. (1996). *Climate Change and Human Health*. Geneva: World Health Organization.

WHO. (2002). *World Health Report*. Geneva: World Health Organization.

WHO. (2005). *Air Quality Guideline Global Update 2005*. Copenhagen: Particulate Matter, Ozone, Nitrogen Dioxide and Sulphur Dioxide.

WHO. (2008). *Outdoor Air Pollution, Children's Health and the Environment*. World health organization.

WHO. (2012). *Mainstreaming gender equality in heallth adaptation to climate change programmes. User's guide*. Geneva, Switzerland: WHO.

WHO. (2013). *Press release N. 221*. International Agency for Research on Cancer, World Health Organization. Retrieved from http://www.iarc.fr/en/media- centre/pr/2013/ pdfs/pr221_E.pdf

WHO. (2014a). *Burden of disease from household air pollution for 2012*. Geneva, Switzerland: World Health Organization.

WHO. (2014b). *Media centre: 7 million premature deaths annually linked to air pollution*. Geneva: World Health Organization.

WHO. (2016a). *Global Health Observatory (GHO) data: mortality from ambient air pollution—situation and trend*. World Health Organization.

Wiebe, K. L., & Gerstmar, H. (2010). Influence of spring temperatures and individual traits on reproductive timing and success in a migratory woodpecker. *The Auk*, *127*(4), 917–925. doi:10.1525/auk.2010.10025

Wilhelm, M., & Ritz, B. (2003). Residential proximity to traffic and adverse birth outcomes in Los Angeles County, California, 1994–1996. *Environmental Health Perspectives*, *111*(2), 207–216. doi:10.1289/ehp.5688 PMID:12573907

Wilhelm, M., & Ritz, B. (2005). Local variations in CO and particulate air pollution and adverse birth outcomes in Los Angeles County, California, USA. *Environmental Health Perspectives*, *113*(9), 1212–1221. doi:10.1289/ehp.7751 PMID:16140630

Wilkinson, C., Saarne, T., Peterson, G. D., & Colding, J. (1998). Strategic Spatial Planning and the Ecosystem Services Concept: an Historical Exploration. *Ecology and Society*, *18*(1), art37. doi:10.5751/ES-05368-180137

Willems, A. (2007). The taxonomy of rhizobia: An overview. In *First International Meeting on Microbial Phosphate Solubilization*. Springer. 10.1007/978-1-4020-5765-6_1

Williams, B. K., Szaro, R. C., & Shapiro, C. D. (2009). *Adaptive management: the US Department of the Interior technical guide*. US Department of the Interior.

Williams, J. W., & Jackson, S. T. (2007). Novel climates, no-analog communities, and ecological surprises. *Frontiers in Ecology and the Environment*, *5*(9), 475–482. doi:10.1890/070037

Williams, S. E., Shoo, L. P., Isaac, J. L., Hoffmann, A. A., & Langham, G. (2008). Towards an integrated framework for assessing the vulnerability of species to climate change. *PLoS Biology*, *6*(12), e325. doi:10.1371/journal.pbio.0060325 PMID:19108608

Willis, C. G., Ruhfel, B., Primack, R. B., Miller-Rushing, A. J., & Davis, C. C. (2008). Phylogenetic patterns of species loss in Thoreau's woods are driven by climate change. *Proceedings of the National Academy of Sciences of the United States of America*, *105*(44), 17029–17033. doi:10.1073/pnas.0806446105 PMID:18955707

Willis, K. J., & Bhagwat, S. A. (2009). Biodiversity and climate change. *Science*, *326*(5954), 806–807. doi:10.1126cience.1178838 PMID:19892969

Wilson, E. O., & MacArthur, R. H. (1967). The theory of island biogeography. Princeton, NJ: Academic Press.

Wilson, D. (1995). Endophyte - the evolution of term, a classification of its use and defin-ition. *Oikos*, *73*(2), 274–276. doi:10.2307/3545919

Wilson, R., & Spengler, J. (1996). *Particles in our air*. Harvard University Press.

Wilson, S. J., Kirby, C. J., Koenigsfeld, A. T., Keisler, D. H., & Lucy, M. C. (1998). Effects of Controlled Heat Stress on Ovarian Function of Dairy Cattle. 2. Heifers. *Journal of Dairy Science*, *81*(8), 2132–2138. doi:10.3168/jds.S0022-0302(98)75789-3 PMID:9749377

Wisheu, I. C., Rosenzweig, M. L., Olsvig-Whittaker, L., & Shmida, A. (2000). What makes nutrient-poor Mediterranean heathlands so rich in plant diversity? *Evolutionary Ecology Research*, *2*(7), 935–955.

Wittig, V. E., Ainsworth, E. A., & Long, S. P. (2007). To what extent do current and projected increases in surface ozone affects photosynthesis and stomatal conductance in trees? A metaanalytic review of the last 3 decades of experiments. *Plant, Cell & Environment*, *30*(9), 1150–1162. doi:10.1111/j.1365-3040.2007.01717.x PMID:17661752

Wolfenson, D., Roth, Z., & Meidan, R. (2000). Impaired reproduction in heat-stressed cattle: basic and applied aspects. *Animal Reproduction Science, 60–61*, 535–547.

Wolfenson, D., Bachrach, D., Maman, M., Graber, Y., & Rozenboim, I. (2001). Evaporative cooling of ventral regions of the skin in heat-stressed laying hens. *Poultry Science*, *80*(7), 958–964. doi:10.1093/ps/80.7.958 PMID:11469662

Wolfenson, D., Thatcher, W. W., Badinga, L., Savio, J. D., Meidan, R., Lew, B. J., ... Berman, A. (1995). Effect of heat stress on follicular development during the estrous cycle in lactating dairy cattle. *Biology of Reproduction*, *52*(5), 1106–1113. doi:10.1095/biolreprod52.5.1106 PMID:7626710

Woodcock, J., Edwards, P., Tonne, C., Armstrong, B.G., Ashiru, O., Banister, D., & … Cohen, L. (2009). Public health benefits of strategies to reduce greenhouse-gas emissions: Urban land transport. *Lancet*, 374. PMID:19942277

Woodruff, T. J., Darrow, L. A., & Parker, J. D. (2008). Air pollution and postneonatal infant mortality in the United States, 1999–2002. *Environmental Health Perspectives*, *116*(1), 110–115. doi:10.1289/ehp.10370 PMID:18197308

World Bank. (2011). *Middle-East and Northern Africa water outlook*. Prepared by Future Water.

World Fish Center (WFS). (2009). *Climate Change: Research to Meet the Challenges Facing Fisheries and Aquaculture*. Issues Brief 1915 World Fish Center.

World Health Organization (WHO). (2014). *Food Security: Trade, foreign policy, diplomacy, Heal*. Retrieved from http://www.who.int/trade/glossary/story028/en/

Wu, J., Ren, C., Delfino, R. J., Chung, J., Wilhelm, M., & Ritz, B. (2009). Association between local traffic-generated air pollution and preeclampsia and preterm delivery in the South Coast Air Basin of California. *Environmental Health Perspectives*, *11*(11), 1773–1779. doi:10.1289/ehp.0800334 PMID:20049131

Wu, S., Mickley, L. J., Jacob, D. J., Rind, D., & Streets, D. G. (2008). Effects of 2000–2050 changes in climate and emissions on global tropospheric ozone and the policy-relevant background ozone in the United States. *Journal of Geophysical Research*, *113*(D18), D18312. doi:10.1029/2007JD009639

Xenopoulos, M. A., Lodge, D. M., Alcamo, J., Märker, M., Schulze, K., & Van Vuuren, D. P. (2005). Scenarios of freshwater fish extinctions from climate change and water withdrawal. *Global Change Biology*, *11*(10), 1557–1564. doi:10.1111/j.1365-2486.2005.001008.x

Xia, S., Liu, Y., Yu, X., & Fu, B. (2018). Challenges in coupling LTER with environmental assessments: An insight from potential and reality of the Chinese Ecological Research Network in servicing environment assessments. *The Science of the Total Environment*, *633*, 1302–1313. doi:10.1016/j.scitotenv.2018.03.284 PMID:29758883

Xie, W., Huang, J., Wang, J., Cui, Q., Robertson, R., & Chen, K. (2018). Climate change impacts on China's agriculture: The responses from market and trade. *China Economic Review*. doi:10.1016/j.chieco.2018.11.007

Xoplaki, E. (2002). *Climate variability over the Mediterranean* (PhD thesis). University of Bern, Switzerland.

Xu, X., Ha, S. U., & Basnet, R. (2016). A review of epidemiological research on adverse neurological effects of exposure to ambient air pollution. *Frontiers in Public Health*, *4*, 157. doi:10.3389/fpubh.2016.00157 PMID:27547751

Xu, X., Sharman, R. K., Talbott, E. O., Zborowski, J. V., Rager, J., Arena, V. C., & ... . (2011). PM10 air pollution exposure during pregnancy and term low birth weight in Allegheny County, PA, 1994–2000. *International Archives of Occupational and Environmental Health*, *84*(3), 251–257. doi:10.100700420-010-0545-z PMID:20496078

Yadav, K. K., Gupta, N., & Kumar, V. (2015). *Environmental challenges: Problem and management*. Lap Lambert Academic publishing.

Yaghoobi-Ershadi, M. R., Akhavan, A. A., Zahraei-Ramazani, A. R., Jalali-Zand, A. R., & Piazak, N. (2005). Bionomics of Phlebotomuspapatasi (Diptera: Psychodidae) in an endemic focus of zoonotic cutaneous leishmaniasis in central Iran. *Journal of Vector Ecology*, *30*(1), 115. PMID:16007964

Yaghoobi-Ershadi, M. R., Hakimiparizi, M., Zahraei-Ramazani, A. R., Abdoli, H., Akhavan, A. A., Aghasi, M., ... Ranjbar, A. A. (2010). Sand fly surveillance within an emerging epidemic focus of cutaneous leishmaniasis in southeastern Iran. *Journal of Arthropod-Borne Diseases*, *4*(1), 17. PMID:22808384

Yaghoobi-Ershadi, M. R., Javadian, E., & Tahvildare-Bidruni, G. H. (1995). Leishmania major MON-26 isolated from naturally infected Phlebotomus papatasi (Diptera: Psychodidae) in Isfahan Province, Iran. *Acta Tropica*, *59*(4), 279–282. doi:10.1016/0001-706X(95)92834-3 PMID:8533663

Yang, C. Y., Tseng, Y. T., & Chang, C. C. (2003). Effects of air pollution on birth weight among children born between 1995 and 1997 in Kaohsiung, Taiwan. *Journal of Toxicology and Environmental Health. Part A.*, *66*(9), 807–816. doi:10.1080/15287390306385 PMID:12746128

Yang, G., Ye, B., Xie, X., & Zhou, L. (2010, July). Dynamics of oasis landscape in inland Shule River basin in arid northwest China. In *Geoscience and Remote Sensing Symposium (IGARSS), 2010 IEEE International* (pp. 922-925). IEEE. 10.1109/IGARSS.2010.5653756

Yari, A. R., Goudarzi, G., Geravandi, S., Dobaradaran, S., Yousefi, F., Idani, E., ... Mohammadi, M. J. (2016). Study of ground-level ozone and its health risk assessment in residents in Ahvaz City, Iran during 2013. *Toxin Reviews*, *35*(3-4), 201–206. doi:10.1080/15569543.2016.1225769

Yates, D., Purkey, D., Sieber, J., Huber-Lee, A., & Galbraith, H. (2005). WEAP21: A demand, priority, and preference driven water planning model: Part 2, Aiding freshwater ecosystem service evaluation. *Water International*, *30*, 501–512. doi:10.1080/02508060508691894

Yates, D., Purkey, D., Sieber, J., Huber-Lee, A., Galbraith, H., West, J., & Rayej, M. (2009). Climate driven water resources model of the Sacramento Basin, California. *Journal of Water Resources Planning and Management*, *135*(5), 303–313. doi:10.1061/(ASCE)0733-9496(2009)135:5(303)

Yates, T. L., Mills, J. N., Parmenter, C. A., Ksiazek, T. G., Parmenter, R. R., Vande Castle, J. R., ... Peters, C. J. (2002). The ecology and evolutionary history of an emergent disease: Hantavirus pulmonary syndrome. *Bioscience*, *52*(11), 989–998. doi:10.1641/0006-3568(2002)052[0989:TEAEHO]2.0.CO;2

Yavar, R., Abedin, S., Reza, A. M., Ali, O. M., Sina, R., Mehdi, M., ... Babak, F. (2011). Phlebotomus papatasi and Meriones libycus as the vector and reservoir host of cutaneous leishmaniasis in Qomrood District, Qom Province, central Iran. *Asian Pacific Journal of Tropical Medicine*, *4*(2), 97–100. doi:10.1016/S1995-7645(11)60045-X PMID:21771429

Yost, M. G., Gadgil, A. J., Drescher, A. C., Zhou, Y., Simonds, M. A., Levine, S. P., ... Saisan, P. A. (1994). Imaging indoor tracer-gas concentrations with computed tomography: Experimental results with a remote sensing FTIR system. *American Industrial Hygiene Assoc.*, *55*(5), 395–402. doi:10.1080/15428119491018835 PMID:8209844

Young, K. (2009). Omega-6 (n-6) and omega-3 (n-3) fatty acids in tilapia and human health: a review. *International Journal of Food Sciences and Nutrition, 60*(sup5), 203–211. doi:10.1080/09637480903140503

Youn, S. J., Taylor, W. W., Lynch, A. J., Cowx, I. G., Beard, T. D., Bartley, D., & Wu, F. (2014). Inland capture fishery contributions to global food security and threats to their future. *Global Food Security*, *3*(3–4), 142–148. doi:10.1016/j.gfs.2014.09.005

Yousaf, S., Andria, V., Reichenauer, T. G., Smalla, K., & Sessitsch, A. (2010). Phylogenetic and functional diversity of alkane degrading bacteria associate with Italian ryegrass (*Lolium multiflorum*) and Birds foot trefoil (*Lotus corniculatus*) in a petroleum oil-contaminated environment. *Journal of Hazardous Materials*, *184*(1-3), 523–532. doi:10.1016/j.jhazmat.2010.08.067 PMID:20851515

Yue, Y., Li, M., Zhu, A., Ye, X., Mao, R., Wan, J., & Dong, J. (2016). Land Degradation Monitoring in the Ordos Plateau of China Using an Expert Knowledge and BP-ANN-Based Approach. *Sustainability*, *8*(11), 1174. doi:10.3390u8111174

Zanobetti, A., Bind, M.-A. C., & Schwartz, J. (2008). Particulate air pollution and survival in a COPD cohort. *Environmental Health, 10*, 7–48. PMID:18847462

Zarandi, S. M., Alimohammadi, M., Moghaddam, V. K., Hasanvand, M. S., Miranzadeh, M. B., Rabbani, D., ... Tehrani, A. M. (2015). Long-term trends of nitrogen oxides and surface ozone concentrations in Tehran city, 2002–2011. *Journal of Environmental Health Science & Engineering*, *15*, 13–63. PMID:26380092

Zeroual, E. (1995). *Enregistrements climatiques dans les sédiments du Lac Isli (Haut Atlas du Maroc)* (PhD thesis). University of Neuchatel, Switzerland.

Zhang, X. C. (2005). Spatial downscaling of global climate model output for site-specificassessment of crop production and soil erosion. *Agricultural and Forest Meteorology*, *135*(1-4), 215–229. doi:10.1016/j.agrformet.2005.11.016

Zhang, X., Friedl, M. A., Schaaf, C. B., Strahler, A. H., Hodges, J. C., Gao, F., & Huete, A. (2003). Monitoring vegetation phenology using MODIS. *Remote Sensing of Environment*, *84*(3), 471–475. doi:10.1016/S0034-4257(02)00135-9

Zhang, Y., He, L., Chen, Z., Zhang, W., Wang, Q., Qian, M., & Sheng, X. (2011). Characterization of lead-resistant and ACC deaminase-producing endophytic bacteria and their potential in promoting lead accumulation of rape. *Journal of Hazardous Materials*, *186*(2-3), 1720–1725. doi:10.1016/j.jhazmat.2010.12.069 PMID:21227577

Zhang, Y., Li, C., Krotkov, N. A., Joiner, J., Fioletov, V., & McLinden, C. (2016). Continuation of long-term global SO_2 pollution monitoring from OMI to OMPS. *Atmos Meas Tech Discuss*, *10*(4), 1495–1509. doi:10.5194/amt-10-1495-2017

Zhao, C. X., Wang, Y. J., Wang, Y. Q., & Zhang, H. L. (2013). Interactions between fine particulate matter (PM2.5) and vegetation: A review. [in Chinese]. *Shengtaixue Zazhi*, *32*, 2203–2210.

Zheng, D., Ducey, M. J., & Heath, L. S. (2013). Assessing net carbon sequestration on urban and community forests of northern New England, USA. *Urban Forestry & Urban Greening*, *12*(1), 61–68. doi:10.1016/j.ufug.2012.10.003

Zhioua, E., Kaabi, B., & Chelbi, I. (2007). Entomological investigations following the spread of visceral leishmaniasis in Tunisia. *Journal of Vector Ecology*, *32*(2), 371–374. doi:10.3376/1081-1710(2007)32[371:EIFTSO]2.0.CO;2 PMID:18260530

Zhiying, L., & Fang, H. (2016). Impacts of climate change on water erosion: A review. *Earth-Science Reviews*, *163*, 94–117. doi:10.1016/j.earscirev.2016.10.004

Zhou, T., Yi, C., Bakwin, P. S., & Zhu, L. (2008). Links between global CO2 variability and climate anomalies of biomes. *Science in China. Series D, Earth Sciences*, *51*(5), 740–747. doi:10.100711430-008-0024-5

Zhu, T. M. W., Jenkins, M. W., & Lund, J. R. (2005). Estimated impacts of climate warming on California water availability under twelve future climate scenarios. *Journal of the American Water Resources Association*, *14*(5), 1027–1038. doi:10.1111/j.1752-1688.2005.tb03783.x

Ziska, L. H., Morris, C. F., & Goins, E. W. (2004). Quantitative and qualitative evaluation of selected wheat varieties released since 1903 to increasing atmospheric carbon dioxide: Can yield sensitivity to carbon dioxide be a factor in wheat performance? *Global Change Biology*, *10*(10), 1810–1819. doi:10.1111/j.1365-2486.2004.00840.x

Zouirech, M., Belghyti, D., El Kohli, M., & Faraj, C. (2013). Entomological investigation of an emerging leishmaniasis focus in Azilal province, Morocco. *Pakistan Entomologist*, *35*, 11–15.

About the Contributors

Ahmed Karmaoui started his academic career as Research Fellow at Cadi Ayyad University and awarded Ph.D. from Faculty of Semlalia; and as an associate editor at the International Journal of Climate Change: Impacts and Responses, editor at several International Journals such as Journal of Research in Chemistry & Environment (SRI: JRCE), the International Journal of Earth & Environmental Sciences, Nanomedicine and Nanoscience Research Journal; as International Advisory Board member of the International Journal of ICT Research and Development in Africa (IJICTRDA), and the International Journal of Social Ecology and Sustainable Development (IJSESD). He is a Member of the Asian Council of Science Editors. KARMAOUI has reviewed for 61 international journals, graduated from Publons Academy, and awarded Top 1% of international Reviewers (2018) by Publons (part of CLARIVATE, supported by WEB OF SCIENCE). He has published various peer-reviewed international journal articles and book chapters on a range of environmental topics. He is currently the president of the Southern Center for Culture and Science, Morocco. Broad areas of his research interests include ecosystem services assessment, climate change impacts, response, adaptation, disaster risk reduction, environmental health, risk and vulnerability, water management.

* * *

Kalaiselvi Beeman works as Soil Scientist in ICAR-National Bureau of Soil Survey and Land Use Planning, Regional Centre, Bangalore. She completed doctorate degree from Tamil Nadu Agricultural University with specialisation in Soil Fertility. She has four years experience in the field of Land resource inventory (LRI), GIS applications for LRI and Land use planning.

Abdelkrim Ben Salem is a PhD researcher on environment sciences at the University of Marrakesh. His current research interest is on "Vulnerability and Adaptation to Climatical Changes in the Biosphere Reserve of the Oasis in Southern Morocco" This work that was proposing has an objective, to methodically, clearly and objectively assess scientific, technical and socio-economic information which is necessary and useful for understanding the scientific basis of risks related to climate change, identify more precisely the possible consequences of this change and consider possible strategies for adaptation and mitigation. He is also interested in ecosystem services valuation and water management and usage.

Souad Ben Salem is a PhD student Cadi Ayyad University, Morocco. She has a Master degree in Sciences and Technics. She is actually a member of the GRIVAC Reseach group of Impact, Vulnerability and adaptation to climate change in Morocco, Laboratory of hydrology, Ecotoxicology Sanitation and Global change, Faculty of sciences semlalia, University Cadi ayyad. Her research subject is about the Vulnerability of oasis ecosystems in southeastern of Morocco to climate change and integrated management of natural resources.

Banwari Dandotiya is researcher in School of Studies in Environmental Science, Jiwaji University Gwalior and working on health impacts of air pollution, air quality monitoring, environmental pollution and health effects, etc.

Zubair A. Dar is a PhD scholar at Sher-e-Kashmir University of Agricultural Sciences and Technology of Kashmir, J&K, India in the division of Environmental sciences. He had completed his masters degree from university of Kashmir, India. The author has a research experience in soil chemistry, organic farming, soil and water pollution. He has published my research papers in national and international journals. He has qualified national eligibility test (NET). He has attended various national and international conferences.

Mohamed Ait El Mokhtar works at the faculty of science Semlalia, Cadi Ayyad University. Mohamed works on organic and biologic fertilizers, Date palm, Salinity, Climate change, oases.

Fadoua El Qorchi is a Ph.D. student at the Department of Environmental Sciences, Faculty of Science Semlalia, University Cadi Ayyad, Marrakech, Morocco . Her doctoral research investigates agriculture, water and socio economic vulnerability and impact to climate change in upper Draa valley.

Rajendra Hegde is a Principal Scientist in ICAR-National Bureau of Soil Survey and Land Use Planning, Regional Centre, Bangalore and conducted pioneering research in the field of resource inventory (LRI) and Land use planning. He has 25 years experience and published more than 40 research papers in peer reviewed journals.

Abdelilah Meddich is a Professor at the Faculty of Science Semlalia (Cadi Ayyad University), who works on the role of organic and biological fertilizers in improving tolerance of oasis crops including date palm to different biotic and abiotic stresses.

Lalitha M. is a Soil Scientist in ICAR-National Bureau of Soil Survey and Land Use Planning, Regional Centre, Bangalore. She has seven years experience in the field of Land resource inventory (LRI), GIS applications for LRI and Land use planning.

Janani N. works as senior research fellow in ICAR-National Bureau of Soil Survey and Land Use Planning, Regional Centre, Bangalore. Her specialisation is soil and water conservation engineering.

Wilson Okaka is a Ugandan who is dedicated to public policy awareness communication for sustainable development. He was educated and trained locally and from abroad in the fields of: Political Science, Public Administration and Management, Renewable Energy Studies, Project Management, Journalism, and Mass Communication in Uganda, UK, Italy, Sweden, and USA. His research areas of interests include: climate change and environmental policy communication, public health, gender equality and equity, ICT for community media (community radio); journalism and mass communication; media training for Science Journalists; public policy communication campaign; renewable; teaching communication and, environmental education. He is a Lecturer at Kyambogo University, Uganda; from where he presented conference papers, published books, book chapters, and peer-reviewed for many international journals. In addition, Wilson Okaka belongs to many interdisciplinary research, media, scientific, and education networks. He has taught at Kyambogo University, Uganda, since its inception. He has been to many African, Asian, Middle East, USA, and European states for studies, research, consultancies, conferences, training, and tourism.

Srinivasan R. has seven years' experience in land resources characterization, water quality and land quality assessment, and applications of Remote Sensing and Geographic Information System in natural resource management in different part of West Bengal, Odisha, Sikkim, Karnataka and Tamil Nadu. Major activities is Mapping of major soils and land evaluation and suggestion of suitable crops and site specific nutrient management and also actively involving capacity building progrmmes like training and teaching to line department and university staffs.

Dharumarajan S. is a Senior scientist in National Bureau of soil survey and land use Planning and conducted pioneering research in land resource inventorisation, digital soil mapping, desertification/ land degradation mapping and vulnerability analysis . He completed post doc from The University of Sydney. He published more than 30 research papers in peer reviewed journals.

Veeramani S. works as senior research fellow in ICAR-National Bureau of Soil Survey and Land Use Planning, Regional Centre, Bangalore. His specialisation is Remote sensing and he has 4 years experience in both research and teaching.

Samreen Siddiqui is a Ph.D. student at Texas A & M University, Corpus Christi working on a USDA-NIFA funded project towards her dissertation. She has completed her Master's degree in Biology with a thesis focused on ecotoxicology from Valdosta State University, GA, USA. Her earlier Master's degree was in Aquatic Ecosystem Management from Edinburgh Napier University, UK. Samreen Siddiqui served as an author as well as reviewer in various peer-reviewed publications and write blogs for American Fisheries Society. She is a Chair in outreach committee within SETAC- NASAC (North America Student Advisory Council). Her research focuses on impacts of Climate change and pesticides in aquatic organisms.

Zerouali Siham has ten years experience in teaching biological sciences. She is a Research Fellow at Southern Center for Culture and Science, Morocco. She has published various peer-reviewed articles and book chapters on a range of environmental topics.

Vartika Singh holds Ph.D. degree in Hydrogeology and Remote Sensing, M.Phil. degree in Hydrogeology from Vikram University and Master degree in App. Geology from Kurukshtera University. She has GATE Qualified all India rank 47. She has 7 years of research experience with Defence Terrain Research Laboratory (DTRL), DRDO, Delhi. Where she is associated with various important projects of Defence Research & Development Organization, DTRL labs like GIM, Thar, and PISTA. Her specialization in research is in Hydrogeology, Remote Sensing, Environmental Geology & Artificial Intelligence. She has more than 30 publications in various national, international journals. She is one of the founder member of Computer Intelligence Research Group and Magazine. She is an Editorial Board of several national and international journals.

Index

A

abiotic stress 83-85, 94
adaptation 3, 23-24, 37-38, 40, 58, 60, 119, 127-132, 141, 191, 196-197, 199, 209-210, 215, 233, 245, 256, 262, 304-305, 307-308, 310-313, 316-317, 323, 327, 329
Adaptative Practices 329
adaptive practices 316-318, 326
agro-ecosystem 37, 40, 44, 67, 69, 72, 77, 290
air 4, 29-30, 37, 94, 96-102, 115, 121, 160-161, 163-170, 188, 198, 224, 279
air pollution 96-102, 115, 161, 163, 165, 167-168, 170
ambient air 97-98, 100-102, 115
anthropogenic factors 155, 158, 197, 317
anthropogenic pressure 5
aquaculture 38, 198, 249, 252-255, 259-262, 279
arid 7, 11, 25, 28, 44, 46-47, 55, 58, 68-69, 75, 117-119, 121-123, 132, 150, 184-185, 192, 196-200, 204-206, 208, 211, 215, 218, 220, 225, 233, 237, 247-248, 253, 255-256, 282, 292, 305-306, 308, 311-312
arid zone 44, 199, 308
associations 23, 280, 291, 295, 298-299

B

Bayoud 35, 39, 217-218, 236, 245
biodiversity 3-4, 7-8, 11-12, 15, 37, 68, 79, 117-121, 123-132, 141-142, 146-151, 155, 158, 160, 168, 188, 250-251, 257, 288, 304-305, 307, 309, 312, 328
Biodiversity Extinction 141
biodiversity loss 117, 126-127, 141
Biodiversity Richness 119, 141
biological diversity 46, 141-143, 145-147, 151, 155, 158
Bioremediation 79, 86, 94

C

capture fisheries 249, 252-255, 258, 262, 279
Cardiovascular health effects 115
climate 1-5, 8, 10-12, 23-24, 26-31, 35-40, 44-49, 52, 58-60, 68-69, 75, 77, 83, 99, 101, 117-121, 124-130, 132, 141-143, 145-151, 153-155, 158, 160-161, 183-184, 191-193, 196-206, 208-211, 216-219, 221-229, 232-238, 245, 247-249, 251-252, 255, 257-259, 261-263, 279, 281-283, 286-288, 290-292, 304-313, 316-318, 321, 323-324, 326-330
climate change 1-3, 5, 8, 10-11, 23-24, 26-27, 29-31, 35-38, 40, 44-46, 48, 52, 59-60, 75, 83, 99, 101, 117-121, 124-130, 132, 141-143, 145-146, 148-151, 153-155, 158, 161, 183-184, 191-193, 196-201, 203-206, 208-211, 216-218, 221-228, 232-235, 237-238, 245, 247-249, 251-252, 255, 257-259, 261-262, 279, 281-282, 304-313, 316-317, 321, 323, 326-330
climate change adaptation 37-38, 119, 210, 308, 310, 312, 329
climate change mitigation 210, 310, 329
Climate Resilience 330
climatic parameters 37, 245, 286, 292-296, 298
Climatic Scenario 245
conservation 13, 27, 60, 117, 119, 124-125, 128-130, 132, 141, 203, 251, 263, 304-305, 311-312, 328
COP22 238, 245
cultural services 34, 37, 77, 247, 307-308

D

desertification 7, 11-12, 15, 24, 39, 183-185, 188, 193, 251, 327
Disease Transmission 280, 290

E

Economic Valuation 68, 77
economic value 69, 72, 75, 77, 154
ecoservices 305-313
ecosystem 1-2, 5-6, 10-13, 23-24, 26-29, 31, 34-37, 40, 44-46, 58, 67-69, 71-72, 75, 77, 86, 99, 101, 115, 117-118, 121, 126-128, 132, 141-142, 149, 153-155, 158, 160-161, 168, 170, 197, 199, 204-205, 209-210, 217-218, 224, 235, 238, 245, 247, 250-251, 257-259, 279-281, 288, 304-305, 308-309
ecosystem service 45-46, 72, 280
ecosystem services 6, 12, 23-24, 26, 28-29, 31, 34-37, 40, 44-45, 58, 67-69, 71-72, 75, 77, 127, 142, 149, 153, 155, 158, 168, 170, 204, 210, 247, 250-251, 257, 279-281, 288, 304-305
endophytes 78-88, 94
environment 3-5, 7, 15, 26, 30, 38, 44, 48, 58, 63, 69, 78-80, 83, 94, 96-97, 101-102, 115, 119, 121, 141, 160-161, 163-164, 166, 192, 196-197, 204, 220, 236, 245, 261, 263, 279, 281-282, 287, 305, 310, 328
Equilibrium of Ecosystems 217, 245
exposure 81, 96, 98-102, 165, 168, 170, 198, 210, 215, 259

F

FAO 204, 206, 208-210, 221-222, 224, 251-256, 317
food security 209, 247, 249, 251-255, 257, 259, 261-264, 279, 304, 306-309, 311, 313

G

gaseous pollutants 99, 102, 115, 163, 169
GCMs 124, 223, 245
Grassland 249, 279
greenhouse gases 46, 119, 143, 145, 163-164, 170, 197, 201, 203, 211, 247-248, 279, 329
groundwater 15, 24, 28, 34, 36-39, 45-48, 51-52, 55-60, 218, 225-228, 232-235, 238, 253, 255, 259, 263, 280-281, 284, 286, 288

H

habitat fragmentation 148-149, 158
health effects 96, 98-102, 115
heavy metals 78-79, 86, 95, 165-166
human health 98-99, 102, 161, 164-165, 170, 311, 313

I

inland fisheries 196-199, 204, 206, 208-209, 211, 215, 262
IPCC 8, 10, 24, 30, 46, 127, 145, 192, 197-199, 203-204, 206, 208, 227, 245, 248, 251, 257-259, 311
Isingiro district 305-308, 312

L

land degradation 183-185, 187-191, 193, 257
landscape dynamics 316, 321, 330
leishmaniasis 37, 68, 280-288, 290-292, 295, 298-299
LTER Network 1-2, 4-5, 12-14

M

MAB 235, 245
MEA 24, 68, 77, 199-200, 203-204, 250
medicinal plants 153-154, 159, 287, 324
MIOS 49, 232, 245
mitigation 38, 150, 155, 197, 199, 210, 216, 251, 262, 307-308, 310-313, 316-317, 329
MIVAD 226-227, 245
Morocco 1-6, 8, 10-13, 23-25, 30, 33, 38-40, 45-46, 58, 63, 67-68, 217-225, 227-228, 232, 236, 238, 280-282, 288, 291-295, 299, 316-317, 319

N

NAO 219, 245
NDVI 28, 33, 44, 189-190
Nitrogen Fixation 81, 95
NPP 28, 33-34, 44, 190

O

oases 8, 12, 15, 24-25, 33, 38-39, 47, 58, 68, 72, 75, 217-218, 221, 224-226, 228-230, 232, 234-238, 245, 281-284, 287-288
oasis 7, 11-13, 23-25, 28, 35, 38, 40, 44, 46, 49, 58, 60, 67-69, 72, 77, 218, 224-225, 231-232, 235-238, 245, 280-282, 286, 290
oasis agro-ecosystem 40, 69
oasis system 68, 280, 290
ocean acidification 247, 251, 258, 260, 262, 279
ozone layer 197

P

parasitic diseases 280-281, 290
particulate matter 97-102, 115, 161, 163, 165-166, 169-170
particulate pollutants 96
Perception 306, 316, 318, 320-323, 330
phytoremediation 78-79, 86-88, 95, 168
pollution 4, 68, 96-102, 115, 124, 146, 161, 163-168, 170, 197, 247
potential risk 146, 155, 159
provisioning services 37, 69, 77, 153, 247

R

rainfall 24-25, 27-31, 33, 38-39, 44, 46-47, 55, 58, 68-69, 72, 75, 118-119, 184, 188, 191-193, 197, 199, 208, 215-216, 218-223, 225, 227, 229, 231-232, 248, 251, 257-258, 264, 281-282, 293, 309, 318, 321
Regulating Services 37, 77, 247
REMO 222, 224, 227, 245
resilience 23-24, 37-38, 40, 118, 126, 128-129, 132, 149, 208, 210, 251, 257, 262, 304-305, 309-310, 316, 322-323, 328, 330
Respiratory health effects 115

S

semi-arid 46, 117-119, 123, 132, 150, 184, 192, 196-200, 204-206, 208-209, 211, 216, 247-248, 251-253, 256-257, 264, 304-305, 310, 312-313
Semi-Arid or Hot Steppe 216
sensitivity 24, 27, 198-199, 216, 259, 262, 306
soil salinity 192, 217
supporting services 24, 77, 247
sustainability 45, 102, 154, 184, 210, 218, 224, 232, 237, 245, 249, 252, 257, 310-311, 316, 328, 330

T

Tleta watershed 316-319, 321, 323, 328
tree 71, 85, 102, 149

U

Uganda 199, 209, 304-310, 312
UNFCCC 197-198, 224, 245, 252, 311
urban area 25, 96, 115
urban environment 96-97, 101-102, 115
Urban vegetation 101-102, 115

V

vector-borne disease 280, 282, 287-288, 290
vegetation 6, 11, 24, 27-28, 31-34, 37, 39, 44, 72, 99, 101-102, 115, 117, 119, 124, 143, 145, 148, 163, 185, 187, 189-190, 203, 229, 245, 280-282, 284, 286, 288, 306, 312, 317
vulnerability 1-6, 8, 11-13, 23-24, 37, 40, 60, 72, 125-129, 131-132, 152, 193, 198, 209-211, 216, 259, 262, 287, 306, 308-311, 329
vulnerability assessment 2, 4
Vulnerable Area 217, 245

W

water scarcity 35, 40, 59, 197, 225, 237
watershed 38, 52, 208, 316-319, 321, 323, 327-328, 330
WEAP 26, 29, 31, 45-46, 48-55, 60, 63, 232, 245

Z

Ziz 25, 28, 31, 33, 35, 38, 45-46, 48-50, 52, 55-59, 61, 228-229, 232
Zoonotic Cutaneous Leishmaniasis 280-281, 286-287, 290-291

www.ingramcontent.com/pod-product-compliance
Lightning Source LLC
LaVergne TN
LVHW081056180926
848390LV00005B/195